PHYSICOCHEMICAL METHODS OF MINERAL ANALYSIS

Edited by

Alastair W. Nicol

Department of Minerals Engineering
University of Birmingham
Birmingham, England

PLENUM PRESS • NEW YORK AND LONDON

Library of Congress Cataloging in Publication Data

Main entry under title:

Physicochemical methods of mineral analysis.

Includes bibliographical references and index.
1. Mineralogy, Determinative. 2. Materials–Analysis. I. Nicol, Alastair W.
QE367.2.P49 549'.1 72-95070
ISBN 0-306-30739-1

A Division of Plenum Publishing Corporation
227 West 17th Street, New York, N.Y. 10011

United Kingdom edition published by Plenum Press, London
A Division of Plenum Publishing Company, Ltd.
Davis House (4th Floor), 8 Scrubs Lane, Harlesden, London, NW10 6SE, England

Printed in the United States of America

PHYSICOCHEMICAL METHODS OF MINERAL ANALYSIS

To Muriel and Laki

CONTRIBUTORS

H. Bennett — *British Ceramic Research Association*
Queens Road, Penkhull
Stoke-on-Trent ST4 7LQ, England

K. G. Carr-Brion — *Warren Spring Laboratory*
P.O. Box 20, Gunnelswood Road
Stevenage, Hertfordshire SG1 2BX, England

V. C. Farmer — *Department of Spectrochemistry*
Macaulay Institute for Soil Research
Craigiebuckler, Aberdeen AB9 2QJ, Scotland

G. L. Hendry — *Department of Geology*
University of Birmingham, P.O. Box 363
Birmingham B15 2TT, England

V. I. Lakshmanan — *Department of Minerals Engineering*
University of Birmingham, P.O. Box 363
Birmingham B15 2TT, England

G. J. Lawson — *Department of Minerals Engineering*
University of Birmingham, P.O. Box 363
Birmingham B15 2TT, England

M. H. Loretto — *Department of Physical Metallurgy and Science of Materials*
University of Birmingham, P.O. Box 363
Birmingham B15 2TT, England

R. C. Mackenzie — *Macaulay Institute for Soil Research*
Craigiebuckler, Aberdeen AB9 2QJ, Scotland

G. D. Nicholls — *Department of Geology*
University of Manchester
Manchester M13 9PL, England

A. W. Nicol — *Department of Minerals Engineering*
University of Birmingham, P.O. Box 363
Birmingham B15 2TT, England

H. N. Southworth — *Department of Physical Metallurgy and Science of Materials*
University of Birmingham, P.O. Box 363
Birmingham B15 2TT, England

M. Wood — *Department of Geology*
University of Manchester
Manchester M13 9PL, England

FOREWORD

This book has developed from a short residential course organised by the Department of Minerals Engineering and the Department of Extra Mural Studies of the University of Birmingham. The course was concerned mainly with physical methods of analysis of minerals and mineral products, and particular regard was given to 'non-destructive' methods, with special emphasis on newly available techniques but with a review of older methods and their recent developments included therein.

Mineral analysis is obviously of great importance in all the stages of mineral exploration, processing, and utilisation. Selection of a method for a particular mineral or mineral product will depend upon a number of factors, primarily whether an elementary analysis or a phase or structure analysis is required. It will also depend upon the accuracy required. The chapters in the book covering the different methods show the range of useful applicability of the methods considered and should prove valuable as an aid in selecting a suitable method or methods for a given set of circumstances.

The book, referring as it does to the majority of the instrumental methods available today (as well as, for comparison, a useful contribution on the place of classical wet chemical analysis) will be valuable to the student as well as to those analysts, research workers, and process engineers who are concerned with the winning, processing, and utilisation of minerals and mineral products.

Stacey G. Ward

PREFACE

The past decade has seen great strides being made in all branches of science, and nowhere more than in the field of analysis and characterisation of materials, both in the number and the variety of techniques that have become available. In the specific area of physicochemical methods of analysis, based on monitoring the interaction of beams of electrons or electromagnetic radiation with matter, this has resulted not only in new and powerful additions to the analysts' repertoire but also in the upgrading of older methods to give them improved accuracy and flexibility, and a new lease on life for some. We may cite, in particular, the electron probe microanalyser, the scanning electron microscope, Auger spectroscopy and non-dispersive X-ray fluorescence analysis, the development of which has allowed us not only to determine the elemental and phase compositions of a material, but also to look in detail at the distribution of the elements among the phases present, or study elemental concentrations in the extreme surface layers in a way that was quite impossible in previous years. Such techniques are very appropriate to the particular problems encountered in the study and analysis of minerals and mineral products, such as glass, ceramics, cement, etc., and the information that they can give may prove crucial in explaining, for example, why a given lead-zinc ore is not amenable to beneficiation by froth flotation; investigation of one such 'problem' ore with the electron probe microanalyser showed that the galena was heavily contaminated by zinc at the sub-micron level and comminution could not separate the two phases. The methods also are important as the basis of sensors for automatic control systems which are currently being developed. But, as ever, the main problem in applying all these techniques lies in translating the methods from the research laboratories, where they have been developed, to the industrial environment, where they are needed.

It is, moreover, a basic tenet of the Editor's method of teaching that a student will be able to understand a process better or apply a technique more sensibly and effectively if he is familiar with the scientific principles and the basic theory that underlie the process or technique. Too often in analysis does the 'black box syndrome' raise its ugly head as the operator, working by rote, pushes button A, turns knob B until the pointer C reaches the line, and copies a number from the dial D, without ever really knowing how the reading is obtained or what factors may intrude to spoil the accuracy of the final figure. This lack of knowledge of basic principles, especially in conjunction with an illuminated digital read-out, may result in a touching, if sometimes disastrous, faith in the magical properties of the numbers that appear in the box, with no thought of what the number actually signifies in terms of the parameter being measured, of how this value is related to the required parameter, or of the degree of confidence that can be placed in the accuracy of the number. It is all too easy to forget that the instrument records the signal that it receives from the test material, and not necessarily the signal that the analyst wants it to record. Also, errors, and sometimes gross errors, can creep into an analysis if the interference caused by an apparently innocent 'other ion' is not identified and allowance made for it. Reproducibility is so often confused with accuracy because people forget that high precision can mean simply that the instrument is making the same mistake on each reading!

Therefore, this book, as was the Residential Course from which it sprang, has been planned to try to present an account of these new methods with particular reference to their use in mineral analysis. It discusses the application of physicochemical methods of analysis, using principally electromagnetic and electron beam stimulation and sensing techniques, to materials of especial interest to the minerals engineer, and puts particular emphasis on the so-called 'non-destructive' methods of analysis. Throughout the book, the aim has been two-fold, to introduce the various techniques and give a description of the type of information each provides together with an account of the good and bad points of the method, its problems, etc., and to show how each method works in terms of the basic scientific principles involved. The book begins, therefore, with a chapter on basic principles, atomic theory, bonding, crystal field theory, the interaction of energy with matter, and an introduction to the detectors used in physicochemical analysis. The next four chapters discuss elemental analysis by optical and X-ray fluorescence methods, radiotracer techniques, and spark source mass spectrometry. A chapter on the application of X-ray methods to automatic control follows, then a section on phase analysis using X-ray diffraction, electron microscopy, thermal methods and infra-red spectroscopy. The last two chapters present an account of some of the very new techniques for analysis, including electron probe microanalysis, scanning electron microscopy, Auger spectroscopy, and the field ion microscope, plus a review of analytical methods which relates the position of physicochemical analysis to absolute, wet chemical techniques and assesses the usefulness of these new methods in a variety of situations. The original Course also included reflected light microscopy as one of its topics, but circumstances outside the Editor's control have made it impossible to include this in the book. Readers will find that each chapter contains a section on the basic theory particularly relevant to that topic, which may be omitted on a cursory read-through, but which is intended to improve the reader's understanding of the method, by supplementing the treatment given in Chapter 1.

Of course, a book such as this is not the work of one person, and I wish to record my most sincere thanks to all who helped me in its preparation.

Firstly, my authors, who patiently bore every request made of them and allowed me to recast their chapters often to a considerable extent in a search for uniformity of coverage of the various methods. Secondly, the many firms who supported the original Course and supplied photographs and figures for the book; acknowledgements are made separately throughout the chapters. Thirdly, Professor Ward and Dr. Lawson for their continuing help and support throughout the gestation period, and particularly Dr. Lakshmanan for being my conscience at all times, and for providing much needed encouragement when it looked as if the end would never come! Fourthly, Dr. J. I. Langford of the Department of Physics for performing the invaluable service of editing my own chapter, on X-ray diffraction, and the office staff of the Department of Minerals Engineering for their help in preparing parts of the typescript. And finally, my wife for bearing the total chaos that reigned in our study while the magnum opus was becoming a reality. Truly, without their help this book would never have been.

Alastair W. Nicol
University of Birmingham

CONTENTS

CHAPTER 1

Introduction, Basic Theory and Concepts

A. W. Nicol and V. I. Lakshmanan

Department of Minerals Engineering
University of Birmingham
Birmingham B15 2TT
England

1.1 BASIC THEORY

The analysis and characterisation of materials by physicochemical methods depends almost entirely on our ability to detect and measure the interaction between the substance under study and some form of electromagnetic radiation, and to relate this interaction to the various processes that can occur within the material. In general we may monitor either the emission of radiation from the material, as in emission spectroscopy, or the absorption of radiation by the material, as in absorption spectrography and thermal methods, or the conversion of one type of radiation into another, as in optical and X-ray fluorescence, or the diffraction of radiation, as in X-ray and electron diffraction. Spark source mass spectroscopy differs somewhat from the other techniques discussed herein, because the measured effect results from the interaction of charged particles with the magnetic field through which they move. The form of the interaction clearly varies for the different techniques which will be considered in other chapters, and it is the aim of this book not only to give an introduction to a group of the more important physicochemical methods currently available, but also to provide some of the theoretical background to the methods in the hope that this will permit a more reasoned and efficient use to be made of them.

At the simplest level, a material object can intercept and absorb all the energy in a beam of electromagnetic radiation and so cast a 'shadow', which we can observe directly, if the radiation lies in the visible portion of the spectrum, or indirectly by, as in the case of a beam of electrons, making the beam 'visible' through its action on a fluorescent screen. The resulting shadowgraphs are of limited diagnostic use.

Alternatively, the object can intercept and absorb only part of the energy in the incident beam, the remainder being transmitted but with a lower intensity. This effect, again considered in its simplest form, lies at the basis of transmission optical and electron microscopy, in which the internal structure of the material under study can be observed by virtue of the varying extents to which the beam of incident light or electrons is absorbed or scattered by the different features in the sample. But much more subtle interactions can occur between matter and electromagnetic radiation, involving not only partial absorption of an incident beam of radiation but also differential absorption or scattering of radiation of different wavelengths, in the beam, and it is these interactions that underlie the methods discussed in the subsequent chapters of this book.

1.1.1 Quantum Theory

In the years before 1900 it had been assumed that energy was absorbed or emitted by a substance as a continuum, despite the well known interrelationship between the intensity of the energy emitted by a so-called 'blackbody' and the wavelength of the emitted radiation. Planck [1] realised that the classical laws of energy transfer could not be applied to the interactions involved in this type of emission or absorption, since they involved the behaviour of the separate atoms in the material and not the macroscopic effect of all the atoms taken together. He showed that the observed energy distribution could be completely explained by postulating, firstly, that all materials consist of a large number of oscillators vibrating with a wide range of frequencies, from zero upwards, with a Maxwell-Boltzmann distribution having a preferred frequency which depends on the temperature of the body, and, secondly, that energy is emitted or absorbed by the vibrators discontinuously in discrete amounts, or 'quanta', whose values are related to the frequency of the vibrator. Mathematically the relationship is given by

$$E = nh\nu \tag{1.1}$$

where E is the energy of the quantum, h is a universal constant, the Planck's constant, equal to 6.625×10^{-34} J.sec, ν is the frequency in $\sec^{-1}$, and n is an integer which is normally taken to be unity.

The introduction of this new concept, that energy could be transferred only in discrete quanta of well defined values, revolutionised the thinking of the time, especially concerning the nature of materials at the atomic level, and provided the impetus for the vast increase in our understanding of the physical world that has occurred during this century, as well as providing the basis for virtually all of the techniques to be discussed in subsequent chapters.

The first development from Planck's original idea was made by Einstein [2], who extended the idea of the quantisation of energy to include the propagation of energy, particularly by the medium of electromagnetic radiation. He showed that such radiation could propagate energy through space also in definite quanta, or 'photons', of value

$$E = h\nu = \frac{hc}{\nu} \tag{1.2}$$

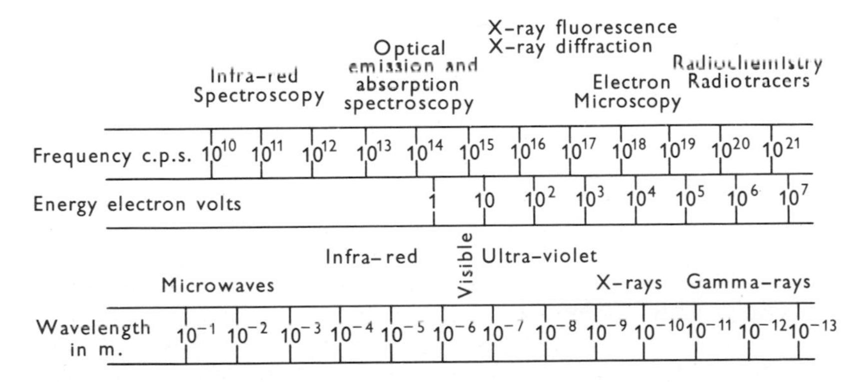

Figure 1.1 The electromagnetic spectrum, with the ranges applicable to the various techniques indicated.

where c is the velocity of light and ν and λ are the frequency and wavelength of the radiation. Such photons can interchange energy with other oscillators capable of vibrating at the same frequency and so can be emitted or absorbed by these vibrators in a very selective manner. Note that the photon energy increases as the wavelength of the associated radiation decreases and figure 1.1 shows the electromagnetic spectrum in diagrammatic form, with the photon energies and wavelengths corresponding to the various commonly named regions.

1.1.2 Bohr-Rutherford Model of the Atom

The next major development followed with Bohr's application [3] of the quantum theory to the model of the atom which had been proposed by Rutherford [4]. Rutherford's picture, in which a small, dense, positively charged nucleus was surrounded by a cloud of dispersed, negatively charged electrons, suffered from the major criticism that if the postulated electrons were assumed to be stationary they would inevitably be attracted to the nucleus and hence be annihilated whereas if they were assumed to be in motion around the nucleus classical theory demanded that they should radiate energy, since the system would then comprise an electric charge moving in a non-uniform potential field, in which case the orbit would decay and the electron should again spiral into the nucleus. Bohr showed that this apparently insoluble situation could be explained if the electrons moved in orbits around the nucleus which corresponded to states in which the angular momentum of the electron was an integral multiple of some fundamental energy, i.e. the energy of the electron was 'quantised'. The electron did not radiate energy when moving in such an orbit, which was therefore a stable or stationary state. Bohr further showed that several such orbits, or shells, could exist for any atom and that movement of an electron from one stationary state to another involved a definite, quantised amount of energy, corresponding to the difference in the energies of the two states involved.

Using this model he successfully accounted for the mathematical representation of the emission spectrum from hydrogen, given by Ritz [5] in the form

$$\bar{\nu} = R\left(\frac{1}{n_2^2} - \frac{1}{n_1^2}\right) \tag{1.3}$$

where $\bar{\nu}$ (= $1/\lambda$) is the wave number for the emission line in cm^{-1}, R is the Rydberg constant for hydrogen, and n_1 and n_2 are integers such that $n_1 > n_2$. Using a modified Planck notation, Bohr showed that the emission lines would be generated by an electron jumping from one energy level in the atom to another of lower energy, thus

$$\bar{\nu} = \frac{E'' - E'}{hc} \tag{1.4}$$

where E'' and E' are the energies of the levels involved. This equation is, in fact, true for all energy transitions whether they involve electron transitions or not, and it will be applied throughout the discussions of virtually every technique in this book.

Today, the Bohr-Rutherford model of the atom has been further modified by the later work of such people as Schrödinger, Heisenberg, Pauli and de Broglie, who together have shown that the solid, particulate electrons moving in exactly defined orbits, postulated by Bohr, must be replaced by rather more vague particles, which may be thought of as very short wavelength electromagnetic radiations under certain circumstances, contained within regions of space roughly corresponding to Bohr's orbits but with a much more complex fine structure involving sub-levels and separate orbitals within each Bohr shell. The energy of a given electron is defined by a set of four 'quantum numbers', and Pauli's principle [6] states that no two electrons in the same atom may possess the same set of quantum numbers. Detailed discussions of the modified Bohr-Rutherford model for the atom may be found in any good textbook on physical chemistry, and the treatments by Glasstone [7] and Mahan [8] may be particularly mentioned.

From the point of view of the analyst, however, the most important contribution that these workers made was to introduce the concept of the atomic orbital into our picture of the atom. Not only did its introduction provide an explanation for the fine structure seen in atomic emission line spectra, it also opened up the possibility of understanding and explaining molecular spectra, in which several additional features not seen in atomic spectra are found. In particular, the spectra from atoms are 'line spectra' with very sharp emission or absorption lines at well defined wavelengths or frequencies, but those arising from molecules are 'band spectra' which extend over a range of wavelengths and which, on very close examination, can be seen to comprise a large number of closely spaced but quite discrete lines. The explanation lies in the realm of valence theory, i.e. in how atoms are held together.

1.1.3 Valence Theory

It is generally accepted today that the electrons in an atom are contained in orbitals, which in turn may be combined in groups to form sub-levels, which finally combine to give the shells that Bohr originally proposed. Each orbital can contain up to two electrons and the orbitals are grouped according to the values of their quantum numbers, the principal quantum number, to denote the Bohr shell, the angular momentum quantum number, to denote the sub-level within the shell, and the magnetic quantum number, to denote the orbital. The fourth quantum number is the 'spin quantum' and is assigned the values +½ and –½. Certain rules exist relating each type of quantum number to the one above it in the hierarchy of the levels, so that the possible values which the angular momentum quantum number can take depend on the value of the principal quantum number for that level, and the possible values of the magnetic quantum number depend, in turn, on the value of the angular momentum quantum number for the sub-level. Again, detailed discussions are given by Glasstone and Mahan. Briefly, however, this treatment has produced a system of nomenclature, based largely on spectroscopic symbols, to denote the energy levels in an atom and the degree to which these levels and sub-levels (and orbitals, by implication) are filled in the atom in a given state. Figure 1.2 shows diagrammatically the sequence of sub-levels in increasing order of energy.

The distribution of the electrons among these energy levels is denoted by adding a superscript to the sub-level designation to show the number of electrons present in the orbitals in that sub-level. Thus, the lowest energy, or ground state, electronic configuration of the hydrogen atom may be represented by $1s^1$, showing that the atom contains one electron in its 1s shell. Similarly, helium may be represented by $1s^2$, denoting that the single orbital in the s-shell contains its maximum number of two electrons, and a representative selection of ground state electronic configurations in atoms is given in table 1.1. Note how the available levels are filled in a regular manner, from the lowest energy upwards, and that the energies of the orbitals in a sub-level, i.e. with the same angular momentum quantum number, are equal. Differences in their energies show up only in the presence of a uni-axial magnetic field.

Formation of Bonds

Bonds form between atoms to reduce the total free energy of the system and so make it chemically more stable. The way in which bonds form can be described mathematically by using either the ionic approximation, which involves transfer of electrons between the atoms concerned, or the covalency picture, in which the electrons are shared by the atoms. In both cases, the basis for the formation of bonds appears to be that each atom is trying to achieve the so-called rare gas configuration with a closed shell of, usually, eight electrons in its outer orbital shell.

In ionic compounds, which comprise compounds between a metal and a non-metal such as NaCl or MgO, this situation is achieved by the metal atom donating its outer electron or electrons to the non-metal to give a positively charged cation and negatively charged anion, with the electronic configurations, in NaCl, $Na^+ = 1s^2; 2s^2, 2p^6$ and $Cl^- = 1s^2; 2s^2, 2p^6; 3s^2, 3p^6$, and in MgO, $Mg^{2+} = 1s^2; 2s^2, 2p^6$ and $O^{2-} = 1s^2; 2s^2, 2p^6$. In covalent compounds, which principally mean compounds of non-metals, the situation is achieved by the atoms sharing the available electrons so that each atom is apparently surrounded by the required eight electrons, at least on a time-average basis. Thus we can picture the electronic configurations of methane, CH_4, and water, H_2O, as shown in figure 1.3, and note that covalent bonds tend to form between atoms which already possess nearly filled outer shells. The case of the bonding in metals partakes of some of the features of both ionic and covalent bonds, since the electrons are thought of as being shared

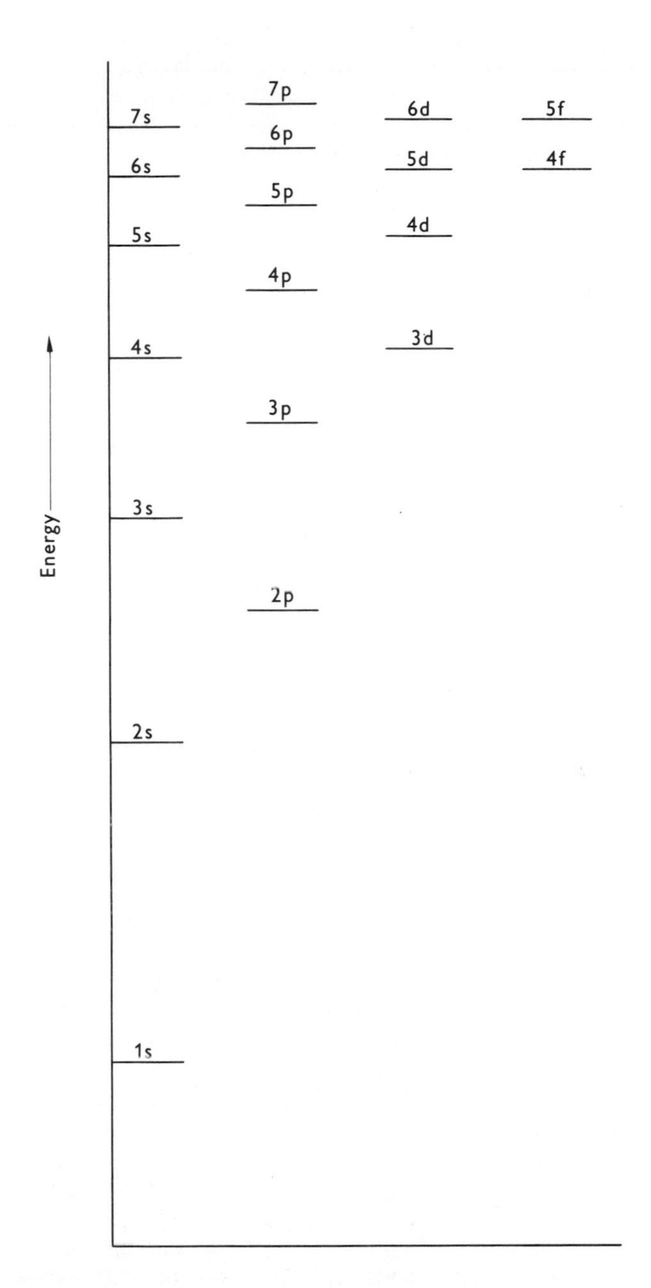

Figure 1.2 Diagrammatic representation of the electronic energy levels in a typical atom.

TABLE 1.1

Atomic No.	Element	Electronic Configuration
1	H	$1s^1$
2	He	$1s^2$
3	Li	$1s^2; 2s^1$
4	Be	$1s^2; 2s^2$
5	B	$1s^2; 2s^2,2p^1$
6	C	$1s^2; 2s^2,2p^2$
7	N	$1s^2; 2s^2,2p^3$
8	O	$1s^2; 2s^2,2p^4$
9	F	$1s^2; 2s^2,2p^5$
10	Ne	$1s^2; 2s^2,2p^6$
11	Na	$1s^2; 2s^2,2p^6; 3s^1$
12	Mg	$1s^2; 2s^2,2p^6; 3s^2$
13	Al	$1s^2; 2s^2,2p^6; 3s^2,3p^1$
14	Si	$1s^2; 2s^2,2p^6; 3s^2,3p^2$
15	P	$1s^2; 2s^2,2p^6; 3s^2,3p^3$
16	S	$1s^2; 2s^2,2p^6; 3s^2,3p^4$
17	Cl	$1s^2; 2s^2,2p^6; 3s^2,3p^5$
18	Ar	$1s^2; 2s^2,2p^6; 3s^2,3p^6$ (Argon core)
19	K	$1s^2; 2s^2,2p^6; 3s^2,3p^6; 4s^1$
20	Ca	$1s^2; 2s^2,2p^6; 3s^2,3p^6; 4s^2$
21	Sc	$1s^2; 2s^2,2p^6; 3s^2,3p^6,3d^1; 4s^2$
22	Ti	$1s^2; 2s^2,2p^6; 3s^2,3p^6,3d^2; 4s^2$
23	V	$1s^2; 2s^2,2p^6; 3s^2,3p^6,3d^3; 4s^2$
24	Cr	$1s^2; 2s^2,2p^6; 3s^2,3p^6,3d^5; 4s^1$
25	Mn	$1s^2; 2s^2,2p^6; 3s^2,3p^6,3d^5; 4s^2$
26	Fe	$1s^2; 2s^2,2p^6; 3s^2,3p^6,3d^6; 4s^2$
27	Co	$1s^2; 2s^2,2p^6; 3s^2,3p^6,3d^7; 4s^2$
28	Ni	$1s^2; 2s^2,2p^6; 3s^2,3p^6,3d^8; 4s^2$
29	Cu	$1s^2; 2s^2,2p^6; 3s^2,3p^6,3d^{10}; 4s^1$
30	Zn	$1s^2; 2s^2,2p^6; 3s^2,3p^6,3d^{10}; 4s^2$
31	Ga	Argon core,$3d^{10}; 4s^2,4p^1$
32	Ge	Argon core,$3d^{10}; 4s^2,4p^2$
33	As	Argon core,$3d^{10}; 4s^2,4p^3$
34	Se	Argon core,$3d^{10}; 4s^2,4p^4$
35	Br	Argon core,$3d^{10}; 4s^2,4p^5$
36	Kr	Argon core,$3d^{10}; 4s^2,4p^6$ (Krypton core)
37	Rb	Argon core,$3d^{10}; 4s^2,4p^6; 5s^1$
38	Sr	Argon core,$3d^{10}; 4s^2,4p^6; 5s^2$
39	Y	Argon core,$3d^{10}; 4s^2,4p^6,4d^1; 5s^2$
40	Zr	Argon core,$3d^{10}; 4s^2,4p^6,4d^2; 5s^2$
41	Nb	Argon core,$3d^{10}; 4s^2,4p^6,4d^3; 5s^2$
42	Mo	Argon core,$3d^{10}; 4s^2,4p^6,4d^5; 5s^1$
43	Te	Argon core,$3d^{10}; 4s^2,4p^6,4d^5; 5s^2$
44	Ru	Argon core,$3d^{10}; 4s^2,4p^6,4d^6; 5s^2$
45	Rh	Argon core,$3d^{10}; 4s^2,4p^6,4d^7; 5s^2$
46	Pd	Argon core,$3d^{10}; 4s^2,4p^6,4d^8; 5s^2$

TABLE 1.1 cont.

Atomic No.	Element	Electronic Configuration
47	Ag	Argon core,$3d^{10}$; $4s^2$,$4p^6$,$4d^{10}$; $5s^1$
48	Cd	Argon core,$3d^{10}$; $4s^2$,$4p^6$,$4d^{10}$; $5s^2$
49	In	Krypton core,$4d^{10}$; $5s^2$,$5p^1$
50	Sn	Krypton core,$4d^{10}$; $5s^2$,$5p^2$
51	Sb	Krypton core,$4d^{10}$; $5s^2$,$5p^3$
52	Te	Krypton core,$4d^{10}$; $5s^2$,$5p^4$
53	I	Krypton core,$4d^{10}$; $5s^2$,$5p^5$
54	Xe	Krypton core,$4d^{10}$; $5s^2$,$5p^6$ (Xenon core)
55	Cs	Krypton core,$4d^{10}$; $5s^2$,$5p^6$, $6s^1$
56	Ba	Krypton core,$4d^{10}$; $5s^2$,$5p^6$; $6s^2$
57	La	Krypton core,$4d^{10}$; $5s^2$,$5p^6$,$5d^1$; $6s^2$
58	Ce	Krypton core,$4d^{10}$,$4f^1$; $5s^2$,$5p^6$,$5d^1$; $6s^2$
59	Pr	Krypton core,$4d^{10}$,$4f^2$; $5s^2$,$5p^6$,$5d^1$; $6s^2$
60	Nd	Krypton core,$4d^{10}$,$4f^3$; $5s^2$,$5p^6$,$5d^1$; $6s^2$
61	Pm	Krypton core,$4d^{10}$,$4f^4$; $5s^2$,$5p^6$,$5d^1$; $6s^2$
62	Sm	Krypton core,$4d^{10}$,$4f^5$; $5s^2$,$5p^6$,$5d^1$; $6s^2$
63	Eu	Krypton core,$4d^{10}$,$4f^7$; $5s^2$,$5p^6$,$5d^0$; $6s^2$
64	Gd	Krypton core,$4d^{10}$,$4f^7$; $5s^2$,$5p^6$,$5d^1$; $6s^2$
65	Tb	Krypton core,$4d^{10}$,$4f^8$; $5s^2$,$5p^6$,$5d^1$; $6s^2$
66	Dy	Krypton core,$4d^{10}$,$4f^9$; $5s^2$,$5p^6$,$5d^1$; $6s^2$
67	Ho	Krypton core,$4d^{10}$,$4f^{10}$; $5s^2$,$5p^6$,$5d^1$; $6s^2$
68	Er	Krypton core,$4d^{10}$,$4f^{11}$; $5s^2$,$5p^6$,$5d^1$; $6s^2$
69	Tm	Krypton core,$4d^{10}$,$4f^{12}$; $5s^2$,$5p^6$,$5d^1$; $6s^2$
70	Yb	Krypton core,$4d^{10}$,$4f^{14}$; $5s^2$,$5p^6$,$5d^0$; $6s^2$
71	Lu	Krypton core,$4d^{10}$,$4f^{14}$; $5s^2$,$5p^6$,$5d^1$; $6s^2$
72	Hf	Krypton core,$4d^{10}$,$4f^{14}$; $5s^2$,$5p^6$,$5d^2$; $6s^2$
73	Ta	Krypton core,$4d^{10}$,$4f^{14}$; $5s^2$,$5p^6$,$5d^3$; $6s^2$

Subsequent elements fill the 5d, 6p, and 7s sub-shells in a manner analogous to the filling of the 4d, 5p, and 6s sub-shells in the series molybdenum through barium, and finally the trans-actinide and trans-uranium elements probably form a series very similar to the rare earth series, as the 5f sub-shell fills, although doubt still exists on this point.

Electronic configurations for the ground states of the atoms. Note (a) the way in which the levels and sub-levels fill from the lowest available energy upwards, as shown in figure 1.2; (b) how the spherical symmetry that can be obtained with d^5, d^{10}, f^7, and f^{14} configurations stabilise these configurations relative to the d^4; s^2, etc., configurations; (c) that precise definition of ground state configurations becomes more difficult for the heavier atoms, since the available levels differ by only very small amounts of energy (figure 1.2).

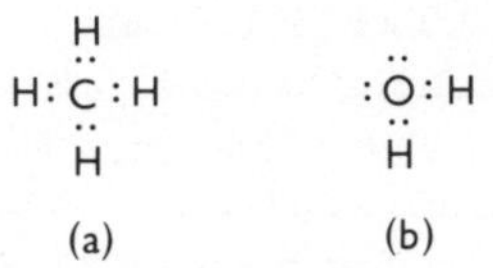

```
   H
   ..            ..
H: C :H        :O: H
   ..            ..
   H             H

  (a)           (b)
```

Figure 1.3 Formal representations of the electronic configurations in (a) methane, CH_4, and (b) water, H_2O. Note the highly symmetrical nature of the methane molecule, the presence of two 'lone pairs' of electrons in water, i.e. pairs of electrons on the oxygen atom not directly involved in bonding with the hydrogen atoms, and the way in which each atom is associated with its 'rare gas number' of electrons (2 for hydrogen, 8 for carbon and oxygen).

between all the atoms in the metal, rather than between pairs of atoms as in covalently bonded compounds, but without giving the formal separation into cations and anions of ionic theory. Metallic bonding is usually treated in terms of the band theory, about which more in the next section. Our interest will lie mainly with the covalent and metallic representations, in laying the theoretical basis for the analytical methods to be discussed in the subsequent chapters of this book.

Molecular Bonding

The formation of a covalent bond may be described by postulating that the two atoms involved in the bond combine their atomic orbitals to give a set of 'molecular orbitals' associated with both atoms. Mathematically, this is described as the 'Linear Combination of Atomic Orbitals', and, according to LCAO theory, occurs between atomic orbitals of similar energy in the two atoms, in such a way that each pair of atomic orbitals, one from each atom, gives rise to one molecular orbital of lower energy than the original atomic orbitals, the so-called 'bonding orbital', and one of higher energy, the so-called 'anti-bonding orbital'. Thus, in the homopolar H_2 molecule, the two 1s atomic orbitals from the hydrogens combine to give a bonding and an anti-bonding orbital in the molecule, and the two electrons associated with the atoms enter the bonding orbital, to minimise the energy of the system and so make it stable. In the heteropolar methane molecule, however, combination occurs between a $2(sp^3)$-hybrid atomic orbital in the carbon [9] and the 1s orbital of a hydrogen, since the 1s orbital in carbon is at a much lower energy than that in hydrogen. Again, bonding and anti-bonding molecular orbitals are set up in the system to correspond to the four C–H bonds formed, but in this case the 1s orbital of carbon also exists in the system as a separate atomic orbital with its electron pair. The energy of this orbital is almost the same as in the free atom. Note that, once again, the Pauli exclusion principle applies to the molecular orbitals in a molecule as well as to the atomic orbitals in a free atom, and so no two electrons in the molecule may have identical quantum numbers. It follows, therefore, that no two of the four C–H bonds in methane can possess the same set of principal, etc., quantum numbers, and so the four bonds must have very slightly different energies to satisfy this rule, at any instant.

This may constitute a point of difficulty, since we are taught that the four bonds in CH_4 are identical, but the problem can be resolved by distinguishing between the instantaneous bond energy and the time average bond energy. The time average energy is the same for all four bonds, but the energies of the four separate bonds are different at any one given instant. Another way of saying this is that the four bonds vary in energy over a range and that no two bonds have the same energy, within this range, at the same time. This is both a cause and a consequence of the fact that materials are not static, but the atoms are in constant movement relative to one another, a fact which will be of great importance in subsequent discussion.

The LCAO method is especially applicable to covalently bound atoms in simple molecules, but it is also applicable to three-dimensional structures, such as diamond, where each bond between two atoms can be treated in virtual isolation, and it can be extended to include metals, and even ionically bonded materials [10]. In metals the LCAO method must be modified slightly to fit the somewhat different conditions which apply in these three-dimensional molecules. Bonding in metals is considered in terms of the band theory [11], wherein all the outer, or valency, orbitals in the atoms of the metal are thought of as contributing to molecular orbitals which cover all the atoms in the crystal, instead of covering pairs of atoms as above. The resulting bands of molecular orbitals comprise large numbers of separate bonding and anti-bonding molecular orbitals, one bonding/anti-bonding pair arising from every atom that contributes, all with very slightly different energies, as demanded by the exclusion principle. The available electrons enter the bonding orbital band and normally only partially fill it, since metals are very electron deficient with respect to the next heavier rare gas, and it is the virtual continuum of energy levels that is generated that gives rise to the high thermal and electrical conductivities of metals. In such a system, electronic transitions can occur between bands, as between the levels in an atom, but now there will be a range of energies over which the transitions can occur if they involve the bonding electrons, although the presence of non-bonding electrons in virtual atomic orbitals will provide sharp transitions also.

To summarise, the properties of materials, and particularly their mode of interaction with radiant energy, can be understood quite well in terms of the atomic and molecular orbitals which current theories of bonding invoke. Such a simplified picture, however, does not explain all the features shown by materials; for example it does not explain why the atomic environments of certain cations are unsymmetrical [12] while simple ionic theory would predict them to be symmetrical, or why certain cations are colored in solution.

To understand these and other effects we must improve our mathematical description of materials and this we can do by considering the application of ligand field theory, which considers in more detail the interaction which can occur between the electric fields of a group of atoms and a central atom which they surround.

1.1.4 Ligand Field Theory

Let us begin with some observations. It is well known that copper metal is a reddish color, that anhydrous copper sulphate ($CuSO_4$) is white, that hydrated copper sulphate ($CuSO_4$) is light blue, that a solution of copper sulphate in water is light blue in color, that addition of excess ammonium hydroxide gives a dark blue color, but that addition of concentrated hydrochloric acid gives a green color. In like manner, aqueous nickel solutions are green and dimethylglyoxime in methanol is colorless but together they produce a dark red complex, or colorless aluminium and aluminon solutions give a characteristic bright red lake, and these examples may be reduplicated many times. Two questions, in particular, arise from these observations, firstly why do we observe colors at all, and secondly why do we observe different colors for the same cation in contact with different anions?

The first question may be answered by noting that we observe an object to be white or colored depending on whether, in the light reaching our eyes from the object, the intensities of all the wavelengths in the visible region of the electromagnetic spectrum are equal, or unequal, due to preferential emission or absorption of certain wavelengths by the object. If excess intensity is emitted by the object the eye 'sees' the emission color, e.g. a sodium flame emits at 5893Å and so appears yellow, but if light is preferentially absorbed the eye 'sees' the complementary color, so that copper ions in aqueous solution absorb at about 8100Å, in the orange region, and impart a blue coloration to the liquid. Hence the formation of a colored compound implies that this compound can emit or absorb electromagnetic radiation of specific wavelengths preferentially, and this, in turn, indicates the presence in the compound of quantised vibrators with energy levels separated by a gap equivalent to the photon energy of the light involved.

There remains the problem of the different colors exhibited by the same cation, and we may illustrate this with copper. As we have seen, anhydrous copper sulphate is white whereas the hydrated compound is blue. Crystal structure analysis has shown that the principal difference in the two compounds is centered around the copper ion, which is surrounded by five oxygen ions in a very unsymmetrical arrangement in the anhydrous form but by six oxygens in a distorted octahedral arrangement in the hydrated compound [13]. This octahedral arrangement is also found in the hydrated ion in solution and it is reasonable to suppose that the color is somehow associated with this atomic arrangement.

The absorption of energy in the visible region, i.e. of a relatively low photon energy radiation, implies the existence in the absorber of levels which are energetically quite closely spaced, a situation which does not obtain in the majority of simple ions on the basis of straightforward ionic theory based on the Bohr-Rutherford model. According to the simple theory of filling orbitals (section 1.1.3), the electron configuration in the Cu^{2+} ion is $1s^2$; $2s^2$, $2p^6$; $3s^2$, $3p^6$, $3d^9$ and the energies of the five 3d-orbitals are the same. As a consequence of this, there will be one electron unpaired in the 3d-orbital set, and the ion will be paramagnetic with the corresponding magnetic moment of 1.73 Bohr magnetons [14]. The experimental value for cupric ions lies very close to this theoretical value, but other cations belonging to the transition metal and rare earth sub-groups show abnormal magnetic properties, so that Co^{2+} ions (theoretical moment 3.87 B.M.) exhibit magnetic moments in the range 4.1-5.2 B.M., whereas Cr^{3+} (theoretical moment 3.87 B.M. also) has a moment of about 3.8 B.M. Such magnetic anomalies are often associated with colored compounds.

Bethe [15] was the first to suggest a model to explain these observations, which he postulated in the form that is now conventionally called the crystal field theory, and is based on a purely electrostatic or ionic approach. The idea that the ligands surrounding an atom or ion could form covalent bonds with it by donating electron pairs to it had been developed by Pauling [9], and was applied a few years later, in the mid-1930's, by Van Vleck [16] to the same problem and is now referred to as the molecular orbital treatment. This treats the problem from a covalent bonding point of view, but has a close fundamental relationship with the crystal field treatment, since both refer to the symmetry of the atoms in the complex surrounding the central atom or ion. Crystal field theory, in its original form, suffers from not making allowance for the partly covalent nature of the metal-ligand bonds involved, but it provides a simple treatment of many aspects of the electronic structures of complexes and is more convenient to use than the more complex molecular orbital theory. Today, the crystal field theory has been modified by introducing empirical adjustments to certain parameters to allow for this partly covalent nature, without introducing the complications of covalency. Readers should note, however, that ligand field theory is also used to denote any of the gradation of theories ranging from crystal field to molecular orbital. Cotton and Wilkinson [17] have discussed nomenclature and give a rigorous discussion of these theories.

Basically, Bethe pointed out that it was unreasonable to consider an ion in a material to be completely isolated and that it must be considered as part of a unit with its surrounding atoms and groups. He showed that, since electric and magnetic fields are associated with all atoms, if two atoms are placed contiguously, their electrical and magnetic fields will overlap and interact and the net effect of this interaction will be to split the five equi-energetic d-orbitals of ions in the transition metal series into sub-groups with dissimilar energies. The manner of this splitting depends critically on the symmetry of the atomic arrangement about the ion and on the strength of the ligand involved. In practice, we can distinguish between regular octahedral and tetrahedral, distorted octahedral and tetrahedral, and square co-planar arrangements, which last can also be considered as an extremely distorted octahedral case. As shown diagrammatically in figure 1.4a, a regular octahedral symmetry produces two groups of orbitals, the three t_{2g}-orbitals with energy less than the original d-orbitals, and the two e_g-orbitals with higher energy. Progressive distortion of the octahedron further splits these groups until, in the square co-planar configuration, the energy of one orbital has increased to a level far above those of the other four, which are nearly equal. Figure 1.4b shows that the situation is similar in tetrahedral symmetry, except that here the splitting gives two e-orbitals with lower energy and three t_2-orbitals with higher, and distortion of the

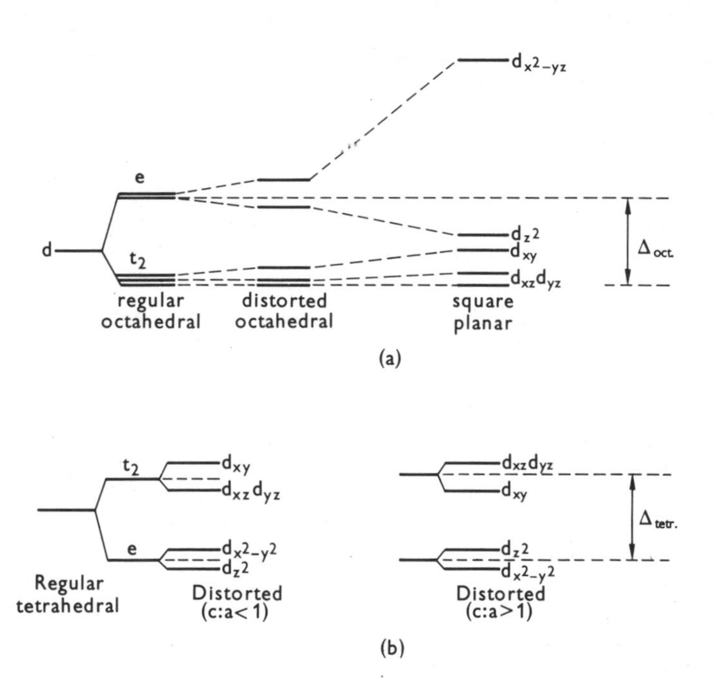

Figure 1.4 Splitting of d orbital energy levels due to crystal field effects in a number of environments. (a) in an octahedral field which is, from left to right, regular about the central atom, distorted towards a tetrahedral shape, and grossly distorted to give a square planar arrangement; (b) in a tetrahedral field which is regular or distorted in two senses about the central atom. (Reproduced from ref. 12, by kind permission of Oxford University Press.)

tetrahedral arrangement results in interchanging the energy levels within these groups rather than further splitting of the levels. The available electrons in the central ion then distribute themselves among these new orbitals, according to the usual rules [12]. Balhausen [18] and Orgel [19] have written at length on ligand field theory and its applications, for those readers who wish to pursue the topic further.

1.1.5 Applications to Physicochemical Analysis

From the point of view of the techniques discussed in later chapters, a major importance of crystal or ligand field theory lies in its ability to account for colors. As we have seen, the effect of the surrounding ligands is to split the d-orbitals of the central cation into two groups with energy separations of the order of 200 kJ.mole^{-1}, which is comparable with the energies of chemical bonds and is equivalent to the energy of a photon of electromagnetic radiation with a wavelength of about 5000-6000Å, i.e. within the visible range. Moreover, the magnitude of the splitting depends on the strength of the ligand involved, and studies have shown that the more common ligands can be arranged in their order of ability to cause d-orbital splitting, with the same cation in a given oxidation state, as

$$I^- < Br^- < Cl^- < F^- < OH^- < H_2O < -NCS^- < NH_3 < NO_2^- < CN^-$$

Thus, in the copper series quoted above, $CuCl_4^{2-}$ absorbs in the low energy, red region of the spectrum at 10,500Å, $Cu(H_2O)_6^{2+}$ absorbs in a higher energy region at 8100Å, and $Cu(NH_3)_4^{2+}$ absorbs in the highest energy region at 6500Å. The electronic transitions are, of course, between the t_{2g} and e_g levels in octahedral complexes and between e and t_2 in tetrahedral, i.e. between levels which do not exist in the absence of the ligands in the required symmetry about the ion. Note that the ligands quoted in the above list include both anions, in which the charge field and the free electron pairs are active, and uncharged molecules, in which the molecular dipole and again the free electron pairs play a role.

But, important as ligand field theory is in treating the optical spectrum of an element in its various compounds, it is also important to realise that ligand field effects are not confined to optical spectra. The effect of the electromagnetic field of contiguous atoms affects all the energy levels of all the electrons in an atom or ion, and the effect is simply more immediately noticeable in the case of optical spectra. In particular, it is vital to realise that the energies of the *K*- and *L*-shells will be modified by the crystal field surrounding the atom, and so the energy difference between them will be affected. But, as we shall see, the wavelength of the characteristic X-rays produced by an element depends on the energy difference between the *K*- and the *L*-shell, and so the effect of the ligand field at this level is to shift the wavelength of the $K\alpha$ emission peak depending on the ligand field.

White [21] has shown that this can constitute a quite noticeable effect which is particularly important when trying to determine elements quantitatively, since its effect is to make the required peak wander with respect to the measuring system and so give spuriously low readings. In electron probe microanalysis studies, particularly, he has also shown that this effect can be used to indicate the coordination shell around an atom or ion, and he can distinguish between, for example, Al^{3+} in 4-coordination and in 6-coordination with oxygen, and even between these and Al^{3+} in 5-coordination with oxygen in favorable cases. The effect, therefore, is one which the analyst must beware of at one level, and use with gratitude at another since it can give information unavailable from any other source.

1.2 PRODUCTION OF SPECTRA

Having discussed the factors within an atom which can be involved in energy transitions leading to the absorption and emission of electromagnetic radiation, we must now turn our attention to the ways in which these transitions can be brought about and to the fraction of atoms or groups of atoms actually involved in the observed transitions occurring in a system under study. We shall do this by first considering the significance of the temperature of the system and the mechanisms whereby transitions can occur, and then applying these to high energy events involving the nucleus, to intermediate energy events involving the extranuclear electrons, and to low energy events involving atomic and molecular movements.

Let us first consider the states in which a system may exist at different temperatures but in the absence of major perturbing forces or energy sources. Thermodynamic theory requires that all materials be in their 'zero point energy' state at 0°K, i.e. that at this temperature all the atoms in the body be at rest relative to one another and that the electronic configuration be that of the 'ground state' or lowest energy state. For a free atom this corresponds to the configuration given in table 1.1 and for a bonded atom to a configuration in which all the electrons are in the lowest energy atomic or molecular orbitals available. At temperatures above 0°K, materials possess energy in excess of their zero point energies, which energy must be stored by the atoms in the body and corresponds to the heat capacity. Einstein [2] showed that storage occurs principally as quantised mechanical vibration of the atoms, although a certain amount of energy is also stored by promoting electrons into higher, or 'excited', states in the atoms or molecules.

Now, Maxwell [22] had earlier shown that the thermal energy of a gas is stored as kinetic energy of motion of the individual gas molecules and that their velocities were spread over a large range of values, with a distribution which could be represented by curves such as those shown in figure 1.5. Note that, at a given temperature, there exists a most probable velocity and a spread of velocities and that, as the temperature rises, the most probable velocity increases in value and the spread of velocities moves to a higher range, so that the fraction of molecules with a velocity in excess of a given

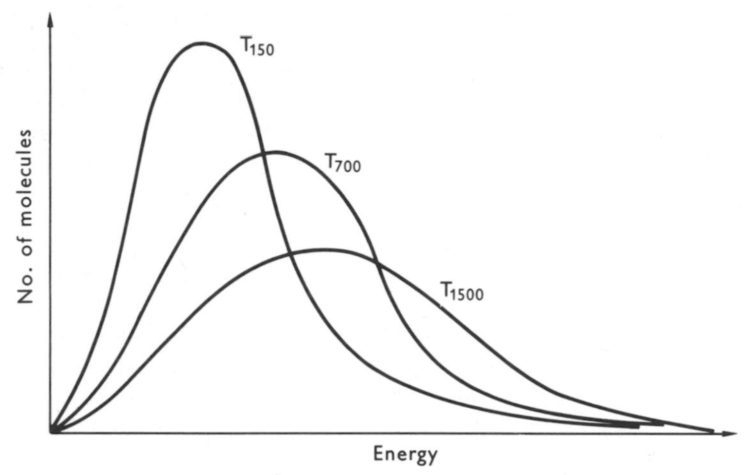

Figure 1.5 Maxwell-Boltzmann curves relating the fraction of molecules with a given energy to increasing energy at different temperatures. Note how the average energy and the range of energies increase with increasing temperature.

value increases markedly with temperature. This fraction is given by the integrated area under the curve from the chosen energy to infinity, divided by the total area. Application of the more general Boltzmann equation [23] shows that this fraction can be written in the form

$$N = N_0 \exp\{-E/RT\} \tag{1.5}$$

where N is the number of molecules with velocities equal to or greater than the chosen velocity v, N_0 is the Avogadro number, E is the energy corresponding to the chosen velocity ($E = \frac{1}{2}N_0 mv^2$ where m is the molecular weight), R is the gas constant, and T is the absolute temperature.

This Maxwell-Boltzmann distribution law was derived in terms of non-quantum statistics for the system involved, and so is formally inapplicable to systems of atoms or electrons, which must be considered as indistinguishable particles. Bose [24] and Einstein [25], however, showed that the corresponding distribution law for such a system takes the form

$$n_i = \frac{p_i}{C^{-1}\,e^{E_i/RT} - 1} \tag{1.6}$$

where n_i is the fraction of particles possessing energy E_i/N_0, p_i is a statistical weight factor required to allow for the possibility of a number of quantum levels of almost identical energies, and C is a constant. However, for atoms and molecules, $C^{-1}e^{E_i/RT}$ is very large compared with 1, and so equation (1.2) reduces to the form

$$N = p_i C \exp\{-E/RT\} \tag{1.7}$$

for the fraction of atoms or molecules with energy greater than E/N_0. This is of the same form as the original Maxwell-Boltzmann equation, with the addition of the statistical weight factor, and the simpler equation will suffice for the present discussion. Also, although Fermi-Dirac statistics [26] show that the above approximation cannot be made rigorously for electrons, again the Maxwell-Boltzmann relationship is sufficiently good for an elementary discussion.

Thus we find that, in any system at any but the very highest temperatures, the vast majority of atoms, molecules and electrons are in lower energy states, although there will always be a fraction in higher energy states. Moreover, the fraction of particles in higher energy states increases with temperature, and this relationship holds both for 'classical' systems involving non-quantised energies and for quantised energy systems. We can now proceed to consider how these high and low energy populations can be used in the detection and determination of the whole population present.

1.2.1 Nuclear Reactions

The energy levels available to the nucleons are quantised in a manner identical to any other system, but the energies involved in the transitions between energy states are appreciably greater than those for extranuclear processes. The ground state in this case corresponds to the stable nucleus and excited states to radioactive isotopes, to a first approximation. An activated nucleus can be formed from a stable one by increasing its energy, which involves bombarding the latter with a beam of moderately high energy nucleons, high energy electrons, or photons. So-called 'thermal' neutrons, with energies of the order of about 4 electron-volts, are often used, particularly in 'Neutron Activation Analysis' (Chapter 4), and are readily available as a by-product of a nuclear pile or from an accelerator, such as the Dynamitron [27].

The resulting activated nucleus can shed its excess energy by emitting either a photon of γ-radiation, in which case the atom retains its chemical identity, or a nucleon, electron, or α-particle, in which case the nucleus may transmutate into that of another element. The 'daughter' element will be one higher in the periodic classification if an electron is emitted, one lower if a proton or positron is emitted, two lower for an α-particle, but an isotope of the same element with one unit less of atomic weight for a neutron. Alternatively, the nucleus can capture an extranuclear electron from the K-shell, to give the element one lower in the classification, with the emission of the X-rays characteristic of the daughter nucleus. Hence there are quite a large number of possible routes whereby the excited atom can lose its excess energy, but all share the feature that the quantised energy of the emitted photon, electron, or nucleon depends on the nucleus from which it originates. In practice, the radiation may come either from the original, 'parent', nucleus or the 'daughter' nucleus derived therefrom by one or other of the nuclear reactions, and so precise characterisation of the radiation in terms both of its energy and its total intensity, in conjunction with a knowledge of the relevant decay series, can not only identify the 'parent' nucleus, but also give an estimate of its abundance.

It is, however, important to remember, in discussing the observed intensity of the radiation from a sample and hence the rate of decay of the excited nuclei, that only a fraction of the atoms present in an irradiated sample are ever converted into the excited form and so become capable of taking part in the emission process. Moreover, of these activated nuclei, only a small fraction is emitting at any given instant and so contributing to the observed signal. Since the observed intensity can be used as a measure of the total number of atoms of an element present, it follows that the total population of an element is determined from the signal generated by only a small fraction of the atoms present. Complete standardisation of conditions is thus necessary in order to ensure that the same fraction of atoms is 'seen' for different samples in a series, or in the samples and the calibration standards used. This point will be discussed again in the next section.

1.2.2 Extranuclear Electronic Transitions

Many quantised transitions, with a wide range of energies, are possible within the extranuclear electronic structure of an atom. As we saw in section 1.1, these transitions are associated with movements of electrons between available atomic or molecular orbitals within the atom or molecule and, since the energy differences between orbitals vary greatly depending on whether the inner or the bonding orbitals are involved (cf. figure 1.2), many different energy jumps are possible. Which jump or jumps occur depends on the amount of energy supplied to the atom.

Energy to stimulate transitions from low to high energy states can be supplied by heat energy, by electromagnetic radiation with wavelengths in the region between γ-rays and visible light, or by a beam of accelerated electrons of intermediate energy. We may then choose to monitor either the amount of energy re-radiated by the system as its atoms return to their ground state levels, the emission or fluorescence spectrum, or we may monitor the amount of energy abstracted from the stimulating medium, in absorption spectrometry.

Heat and Electromagnetic Stimulation

Stimulation by heat energy involves relatively small energy increments, and so such stimulation tends to excite only the electrons in the outer orbitals, where the energies of transition between adjacent states are small. Subsequent decay to the ground state is accompanied by emission of photons in the visible region to give the characteristic visible light emission spectrum (Chapter 2). Since only a very limited number of electrons are involved, the spectra produced are relatively simple, since there exist only one or two routes whereby the atom can lose its excess energy during the decay process. In common with all other methods, only a small fraction of atoms, governed by equation (1.7), are actually excited and radiate, and so stimulation by heat is very dependent on the temperature of the flame used.

Incident electromagnetic radiation can be absorbed by an atom or molecule and the photon energy re-radiated either without any alteration in energy, or after some of the energy has been dissipated within the atom, when a photon of lower quantum energy is produced, or the energy may be dissipated wholly within the atom or molecule and be degraded to heat energy before re-radiation. Radiation scattered without a wavelength change may contribute to diffraction effects, particularly in the case of X-rays (Chapter 7), but is not useful for identifying atomic species present. Radiation scattered with a wavelength longer than the incident wavelength, due to the photon energy loss, appears as fluorescent radiation with a wavelength characteristic of the atom or molecule involved and so may be used for elemental analysis (Chapter 3). Moreover, the energy so abstracted from the incident beam of radiation can be measured, as it can in the case of complete degradation to heat, and also used for elemental analysis, in absorption spectrometry (Chapter 2).

In most of the above cases, then, the method used involves first stimulating some of the atoms or molecules into excited states and then detecting the energy emitted as they drop back into their low energy states again. We have seen, in section 1.2, how the energies of the particles in a system are distributed over a range of values, and equation (1.7) gives the form which is useful for electrons. It will be clear from this that the number of atoms, etc., which can exist in an excited state at any temperature under the different stimulation regimes is a fraction, and often a very small fraction, of the total number present. Hence, only a small fraction of the particles present actually generate a signal at any time, and this fraction will depend critically on the conditions obtaining in the system at the time of measurement. Temperature is particularly important in this context, since it controls the number of atoms, etc., which can be promoted to the high energy state, and quite small changes in temperature can noticeably affect the fraction of particles going into the excited state and subsequently radiating. Hence temperature changes affect the observed signal appreciably.

This is not usually very important in qualitative work, where a stronger signal may be a positive advantage, but it is very important in quantitative measurements since the amount of an element present is normally measured by comparing the observed intensity of the radiation from the sample with the intensities of the same radiation from standards containing known concentrations of the element, measured under identical conditions. It follows that, since the intensity of the signal depends on the fraction of atoms, etc., generating the signal, these methods rely on the identical fraction of atoms being excited to their higher energy states in any specimen studied under a given set of conditions. Naturally, the total population is being estimated on the basis of the reaction of a part population, but a constant part population, to the stimulation medium, and it will be clear that, since this part usually represents only a small fraction of the whole, even slight changes can give rise to large errors in the overall estimate. Thus a change from 1% to 1.1% of the population becoming excited will represent only a small change in the number of atoms taking part in the emission process but a 10% increase in the observed signal, and hence a 10% overestimate in the derived concentration of the element. Close control over the conditions under which emission and fluorescence spectra are measured, and especially the temperature of the system, is critically important.

It also follows from the above discussion that, however many atoms, etc., are excited and emit their characteristic radiation, very many more atoms remain in their low energy states, even at temperatures around 1000°C. These atoms are capable of absorbing radiation, usually radiation with an energy equal to or slightly greater than the emission energy. Moreover, their greater numbers can give a proportionately greater absorption signal in many situations, compared with the emission signal, particularly when the energy-equivalent wavelength lies in the visible region. This has been put to good use in the relatively new technique of atomic absorption spectrometry (Chapter 2) wherein highly monochromatic, visible radiation is passed through the vapour of the sample in a flame and is absorbed specifically by one atomic species with reduced spectral interference from other atoms in the sample.

Absorption of emitted radiation by the atoms in a sample can, however, interfere noticeably with the observed signal strength by reducing its intensity at the surface of the specimen, fortunately in a systematic, if complex, manner. This phenomenon applies particularly to studies involving X-rays, and the effect of atomic absorption will be discussed in Chapters 3 and 7. Techniques are currently being developed to utilise such absorptivities in the determination of heavy elements, such as lead, or in the measurement of particle sizes, especially in suspension [28], but such applications are outside the scope of this book.

Electron Beam Excitation

Stimulation by electrons occupies a somewhat special position in the list of energy sources for excitation, due to the dual wave-particle nature of the electron. This has been discussed by Glasstone [7]. Most importantly, if an electron is accelerated through a potential gradient of V volts, the associated wavelength is given by

$$\lambda = \sqrt{\frac{150}{V}} \times 10^{-8}\,\text{cm} \tag{1.8}$$

provided that the potential is sufficiently low that the rest mass may be assumed for the electron. For voltages above 10kV this assumption no longer holds and the more accurate form, which allows for relativistic effects,

$$\lambda = \sqrt{\frac{150}{V}} \times \frac{1}{1 + 4.95 \times 10^{-7}\,V} \times 10^{-8}\,\text{cm} \tag{1.8a}$$

must be used. Thus the electron is capable of transporting considerable photon energies and is able to stimulate all kinds of electronic transitions within the atom, and hence to generate a wide range of spectral radiations from it. At these energy levels there exist a variety of routes whereby an atom may shed its excess energy, and each route will give rise to one or more types of radiation to produce complex spectra. The different measurement techniques based on, for example, the characteristic X-rays generated (Chapter 3) or the Auger electrons liberated (Chapter 11) thus operate simply by concentrating attention on the one or the other effect, and readers should realise that all possible types of radiation are generated simultaneously when a sample is irradiated by an electron beam.

Accelerated electrons can be absorbed and re-radiated without energy change, and such coherently scattered electrons contribute to electron diffraction patterns (Chapter 8) or to scanning electron micrographs (Chapter 11), or they may be absorbed and re-emitted as lower energy electrons, after dissipating energy within the atom or solid, to give the characteristic Auger and the semi-characteristic secondary backscattered electrons, also discussed in Chapter 11.

Alternatively, the electron beam may displace electrons from the various shells in the atom, and this can give rise to a wide range of electromagnetic radiations. If the innermost electrons are fully ionised, the subsequent transitions from, for example, the *L*- to the *K*-shell or the *M*- to the *L*-shell produce the *K* and *L* series X-rays characteristic of the element. These relatively high energy transitions are then followed by a series of lower energy transitions, resulting in the emission of longer wavelength radiations, as the outer electrons move into the inner shells to re-establish the ground state configuration. These wavelengths lie in the ultra-violet and visible regions of the spectrum but, as the discussion of the outer, valence electrons given in section 1.1.3 shows, they will be of less use for identifying separate elements because, on the one hand, the relative closeness of the atomic orbitals makes it less easy to assign a given wavelength to a single transition between atomic orbitals and, on the other, and more importantly, the existence of molecular orbitals will spread the emission lines into bands which are more difficult to monitor. Cathodoluminescence is, however, used in the identification of certain minerals, especially in the scanning electron microscope (Chapter 11). Recently there has been an upsurge in interest in utilising as much of the information generated by electron beam bombardment of the specimen as possible, and so instruments are being built which combine transmission and scanning electron microscopy with X-ray fluorescence analysis and ultra-violet and visible cathodoluminescence studies.

Stimulation by electrons suffers from the absorption problems mentioned in the previous sub-section, and results in electron beam studies being confined generally to the outer layers of the specimen, because of the difficulty both of getting the electron beam into the material and of getting the generated signal out. Temperature is not quite so critical, since the energies involved are very much higher than thermal energies. Electron beam stimulation is, at times, a somewhat mixed blessing, since the beam can be focused onto a very small area of the specimen to give precise information about, for example, variations in composition over short distances in the sample, but the very range of wavelengths generated simultaneously can prove to be an *'embarras de richesses'* in some situations (Chapter 3, section 3.2.1). In the main, however, the advantages far outweigh the disadvantages and electron beam stimulation constitutes the basis for some of the most powerful tools presently available for investigating materials.

1.2.3 Atomic and Molecular Movements

It was shown, in section 1.2, that the thermal energy of a body is principally stored as quantised atomic vibrations. These vibrations are subject to the rules of quantised states and to the modified Maxwell-Boltzmann energy distribution pattern, as are electrons, but the quantised energy transitions involved in going from low to high energy states are much smaller than for electrons. Hence these atomic and molecular movements correspond to longer wavelength electromagnetic radiations, in fact to radiations lying in the infra-red regions (figure 1.1), and form the basis for infra-red spectroscopy (Chapter 9).

Movement of the atoms in a molecule or crystal relative to one another clearly results in variations in the instantaneous distances between pairs of atoms, the bond lengths, and in the angles between atom-atom vectors from a given atom, the bond angles. The overall motions of the atoms will be complex and controlled by the strengths of the separate bonds in the structure, but they may usually be resolved into motions which result in changes in the quantities quoted above, i.e. motion along the line of centres between atom pairs changing the bond length, the so-called stretching vibrations, and motions in the planes of adjacent bonds changing the bond angle, the so-called bending vibrations. These motions have been illustrated in figure 1.6, by reference to the carbonate ion, CO_3^{2-}.

As shown, the ion possesses four basic modes of vibration, of which two are doubly degenerate and one represents movement of the electron cloud only. Note how modes V_1 and V_3(a) represent stretching vibrations only, whereas mode V_4 represents pure bending vibration, and mode V_3(b) represents mixed stretching and bending. Of the four modes, all except V_1 produce a change in the dipole moment and so are infra-red active. V_1 is

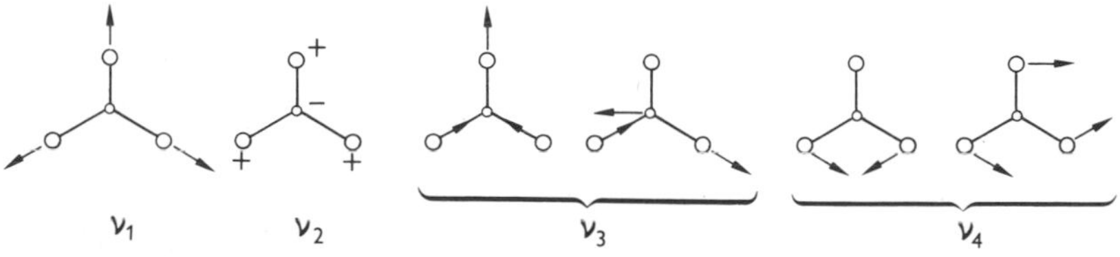

Figure 1.6 Modes of vibration in the carbonate ion, CO_3^{2-}. Modes V_2, V_3 and V_4 are infra-red active, mode V_1 is Raman active. (Courtesy of Dr. V. C. Farmer.)

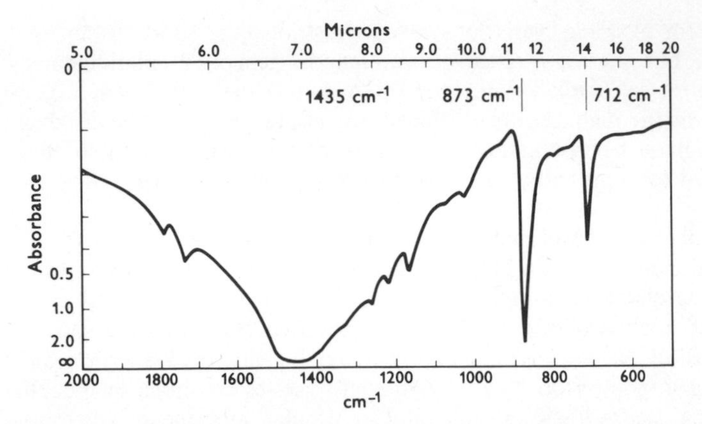

Figure 1.7 Infra-red spectrum for calcium carbonate showing the three absorption bands for the CO_3^{2-} anion. The lines corresponding to bending and electron cloud movement can be seen clearly at 712cm^{-1} and 873cm^{-1}, but the stretching band at 1435cm^{-1} is somewhat hidden within the general Ca-O absorption band between 1000cm^{-1} and 1800cm^{-1}.

infra-red inactive but is active in the related Raman spectrum (Chapter 9). Hence the simple carbonates absorb at three quite well defined wavelengths to give an absorption spectrum with three distinct absorption bands. Stretching vibrations involve larger energies than do bending motions and so vibrational modes involving stretching occur at higher frequencies, shorter wavelengths, than do bending modes. Movement of the electron cloud is intermediate between stretching and bending. The spectrum of calcite, $CaCO_3$, is shown in figure 1.7.

For the particular case of the stretching mode, the theoretical frequency at which absorption of infra-red radiation occurs is given by

$$\nu = \frac{1}{2\pi c}\left(\frac{k}{\mu}\right)^{1/2} \tag{1.9}$$

where ν is the wave number in cm^{-1}, k is the force constant, in dynes$\cdot$cm^{-1}, acting between the atoms; c is the velocity of light, and μ is the reduced mass given by

$$\mu = \frac{m_a \cdot m_b}{m_a + m_b} \tag{1.10}$$

where m_a and m_b are the isotopic weights of the atoms involved.

Note that there is, once again, a need for close standardisation of procedure, especially when making comparative measurements on two samples. Once more, temperature control is important, although some latitude is allowable for purely qualitative measurements.

If the temperature is raised by appreciable amounts, especially by tens or hundreds of degrees, the atomic movements thereby induced may become so great that changes may occur in the structure itself. These changes may take the form of loosening hydrogen bonds in certain silicates [29], or of polymorphic transformations, as in the α- to β-quartz transformation at 578°C, or they may involve loss of volatiles, as in dehydration or carbonate decomposition reactions. These last quoted changes bring us into the range of thermal analysis, wherein weight losses or changes in the rate of heating of a body relative to a standard body are monitored and give information relating to the phases originally present or to the reactions occurring. Thus, as will be seen in Chapter 10, the amount of weight lost from a hydrated mineral can indicate its water content, and the temperature range over which the loss occurs can suggest the form in which the water is present in the crystal structure. Differential thermal analysis is useful for giving further information relating to the amount of heat energy required to bring about changes, both those which also involve weight changes and those which involve no change in weight but only a change in crystalline structure. Thermogravimetry and differential thermal analysis differ somewhat from the other methods discussed, however, in that they do not involve directly monitoring absorbed or emitted electromagnetic radiation or electrons.

1.2.4 Ionisation of Atoms

At the very high temperatures found in an electric spark, minerals are not only vaporised but many of the atoms are also converted into ionic species by the loss of one, or occasionally more, electron. These positive ions can be accelerated through a potential gradient of some tens of kilovolts, passed through an energy selector, and the resulting beam of ions, all with the same kinetic energy, used as input to a mass spectrometer. Here they are subjected to a homogeneous, strong magnetic field at right angles to their direction of motion, under whose influence the ions are forced to move in circular paths with radii of curvature determined by the mass-to-charge ratio (in effect to the value of $\sqrt{m/e}$) for each ion in the beam. The result is to spread the ions out into a spectrum, with the lightest elements detectable nearest to the inlet slit and heavier elements detectable at systematically increasing distances. The position of a beam of ions relative to the slit thus identifies the nature of the ion, and its intensity gives the concentration, in the usual way. The method is discussed in Chapter 5, and again does not involve direct monitoring of electromagnetic radiation.

1.3 DETECTORS FOR ELECTROMAGNETIC RADIATION

Various forms of detector are used to monitor and measure the radiations emitted, absorbed, or diffracted in the course of physicochemical methods of analysis. They include photographic film or plate, particularly useful for qualitative or semi-quantitative methods, and electronic detectors of several kinds, of especial use in situations where fast detection rates are needed and in many quantitative analytical methods. Such detectors may conveniently be sub-divided into the low energy group, used for longer energy wavelengths in the visible and infra-red regions of the spectrum, and the high energy group, used for γ-rays, X-rays, and electrons. We may consider the three groups separately.

1.3.1 Film Methods

Photographic emulsions, deposited either on celluloid or glass substrates, provide an easily used, flexible, relatively inexpensive, and quite accurate detection system for electromagnetic radiations over a large range of the electromagnetic spectrum. They are particularly useful for wavelengths lying between the visible and the X-ray regions, although they find some uses in detecting γ-rays directly and indirectly (Chapter 4, sections 4.6 and 4.7). They are eminently suitable for qualitative studies, but they can be used for quantitative purposes provided that sufficient care is taken in their exposure and subsequent processing. Film methods possess the undoubted psychological advantage of providing a visible record of the phenomenon being investigated, although it is important to remember that the film will record only those emanations which will affect it, and this may result in the loss of information or the addition of spurious information if the wrong conditions are chosen.

The principles underlying the formation of a photographic image through the action of a beam of electromagnetic radiation or ionised species are probably quite familiar, and have been well discussed in many books [30]. Briefly, however, the emulsion consists of a large number of basically silver bromide and iodide crystals, diameter 0.1-1μm, distributed evenly throughout a gelatine base. When this emulsion is exposed to a beam of electromagnetic radiation or ions, a very few silver ions in specific crystallites are reduced to silver atoms, and these act as catalysts to assist in the reduction of the remaining silver in these affected grains during the subsequent development stage, and as nuclei for the resulting silver particles. It is important to remember that the permanent image is composed of silver particles and that the light and dark areas comprise regions of low and high particle densities, depending on the pattern of low and high beam intensities incident on the different regions of the film. Note that silver in unaffected crystallites is not normally reduced in the development process, and is removed in the fixing step, but excessive development can result in this silver being reduced, and so contribute to the background 'fog' on the film. Insufficient development, on the other hand, can result in not all the affected particles being fully reduced, and the production of a weak image. Stray radiation incident on the film also contributes to fogging, by sensitising crystallites not contributing to the image. Such radiation can arise from a variety of sources and its effect generally increases with the exposure time.

Thus the position and the darkness of the image is determined primarily by the position and intensity of the beam incident on the film, but the intensity is also noticeably affected by the conditions of development. Moderate care in choosing the length of exposure and the development time, in order to give a sufficiently dark image without excessive fogging, usually suffices for qualitative purposes, but great care must be taken to standardise both exposure time and development conditions for any quantitative applications.

A further property of the photographic image must be considered in quantitative work, however, arising from the particulate nature of the photo-sensitive crystallites in the emulsion. As we have seen, the image is generated by photons or ions sensitising individual crystallites, which subsequently become silver particles whose density determines the degree of darkening at a given point in the film. Since one photon or ion gives rise to one silver atom, and as few as ten silver atoms can sensitise a single crystallite, it would appear reasonable to suppose that a straightline relationship would be found between the number of photons incident on the film and the degree of blackening at that point. Hence the degree of blackening could be used as a measure of the intensity of the radiation and so ultimately of the concentration of the element or crystalline phase generating the beam. In practice this condition holds over a limited, if quite wide, range of intensities, as shown in figure 1.8. Very low intensities suffer from the graininess of the film, although this is not a serious problem with modern fine-grained emulsions, but high intensities suffer from the much more important problem of saturation. As the crystallites in a given region

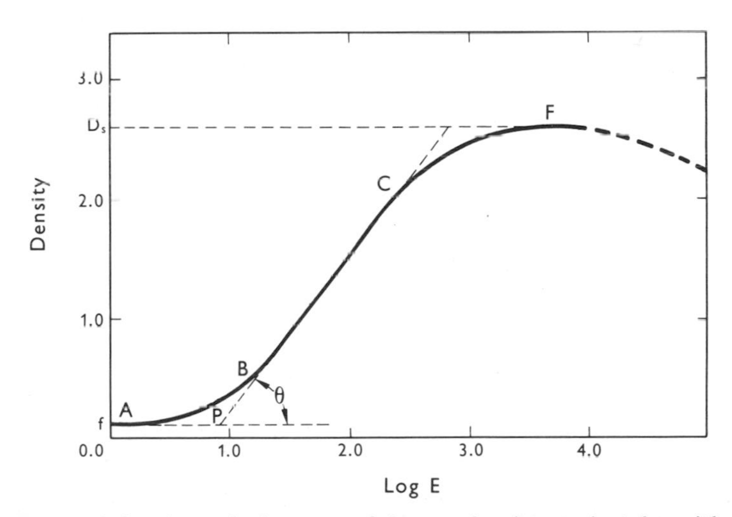

Figure 1.8 A typical curve of image density against logarithm of exposure, showing the toe (AB), the region of direct proportionality (BC), the shoulder (CF) where proportionality no longer holds, and the saturation density, D_S. (Reproduced from ref. 30, by kind permission of Macmillan Publishing Co.)

of the film, subject to a very intense beam, become sensitised, fewer are left unsensitised for later photons to affect and a stage is eventually reached where the probability that an incoming photon will interact with an unsensitised crystallite is very small compared with the probability of its interacting with a previously sensitised one. Such photons, therefore, do not contribute to the final degree of blackening of the film, and hence the observed intensity will be lower than that expected on the basis of extrapolation from regions not subject to saturation.

These problems can be circumvented by making a set of exposures for different times and using the shortest times to give intensity data for the most intense beams, the longest for the least intense, and features common to several films to provide the necessary normalisation factors between the different exposures.

Film techniques also suffer somewhat from the fact that photographic emulsions are largely non-discriminatory for wavelengths over quite wide ranges, but respond to virtually all radiations with wavelengths shorter than the limiting value for the type of film being used. Moreover, the fogging, which corresponds to 'noise' in electronic detectors (next sections), can make measurement of quantitative values more difficult, especially when the fogging is not uniform across the film. These effects are particularly noticeable in X-ray film work, when air-scattered X-rays cause general fogging of the film, and the presence of iron in a sample irradiated with copper *K* radiation gives rise to unwanted iron *K* fluorescent X-rays which cause general blackening of the film and can completely obliterate the diffraction image (Chapter 7).

Despite these difficulties and drawbacks, film methods constitute an important form of detector in such techniques as arc spectroscopy, spark source mass spectrometry, X-ray diffraction, and electron microscopy.

1.3.2 Low Energy Detectors

Electronic detectors for longer wavelengths depend on the use of sensitive materials to convert the incident radiant energy into an electrical signal, plus the necessary circuitry to measure and record the voltage or the current generated. The detectors used for ultra-violet and visible radiation depend on the generation of electrons from photo-sensitive materials, whereas those for infra-red wavelengths depend on the heating effect of such radiation on a thermocouple junction. Since all the methods depend ultimately on measuring the numbers of electrons set free by the incident radiation, the strength of the signal, voltage or current, will be a measure of the amount of radiant energy incident per unit time, and so of the intensity of the radiation.

Photoelectric Cells

Electrons are liberated from certain materials when radiation of energy greater than a minimum quantum value is incident on them, in a manner akin to the stimulation procedures discussed in section 1.2.2. The electrons

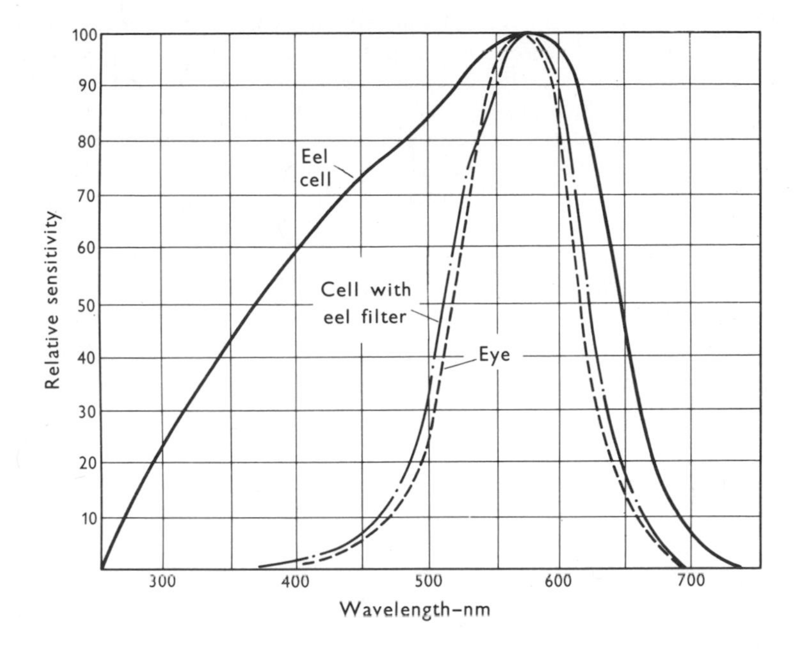

Figure 1.9 Spectral response curve for '*EEL*' selenium cell with and without a correction filter to match with the response curve of the normal eye. (Courtesy of Evans Electroselenium Ltd.)

so generated will come from the outer, valence levels, and usually involve ionisation from a molecular orbital band in the substance. We showed, in section 1.1.3, that transitions from such an orbital band can occur over a range of energies, and so photoelectric cells normally respond to a range of wavelengths, as shown in figure 1.9. It follows that different photoelectric cells should be chosen for work in different regions of the spectrum, since the signal generated by a cell operating at the extreme high or low wavelength end of its range will be very small indeed. Hence, most optical spectrometers are fitted with two photocells, one sensitive to wavelengths in the red portion of the spectrum, 4000-9000Å, and the other to the blue region, 3000-5000Å. Photoelectric cells fall into two main classes, the photoemissive and the barrier layer types.

The photoemissive cell consists basically of an evacuated bulb containing a cathode, comprising a layer of caesium or potassium oxide doped with silver oxide deposited on a polished metal sheet, and a ring anode maintained at some tens of volts positive relative to the cathode. A typical cell and the basic circuit used are shown in figure 1.10. Electrons set free by the incident radiation are attracted to the anode and allow a current to flow through the upper circuit in figure 1.10, and also change the potential drop across the resistance R. The changes in the current or the potential drop can be amplified, detected, and measured using a simple meter for simple applications, but a potentiometric circuit for accurate values, and related to the intensity of the light falling on the cathode. Theoretically the relationship should be linear, since the greater the number of quanta incident on the detector the greater the number of electrons emitted, and so the larger the signal, but in practice the signal tends to fall off for higher intensities, largely due to a saturation effect similar to that observed with film. This non-linearity is usually corrected for in the subsequent measuring circuits.

The photomultiplier tube, shown in figure 1.11, is a development of the simple photoemissive cell. The cathode is again coated with a photosensitive layer to provide the primary electrons, but the anode consists of a series of

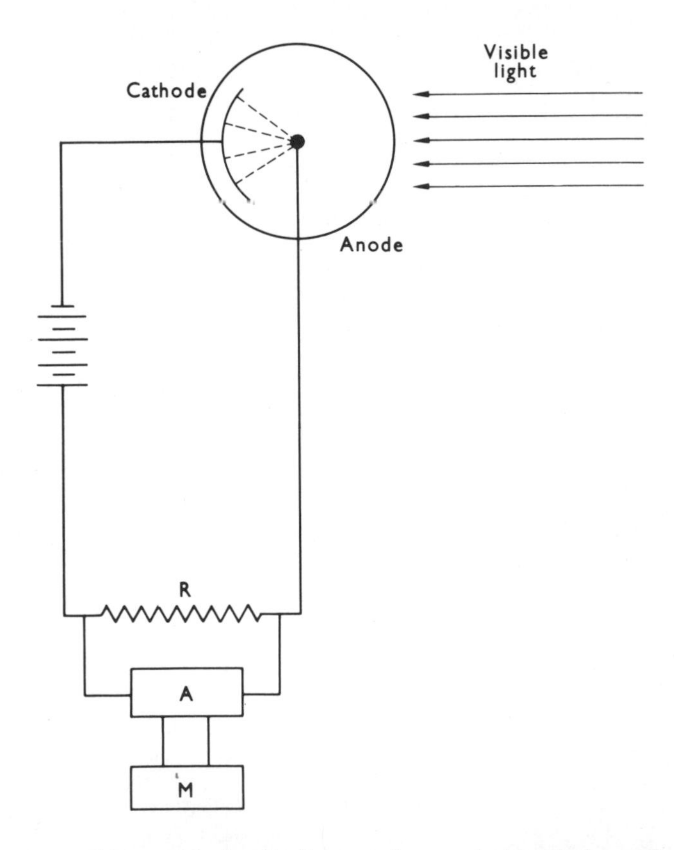

Figure 1.10 Schematic circuit diagram for a photoemissive cell. The dotted lines show the paths of the electrons emitted from the cathode and collected by the anode, R is a circuit resistance, A an amplifier, and M the voltage measuring unit.

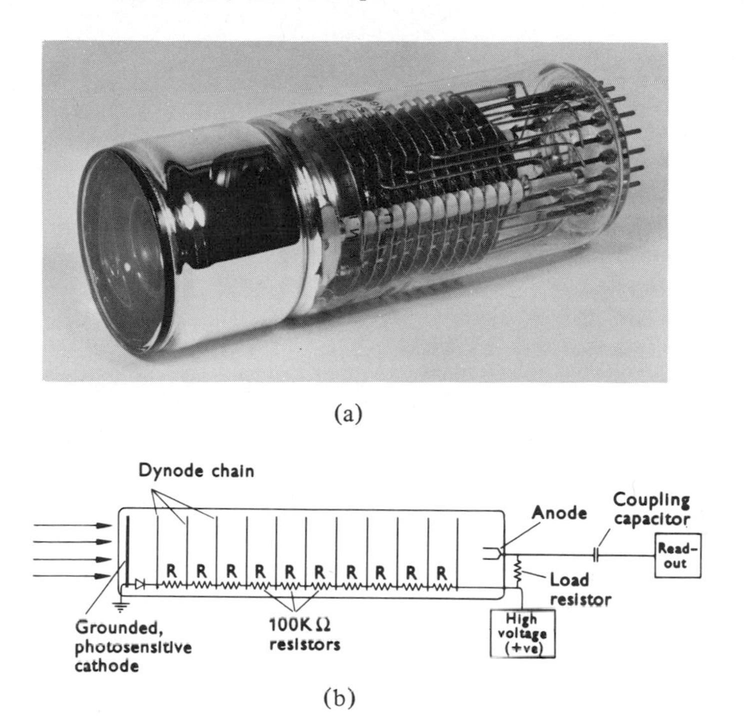

Figure 1.11 The photomultiplier tube (a) and block schematic of its electronic circuitry (b). (Photograph by courtesy of EMI Electronics Ltd.)

plates at successively higher voltages, coated with a material which emits electrons on bombardment with electrons. Hence the primary electrons are accelerated across the gap to the first anode plate where they release further electrons, which repeat the process at the second plate, and so on down the tube in a cascade effect. Thus, for a tube containing 10 plates coated with a material emitting 4 electrons per incident electron, the multiplication factor for the electrons set free will be about 10^6x. Output currents are limited to the milliamp range, however, and so the photomultiplier tube can measure only low incident intensities, and high intensities can damage it. It can measure intensities of about 1/200 of the strength of those measurable with the ordinary phototube.

The barrier layer, or photronic, cell is different in construction and mode of operation from the photoemissive cells, although it again depends on the ability of light to set free electrons from a material. The photosensitive element comprises a photosensitive semiconductor layer, selenium is often used, deposited on a metal plate and covered with a thin, transparent metal layer. Electrons set free in the selenium pass through the theoretical 'barrier' layer between it and the metal layer, to set up a cell with the selenium layer and its backing plate as the positive pole, and the transparent metal cover as the negative, as shown in figure 1.12. The resulting current is usually measured directly, since it is not very easy to amplify the direct current signal accurately. Barrier layer cells are rugged and give good linearity of intensity response with a low resistance in the measuring circuit. The inability to amplify the signal, however, makes them a poorer choice than photoemissive cells for the most sensitive applications. High levels of illumination, however, merely produce a temporary fatigue effect, which can be corrected by storing the cell in the dark.

Detectors for infra-red wavelengths are either a simple bimetallic thermocouple or a series of thermocouple beads in series, a thermopile, to give a greater signal. Chromel-alumel junctions are normally used, to give the maximum signal for a given amount of incident radiation. The voltage generated against a cold junction is amplified, measured and recorded in the usual way.

Low energy detectors show very poor discrimination of wavelengths, but respond to values over quite a wide range, as shown in figure 1.9. Their response can be varied by the use of filters, and figure 1.9 shows how the response of a barrier layer cell may be modified by the use of the relevant absorption filter to match more closely the response of the human eye, for applications in reflected light microscopy. Electronic discrimination by pulse height analysis cannot be used, however, and so the resolution obtained in an optical spectrometer by using a prism or grating cannot be backed up in the manner possible for high energy detectors (section 1.3.3).

1.3.3 High Energy Detectors

These methods are again based on the interaction of the radiation being measured with matter, but this time they depend on the ionising power of the radiation involved. The ionisation, which produces free electrons and positive ions or positive holes, may occur either in a low pressure gas in contact with charged electrodes, as in the gas proportional and Geiger-Muller counters, or in a crystal such as sodium iodide activated by thallium, the scintillation counter, or in a solid state semi-conductor crystal, usually silicon or germanium doped with lithium and operated at liquid nitrogen temperatures. Each detector produces signal pulses in numbers proportional to the intensity of the incident radiation, and these are amplified and passed through a pulse height discriminatory circuit, if necessary, for display on a paper chart recorder or as a signal integrated over a pre-determined time interval.

Gas Detectors

These are based on, and have developed from, the ionisation chamber which, in its simplest form, comprises a pair of electrodes, one maintained at earth potential and the other at a potential of some volts positive relative to earth, contained in a partially evacuated vessel at a constant, low pressure of gas. Ion pairs, formed by the interaction of the incident radiation with

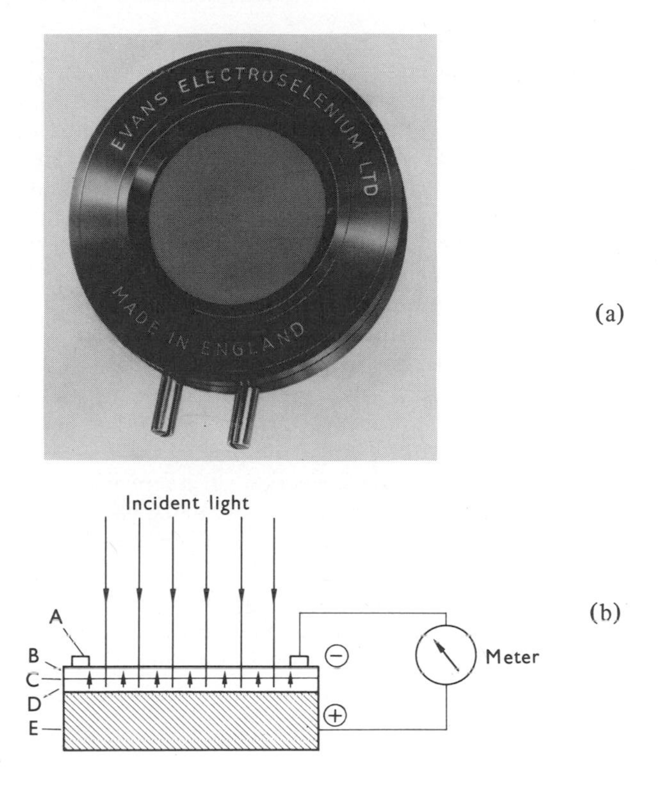

Figure 1.12 Barrier layer photocell (a) and schematic circuit diagram, showing A, collecting ring; B, transparent metal layer; C, hypothetical barrier layer; D, selenium layer; E, metal base. Note how the incident light penetrates the transparent metal layer and sets free photoelectrons in the upper part of the selenium layer which pass through the barrier layer, as shown by the small arrows, to be collected by the metal layer and ring. (By courtesy of Evans Electroselenium Ltd.)

the gas atoms, are collected by the electrodes and the resulting voltage pulse produces the output signal, which may be amplified as needed and recorded. Such an instrument is capable of detecting α-particles and electrons or β-rays quite readily, but the ionising effect of electromagnetic radiations is generally insufficient to generate a detectable signal. Increasing both the physical size of the electrodes and the voltage between them can overcome this difficulty, and makes the Geiger-Muller and gas-proportional counters suitable not only for X-rays but also for lower energy γ-rays.

We may understand why this should be so by considering the series of events which may occur when a high energy particle or photon interacts with a gas atom in a counter tube. The initial interaction knocks an electron from the atom and forms a positive ion plus a free electron. In the absence of an appreciable electric field, the electron will drift about, undergoing minor collisions with other atoms, until either it is recaptured by an atom or ion, or it reaches the anode plate. If a significant electric field exists, the electron will tend to move rather more rapidly towards the anode, being accelerated under the influence of an ever increasing force the nearer the electron comes to the electrode, and will again undergo collisions with gas atoms. As the voltage across the counter increases, so the energies of the free electrons increase until they reach a level at which their interactions with the gas molecules are sufficiently energetic to eject secondary electrons from these particles, forming further ion pairs. If the field gradient is so high that an electron can regain the energy it lost in forming one ion pair before it collides with another gas molecule, it will be clear that this process will become cumulative and that an 'avalanche' of electrons, and positive ions, will be produced, the so-called 'Townsend avalanche' [31] . Thus the end result of the formation of a single ion pair by the interaction of a photon with the counter gas is the collection of a very much larger number of electrons at the anode, and hence of a very much larger output signal. If the incident particle or photon produced n primary ion pairs by direct ionisation, and Zn electrons are collected at the anode, the factor Z is the gas multiplication factor and clearly depends on the operating voltage across the electrodes, while the voltage of the output pulse is $-Zne/C$, where e is the charge on the electron and C is the capacitance of the detector.

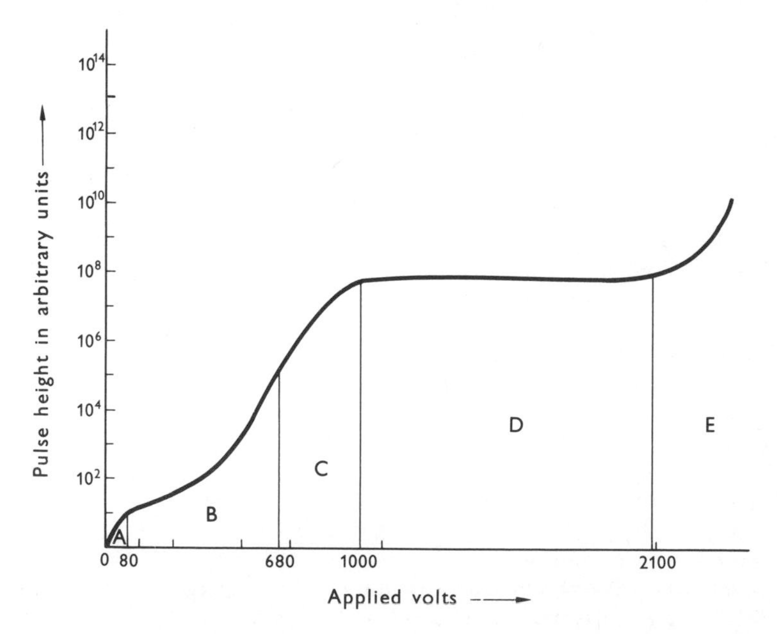

Figure 1.13 Variation of the gas multiplication factor, Z, with increasing applied voltage across the electrodes of a gas-filled ionisation detector tube.

Figure 1.13 shows the number of ion pairs received by the electrodes per quantum of incident radiation from a fixed source, plotted against the operating voltage for a typical gas-filled detector. In effect we are plotting Z against the operating voltage. The marked regions of the curve may be identified as follows:

A – the ionisation chamber region, in which the applied voltage is so low that virtually no gas multiplication occurs ($Z \approx 10$) and the number of electrons collected at the anode is of the same order of magnitude as the number of ion pairs produced.

B – the proportional region, in which the applied voltage is sufficient to permit an appreciable but limited amount of gas multiplication, the value of Z lying in the range $10^2 - 10^5$. Note that the value of Z depends markedly on the exact value of the operating voltage.

C – the limited proportionality region, in which the value of Z has become large, but is still obviously voltage dependent.

D – the Geiger plateau region, in which the value of Z has become so large that saturation effects have set in, for reasons basically the same as those for film (section 1.3.1), and Z has become effectively voltage independent.

E – the emission region, in which the operating voltage is so high that a continuous discharge is set up between the electrodes, rendering the system useless as a counter.

Regions *B* and *D* represent the conditions used for gas proportional and Geiger-Muller counters respectively. Region *A*, the ionisation chamber, represents the conditions used for the detection of α- and β-particles.

The above discussion has been developed in terms of the electrons only, since these are by far the fastest moving particles in the system, but it must be remembered that the avalanche process necessarily forms a sheath of positive ions near to the anode, where the field gradient is highest, and this has to be dissipated before another avalanche can develop, i.e. before another incoming photon can be detected. This 'quenching' involves, basically, discharging the positive ions at the cathode, a process which is beset by two main difficulties. In the first place, the heavy positive ions move relatively slowly to the cathode, resulting in long 'dead times' when the counter tube is inoperative, and, in the second place, the positive ions can accumulate sufficient energy during their approach to the cathode that their impacts thereon can liberate electrons, which can proceed to trigger a new avalanche and so give rise to spurious counts. The quenching is speeded up in most counters, and the energy of the ions incident on the cathode is minimised, by adding small amounts of organic or halogen gases to the detector to absorb the positive charge and redistribute the energy of the positive ions by collisions which result either in the dissociation of the polyatomic molecules or the dissipation of the energy among the available bonds. Note that the lower value of Z for a gas proportional counter means that less positive ions are formed and so the dead time for a gas proportional counter is much less than that for a Geiger-Muller tube.

(a)

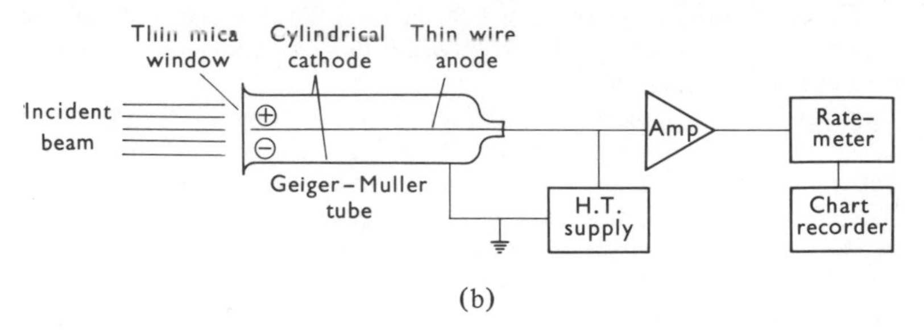

(b)

Figure 1.14 Geiger-Muller counter tube (a) and its associated electronic circuitry (b) in block schematic form. Note that the amplifier (Amp) may be omitted in some situations.

The Geiger-Muller Counter

This is the simpler and cheaper counter and is shown in figure 1.14. The metal cylinder, typically about 10cm long and 2cm diameter, is earthed and acts as the cathode while the coaxial, centrally mounted tungsten wire forms the anode at a positive potential of 1–1.5kV. This potential ensures that a sufficiently high electrical field exists in the argon or krypton gas filling the tube to ensure that a continuous discharge is maintained briefly throughout the length of the tube by the avalanche process, once the primary ion pair has been formed by an incoming photon or electron. Under the prevailing conditions, the gas multiplication factor is so large that the same number of electrons are received at the anode irrespective of the number of ion pairs initially formed. The counter, therefore, cannot discriminate between photons of different quantum energies, but counts all incident photons equally. Note that the counter is operated in the region in which small drifts in the operating voltage do not appreciably affect the value of Z, i.e. the counting rate of the tube is intrinsically stable and it does not require a well stabilised potential source. Moreover, the counter produces large voltage pulses which require little or no amplification before being measured and recorded.

The incident radiation enters the counter through a thin mica end window, and ion pairs formed in the 'dead space' between the window and the end of the anode wire are not detected. This, together with the dead time, results in the counter not responding to all the radiation incident on it, but only to a fraction. Typically the dead time for a Geiger-Muller counter is of the order of 50μsec for a self-quenched tube containing halogen gas, while the time to form the avalanche can be measured in nanoseconds at most. Thus the tube is 'off' for an appreciable fraction of the time, and theoretically cannot register rates greater than about 2×10^4 counts per second. In practice, photons do not arrive at the tube in the evenly spaced manner that the theoretical count rate would demand, but tend to be bunched, so that many of the photons are not registered. It is usual to make allowance for this by applying the correction formula

$$R_t = \frac{R_m}{1 - R_m t} \qquad (1.11)$$

to give the true count rate, R_t, from the measured count rate, R_m, for a counter with a dead time t. Thus the efficiency of the counter falls as the count rate increases, and there is an appreciable deviation from linearity at rates above 10^3 counts per second. Note that the dead time used in equation (1.11) contains a contribution from the lag in the measuring circuits for the tube.

The Geiger-Muller counter is thus a robust, cheap, and easy to use instrument. The associated electronics are not sophisticated and the output signal requires no pre-amplification before it is measured. The tube does suffer from its inability to discriminate between radiations with different energies, and this makes it less useful for precise measurements than the other high energy counters. Geiger-Muller tubes find their major uses in simple β-radiation detectors and some of the cheaper X-ray diffractometers.

The Gas Proportional Counter

The gas proportional counter, shown in figure 1.15, is basically very similar to the Geiger-Muller tube and, in principle, the two are interchangeable since the only formal difference is the operating voltage. In practice, however, the proportional tube differs from the Geiger-Muller tube in that the incident radiation enters the chamber through a window in the side of the tube and the window is electrically continuous with the walls of the cylinder, to ensure a symmetrical electrical field within the tube. A thin

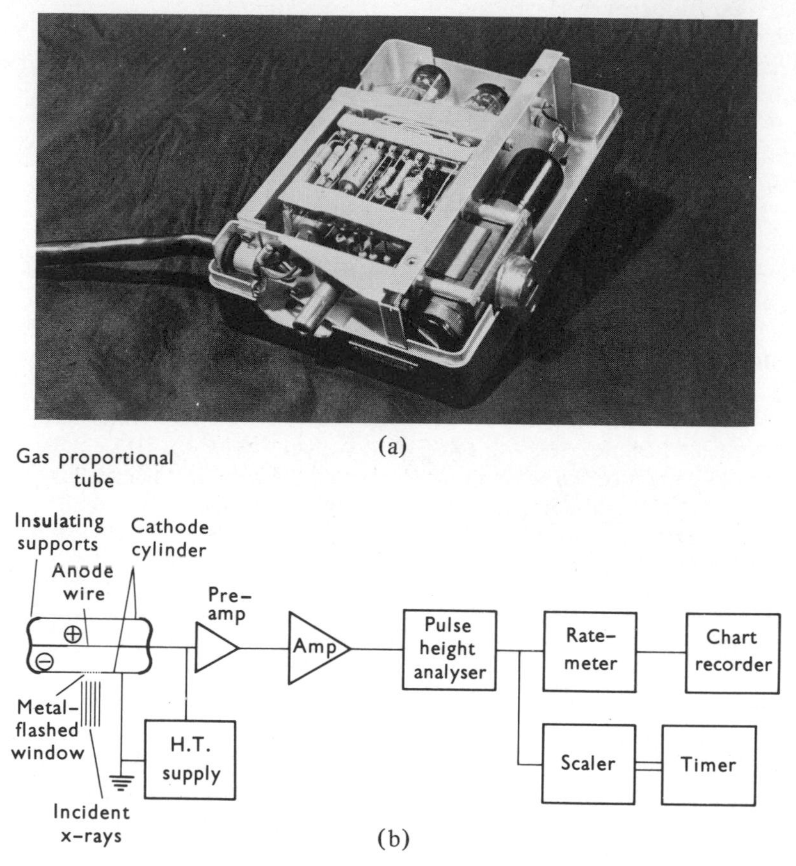

Figure 1.15 Gas proportional counter tube (a) and block schematic of its electronic circuitry (b). Note the need for a pre-amplifier located physically close to the counter tube, to boost the very weak signal produced to a sufficiently high value to allow it to be passed down a longer connection to the main amplifier and counting chain without unacceptably high losses.

beryllium sheet can be used for more energetic radiations, e.g. β-rays but long wavelength X-rays require an even less strongly absorbing window, and Mylar, polypropylene, or polycarbonate film is used, flashed lightly with aluminium to provide the necessary electrical properties. The filling is normally a mixture containing 90% argon and 10% methane, to give a self-quenching system. The quenching action, however, results in the gradual decomposition of the organic molecules, and the argon can diffuse through the thin window, so sealed gas proportional tubes have a limited life. The gas flow proportional counter, in which the atmosphere is continuously renewed by slowly bleeding in new gas from a cylinder to preserve the same quenching characteristics, is preferred for many applications requiring very stable counting characteristics over a long period of time.

The operating voltage for the tube is chosen such that the gas magnification factor is of the order of $10^4 - 10^5$x, i.e. formation of a single primary ion pair does not result in the tube becoming saturated by the subsequent Townsend avalanche. Instead, at a given operating voltage, the size of the output pulse ($-Zne/C$ volts) is controlled by n, the number of primary ion pairs formed by the incident photon, which is, in turn, controlled by the quantum energy of the photon. It follows that the proportional counter is able to discriminate between photons of different energies, from the voltages of the separate output pulses, provided that the operating voltage of the tube is kept absolutely constant. This, of course, demands a very much better stabilised power source than the Geiger-Muller tube, and the lower value of Z means that the output signal is much weaker and requires pre-amplification before it can be measured and recorded by rather more sophisticated circuitry.

The design of the proportional counter eliminates the dead space and the smaller value of Z reduces the dead time to about 0.2μsec. This is generally shorter than the counting time of the associated circuitry, and so the effective dead time is controlled by the electronics, and is typically of the order of 2μsec. The proportional counter is thus capable of much higher counting rates than the Geiger-Muller type, and 5×10^4 counts per second are possible without excessive deviation from linearity, although equation (1.11) must be used for precise rates.

The gas proportional counter is less rugged than the Geiger-Muller, and its associated electronics are very much more complex and expensive. It also requires rather more servicing, since the window can become damaged, the gas atmosphere can deteriorate, and dirt on the central wire can markedly affect the performance of the tube. Despite these operational difficulties, however, the gas proportional tube is widely used, especially in X-ray spectrometry and diffractometry because of its good discriminatory powers and its basic simplicity.

The Scintillation Counter

In the scintillation counter, shown in figure 1.16, the interaction of the incident radiation takes place in a solid medium, a crystal of sodium iodide activated by the addition of about 1% thallium, to generate an electrical signal by a two stage process. The incident photon first promotes an electron from the bonding molecular orbital band of the sodium iodide into a conduction band, then the 'positive hole' left in the valence band migrates to a thallium ion, and finally the promoted electron and the positive hole recombine at the thallium atom with the emission of a photon of visible light. The minimum energy required for the initial stimulation is 3eV, although the losses which occur in both stages of the overall conversion

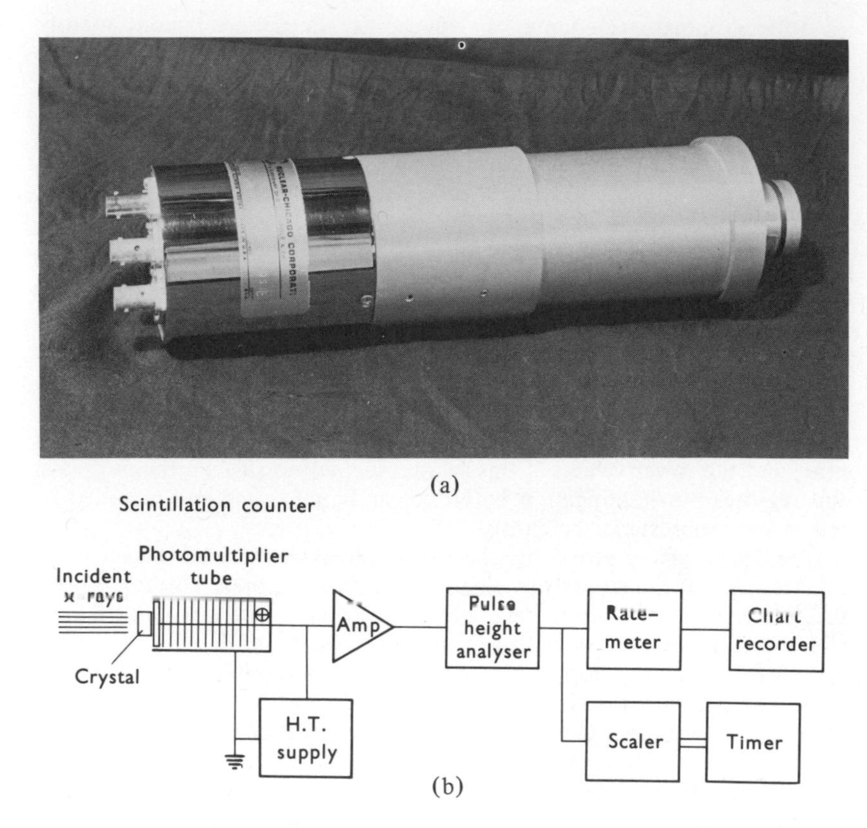

Figure 1.16 Scintillation counter (a) and block schematic circuitry (b).

process raises this effectively to about 300eV, and photons of higher energies can stimulate proportionally more electrons, to generate more photons of visible light. The light produced next activates the cathode of a photomultiplier tube (section 1.3.2) to produce signal pulses with voltages proportional to the total light intensity generated in the crystal, i.e. proportional to the quantum energy of the photon generating the light, again at a rate proportional to the intensity of the incident radiation. It follows that the scintillation counter is again able to discriminate between photons of different energies, although its low conversion efficiency results in a resolution much below that of the gas proportional counter.

The scintillation counter has a very short dead time, of the order of 0.1μsec, and the counting rate is again controlled by the characteristics of the counting circuit, but rates up to 10^5 counts per second may be handled. The output signal is small and again needs to be pre-amplified, and the photomultiplier tube also needs a stable power source.

The scintillation counter is another robust and easily used instrument. It requires little maintenance, although care must be taken to prevent any bright light from falling on the crystal while the photomultiplier tube is at its operating voltage otherwise it may burn out due to the size of the resulting avalanche, and so the hygroscopic sodium iodide crystal must be sealed against visible light as well as against moisture, as shown in figure 1.15.

As mentioned earlier, the conversion efficiency of the system is low, only about 1% of the incident photons being actually counted due to the losses which occur both in the initial stimulation step and the subsequent stage of detecting the light generated, typical efficiencies being 20% and 5% detected in the respective processes. Unfortunately, the two stages clash in their requirements for improved detection efficiency, since the efficiency with which the initial quantum energy of the incident radiation is converted into light photons increases as the crystal thickness increases, but the efficiency of the transmission of this light to the photomultiplier tube decreases with increasing crystal thickness. Moreover, the higher the energy of the radiation, the thicker the crystal needed for efficient conversion. Compromise crystal thickness are thus used, depending on the application, crystals being 2-3mm thick for X-ray work, but at least 2.5cm thick for γ-ray studies. The scintillation counter is widely used in both X-ray and radiochemical applications.

Semiconductor Detectors

The other family of solid state detectors, recently developed but rapidly gaining in importance, is based on semiconductors and depends on the direct measurement of electrons set free by the ionising effect of the incident radiation. Early varieties were based on diamond and cadmium sulphide crystals, used as solid state equivalents of the gas proportional detector, and the junction barrier type, which incorporated a reverse biased n-p junction between gold and silicon as the pulse counter [32]. Today these have been largely superseded by the lithium-drifted silicon and germanium instruments shown in figure 1.17.

These detectors are based on the fact that ionising radiation can generate free electron-positive hole pairs in the intrinsic region which forms by compensation in a p-n junction, and that these free charged species can be separated by a reverse voltage across the junction and collected at the electrodes. In a normal p-n junction, the i-region, where the p-type dopant just balances the n-type dopant, is too thin to be used for detection, but its thickness can be increased manyfold by initially diffusing a limited amount of lithium into the surface layers of specially chosen, high purity, p-type silicon and germanium single crystal slices, then allowing the monovalent ion to drift through the slice, now containing a p-n junction, under reverse bias voltage conditions at about 100-200°C. Dearnaley and Northrop [32] have shown that this treatment spreads the n-type lithium evenly throughout the slice, exactly compensating for the p-type boron which it is virtually impossible to remove, and the resulting slice acts as an intrinsic

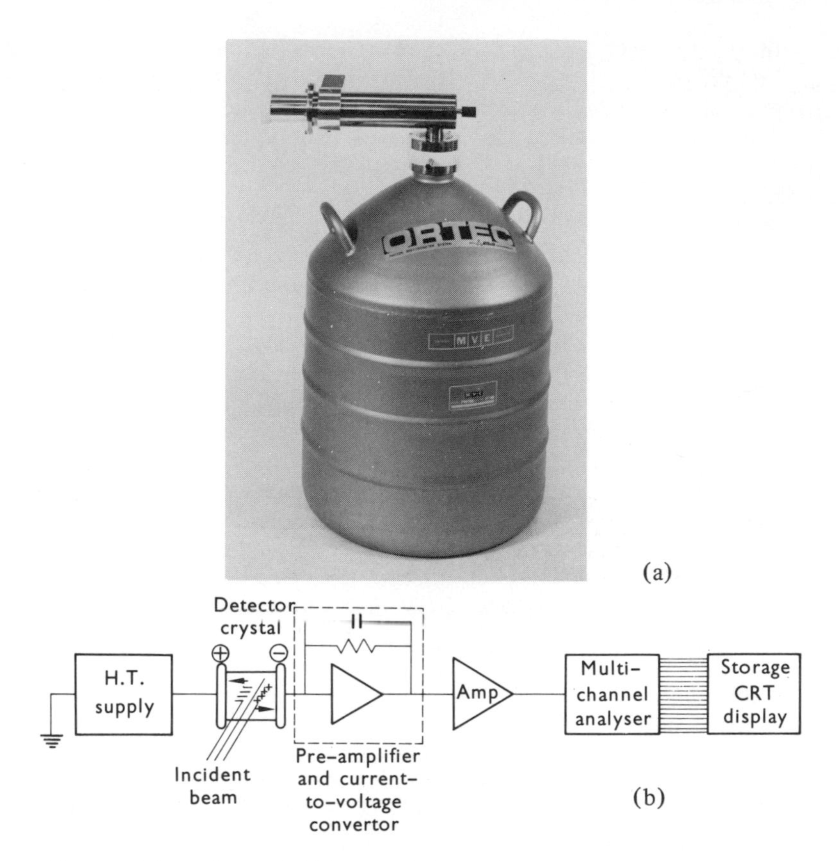

Figure 1.17 Typical lithium-drifted silicon or germanium detector system (a), and the associated circuitry in block schematic form (b). The photograph shows a common dewar-crystal unit with the tube housing the silicon or germanium detector and the preamplifier, both operated at liquid nitrogen temperatures to safeguard the crystal and to reduce electron noise to a minimum, mounted above or below the liquid nitrogen reservoir. (By courtesy of Ortec Inc.)

semiconductor. Thin n- and p-type layers on opposite faces complete the n-i-p sandwich and provide electrode contacts for the reverse bias voltage and for collecting the free charges generated in the thick intrinsic layer. Intrinsic layers up to 10mm thick can be prepared in this way but, while such thicknesses are needed to absorb very high energy radiations, layers 1–2mm thick suffice for the X-rays, or up to 5mm for the γ-rays, used in analysis. As the spectral response curves for lithium-drifted silicon [Si(Li)] and germanium [Ge(Li)] detectors suggest (figure 1.18), Si(Li) detectors are preferred for X-ray studies and Ge(Li) for γ-rays and some more energetic X-rays, operated at reverse bias voltages of the order of $100\text{V}\cdot\text{cm}^{-1}$.

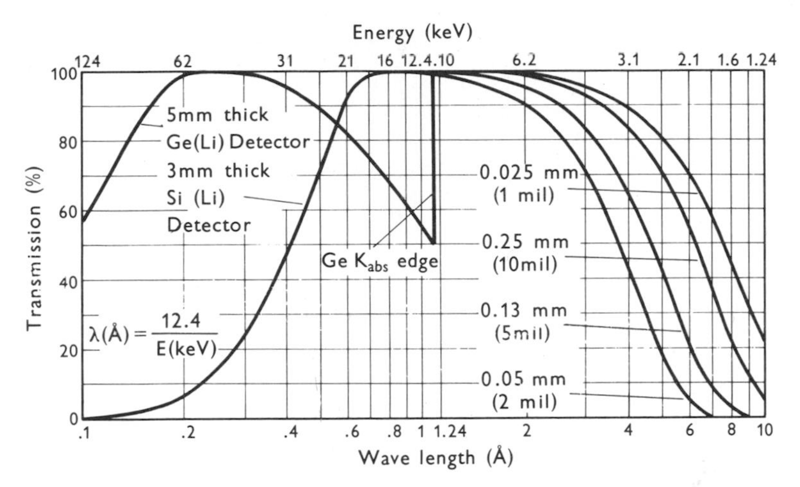

Figure 1.18 Spectral response curves for lithium-drifted silicon and germanium detectors plotted as a function of wavelength on a logarithmic scale. Note how silicon can be used over the whole of the X-ray range, whereas germanium is better suited to detecting high energy γ-radiation. (From Rao, Wood, Palms, and Fink, *Phys. Rev.*, **178**, 1997 (1969), by kind permission of the American Physical Society.)

The energy involved in forming an electron-positive hole pair is again about 3eV, with almost total conversion of the energy of the incident photon, and the pairs can be separated and collected also with nearly 100% efficiency, so that the ultimate resolution capability of lithium-drifted detectors is close to their theoretical value and at least an order of magnitude better than those of gas proportional and scintillation detectors. The resulting signal, whose voltage depends on the number of pairs formed as before, after pre-amplification is thus ideally suited for pulse height discrimination, and the extreme narrowness of the voltage spread for each line permits the use of a multi-channel analyser (next section) in conjunction with an integrating counter array, to provide directly the spectrum of wavelengths present in any beam of incident radiation (cf. Chapter 3, section 3.3.2). Figure 1.19a compares the resolution capabilities of gas proportional, scintillation and semiconductor detectors for the silver $K\alpha$ line at 0.5608Å, 22.1keV, and figure 1.19b those of a scintillation and a Ge(Li) detector for the 1.19MeV and 1.36MeV γ-rays from Co^{60}.

Lithium-drifted silicon and germanium detectors are clearly of great use in monitoring high energy radiations, and their very high resolving powers make it virtually certain that they will take over from gas and scintillation counters in many fields during the next few years. In particular, their ability to present a complete spectrum of incident wavelengths simultaneously makes them of immense use in X-ray applications, since they are able to accept all the radiation emanating from a sample and analyse and count it,

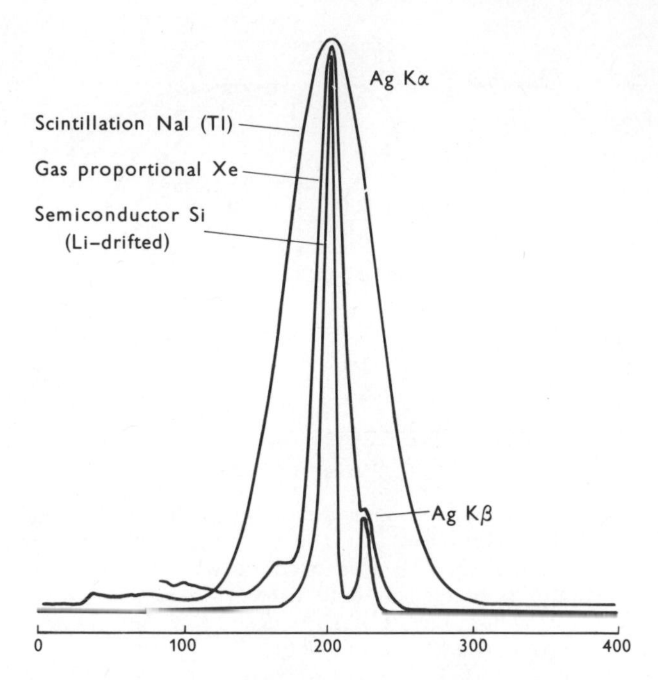

Figure 1.19(a) Demonstration of the resolution capabilities of the three common types of X-ray detector for the silver *K* spectrum obtained from a ^{109}Ag source. (By courtesy of P. G. Burkhalter and W. J. Campbell, U.S. Bureau of Mines, College Park, Md.)

instead of losing a large fraction during diffraction from the analysing crystal. This is particularly important when low intensity X-rays are involved, as in the scanning electron microscope (Chapter 11), but the technique has also been applied to on-line control systems, especially in North America (Chapter 6). The initial cost of the system is moderately high, although by no means excessively so in comparison with other similar systems and the cost can easily be recouped through subsequent savings, but the major disadvantage to the system is that both detectors must be operated at about 100°K, and the germanium variety must be stored at this temperature also since allowing it to stand at room temperature for even a few hours will result in out-drifting of the lithium and loss of its detecting properties. However, even this last is less of a problem now that dewar vessels are available which can store sufficient liquid nitrogen coolant to maintain the required temperature for periods of up to one month, and so ease maintenance of the equipment. Otherwise the instruments are quite robust and present few major problems in use.

In addition to Dearnaley and Northrop [32], Jenkins and De Vries [33], Miller and co-workers [34], and Tavendale [35] have reviewed the uses of high energy detectors generally in X-ray spectrometry and nuclear studies,

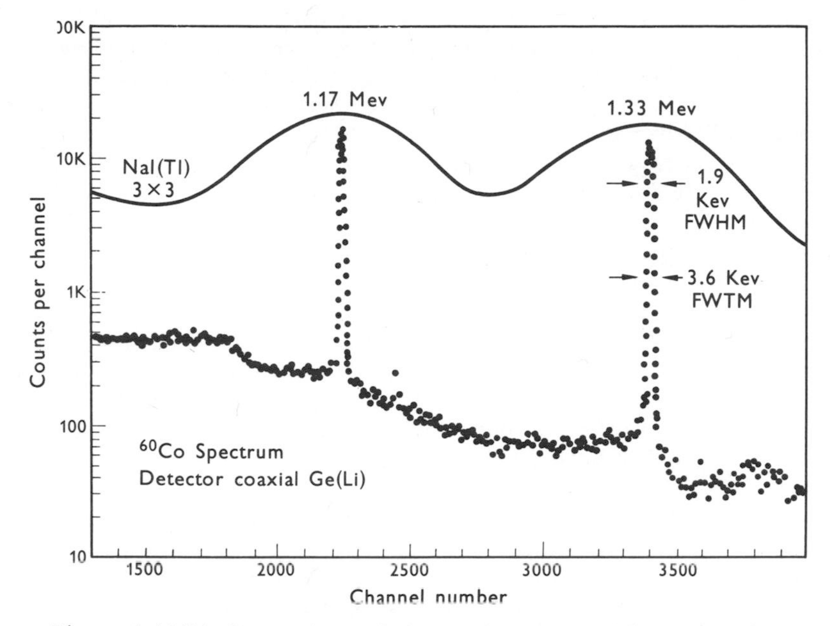

Figure 1.19(b) Comparison of the resolution capabilities of a lithium-drifted germanium and a thallium-activated NaI scintillation detector for the 1.19MeV and 136MeV γ-rays from ^{60}Co. (By courtesy of Nuclear Enterprises Ltd.)

and Brownridge [36] has assembled a bibliography on lithium-drifted germanium detectors.

1.3.4 Electrical Circuitry

Since most manufacturers supply the relevant electrical circuitry for the detector used, the comments in this section will be made in the most general sense, but it may be profitable to mention one or two recent developments, and to sound some words of warning. In the first place, it must be remembered that the eventual resolving power of a detection system is a function not only of the resolving power of the detector but also of the resolving power of the associated electronics, so that the gross resolution includes contributions from all the high voltage sources, amplifiers, and measuring chains in the complete system, and economy in these areas will be very definitely false economy since this may produce results that are, at best, poorer than they might be and, at worst, completely misleading. Electronic 'noise', or random fluctuations in the signal, is the great enemy of resolution and it is important to ensure that all components of the system should be as stable as possible, should show a minimum tendency to drift, and should introduce as little noise as possible into the signal.

Thus, highly stabilised voltage sources are readily available for all detectors, but the source for a Geiger-Muller detector is less stable, and so cheaper, than those for the other detectors, since voltage fluctuations are not so critical in this instrument, but such a source must not be used, for example, for a gas proportional detector since fluctuations in the high voltage will alter the electrical field in the detector and so change the avalanche characteristics and hence the 'Z' factor in the output voltage pulse. The effect of this will be to generate two signals of different voltages from identical photons, and hence give errors in the final energy spread.

Most detectors generate low voltage signals which must be amplified to a level at which they can be handled, and clearly it will be this amplified signal that is actually measured. It follows that the pre-amplifier used must be capable of amplifying the input pulses accurately and stably, within the limitations of the input signal. Dearnaley and Northrop [32] have discussed the generation of noise within the detector itself, and point out that the inherent levels in gas proportional and scintillation detectors are such that relatively unsophisticated solid state circuits suffice for these. With the advent of semiconductors with much lower noise levels and higher intrinsic resolutions, however, reduction of the pre-amplifier noise has become much more important, and circuits including field effect transistors operating at temperatures around 100°K are used to minimise the noise introduced into the signal. The effect of this has been to allow manufacturers to guarantee resolutions of better than 250eV for the Fe $K\alpha$ line at 6.40keV with a Si(Li) detector, and resolutions of around 150–120eV can be achieved, compared with about 1keV for a gas proportional system and about 2.8keV for a scintillation unit.

The signal from a gas proportional, scintillation, or semiconductor detector can next be passed through a pulse height analyser, or pulse height discriminator, circuit in which amplified pulses with voltages lying within a predetermined range only are permitted to pass [37]. All others are electronically blocked. Pulse height analysers can be set to accept either all pulses with energy greater than a minimum value, or all pulses less than a maximum value, or all pulses lying between a lower and an upper limit, depending on how the controls are set. The advantages of such a system will be obvious since, especially in the 'window' mode, it will allow the detector to accept all the radiation incident upon it, but only pulses generated by radiation of the desired energy are allowed through to the final stages of counting. As an example, we may consider the use of a pulse height analyser in X-ray diffractometry. One of the problems associated with the use of copper $K\alpha$ X-rays in diffractometry is that they generate fluorescent X-rays from any sample containing iron, and these fluorescent X-rays are detected along with any diffracted radiation, giving rise to spuriously high background signals which may completely swamp the diffracted signal. Fortunately, the energies of Cu $K\alpha$ and Fe $K\alpha$ X-rays are sufficiently different that they can be resolved using a gas proportional detector and a pulse height analyser will remove the signal due to the iron radiation. The dramatic improvement in the diffraction trace is shown in figure 1.20, and

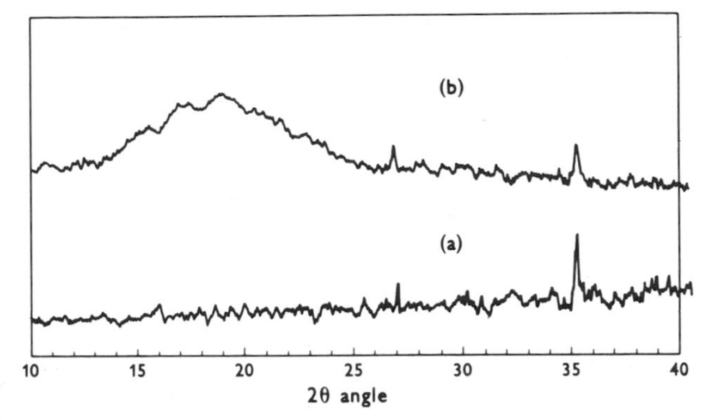

Figure 1.20 Comparison of the X-ray powder diffraction trace obtained from a sample of hematite (a) with and (b) without using a pulse height analyser in the counting chain to remove the signal from the fluorescent Fe K radiation stimulated by the Cu $K\alpha$ radiation incident on the sample.

the same technique can, of course, be used to reduce the overall background signal and again improve the signal-to-noise ratio.

The multi-channel analyser is the logical extension of the pulse height analyser, and comprises a large number, typically between 512 and 4096, of pulse height discriminator units so arranged that each passes signals lying within a narrow voltage range, the whole array covering a wide range of voltages in a series of small, contiguous voltage steps. The incoming signal is thus sorted into a large number of separate signals, each corresponding to a different wavelength in the incident beam, with a resolution depending on the resolving power of the detector and pre-amplifier assembly and on the number of channels available. Each signal may be counted separately in the subsequent circuits. The use of the multi-channel analyser has, of course, increased greatly with the advent of the semiconductor detector, since only this instrument has a sufficiently high resolving power to warrant its application.

The final measurement of the signal can be made using either a ratemeter, to indicate the intensity of the signal, or a scaler, to indicate the integrated signal intensity during a preset time. The actual measurement can be made using a current sensing device, such as a meter, or, better, a potentiometric circuit and a null indicating device. Current sensing is quite sufficient for many applications, but potentiometric methods are very much to be recommended for very accurate work, because of their basically higher accuracies. The multi-channel analyser is often used in conjunction with an integrating cathode ray display, in which the separate signals from the analyser are accumulated in a bank of condensers for a fixed time and their contents subsequently displayed as a trace either on a cathode ray oscilloscope or on a paper chart.

The measured signal level can finally be displayed either simply on a meter, or as a trace on a paper chart, or in digitised form. While some instruments, such as certain recording infra-red and visible spectrophotometers, couple the motion of the paper chart on its drum to the motion of the dispersing prism in the optical path, other instruments, such as the X-ray diffractometer, use time as the connection between the measuring device and the paper chart and depend on a constant slewing speed on the diffractometer head and a constant paper feed rate on the chart recorder. For this reason, paper chart recorders used in diffractometry must use a synchronous motor and must be gear driven, and recorders in which belt drives are used, or in which the paper speed is altered by altering the signal to the motor, do not have sufficient stability and should be avoided. The Philips Company has begun to use stepped motors fed by a synchronous pulse on the diffractometer and the paper drive to avoid this difficulty.

1.4 UNITS

Only a brief mention will be made of units here, since they are discussed elsewhere, at the relevant points in the book. Système International (SI) units are used generally, but not exclusively, in the book, and wavelengths are quoted both in nanometers, nm, and angstrom units, Å, where 1Å = 0.1nm. Some confusion, however, may arise in the use of frequency and wavenumber, and in the practice of quoting radiations in terms of their wavelengths and of their equivalent energies in electron-volts (eV) or kilo-electron-volts (keV).

The frequency of a waveform is simply its velocity divided by its wavelength, and represents the number of wave cycles occurring in a second, whereas the wavenumber is the reciprocal of the wavelength expressed in centimeters, and represents the number of wave cycles occurring in one centimeter. Frequency is quoted in hertz (Hz), or less correctly now in cycles per second, and kilohertz (kHz), and wavenumbers in reciprocal centimeters (cm^{-1}). Clearly, for electromagnetic radiation, the frequency is equal to the wavenumber multiplied by the speed of light, in centimeters per second ($2.99 \times 10^{10} cm \cdot s^{-1}$.

The electron-volt is a measure of energy, and represents the amount of energy gained by an electron when it is accelerated through a potential difference of 1 volt. Numerically it is equal to 1.602×10^{-19} joules, and the multiple units keV and MeV represent 10^3x and 10^6x this basic unit respectively. From equation (1.2) we find that the photon energy of radiation of wavelength λ can be given as

$$E = \frac{hc}{\lambda} \tag{1.2a}$$

and so, since $h = 6.626 \times 10^{-34}$ joule seconds, the wavelength of the electromagnetic radiation corresponding to an energy of 1eV is

$$\lambda = \frac{6.626 \times 10^{-34} \times 2.998 \times 10^{10}}{1.602 \times 10^{-19}} \text{ cm}$$

$$\lambda = 1.24 \times 10^{-4}\text{cm or } 12{,}400\text{Å}$$

In like manner, the wavelength corresponding to 1keV is 12.4Å, and for any general energy, NeV, the corresponding wavelength is given by

$$\lambda = \frac{12{,}400}{N} \text{ Å} \qquad (1.12)$$

1.5 SAFETY

The use of physicochemical methods of analysis for minerals avoids many of the chemical hazards associated with wet chemical techniques, but it introduces other sources of hazard, particularly associated with the use of electrical equipment and with the need to work with ionising radiations or radioactive materials. In addition, the development of flame spectroscopy, and particularly atomic absorption spectrometry, has involved the use of hotter and more intense flames which can pose distinct hazards. It should go without saying that such flames must be treated with great care and never left unattended. This applies especially to the nitrous oxide-acetylene flame and the maker's recommendations in the case of an emergency should be clearly displayed on the instrument and strictly adhered to, should an accident occur, to prevent the shock waves generated by the flame from igniting the cylinder of acetylene gas.

Electrical risks fall into two main categories, those associated with 'normal' mains voltages and those arising from the very high voltages used in some techniques. The low voltage risks can be largely eliminated by careful and thoughtful use of the equipment, for example by ensuring that the 'obvious' precautions are taken, namely that all mains connections are correctly made and that the equipment is properly earthed, that the equipment is properly maintained and kept clean at all times, and that all operators are fully trained in its use. The 'casual' operator is a major laboratory menace, since any piece of equipment works better with a single operator, but where he is unavoidable a code of practice should be instituted and adhered to by everyone, to ensure that the equipment is always operated safely and left in a safe condition after use. This usually pays dividends in improved utilisation of the apparatus and better results. A reference book, such as the 'Industrial Safety Handbook' [38], should be available and the relevant sections read before anyone is allowed to work in the laboratory.

High voltage dangers are usually minimised on the equipment by the manufacturers, by the proper use of screening, interlocks, etc., and it is an act of supreme folly to interfere with the covers or interlocks which are provided for the safety of the operator, however inconvenient these may make the operation of the equipment. Naturally, at least one dry powder fire extinguisher must be provided in every laboratory containing electrical equipment, and clearly labelled for use on electrical fires. Again, all users of the laboratory must be acquainted with its presence and fully conversant with its use.

The dangers from electrical equipment are usually fairly obvious, however, and so tend to be relatively minor in comparison with the much more insidious hazards arising from the use of ionising radiation and radioactive materials. Here the dangers are invisible and virtually undetectable until after the damage has been done and prevention is the only real safeguard. Exposure to ionising radiation can cause moderate to severe superficial burns after one or two seconds, depending on whether X-rays or γ-rays are involved, and prolonged exposure can produce more serious, long-term internal damage.

Figure 1.21 An X-ray diffractometer totally enclosed in a metal box to afford maximum operator protection during use. Note the interlocked micro-switch which prevents the X-ray beam from being energised while the top of the box is open. (Courtesy of Dr. I. R. Harris.)

It is important to remember that ionising radiation is not only produced by instruments labelled as X-ray or γ-ray sources, but that almost any apparatus which involves high energy stimulation, by electron beam bombardment or spark ionisation, will probably generate X-rays as a by-product, and it is important to provide adequate shielding for the operator in any such case. Hence the use of lead glass in the port of the spark source mass spectrometer or in electron microscopes. Radiochemical techniques carry the additional hazards of absorbing radioactive materials by ingestion, in addition to the grave dangers of exposure to γ-rays.

Most countries today operate codes of practice for the protection of persons working with ionising radiation and radiochemicals [39] and the International Union of Crystallography has published its recommendations for X-ray equipment [40]. Particular mention may be made of the recommendation for providing fail-safe devices on all equipment generating X-rays, especially the provision of interlocked safety shutters on all ports to ensure that a beam of X-rays cannot accidentally be withdrawn from the source without having a camera or other device correctly placed to absorb it. Modern diffractometers tend to have completely enclosed beam paths, but further protection can be afforded by a metal box which totally encloses the X-ray tube and goniometer or camera, as shown in figure 1.21. The higher penetrating power of γ-rays require more stringent protection, and lead castles and lead walls must be provided, preferably within a fume cupboard with a good 'normal' extract plus an 'emergency' high flow rate extract and certainly when volatile materials are involved, for all applications using such materials. The use of rubber gloves to minimise possible skin contact and scrupulous care in washing before eating are the only ways to guard against accidental ingestion of non-volatile radioactive chemicals, and the laboratory should be provided with a seamless floor covering which reaches some distance up the walls, to simplify the removal of spills, and have adequate ventilation. Cleanliness at every stage is imperative in all radiochemical work.

In addition to taking precautions, however, all workers who are likely to be exposed to appreciable amounts of radiation over a long period, more than 1.5rem/yr, should be 'designated' for such work and their dose rates monitored continuously. This designation takes the form of a medical examination before the work is begun plus a continuous check on the integrated dose which the worker is receiving during the course of the work. This monitoring may be carried out either by means of film badges which are worn by the operators, often backed up by film badges mounted close to the source of the radiation, or by urine sampling. Film badges, which are normally worn for periods of up to four weeks then sent for processing to a central laboratory, are perfectly adequate for the great majority of applications associated with physico-chemical methods of analysis, but urine sampling may be indicated if, for example, appreciable amounts of Ca^{45} or H^3 are involved in the work. Many governments provide monitoring services and, while it is a legal requirement that workers with ionising radiations or radiochemicals must be designated in certain forms of employment, it is strongly recommended that all workers exposed to these dangers should be designated, whether the legal requirement applies or not.

1.6 SUMMARY

Physicochemical methods of analysis are based on the basic energy transition processes which can occur within materials at the atomic and molecular levels, or on the diffraction or reflection of electromagnetic radiation by substances. Since they involve fundamental processes in the materials studied they provide very powerful methods for characterising substances, provided always that care is taken both in the application of the chosen method to the sample under study and in the interpretation of the results obtained. Readers will find this point, about thoughtful application of each method discussed, stressed throughout the subsequent chapters of the book since it will be clear that accurate and reliable results are impossible to obtain unless the methods are used sensibly, in fact careless application may give totally misleading results.

REFERENCES

1. M. PLANCK, *Verh. Deut. Phys. Ges.*, **2**, 237 (1900); *Ann. Phys.*, **4**, 553 (1901).
2. A. EINSTEIN, *Ann. Phys.*, **17**, 132 (1905).
3. N. BOHR, *Phil. Mag.*, Ser. 6, **26**, 476, 857 (1913).
4. E. RUTHERFORD, *Phil. Mag.*, Ser. 6, **21**, 669 (1911).
5. W. RITZ, *Phys. Zeitschr.*, ix, 521 (1908).
6. W. PAULI, *Z. Physik*, **31**, 765 (1925).
7. S. GLASSTONE, 'Textbook of Physical Chemistry', Macmillan, London, 1962.
8. B. H. MAHAN, 'University Chemistry', Addison-Wesley, Reading, Mass., 1967.
9. L. PAULING, 'The Nature of the Chemical Bond', 3rd ed., Cornell University Press, Ithaca, N.Y., 1960.
10. N. MOTT, 'Materials (A Scientific American Book)' (G. Piel, D. Flanagan, *et al.*, eds.), Chap. 2, W. H. Freeman & Co., San Francisco and London, 1967.
11. H. J. EMELEUS and J. S. ANDERSON, 'Modern Aspects of Inorganic Chemistry', 3rd ed. Routledge and Kegan Paul, London, 1960.
12. A. F. WELLS, 'Structural Inorganic Chemistry', 3rd ed. Oxford, Clarendon Press, Oxford, 1962, p.86.
13. A. F. WELLS, *Ibid.*, p.881.
14. J. H. VAN VLECK and A. FRANK, *Phys. Rev.*, **34**, 1494, 1625 (1929).
15. H. BETHE, *Ann. Phys.*, **5**, 135 (1929).
16. J. H. VAN VLECK, *Phys. Rev.*, **41**, 208 (1932); 'Theory of Electric and Magnetic Susceptibilities', Oxford, Clarendon Press, Oxford, 1932.

17. F. A. COTTON and G. WILKINSON, 'Advanced Inorganic Chemistry', 2nd ed., Interscience, New York, 1966, Chapters 3 and 26.
18. C. J. BALHAUSEN, 'Ligand Field Theory', McGraw-Hill, New York, 1962.
19. L. E. ORGEL, 'Introduction to Transition Metal Chemistry', Methuen, London, 1960.
20. R. G. BURNS, 'Mineralogical Applications of Crystal Field Theory', Cambridge University Press, Cambridge, 1970.
21. E. W. WHITE, *Amer. Mineral.*, **49**, 196 (1964).
22. J. C. MAXWELL, *Phil. Mag.*, Ser. 4, **19**, 19; **20**, 21 (1860); **32**, 390 (1866).
23. L. BOLTZMANN, *Sitz. Acad. Wiss., Wien,* **66**, 275 (1872); *Wiss. Abhandl.*, I, J. A. Barth, Leipzig, 1909, p.316.
24. S. N. BOSE, *Z. Physik,* **26**, 44 (1924).
25. A. EINSTEIN, *Sitz. preuss. Akad. Wiss.*, **1924**, 261; **1925**, 1, 18.
26. P. A. M. DIRAC, *Proc. Roy. Soc.* London, **A112**, 661 (1926).
27. M. R. CLELAND, P. R. HANLEY and C. C. THOMPSON. *I.E.E.E. Trans. on Nucl. Sci.* NS-16, No. 3, 113 (1969).
28. K. J. SCOTT, *Trans. I.M.M., C,* **77**, 85 (1968).
29. E. McIVER, *Ph.D, Thesis,* Cambridge, 1960.
30. C. E. KENNETH MEES and T. H. JAMES (editors), 'The Theory of the Photographic Process', Macmillan, New York, 1966.
31. F. H. SANDERS, *Phys. Rev.*, **41**, 667 (1932).
32. G. DEARNALEY and D. C. NORTHROP, 'Semiconductor Counters for Nuclear Radiations', E. & F. N. Spon, London, 1966.
33. R. JENKINS and J. L. DE VRIES, 'Practical X-ray Spectrometry', 2nd ed., Philips Technical Library, Macmillan, London, 1970.
34. G. L. MILLER, W. M. GIBSON and P. F. DONOVAN, *Ann. Rev. Nucl. Sci.*, **12**, 189 (1962).
35. A. J. TAVENDALE, *I.E.E.E. Trans. Nucl. Sci.*, **NS-11**, No. 3, 191 (1964).
36. I. C. BROWNRIDGE, 'Lithium drifted Germanium Detectors, an annotated bibliography', IFI/Plenum, New York and London, 1972.
37. D. MAEDOR, 'Methods of Experimental Physics', Vol. 2 (E. Bleuler and R. O. Haxby, eds.), Academic Press, New York and London, 1964; Chapter 9, p.593.
38. W. HANDLEY (editor), 'Industrial Safety Handbook', McGraw-Hill, Maidenhead, 1969.
39. DEPARTMENT OF EMPLOYMENT AND PRODUCTIVITY, 'Code of Practice for the protection of persons exposed to ionising radiations in research and teaching', H.M.S.O., London, 1968.
40. Report of International Union of Crystallography, *Acta Cryst.*, **16**, 324 (1963).

CHAPTER 2

Optical Spectrometry

G. J. Lawson

Department of Minerals Engineering
University of Birmingham
Birmingham B15 2TT
England

2.1 INTRODUCTION

Optical spectrometry, as has been indicated in Chapter 1, is concerned with the detection and measurement of radiant energy, particularly that associated with electromagnetic radiations lying in or near the visible portion of the spectrum, absorbed or emitted in the course of certain electronic transitions occurring in atoms or molecules. The term 'optical' is convenient but not strictly correct, since the wavelength range corresponding to the transitions of interest extends through the visible and into the ultra-violet region of the spectrum. Also 'spectrophotometry' is perhaps a more suitable term than 'spectrometry', since all modern instruments employ photoelectronic devices for measuring light intensities. The older name will be retained, however, for convenience and for continuity with older work.

After a brief recapitulation of the underlying theory, three general analytical methods, based respectively on optical absorption, flame emission, and flame (atomic) absorption, will form the basis of the techniques discussed in this chapter. The instruments used will be considered, ranging from simple spectrometers, confined largely to the visible spectrum, to complex instruments covering a very much wider range of frequencies and performing a variety of functions, and the basic techniques used will be reviewed generally but without reference to specific analyses. As far as the instrumentation is concerned, extension from the visible to the ultra-violet region of the spectrum really involves little more than a modification of the optical system of the spectrometer by substitution of gratings and reflectors or silica transmission optics for the glass prism and lenses of the visible range instrument, and provision of a suitable light source. Combined visible–ultra-violet instruments, however, usually also offer improved performance levels in respect of sensitivity, accuracy and spectral discrimination.

Since the sensitive interaction between the material under study and the radiation being monitored occurs at the atomic level, or at the molecular level particularly in materials studied by optical absorption methods, at well defined frequencies characteristic of the element or molecular compound present, optical spectrometry can be used to provide both qualitative and quantitative information about the elements present in a sample It can deal with single elements or with mixtures of elements and compounds, provided that the absorption or emission frequencies for the various species are not coincident or so close together that the instrument cannot resolve the lines involved. When this condition is satisfied, however, optical spectrometric techniques provide rapid, accurate and convenient methods of elemental analysis for minerals [1-3].

2.2 BASIC THEORY

The basic theory underlying all the methods of optical spectroscopy has been largely discussed in Chapter 1, section 1.1, but it is worth reiterating certain points made in that chapter, especially those dealing with the electronic levels involved in the relevant transitions and the fractions of atoms or molecules actually taking part in any absorption or emission process.

As shown in Chapter 1, electronic transitions between permitted energy levels in the atom or molecule are responsible for the observed absorptions or emissions of light which are measured in optical spectrometry. Since light in the visible–ultra-violet region has a relatively low photon energy associated with it, the corresponding electron 'jumps' will be relatively small and do, in fact, correspond to transitions between the outer levels in the atom or molecule, the levels usually involved in bonding. It is the highly characteristic nature of these energy transitions and of their associated light wavelengths that permits the resulting colours to be used to identify atoms and compounds. The ability to identify an atom depends crucially on its chemical and physical environment, since this may affect both the energy levels available for measurement and the electronic populations in these levels.

Very few atoms themselves possess electronic levels with permitted transitions that correspond to visible light energies at room temperature, and so it is generally necessary either to incorporate the element into a complex in which it forms bonds with, for example, oxygen or nitrogen atoms and so set up new electronic levels to provide the required transitions, or to raise the element to a high temperature in a flame. The first method forms the basis of optical absorption spectrometry and the second that of flame emission spectrometry or of flame (atomic) absorption spectrometry depending on the property monitored. In some cases the element under investigation may be combined with a reagent that already exhibits suitable measurable electronic transitions, and which acts as a quantitative 'label' for the element rather than forming new bonds with useful transitions.

A given atom or molecule can absorb or emit radiation depending on whether its intrinsic energy is less than or greater than some limiting value. As has been pointed out in Chapter 1, the systems considered here are in a state of dynamic equilibrium and the statistical distribution of energies among the separate atoms or molecules follows a Maxwell-Boltzmann distribution, which varies with temperature as shown in figure 1.5. Therefore, at any temperature there will be a certain fraction of atoms with low energies, i.e. capable of absorbing radiation, and a certain fraction with high energies, capable of emitting radiation. In either case, however, the actual number of atoms which do absorb or emit radiation at any time is only a fraction of the atoms capable of so doing, although fortunately this fraction is always the same for a given system under a particular set of conditions. It is this constancy that allows us to use optical absorption or emission values in a comparative manner to determine quantitatively the amount of a material present in a sample under investigation. It follows that, in all optical spectrometry, it is important that the chemical and physical environment must be standardised for all measurements.

Figure 1.5 shows that at room temperature, around 300°K, the great majority of atoms are in their low energy states and that absorption of light of the correct quantum energy can readily occur, giving the basis for optical absorption spectrometric methods. On the other hand, at high temperatures, around 1200°K, an appreciable fraction of the atoms possess sufficient energy to promote electrons into high energy electronic levels, from which they can return to a lower energy level, with the emission of a photon of characteristic light. The wavelength and intensity of this light is monitored in flame emission spectrometry. Even at high temperatures, however, the Maxwell-Boltzmann distribution curve shows that more atoms are in relatively low energy states than are in the fully excited state and these are again capable of absorbing radiation of the correct wavelength. The absorption of light of a very closely specified wavelength, generated from a special lamp, by a flame containing the vapour of the element to be determined, forms the basis of the relatively new method of flame, or atomic, absorption spectrometry.

2.3 SAMPLE PREPARATION

Optical spectrometric methods differ from the other techniques discussed in this book in that, at present in every case, the material to be analysed must be brought into solution in a liquid phase at some stage in the procedure. It follows that the methods cannot be applied directly to a mineral in its as-received form, but only to a sample from the mineral following a dissolution process and possibly some concentration and separation procedures. Elemental selection may of course occur at each step in such a preparation scheme, possibly to a greater extent than in the fusion schemes used in X-ray fluorescence and mass spectrographic analysis (Chapters 3 and 5 respectively). Accordingly, to the necessity of ensuring that the primary sample is representative of the bulk from which it was drawn is added the need to ensure that the final sample presented to the spectrometer is also truly representative of the primary sample. A review of methods for the dissolution of mineral materials will be given in Chapter 12 and so these will not be further discussed in detail here. Suffice it to say that even the most complicated or expensive instrument cannot give a result more accurate in relation to the original mineral material than is allowed by the extent to which the element or molecular group to be determined is recovered in the preliminary treatment.

It is worth noting that while in the preparation of a solution for analysis the final sample must be truly representative of the bulk, it is by no means always necessary to achieve total dissolution of the mineral to obtain this. On the contrary, selective dissolution of a required component, or successive dissolution of several components, may be advantageous in simplifying the analysis of the resulting solutions, by removing potentially interfering substances or by permitting separate determinations to be made of the same element in different chemical forms in the same sample. Thus, carbonate minerals may be selectively leached with dilute acid; extraction with ammoniacal solutions can recover elements capable of forming soluble ammine complexes and may permit differentiation, for example, between the different copper compounds present in a copper ore; and in standard methods of coal analysis extraction with dilute hydrochloric or nitric acid is used to recover sulphur present as sulphate and as pyrites respectively.

It is important to remember that ions in solution rarely exist as the unattached ion but are virtually always associated either with molecules of solvent or with some other complexing or chelating agent. Thus, copper ions in aqueous solution exist as the complex $[Cu(H_2O)_6]^{2+}$, whereas in ammoniacal solution they form the tetrammine complex $[Cu(NH_3)_4]^{2+}$. As shown in Chapter 1, the permitted electronic transitions will be different in the two complexes and so the wavelengths of the light which they can absorb will also differ. The aquo-complex absorbs at 810nm and appears light blue in colour while the tetrammine absorbs at 650nm and is dark blue. This example underlines the importance of appreciating the chemical state of the element or group being determined, because a proportion of the element may go undetected if it is in the wrong chemical state for the method being used. This is particularly applicable to absorption studies at room temperature; at the high temperatures of flame methods molecular complexes tend to be destroyed and elements converted to a common form.

The extraneous substances inevitably introduced during the dissolution steps may interfere with the analysis by forming undesired complexes or in other ways. These complexes may persist if their stability constants are sufficiently high, and their possible effects must always be borne in mind. Thus, copper might form the green cupro-chloride complex in the presence of excess chloride ions with corresponding diminution in the concentration of copper ions present. Sometimes it may be quite easy to remove the unwanted substance(s), for example some acid anions can be eliminated by evaporation and fuming, but other materials may require more complex treatment, although this may often be conveniently incorporated into the overall preparation scheme.

It may be necessary to carry out a concentration step during the preparation of the solution to ensure that the concentration of a required element is at an appropriate level. This might be done by simple evaporation, but it may be more advantageous to use a technique such as solvent extraction, since in this way the required element can often be removed very selectively from the original solution and so be purified simultaneously with concentration. Essentially, solvent extraction involves incorporation of the element into a complex, often as a chelate with an organic reagent, which can be extracted quantitatively into an organic solvent phase and subsequently separated physically from the original aqueous solution. The degree of concentration is controlled by varying the relative amounts of aqueous and organic phases or by subsequently evaporating the organic solvent. Potentially interfering substances will remain in the aqueous phase, provided that the conditions of extraction have been so arranged that they are not easily extracted. The final optical measurements may be made either on the organic solution or on an aqueous back-extract from the organic phase. In both situations, the solutions may be analysed without further treatment, especially in flame emission and atomic absorption spectrometry; however, for optical absorption it is more usual to add a chromogenic reagent to develop a strong colour characteristic of the element being sought,. and thus to improve the accuracy and sensitivity of the method. In some cases the complexing reagent used to facilitate solvent extraction may also act as the chromogenic reagent.

Following the dissolution step, the analysis of material from a mineral source differs little from that of any other material by spectrometric methods, except that more complex mixtures of elements are commonly encountered. The different methods of optical spectrometry are now considered in more detail.

2.4 OPTICAL ABSORPTION SPECTROMETRY

This is one of the older methods [4-7] and depends on measuring the amount of light absorbed by a solution which contains the required element, usually in the form of a coloured complex or chelate, at room temperature and at a specific wavelength corresponding to a characteristic electronic transition in the complex containing the element.

2.4.1 Basic Principles of Absorption Spectrometry

The Single-Beam Optical Spectrometer

The principle of a single-beam optical spectrometer is shown diagrammatically in figure 2.1, and the layout of a typical instrument is shown in figure 2.2. Light from a suitable source, usually a tungsten or a deuterium lamp, for the visible and the ulta-violet regions of the spectrum respectively, passes in turn through a dispersion prism or optical grating and a slit system,

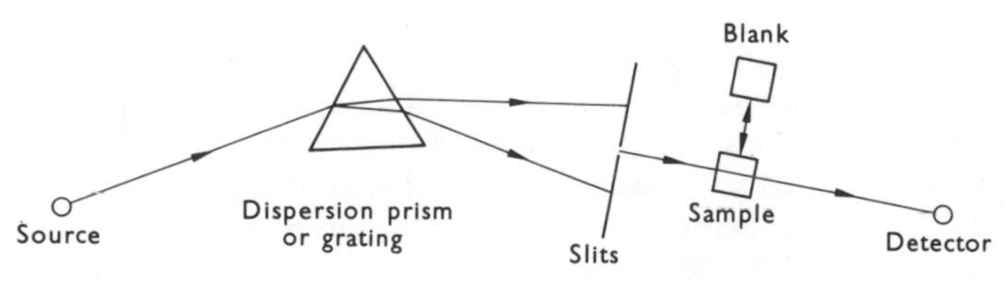

Figure 2.1 Principle of a single-beam spectrophotometer.

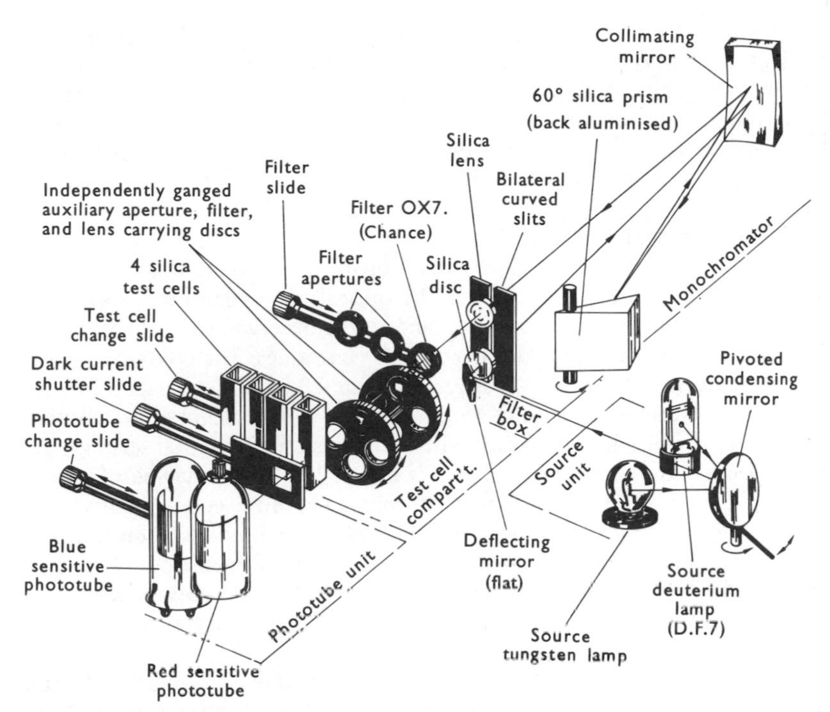

Figure 2.2 Perspective ray diagram of the optical system of the Pye-Unicam SP 500 Absorption Spectrophotometer. (Courtesy of Pye-Unicam Ltd.)

to provide the approximately monochromatic light beam, and then the sample in its special holder, before reaching the detection unit from which a signal is sent to the metering unit to register the intensity of the transmitted light. In practice quite complex and precise optical and mechanical systems are required to ensure the constancy of the light path; a typical arrangement, as used in the Pye-Unicam SP500 precision visible–ultra-violet spectrometer, is shown in figure 2.2. Note that the source beam from either the tungsten or the deuterium lamp is selected by a moveable mirror, again to ensure the constancy of the beam path, rather than by moving the lamps themselves. To ensure constant performance the lamps and all electronic circuits are supplied with current at a constant voltage from a stabilised power pack. The dispersion unit incorporates a silica prism, to accommodate both the visible and the ultra-violet regions, and the slit system performs the dual role of limiting the bandwidth of the transmitted dispersed beam and of collimating the beam both before and after it passes through the dispersion prism. Provision is made for inserting various filters before the light passes through the sample, to absorb heat from the beam or unwanted wavelengths transmitted by the slit system. The sample cell carriage will normally carry up to four cells, each of which may be inserted in the light path at any time. The intensity of the transmitted light is measured by one of two photocells, one being used for measurements at long wavelengths, in the red-green region, and the other for short wavelengths, in the blue–ultra-violet region.

2.4.2 Measurement Techniques

Consideration of figure 2.2 shows that absorption of light will occur in the optical system of the spectrometer, the optical cell which contains the test solution, and the test solution itself, and so the observed total absorption of light will contain contributions from all three sources. The contribution due to the optical system may fairly be assumed to be constant for any particular combination of wavelength and slit width, although its absolute value will change with changes in either of these instrumental settings. The contribution due to the optical cell is controlled by the degree of quality control exercised by the manufacturer, and only good quality cells appropriate to the instrument should be used. Care must be taken to ensure that only silica cells are used for measurements involving ultra-violet wavelengths. It is desirable that each separate cell should absorb exactly the same amount of light as any other, but for the most accurate work it is best to use cells in pre-matched pairs.

The contribution from the solution itself includes, in addition to that from the element to be determined, contributions from the solvent and possibly from other components present, which are usually kept to a minimum. The solvent contribution is very important and allowance must be made for it in every measurement. In practice, the spectrometer is set to a 'zero' or null balance condition at the required wavelength by adjusting the width of the slit system to give a standard response, I_0, on the measuring unit when a reference or 'blank' cell is in the light path. This cell will contain the solvent alone, or the solvent with some of the reagents used in preparing the test solution, but without the test element. To make the test reading, a matching cell containing the test solution now replaces the standard cell and the response, I, is measured using the same instrumental parameters of slit width and wavelength. The dials of most instruments are so calibrated that a direct reading may be obtained of the percentage transmission, $(I/I_0) \times 100\%$, or of the optical density or absorbance, $\log(I_0/I)$.

For the majority of solutes in dilute solution, the optical density is directly and linearly related to the concentration of the solute according to the Beer-Lambert law,

$$\log(I_0/I) = \epsilon cl \qquad (2.1)$$

where ϵ is the molar extinction coefficient, a constant characteristic of the solute; c is the concentration of the solute; and l is the length of the absorption path, i.e. the path length defined by the thickness of the solution layer in the optical cell. Thus, after simple calibration using solutions of quite widely spaced known concentration values, or from a knowledge of ϵ and l, the concentration of a solute in a given solution may readily be determined directly from the optical density. If the solution does not conform strictly to the Beer-Lambert law, a more detailed concentration curve must be drawn, using a more closely spaced series of known solutions, but once this has been prepared unknown concentrations can again be found from the optical density value by interpolation.

Peak Profiles

In common with all techniques which involve the detection and estimation of an emitted or absorbed electromagnetic radiation, absorption by a given substance does not occur at a single wavelength, but rather over a range of wavelengths with varying efficiency. The profiles of typical absorption peaks are shown in figure 2.3, from which it will be seen that the peak is roughly triangular in shape, with a definite width at the half height, and is certainly not a true delta function. Also, the beam of light transmitted by the slit system cannot be ideally monochromatic, with only a single wavelength, but must contain a range of wavelengths which irradiate the sample. Under normal operating conditions, the transmitted bandwidth is kept narrower than the peak width at the half height, as shown in figure 2.3a, in order to maximise the amount of light energy absorbed, and the bandwidth is centred on the wavelength of the maximum absorbance in the peak. The most accurate and sensitive measurements are obtained under these conditions. It is therefore necessary, when first testing a material for use as a

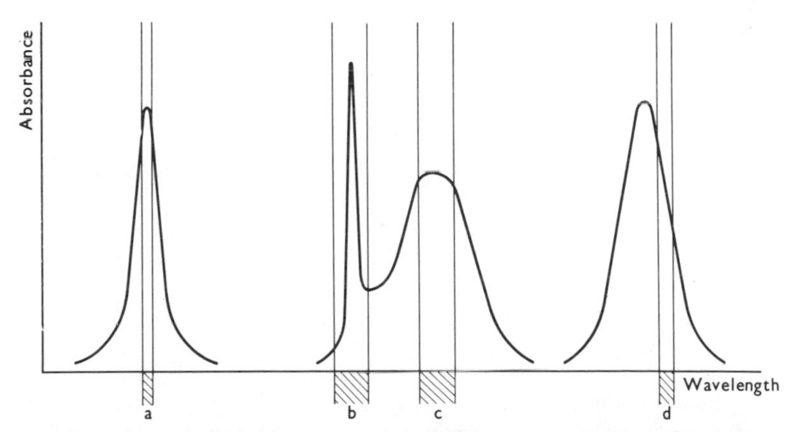

Figure 2.3 Diagrammatic representation of the relationship between absorption peaks and spectral bandwidths, showing (a) the ideal situation, (b) and (c) situations resulting from the use of wide bandwidths with a sharp and a broad peak, and (d) the situation which may arise from the use of an optical filter. In every case the bandwidth is indicated by hatching.

spectrochemical reagent, to define the shapes of the absorption peaks for a given element in a given chemical form, in a given solvent, for a predetermined slit width or bandwidth.

To determine a peak profile with a single beam instrument a series of readings of the absorbance is taken at different wavelengths. This involves resetting the slit system at each wavelength, to take account both of the non-uniform distribution of light in the spectrum of the light source and of the variation in the absorbance of the solvent at the different wavelengths. It must be remembered that changing the slit width alters not only the intensity of the transmitted beam but also the bandwidth of the incident beam, and in certain situations this may cause difficulties. Thus, in regions of low source intensity or high solvent absorption a condition may be reached where the instrument cannot be set to zero with the 'blank' in the beam, or alternatively the transmitted bandwidth may become greater than the width of the absorption peak at the half height, and in either case accurate measurements will be impossible to obtain. Figure 2.3b illustrates the latter situation, which can easily arise when a sharp peak is being measured in a region where the solvent absorption is high. In this case light of wavelengths not absorbed by the test compound can reach the detector and give an anomalously low absorbance reading. The remedy, shown in figure 2.3c, lies in using either a less sharp peak in the spectrum, with a width greater than the bandwidth, or even a plateau region. At these different wavelengths it may, of course, be possible to use a narrower bandwidth.

The high intensity sources used in modern good quality instruments give satisfactory emission levels over a wide spectral range, and the above difficulty will arise more usually from absorption by the solvent than from inadequate illumination. A change of solvent, if possible, may help to overcome the difficulty.

Use of Filters

Some simple absorptiometers employ filters instead of the more expensive prism or grating dispersion systems for wavelength selection. Although every effort is made to give filters with pass bands centred on the peak for the element to be determined, it is unfortunately not uncommon to find that the transmission band for the filter lies to one side of the optimum position for the required absorption peak, so that measurements must be made on the steep gradient of the absorption curve, as illustrated in figure 2.3d. This is not the ideal condition for determining the absorption and can give rise to major uncertainties in the results. Accordingly, if a filter instrument is being used, then unless a procedure is being followed for which a particular filter has been specifically recommended, it is advisable first to check the transmission characteristics of the filter, preferably using a grating dispersion instrument for maximum accuracy, and compare them with the absorption characteristics, measured similarly, of the material to be determined. Although they are somewhat limited in scope, filter instruments are relatively inexpensive in comparison with prism or grating spectrometers, and they can be very useful for routine measurements, particularly where a good match is available between filter transmission and the absorption to be measured. A reasonable case may be made for such an instrument in a situation where it can provide sufficiently accurate results, possibly when a large number of control analyses require to be carried out as a routine; here it may also release a more expensive instrument for work requiring greater flexibility or a higher level of accuracy.

2.4.3 Final Preparation for Measurement

The initial stages of the preparation of a solution for optical absorption measurements have already been discussed in section 2.3. As mentioned therein, some metal ions in solution are coloured in the presence of certain solvating molecules and such solutions could be used directly for absorption measurements. In general, however, with the possible exception of the rare earth metals in aqueous solution, these colours are insufficiently intense for direct use and it is more usual to incorporate the metal into a more intensely coloured complex or chelate form, which affords greater sensitivity and specificity of measurement. Familiar examples of this are the permanganate ion, MnO_4^-, in which the manganese has been oxidised to the Mn^{7+} state and complexed with oxygen to give the intensely purple coloured oxy-anion, and the tetrammine copper(II) ion (or cuprammonium ion), $[Cu(NH_3)_4]^{2+}$, which could be used to determine copper, although better reagents than ammonia are now available.

Ideally the chosen reagent should react quantitatively and specifically with the required element to give a complex with a strong absorption peak at a wavelength remote from any absorption due to the reagent itself or to the solvent used. The complex should be stable over a reasonable period of time, preferably of the order of hours although complexes stable for as short a period as five or ten minutes can be used provided care is taken, and the complex in solution should preferably obey the Beer-Lambert law. In practice, reagents rarely react exclusively with one element and it is often necessary to maintain careful control over the experimental conditions, such as solution pH or the oxidation state of either the required element or of potential interfering elements, to obtain the desired specificity. It was pointed out in section 2.3 that the best way of ensuring specificity involves separating the required element from its accompanying suite of elements, but final elimination of unwanted materials may be achieved by the judicious addition of alternative complexing agents which form strong colourless complexes with the interfering elements but only weak compounds with the required one.

These points may be illustrated by reference to the colorimetric method for the determination of nickel using dimethylglyoxime in alkaline solution in the presence of an oxidising agent, preferably bromine. Under these conditions the colourless dimethylglyoxime forms an intensely red coloured complex with Ni(III), produced by the action of the bromine on Ni(II), which absorbs strongly at 445nm. Some Ni(IV) is thought also to be present in equilibrium amount, and the complex, while not very soluble, is appreciably more soluble than the better known Ni(II) dimethylglyoxime complex. Various cations can interfere with this test but most may be removed from the test solution by utilising the relatively high solubility of the nickel(II) dimethylglyoximate in chloroform to remove the required complex by solvent extraction. Certain elements will still extract with the nickel complex, and so copper present in the chloroform layer must be removed by shaking with ammonia solution, and certain cations such as manganese which form soluble complexes in their higher oxidation states must be treated with hydroxylamine hydrochloride to maintain them in their lower oxidation states. Similarly the addition of citrate or tartrate prevents the precipitation of iron, aluminium, and other trivalent metal hydroxides which would otherwise tend to cloud the solutions and so make accurate determinations impossible.

After this purification step, the nickel complex is decomposed and the Ni(II) returned to an aqueous solution by shaking the separated chloroform layer with dilute hydrochloric acid, and finally the colorimetric solution is prepared. This test solution is stable for only ten minutes, so that careful but fast manipulation is required after the final addition of dimethylglyoxime. Because of the aforementioned limited solubility of the complex, which could lead to precipitation and hence to low readings, the concentration of nickel in the solution must not exceed 6ppm; consequently care must be exercised at all stages of preparation of the solution, in particular at any dilution steps.

Optical Clarity

Before any solution is presented to the spectrophotometer it must be free from suspended solid particles or dispersed droplets of immiscible solvent, which would give rise to scattering of the transmitted light beam and so produce erroneously low concentration readings, if the blank were adulterated, or high values, if the sample were affected. Measurements of absorbance are usually made with solutions in cells of 10mm path length, but the sensitivity may be increased by using larger cells which are available with path lengths up to 100mm. Optical clarity obviously becomes increasingly important as path length is increased; moreover, the advantages gained by using a longer absorption path may be limited if the blank or solvent absorbs appreciably at the chosen wavelength, since the resulting overall attenuation of the light beam may lead to reduction in the instrumental sensitivity or to the necessity for an excessively wide slit setting.

2.4.4 Reagents for Spectrophotometric Absorption Analysis

A wide range of reagents is available for the analysis of many elements, and most produce visibly coloured complexes, principally because spectrophotometric procedures have often developed from simpler colorimetric methods using similar reagents. However, many coloured complexes absorb in the ultra-violet, particularly if they contain aromatic nuclei, and measurement of an ultra-violet peak may offer advantages in increased selectivity or sensitivity. Some methods depend on ultra-violet absorption, and do not involve the formation of a coloured complex. Space does not permit a detailed discussion of the various reagents and methods available, but these have been well documented in the literature, and sources have been listed in the reference list attached to this chapter.

In the majority of spectrophotometric procedures the amount of the required substance present is determined directly from the absorbance of the test solution, but in some a measurement is made of the amount of a reagent remaining after an excess of it has reacted with the required material, in a manner analogous to 'back-titration' in volumetric analysis. The determination of sulphate by quantitative precipitation with an excess of aminochlorodiphenyl followed by determination of the residuum of the reagent, using its absorption in the ultra-violet at 254nm, is a good example of aminochlorodiphenyl followed by determination of the residuum of the in a specific chemical form. It is important that the amount of the excess actually determined should be of the same order as the amount of the required substance present, to minimise the 'conversion errors' involved in the subsequent calculations.

Anyone applying spectrophotometric methods to the analysis of minerals must be aware of one final potential difficulty, namely the complexity of the systems he is likely to be called on to analyse. Most methods described in the literature deal with 'pure' substances, i.e. the required substance mixed with one or two non-interfering materials, and so when analysing complex mixtures it is good practice always to scan the absorption pattern of the test solution over a range of wavelengths and compare the observed peaks, both in terms of position and intensity, with those quoted in the literature in order to identify situations in which the obvious, i.e. strongest, peak suffers overlap from a peak due to another substance, remembering that overlap will occur if significant portions of the two peaks lie within the bandwidth defined by the slit opening. As before (section 2.1.2) the interference may often be avoided by using a different absorption peak, although this usually involves some loss of sensitivity.

Reference has been made to the 'blank' that is used when the spectrophotometer is adjusted to the null position corresponding to a transmitted light intensity I_0, this is essentially the solution obtained by following the recommended sample preparation procedure in the absence of the mineral sample containing the element to be determined. If the resulting complex is extracted into an organic solvent the blank will be that solvent saturated with the final aqueous solution, but if extraction is not part of the procedure the blank will contain the reagents used and substances arising from them and will correct for any absorption so introduced. The blank will however not correct for absorptions due to additional elements in the mineral material examined.

2.4.5 Double-Beam Optical Absorption Spectrometry

The scanning of a spectrum is a tedious operation with a single-beam absorption spectrometer, since the 'zero' condition of the instrument must be checked, with the 'blank' solution in the light beam, at each selected wavelength. This process is more easily carried out using an automatic double-beam spectrophotometer.

In this instrument, the incident light beam from the source is intercepted before it reaches the sample cell by a rotating chopper which alternately directs it through the sample cell or through a reference cell containing the solvent or the 'blank' solution, as shown diagrammatically in figure 2.4. The resulting two separate beams, after transmission through their respective liquid samples, impinge, again alternately, on the measuring photocell to provide a fluctuating output signal so long as the incident energies in the two beams differ, i.e. when the absorptivities of the contents of the two cells are different. The fluctuating output signal is used to drive a servo motor which moves an optical wedge into the reference beam until the signals balance and a constant output is obtained from the photocell. The position of the optical wedge then indicates the absorptivity of the sample. The desired spectral range is scanned automatically by rotating the dispersion prism or grating, using a second motor, and the absorptivity of the sample is recorded automatically either on a chart mounted on a drum which is rotated synchronously with the prism or on a flat-bed recorder used either in an x-y or an x-t mode. In every case, a graph of sample absorption against wavelength results. The correction for the 'blank' absorption is made automatically, and the slit width and/or the degree of electronic amplification are continuously adjusted to regulate the overall output from the measuring unit and keep it within defined limits. Figure 2.5 shows the complete system used in a commercial instrument, the Perkin-Elmer 402 UV double-beam spectrophotometer. Many electronic and other refinements, e.g. facility for scale expansion in the wavelength or absorption axis, are incorporated in currently available commercial spectrometers, but the basic principle outlined above is common to all these instruments.

Double-beam instruments are undoubtedly convenient to use, much more so than single-beam instruments, especially in terms of the automatic

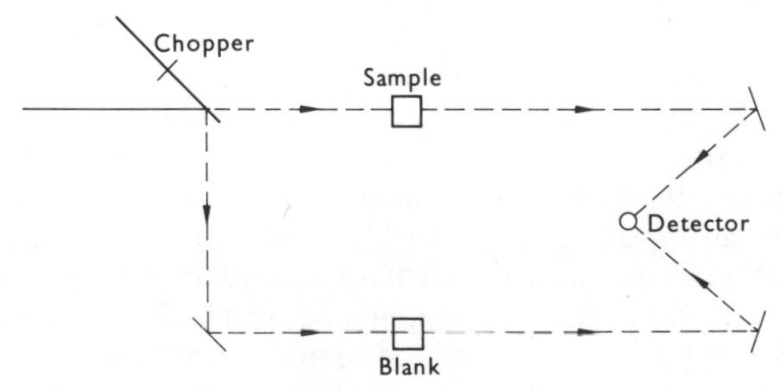

Figure 2.4 Principle of operation of a double-beam spectrophotometer.

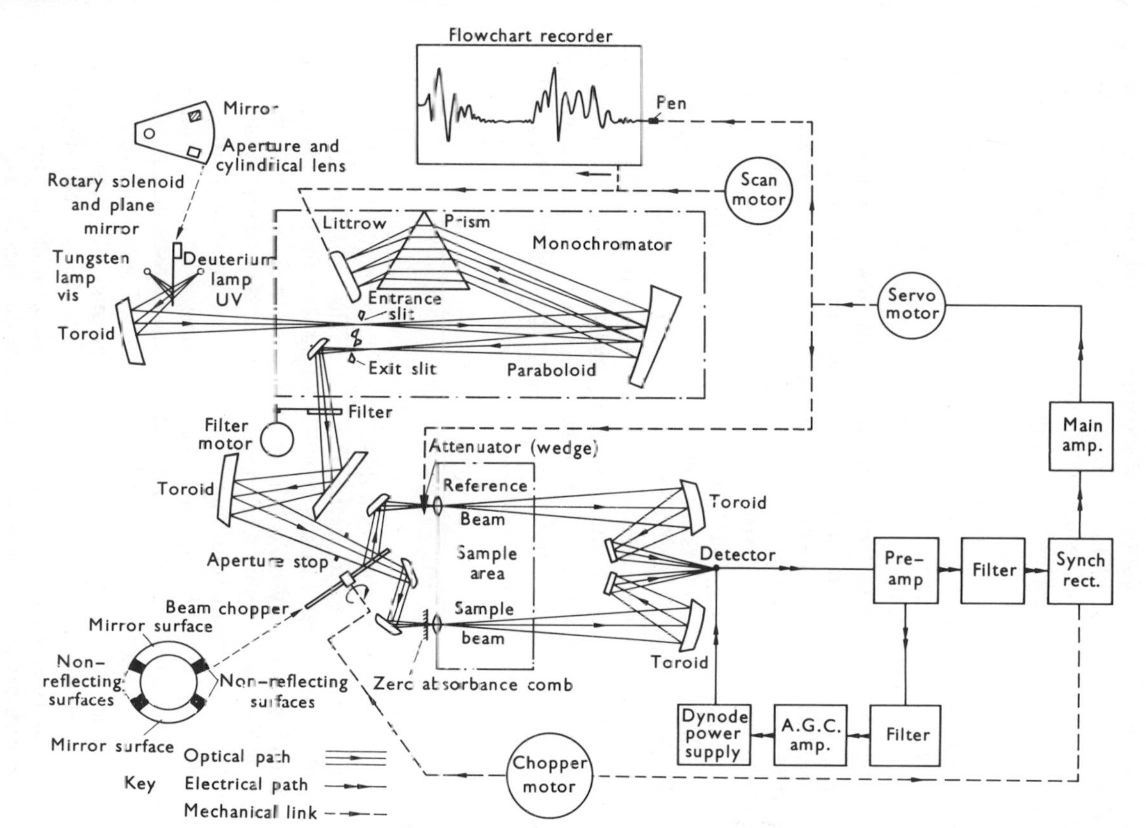

Figure 2.5 Ray diagram and operating system of the Perkin-Elmer 402 recording double-beam absorption spectrophotometer. (Courtesy of Perkin-Elmer Ltd.)

setting of the 'zero' condition, but their very convenience means that they must be used with caution. Mainly this arises because the double-beam instrument gives the operator no immediate indication of the absolute absorptivity of the reference solution at any particular wavelength, and it is quite possible that in regions of high solvent or 'blank' absorptivity the instrument may be attempting the impossible task of distinguishing between two beams of light which are both virtually non-existent. For this reason, the absorption characteristics of the solvent or 'blank' should always be determined separately, either with the double-beam instrument in the 'single-beam mode', i.e. with the 'blank' solution in the sample beam and an empty cell in the reference beam, or with a single-beam instrument. It should be remembered that measuring devices, like the human eye, cannot easily distinguish between two beams of similarly low or similarly high intensity, and so the most accurate results are obtained in the middle range of the absorbance scale, i.e. between $\log(I_0/I)$ values of approximately 0.25 and 0.75 respectively for the low and high values. Both single-beam and double-beam spectrophotometers are built to a high degree of precision, but the discrimination of the double-beam instrument is slightly limited by the response characteristics of its recorder system, and the single-beam spectrophotometer is generally preferred for the most accurate measurements. However, the accuracy of the double-beam instrument is usually quite sufficient for the majority of applications, especially for routine analyses, and its speed and automatic mode of operation give a much greater throughput; consequently it becomes the better choice for most laboratories, provided that the greater cost can be absorbed.

2.5 FLAME EMISSION SPECTROMETRY

At the temperatures reached in a flame appreciable numbers of atoms gain sufficient excess energy to excite their outer electrons into high energy states, from which they can subsequently decay with the emission of radiation of a characteristic wavelength. At its simplest level, this forms the basis of the well-known 'flame test' used to detect easily excited elements, such as sodium, potassium, calcium, strontium, barium, copper, etc., by the colour they impart to a bunsen flame. The same principle is involved in flame emission spectrometry [8–11] but the characteristic light is produced under carefully controlled conditions and analysed spectroscopically, and the intensities at the various wavelengths present are monitored electronically instead of visually as in the simple test. The wavelengths of the emission bands or lines detected provide a qualitative identification of the emitting elements present, and the intensities of the lines give a quantitative measure of each of these elements in a sample. Since compounds are decomposed in the flame, flame emission spectroscopy can be used to measure the amounts of elements only, and not molecular groupings as is possible in optical absorption spectrometry.

2.5.1 Simple Flame Photometers

Simple photometers, suitable for measuring the relatively powerful emissions from elements such as those listed above, comprise a flame into which the sample solution is aspirated via a nebuliser unit, a collimating lens system to collect light from the flame, a filter to eliminate radiation of unwanted wavelengths, and a photocell detector connected to a galvanometer. The flame is usually produced in a special burner, which is supplied with a mixture of town's gas and air at carefully regulated flow rates to maintain the flame conditions as constant as possible. Before reaching the burner the air passes through the nebuliser where the venturi effect aspirates the sample solution through a capillary tube and disperses it as a fine mist so that it is introduced into the flame in a uniform manner. Usually only a proportion of the sample actually aspirated reaches the flame, because the larger droplets are allowed to settle out from the gas flow and are rejected, but provided that this proportion is constant and the rate of flow of the sample solution through the capillary is uniform, the measured intensity of the emission for an element will be steady and directly related to its concentration in the solution. Sometimes 'total consumption' burners are used, in which the whole of the sample solution passes through, and is evaporated in, the flame.

The simple instrument may employ either an optical or an interference filter to select the required wavelength to be transmitted to the photocell detector; the detector is usually a barrier-layer type cell, which has sufficient sensitivity for measurement of strongly emitting elements. The signal from the detector is usually monitored with a sensitive galvanometer or micro-ammeter without amplification.

The flame photometer is an easy and convenient instrument to use and is relatively inexpensive, although its range of applications is somewhat restricted. Quite accurate results can be obtained with it provided that care is taken to standardise the conditions in the instrument, and particularly in the flame. It is important that the operating conditions in the nebuliser and burner units be carefully controlled, especially that the flow rates of the gas and the air streams be kept constant, to ensure that the concentration of the required element in the flame, and hence the intensity of its emission, will remain as constant as the variable nature of a flame will allow. Also, the measured emission must be corrected for the natural emission, or 'baseline', arising from the flame with solvent alone aspirated into it, although when very strongly emitting elements are being determined this baseline emission is relatively small and can often be ignored. As in other similar situations, the correction simply involves measuring the emissivity of the flame at the given wavelength with solvent, and with solvent plus sample, and subtracting the former value from the latter.

Although it is possible to calculate a relationship between the concentration of a metal ion in the examined solution and the corresponding galvanometer reading, it is simpler and more practical to prepare calibration curves for the instrument using standard solutions containing the required element in an appropriate range of concentrations. Ideally, a test measurement should consist of an instrumental comparison of an unknown solution with a prepared standard of the same concentration, but in practice it is found that interpolation into the calibration curve gives good accuracy.

2.5.2 Flame Spectrophotometers

In addition to the group of strong emitters, many elements give rise to weak visible emissions or to emissions in the ultra-violet region. The detection limit for weak emissions may, in general, be improved either by using a hotter flame, to increase the number of atoms emitting, or by using a more sensitive detecting system, or by a combination of the two. The flame temperature may be raised by using hydrogen with oxygen or air, or acetylene with air, oxygen or nitrous oxide, instead of town's gas with air, although these mixtures, especially acetylene-nitrous oxide, are more expensive.

More complex instruments have been developed to measure weak emissions over a wide spectral range and consequently determine poor emitters at moderately high concentrations or strong emitters at very low concentrations. Most of these instruments combine the higher flame temperature with improved optical systems and detectors, and the Pye-Unicam SP 900, for example, employs an acetylene-compressed air flame in conjunction with a dispersion prism and a highly sensitive photomultiplier as detector, for complete spectroscopic analysis. The emitted light beam is chopped mechanically before reaching the detector to produce an a.c. signal which is suitable for electronic amplification, and the level of amplification used may be varied to suit the intensity of the emission being studied. With the prism system the emission spectrum from the sample may readily be scanned to determine peak emission wavelengths and peak profiles, some of which are very sharp indeed, in contrast with the relatively broad bands encountered in optical absorption spectrometry; automatic scanning is possible with a slow motor drive connected to the wavelength selection system, the emitted light intensity being simultaneously recorded on a synchronously driven chart recorder.

With the increase in sensitivity in the instrument, the baseline emission due to the flame becomes more significant as it becomes larger in proportion to the emission from the required element or elements. The baseline emission varies considerably over the full spectral range covered by a normal spectrophotometer; it is therefore usually expedient to scan only short wavelength ranges, over which the baseline variation will not be excessive. Also, the inherent instability of a flame causes the readings for a given sample to fluctuate about the true value, and these fluctuations may become considerable when high amplification levels are used to provide the required sensitivity. The only way in which this 'noise' can be reduced is by applying electronic damping to the amplifying circuit, but this inevitably reduces the speed of response of the instrument and so a balance must usually be found between the sensitivity and the degree of damping necessary to give stable operation. Since the light that is measured in a flame emission

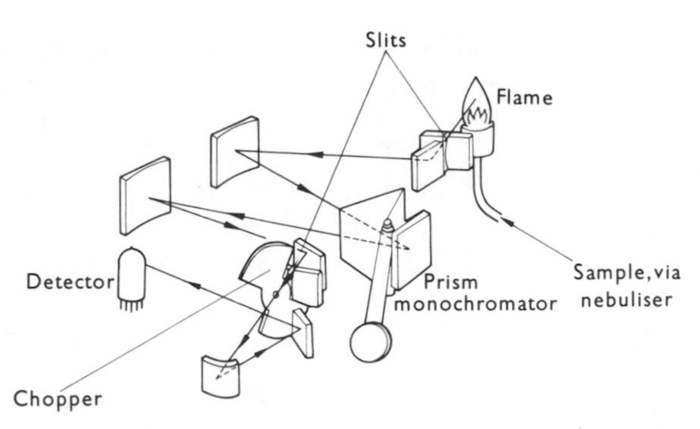

Figure 2.6 Perspective diagram of the operating system and ray paths of the Pye-Unicam SP 900 flame emission spectrophotometer. (Courtesy of Pye-Unicam Ltd.)

spectrometer originates in a single flame that includes atoms of the element being determined, double-beam operation, with its several advantages, is not possible.

Figure 2.6 shows the optical system of the Pye-Unicam SP 900, which is an excellent example of a highly sensitive flame emission spectrophotometer, but which is no longer in production since its function has been combined with that of the atomic absorption spectrophotometer, to be discussed in the next section, in more recent dual purpose instruments such as the SP 90.

2.5.3 Interferences

Many chemical and physical effects can alter the reading obtained with a flame photometer, and it is important that the solutions containing the unknown materials and the standard solutions used for calibration should be as nearly alike as possible, with regard to nature and concentrations of dissolved substances, viscosity, etc., to minimise potential sources of error. The slits of a spectrophotometer are of finite width, and must pass a band of wavelengths, as in optical absorption measurements, so spectral interference may occur if a second element is present having an emission line very close to that chosen for the determination of the required element. An optical system with good spectral discrimination coupled with narrow slit widths can reduce this difficulty provided that the emission lines are sharp, but high concentrations of an unwanted element can give rise to broadened peaks, due to the greater number of excited species present, and so lead to overlapping. A remedy can often be found in using an alternative, but perhaps less intense, emission peak for the required element. The presence of one element may sometimes enhance the emission of another, giving rise to mutual interference.

Chemical interference may result from the differing rates at which different compounds of an element are decomposed in the flame, since free atoms of the required element must be liberated before emission can take place. Thus, calcium sulphate would give a lower emission value than would an equivalent amount of calcium chloride, simply because of its more refractory nature. The addition of a strong complexing agent, such as diaminoethanetetraacetic acid (EDTA), to the test solutions may remove this type of interference by converting all of the required element into a common and relatively easily decomposed state. The test solution may, of course, include an organic solvent, and this is often beneficial, but it must be remembered that such solvents as chloroform or carbon tetrachloride will tend to extinguish the flame unless they are diluted with, for example, a ketone.

The physical conditions in the test solution may also affect the performance of the instrument, and give rise to erroneous readings. For example, if the test solution and the standard differ in viscosity, they will be aspirated through the inlet capillary tube at different rates and so reach the flame in different concentrations relative to the fuel-oxidant mixture. An erroneously low intensity will consequently be obtained for the more viscous solution. Again, solutions which are saturated, or nearly so, with respect to a component may deposit solid when evaporated into the flame, causing partial blockage of the burner and alteration of the flame characteristics and hence of the observed intensities. This may occur during the determination of soluble salts in gypsum, when a sample is extracted with water and the extract, inevitably saturated with calcium sulphate, is analysed for sodium and potassium. In such a case, the difficulty is easily overcome by diluting the test solution so that it is no longer saturated with respect to the interfering compound, since the resulting reduction in the emission intensities does not usually reduce significantly the accuracy of the measurement.

2.5.4 Standardisation and Calibration

As with all optical methods the instrument used must be calibrated using standard solutions of known molarity in the required concentration range. Table 2.1 lists the commonly used emission lines for a range of elements and also their detection limits, expressed in parts per million; working concentrations will generally be about five to ten times these limit values. As mentioned in the previous section, it is important that the chemical and physical characteristics of the standard solution should be as close as possible to those of the test solutions; this may involve making up standard solutions to match the contents of the test samples, but it is often more convenient to isolate the required element from the test solution, e.g. by solvent extraction, or to remove interfering elements, e.g. the use of ion exchange methods to remove alkaline earths in the analysis for alkali metals. The precautions to be taken when using such methods have been discussed in section 2.3.

An alternative, and very useful, technique is the method of standard additions, in which known, increasing amounts of the required element, in an appropriate solution form, are added to aliquots of the test solution.

When the observed intensity readings of the resulting solutions are plotted against the known concentrations due to the additions alone, it is found that these lie on a straight line; when this is extrapolated to zero intensity the 'negative intercept' on the concentration axis corresponds to the original concentration of the element in the test solution. This method of 'internal calibration' has much to recommend it, particularly in its ability to standardise the physical properties of all the solutions used.

In most cases the normal method of plotting a calibration curve of emitted intensity against known concentration and determining unknowns by interpolation suffices, but close intervals should be used to ensure that any test solution is in fact compared with a closely similar standard solution.

2.5.5 Appraisal of Flame Emission Spectrometry

Flame emission spectrophotometry requires only simple instruments for the determination of strongly emitting elements, especially the alkali and alkaline earth elements, and provides a useful and rapid method of routine analysis for such elements, capable of measuring accurately concentrations as low as a few parts per million. Moreover, such elements are often difficult or tedious to determine by other means. With more complex instruments, a wide variety of elements may be determined, many showing detection limits of less than 1ppm with some strong emitters detectable at 0.1ppm or less. The technique permits qualitative analyses to be made easily, but when used under conditions of high sensitivity it suffers from instability due to flame fluctuation, baseline variation, and the several interferences mentioned in section 2.5.3. Flame fluctuations have been reduced, for example by shielding the flame with inert gas, and corrections can be applied for baseline variation and other interferences, but these difficulties have tended to reduce the role of flame emission for the routine analysis of weakly emitting elements to the level of identification only, and accurate determinations of concentrations are now preferentially carried out by atomic absorption spectrometry.

2.6 ATOMIC ABSORPTION SPECTROPHOTOMETRY

Recent advances in flame emission and optical absorption spectrometry have largely been confined to improvements in instrumentation, giving greater sensitivity, improved methods of presentation of the instrumental readings, including digital read-out and scale expansion facilities, and automatic operation, such as the ability to measure the absorption of one or more samples over an extended period of time. Apart from these rather peripheral improvements, however, neither method is capable of a major upgrading in terms of its basic sensitivity, accuracy, and freedom from interferences. The development of atomic absorption spectrophotometry [12–17] has recently provided a powerful new analytical tool which is of particular interest to mineral analysts because it offers a convenient and rapid routine method of analysis for a wide range of elements.

The method is based on the ability of atoms vapourised in a flame to absorb energy from a light beam of a suitable wavelength. Consideration of the Maxwell-Boltzmann distribution curve (figure 1.5) at high temperatures shows that, although an appreciable fraction of the atoms are in excited states, a larger fraction are in low energy states and so are still capable of absorbing energy from electromagnetic radiation of a very specific quantum energy, corresponding once again to a permitted transition, but this time in an atomic species rather than in a molecular species as in optical absorption spectrometry. The absorption peaks exhibited by atoms are narrower than those of molecular compounds, and so the development of atomic absorption spectrophotometry had to await the development of a light source with a correspondingly low bandwidth. Such a source is provided by a discharge lamp fitted with a hollow cathode made from, or containing, the element for whose determination the lamp is to be used. Light emitted from such a lamp comprises the spectral lines of this element only, with no contribution from a continuous spectrum, and so is very selective, being characteristic of the element being measured and virtually no other. In particular, true absorption interference by 'foreign' atoms becomes negligible since the very low bandwidth of the emitted lines, about 0.002nm, effectively precludes any absorption band overlap with other atoms. It may be noted that the narrowest bandwidth achieved with a good slit and monochromator system is of the order of 0.2nm, so that a continuous source is unsuitable for most atomic absorption methods. It is of course possible that a fortuitous overlap may occur between two atoms at a certain common wavelength in their spectra, but it is highly unlikely that overlap will occur at any other line, and so selection of the line at which the absorption is to be measured, using a monochromator, ensures the exclusive character of the absorption measurement. The provision of a monochromator also helps to reduce the 'noise' due to the natural light from the flame, by rejecting the general continuous spectrum to a very great extent.

2.6.1 Instrumentation

The atomic absorption spectrophotometer comprises three main sections, the light source, the absorption flame, and the wavelength selector and detector; these will be considered separately below.

Light Source

The light source is the heart of the atomic absorption spectrophotometer and is normally a hollow cathode lamp which produces a line spectrum specific to the element being determined. As shown in section 2.6 this gives the great advantage of high specificity to the source and leads to high levels of accuracy and sensitivity with a minimum of interference, but this very specificity is a disadvantage in another way. Since each element requires its own lamp, the lamp must be changed for each element being determined in a multi-element analysis; hollow cathode lamps are quite expensive and require appreciable warm-up times, so the user can be involved in a considerable investment both in terms of capital and of time, although the latter may be reduced by storing the lamps in a special rack in which they are kept in a standby condition, plugged into a reduced heater circuit to keep them warm. In addition, the use of a line spectrum makes it impossible to perform a wavelength scan to identify individual elements present, although a check for several specific elements can be made by changing the lamp for each component. This is obviously time consuming, and recent instruments have had turrets fitted to take up to six lamps in a standby condition to facilitate rapid changing. Lamps are now available with several elements in the hollow cathode; these are convenient but costly, and since the life of any hollow cathode lamp is limited they can hardly be justified economically unless most of the elements included in the lamp are required in most of the analyses to be made. Despite the disadvantages attendant on the use of single element hollow cathode discharge lamps, their use has revolutionised optical methods of spectrophotometry and greatly reinforced the techniques available to the analytical spectroscopist.

Absorption Flame

The light from the hollow cathode source passes first through the flame into which the test solution is aspirated and in which the absorption occurs. The flame is normally produced by a slot burner, usually oriented with its long axis parallel to the direction of the light beam to provide the longest possible absorption path. Since a large fraction of the atoms of the required element in the flame are capable of absorbing the relevant radiation, and the absorption path length is long, the magnitude of the absorption signal available for measurement is much greater than the corresponding signal in the flame emission technique.

The test liquid is aspirated into the flame via a nebuliser, as in flame emission spectrometry, and carried forward by the mixture of fuel and oxidant as a fine cloud. Conditions in the flame must be maintained as homogeneous as possible, as described in section 2.5.1. The flame is normally an acetylene-air mixture, but acetylene-nitrous oxide mixtures are used for higher temperatures. Once again, a constant flame is required to produce homogeneous and reproducible conditions for measurement, but a little more latitude is possible here than with the flame in emission work, partly because small changes in the flame condition result in smaller changes in the numbers of absorbing atoms and partly because electronic filtering can be used to remove the effect of such variations, as discussed in the next section.

Wavelength Selection and Detection

After passing through the flame, the light next passes through either a prism or an optical grating and a slit system which selects a narrow band of wavelengths for final transmission to the detector. This selection is carried out in order to minimise the effects due to the general absorptivity of the flame or to fortuitous absorption of certain components of the line spectrum from the lamp by foreign atoms in the flame. The detector unit, which usually incorporates a photomultiplier, transmits a signal either to a meter, in the simpler instruments, or to a digital read-out in the newer models.

The detector of course responds to all the energy incident on it, irrespective of its source, and so will respond to such of the natural radiation from the flame as is of the correct wavelength to pass through the wavelength selector, as well as to the spectral emission lines of the required element, which are generated in the flame and, being effectively coincident with the absorption lines, pass readily through the selection system. The combined effect of these interferences is to increase the general noise level of the detector and to give rise to curved relationships between the concentration and the instrument reading, since the higher emission levels at higher concentrations partly nullify the increased absorption. The effects may be allowed for by careful calibration, and sometimes by choosing an alternative line for measurement, but it is better to modify the measuring system to eliminate flame effects completely and ensure that the detector responds only to light from the hollow cathode lamp. This is achieved by modulating the output from the lamp, either electronically or mechanically, and tuning the detector circuitry to match this modulation and accept only signals with the correct characteristics. The relatively constant light arising in the flame, either due to the combustion process or to the required element, is thus excluded from the measurement and the output signal corresponds only to the degree of absorption of the transmitted light.

Types of Instrument

Since the atomic absorption spectrophotometer possesses a light source distinct from its flame, atomic absorption spectrophotometers may operate either as single-beam or as double-beam instruments. In this they are more akin to optical absorption spectrometers than to flame emission spectrometers, which can operate only as single-beam instruments. Unfortunately, both types of instrument are relatively expensive in comparison with the other two spectrometers.

A simple single-beam spectrophotometer includes a suitable light source, a nebuliser and flame, a slit system and monochromator, and a measuring unit. The light source may be continuous or it may be modulated either electronically or by the addition of a simple mechanical chopper driven synchronously with the detection system modulator. Readings are usually presented on a meter so calibrated that its response is approximately linear with respect to the concentration of the required element. The quantity actually measured is the optical density, $\log(I_0/I)$, as in optical spectrometry, and the Beer-Lambert law [equation (2.1)] holds for low sample concentrations under controlled measurement conditions.

In the more complex, and more expensive, double-beam instrument part of the output from the lamp is diverted by a rotating chopper while the remainder passes through the flame in the normal way, the two beams being recombined subsequently for presentation to the detector system. The chopper automatically modulates the beam passing through the flame, and the electronic circuitry of the detector system rejects signals due to light arising in the flame and separates the signals due to the two parts of the beam. The two signal levels are balanced manually by returning a meter to the null position, first with solvent aspirated into the flame and then with test solution. The difference between the two settings, corresponding to the effect of the required element in the flame, is obtained as a digital presentation of absorbance. Figure 2.7 shows diagrammatically the optical layout of the Perkin-Elmer 303 instrument, which uses the above system.

Double-beam operation does not have the same advantages in atomic absorption as it has in optical absorption, principally because no allowance can be made for the effects of the solvent or of extraneous solutes, since the test solution must be aspirated into the flame and cannot affect the reference beam. In fact, corrections for such effects must always be made by separate calibration of the instrument for the type of solution under examination, in both single and double-beam instruments. The merits usually claimed for double-beam atomic absorption spectrophotometers are greater stability and, particularly, automatic compensation for any variation

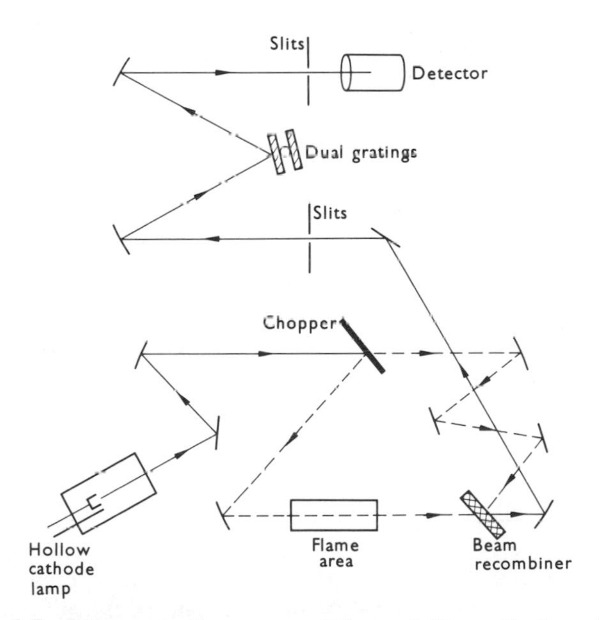

Figure 2.7 Diagrammatic representation of the optical system of the Perkin-Elmer 303 double-beam atomic absorption spectrophotometer. (Courtesy of Perkin-Elmer Ltd.)

in the output from the hollow cathode light source. Since hollow cathode lamps take some time to warm up and reach stable outputs, this means that a double-beam instrument may be put into operation a little more quickly than a single-beam type. On the other hand, the double-beam instrument is considerably more expensive than the single-beam type, and a careful appraisal of all their relative merits is advisable before deciding to purchase one or the other.

2.6.2 Sample Preparation

Sample preparation for atomic absorption is very similar to that for optical absorption or flame emission, except that there is generally less need to isolate the required element because of the spectral resolution available. The method is, however, subject to all the other interferences mentioned in section 2.5.3 for flame emission spectrometry. Consequently it is necessary that all measurements be carried out by comparing unknown solutions with standards which are as nearly identical with them as possible, and the recommendations made in section 2.5.3 apply equally to atomic absorption methods also.

A special technique may be used to deal with solutions which are difficult to handle by conventional methods, e.g. solutions in organic solvents such as chloroform, which would tend to extinguish the flame, or when the volume of solution available is even less than the few millilitres usually necessary for an atomic absorption measurement. A known, small volume of the solution is placed in a tantalum boat which is then held close to the flame to evaporate the solvent without volatilising the required element. When evaporation is complete, the boat is placed in the flame and the contents are volatilised. This procedure gives a 'pulse' reading, similar to that obtained in gas chromatography, rather than the usual steady reading obtained when the sample solution is aspirated into the flame at a constant rate. The pulse is registered on a chart recorder, and the area under the curve is a measure of the amount of the material present. Calibration is carried out in the usual way, using the same volumes of solutions of known molarities.

Work is currently in progress on a method which will permit direct evaporation of solid samples into the light beam, and it may thus be possible, before long, to apply atomic absorption methods to mineral materials without the need for prior dissolution with all its attendant difficulties.

2.6.3 Appraisal of Atomic Absorption Spectrometry

Since the technique is based on light absorption, measurement of a given solution involves first aspirating a portion of the solvent, or of a 'blank' solution, into the flame and adjusting the electronic circuitry to read 100% transmission (zero absorbance), then aspirating the test solution and reading off the reduction in the intensity of the transmitted light. Atomic absorption thus combines the advantages of optical absorption spectrometry, principally the ability to use a steady and relatively powerful light source, measurement of a reduction in a reasonably strong signal rather than the detection of a weak signal as in flame emission spectrometry, and the facility for double-beam operation, with the ability to detect discrete atoms at flame temperatures rather than complex molecules at room temperature. Thus the stability of the measured light is much better than with flame emission, and selectivity is much higher than with optical absorption methods. Photomultiplier tubes are used as detectors, and some instruments allow considerable scale expansion, which increases the readability of the intensity scale, so that the sensitivity attained is comparable with that achieved by the better flame emission instruments.

Table 2.1 includes sensitivities for a representative group of elements, which is by no means exhaustive; these and the detection limits for flame emission are taken from 'Undergraduate Instrumental Analysis' by J. W. Robinson. 'Sensitivity' is usually defined as that concentration which gives an absorption reading of 1%, while the 'detection limit' is that concentration which gives a signal twice the instrumental noise level. Sensitivity limits are usually several times larger than the corresponding detection limits, but both values depend markedly on the instrument and the procedure used. Since the detection limits for flame emission quoted by Robinson are often ten times those quoted for the SP 900 spectrophotometer, it is not unreasonable to use the figures in table 2.1 as a rough comparative guide to the practical sensitivities of the two methods. It is seen that with a few exceptions atomic absorption spectrophotometry provides a sensitivity at least as good as flame emission, and better in many cases. The smallest working concentrations, those at which reasonable practical analyses may be obtained, are usually about five to ten times the sensitivity values.

The use of a flame confines atomic absorption and flame emission methods to the detection of atomic species, but molecular groupings may sometimes be determined by an indirect approach. Thus halides may be determined by quantitative precipitation as the corresponding silver salt, followed by measurement of the silver in the precipitate, and sulphate may similarly be determined after precipitation as barium sulphate.

Atomic absorption spectrometry has proved to be a reliable and flexible technique for the analysis of many elements, particularly metals, and is limited only by the availability of a suitable hollow cathode discharge lamp for the element in question. The technique is simple to apply, involving as it does only dissolution of the sample and aspiration into the flame under the correct experimental conditions, and it is very rapid and accurate. Also, since the atomic absorption spectrometer contains the essential parts of a flame emission spectrometer, many instruments now available are either atomic absorption instruments capable of being converted for use as flame emission instruments by means of attachments, or are completely dual purpose instruments. This overcomes the limitations of atomic absorption methods for qualitative work, since a wavelength scan can be made with the

TABLE 2.1

Element	Flame Emission		Atomic Absorption	
	Line Wavelength (nm)	Detection Limit ($\mu g/cm_3$)	Line Wavelength (nm)	Sensitivity ($\mu g/cm_3$)
Aluminium	396.2	0.5	309.3	1.0
Antimony	252.8	1.0	217.6	0.1
Arsenic	235.0	2.2	193.7	2.0
Barium	553.6	1.0	553.6	0.2
Bismuth	223.1	6.4	223.1	0.1
Boron	518.0	3.0	249.7	250
Cadmium	326.1	0.5	228.8	0.01
Calcium	422.7	0.07	422.7	0.05
Caesium	852.1	0.5	852.1	0.1
Chromium	425.4	5.0	357.9	0.1
Cobalt	242.5	1.7	240.7	0.1
Copper	324.7	0.6	324.7	0.1
Iron	372.0	2.5	248.3	0.1
Lanthanum	442.0	0.7	392.8	75
Lead	405.8	14	217.0	0.01
Lithium	670.8	0.07	670.8	0.03
Magnesium	285.2	1.0	285.2	0.001
Manganese	403.3	0.1	279.5	0.05
Mercury	253.7	2.5	253.7	1.0
Molybdenum	379.8	0.5	313.3	0.1
Nickel	352.4	1.6	232.0	0.1
Niobium	405.9	12	334.9	20
Palladium	363.5	0.1	247.6	0.5
Phosphorus	253.6	1.0	–	–
Platinum	265.9	15	265.9	1.0
Potassium	766.5	0.02	766.5	0.01
Rhenium	346.1	3	346.1	15
Rhodium	369.2	0.7	343.5	0.1
Rubidium	780.0	0.6	780.0	0.1
Silicon	251.6	4.5	251.6	0.8
Silver	338.3	0.6	328.1	0.01
Sodium	589.0	0.001	589.0	0.01
Strontium	460.7	0.06	460.7	0.1
Tin	243.0	1.6	235.5	0.5
Titanium	399.9	5	364.3	1.0
Vanadium	437.9	3	318.4	1.0
Zinc	213.9	77	213.9	0.01

Detection limits and sensitivities in $\mu g/cm^3$ (ppm) for various elements determined respectively by flame emission and atomic absorption spectrophotometry. Working concentrations would be approximately five to ten times the values quoted. Taken from 'Undergraduate Instrumental Analysis' by J. W. Robinson, pp.148 and 183–184, by courtesy of Marcel Dekker, Inc.

instrument in the flame emission mode to identify the elements present, after which it can be used as an atomic absorption spectrophotometer for quantitative measurements of the amounts of the required elements. The Perkin-Elmer 303 can easily be converted from atomic absorption to flame emission operation, the accessories providing for single-beam operation and automatic wavelength scanning, and the Pye-Unicam SP 900 has recently been replaced by the more versatile SP 90, which is a single-beam, dual purpose instrument with a prism monochromator.

The development of atomic absorption – flame emission spectrophotometers has proceeded rapidly over the past few years, and changes are still in progress. The more expensive instruments, costing several thousands of pounds, can provide almost completely automatic analysis; for example, the Pye-Unicam SP 1900, a double-beam instrument with a grating monochromator, employs a turret fitted with up to six hollow cathode lamps each of which can be automatically selected as required, so that up to six elements may be determined simply by adjusting the instrument controls, and the results obtained in printed form.

2.7 SUMMARY

The three optical methods discussed are important in the analysis of mineral materials, and are largely complementary. Thus, optical absorption spectrometry has a very wide applicability, since suitable methods have been elaborated for most elements and for many molecular groups. Selectivity and sensitivity vary with the species to be determined, but in some cases optical absorption offers the highest sensitivity of the three methods, and certainly it is the only one generally applicable to analyses for specific chemical groupings rather than single elements. The instruments available range from inexpensive colorimeters to complex automatic spectrophotometers.

Flame emission spectrometry provides a sensitive method of determining many elements, being particularly useful for strongly emitting elements such as the alkali and alkaline earth elements, which do not readily form coloured compounds suitable for optical absorption methods, and the characteristic nature of elemental spectra allows qualitative analyses to be made by spectral scanning. Selectivity depends mainly on the optical quality of the instrument used, but is not so good as is obtainable with atomic absorption spectrometry. Simple flame photometers for the determination of very strong emitters, such as sodium, potassium, calcium, etc., are inexpensive, but the more precise instruments, used for more weakly emitting elements, have largely been combined into dual purpose spectrometers. For some elements, for example boron, rubidium, caesium, flame emission spectrometry is more sensitive than atomic absorption.

Atomic absorption spectrometry provides a sensitive and very selective method of determining a wide range of elements and, because of the stability and operational simplicity of atomic absorption instruments, is very suitable for routine and automatic analysis. Since specific light sources are needed for the different elements, qualitative analysis by wavelength scanning is not easy to carry out. The instruments involved are expensive, the cheapest costing over £1000, but the capital investment is usually well justified where rapid routine analysis, particularly of metals, is involved.

Mention has been made of instruments by two particular makers; these were chosen as examples of excellent, reliable spectrophotometers, and should not be considered as exclusive of instruments made by other manufacturers. Several companies manufacture a wide range of optical instruments and development in the field is continuous; the mineral analyst contemplating the purchase of an instrument would therefore be well advised to consider all those currently available, particularly when the purchase may represent a large capital investment.

2.8 THE DIRECT READING SPECTROGRAPH

The three methods discussed are, of course, not the only methods of analysis based on the absorption or emission of light that can be used in the determination of the elemental contents of minerals and mineral products, although they are the methods most commonly used. In particular, the characteristic optical spectrum can be excited by using an electric arc or spark instead of a flame. In the spark emission spectrograph this spectrum is analysed using a quartz prism in the usual manner, and the resulting dispersed spectrum focused on to a photographic plate. The positions of the observed lines on the plate serve to identify the element, and their intensity gives a quantitative indication of the amount present in the sample.

A development of the spark emission spectrograph, which appeared in the 1960's, was the direct reading spectrograph, often referred to as the direct reader. In this instrument the photographic recording plate is replaced by a bank of slits and phototubes and a long path length is provided between the prism and the detectors, to ensure sufficient dispersion of the components of the light beam. Each slit and phototube is mechanically positioned to record the intensity for one specific element, and an average instrument can record intensity data for up to 30 elements simultaneously. In practice, the sample is ground, mixed with graphite, and cast into an electrode, in a manner similar to that for spark source mass spectroscopy, described in Chapter 5, section 5.2.1. The resulting electrode is then sparked for a predetermined time, and the intensities accumulated by the different phototubes are printed out automatically and the various elemental concentrations determined by comparison with standard materials.

The instrument is useful, and was widely used during the 1960's, but to some extent has been overtaken by improvements in technique in X-ray fluorescence and spark source mass spectroscopy. It is expensive, costing between about £10,000 and £40,000, and consequently the mineral analyst contemplating purchase of analytical equipment of this type might be well advised to consider as an alternative a good XRF unit, which possesses greater flexibility.

ACKNOWLEDGEMENTS

The author wishes to express his thanks to Professor Stacey G. Ward, Head of the Department of Minerals Engineering, for his interest and encouragement, and to his colleague Dr. C. V. Phillips for his assistance in writing the section on atomic absorption spectrophotometry. Thanks are also due to Perkin-Elmer Limited and to Pye-Unicam for their courtesy in supplying diagrams of their instruments, and to Marcel Dekker Inc. for permission to reproduce the figures used in table 2.1.

REFERENCES

1. G. H. AYRES, 'Quantitative Chemical Analysis', 2nd ed.; Harper and Row, New York, 1968.
2. G. W. EWING, 'Instrumental Methods of Chemical Analysis', 3rd ed.; McGraw-Hill, New York, 1967.
3. J. W. ROBINSON, 'Undergraduate Instrumental Analysis'; Marcel Dekker, New York 1970. This is very useful introduction to a variety of instrumental methods, describing principles, methods and some applications.
4. R. P. BAUMAN, 'Absorption Spectroscopy'; Wiley, New York, 1962.
5. G. F. LOTHIAN, 'Absorption Spectrophotometry', 3rd ed.; Hilger and Watts, London, 1969.
6. E. B. SANDELL, 'Colorimetric Determination of Traces of Metals', 3rd ed.; Wiley-Interscience, New York, 1959.
7. F. D. SNELL and G. T. SNELL, 'Colorimetric Methods of Analysis', Vols. I, II and IIA; Van Nostrand, New York, 1948-59.
8. R. MAVRODINEANU and H. BOITEUX, 'Flame Spectroscopy'; Wiley, New York, 1965.
9. J. A. DEAN, 'Flame Photometry'; McGraw-Hill, New York, 1960.
10. F. BURRIEL-MARTÍ and J. RAMIREZ-MUÑOZ, 'Flame Photometry'; Elsevier, New York, 1964.
11. R. HERRMANN and C. T. J. ALKEMADE, 'Chemical Analysis by Flame Photometry', 2nd ed. (trans. by P. T. GILBERT); Wiley, New York, 1963.
12. G. D. CHRISTIAN and F. L. FELDMAN, 'Atomic Absorption Spectroscopy'; Wiley-Interscience, New York, 1970. This book contains a comprehensive collection of methods, but not applied specifically to mineral analysis.
13. W. T. ELWELL and J. A. F. GIDLEY, 'Atomic Absorption Spectrophotometry', 2nd ed.; Pergamon, Oxford, 1967. Contains detailed methods for many elements, and a useful comparative chart describing available commercial atomic absorption instruments.

14. W. SLAVIN, 'Atomic Absorption Spectroscopy'; Interscience, New York, 1968. Considers the analysis of individual metals, and quotes examples of geochemical and industrial applications, including some from the minerals industry.
15. J. W. ROBINSON, 'Atomic Absorption Spectroscopy'; Marcel Dekker, New York, 1966.
16. I. RUBEŠKA and B. MOLDAN, 'Atomic Absorption Spectrophotometry' (trans. by P. T. WOODS); Iliffe, London, 1969.
17. The periodical *Atomic Absorption Newsletter*, published by the Perkin-Elmer Corporation, is a useful vehicle for recent advances in atomic absorption.

The above list, which is by no means exhaustive, is intended as a guide to further reading. Pye-Unicam Limited produce method sheets which describe in detail many specific analytical procedures worked out for their own various instruments. The Perkin-Elmer Corporation similarly publish methods for their atomic absorption instruments.

CHAPTER 3

X-ray Fluorescence

G. L. Hendry

Department of Geology
University of Birmingham
Birmingham B15 2TT
England

3.1 INTRODUCTION

The status of chemical analysis by X-ray fluorescence has developed from that of a laboratory curiosity in the decade 1940-50 to its current position as an almost universal method with, according to a recent estimate, over 8000 spectrometers in use throughout the world. The rapid growth and the advances in technique during that time are largely due to post-war developments in vacuum technology and electronics, which not only have replaced pumped X-ray tubes by the more stable sealed versions, but also have led to direct methods of detecting and estimating X-ray photons superseding film based methods. The modern X-ray spectrometer is, therefore, a sophisticated electronic instrument capable of accurate, fully automatic analysis, with obvious potential in on-line process control, as discussed in Chapter 6.

During this developmental period a great deal of the theoretical background to the method has been covered in many books and papers. Only brief reference will be made to theory in the chapter, and those interested in pursuing the topic further are recommended to the reading list at the end of this chapter. The classic work by Compton and Allison is particularly recommended as a foundation text. For up to date information on developments in technique and general methodology and equipment, Chemical Abstracts, Mineralogical Abstracts, X-ray Spectrometry and the Advances in X-ray Analysis series should be consulted.

The method possesses the great advantage of being non-destructive, amenable to automation, and rapid. Moreover, with automatic equipment it is possible to analyse for more than one element in a single run, so that a large amount of data can be accumulated quickly, and without operator intervention. The possible disadvantage that a computer is needed to keep pace with the conversion to elemental concentrations is now largely removed by the availability of cheap electronic calculators which may be operated on-line to the instrument.

Inherent problems such as absorption of X-rays by the sample itself, sample preparation, and light element analysis, have largely been overcome by a combination of improved apparatus and methodology. Today, despite the large capital cost of the equipment, the accuracies of the element concentrations obtained by X-ray spectrometry are comparable with values obtained by any other method. Indeed the cost per analysis is generally lower than competing techniques if large numbers of samples are to be processed.

The purpose of this chapter, therefore, is to assess these advantages and disadvantages with particular reference to mineral analysis.

3.1.1 Safety

All workers with X-rays must be aware of the dangers inherent in the use of such high energy radiations and are referred to the discussion of precautions to be taken, in Chapter 1, section 1.5, and also to the publications by the

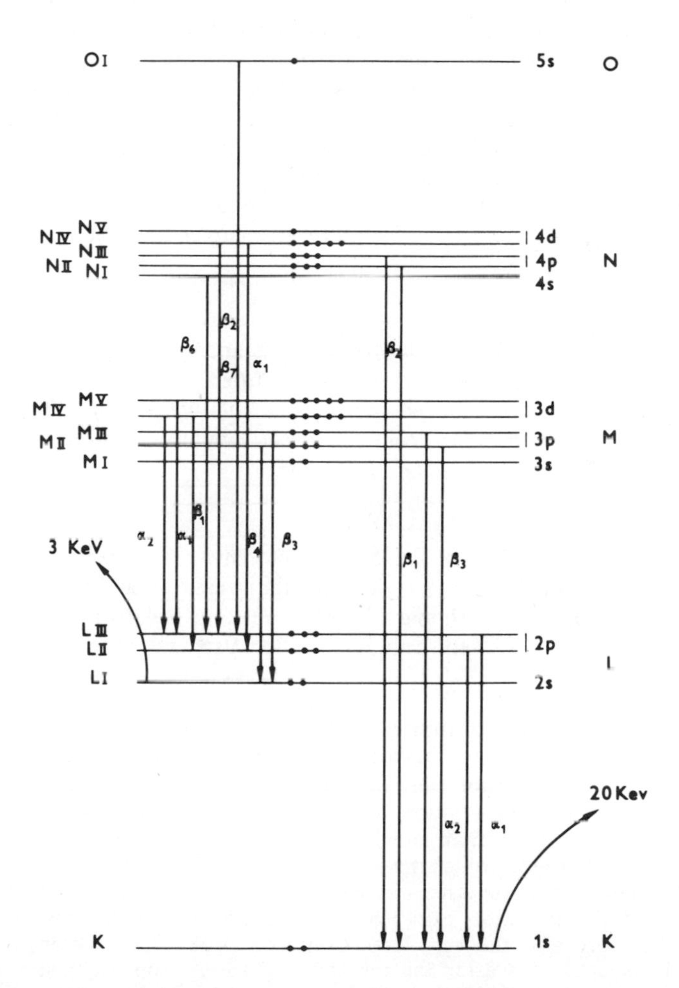

Figure 3.1 Energy levels of electrons in the molybdenum atom. The arrows represent electron transitions producing *K*- and *L*-spectrum X-ray emission lines. The upward curving arrows represent the ionisation of the *K*- and *L*-electrons.

Department of Employment and Productivity [1], the International Union of Crystallography [2], Jenkins and Haas [3], and the information provided by the instrument manufacturers.

3.2 BASIS OF THE METHOD

The model of the atom published initially by Bohr in 1913 and subsequently modified by Rutherford has been discussed in Chapter 1. As has been pointed out, the features of primary interest to the X-ray spectroscopist is the existence of discrete energy levels, the *K, L, M, N* shells, for the electrons surrounding the central nucleus. When sufficient energy is applied to the atom, electrons may be ejected from an inner shell and replaced, within 10^{-15} seconds, by electrons from outer shells. At each step in this process a photon of electromagnetic radiation is emitted with a wavelength in the X-ray region, i.e. in the range 0.1 through 70Å, corresponding to the energy differences between the shells, as shown diagrammatically in figure 3.1. The photon energies and their associated wavelengths are given by

$$E = |E_{\text{outer}} - E_{\text{inner}}| = \frac{hc}{\lambda} = \frac{12{,}395}{\lambda} \tag{3.1}$$

where E is the photon energy, h is Planck's constant, c is the velocity of light, and λ is the wavelength of the X-radiation generated.

The spectrum so produced comprises relatively few lines, and Moseley [4] showed, in 1914, that there is a simple relation between the wavelength and the atomic number, Z, namely

$$\frac{1}{\lambda} = K(Z - 1) \tag{3.2}$$

Moseley realised that the unique, or 'characteristic', line spectra produced by each element could be used to identify that element, and further that a mixture would produce a simple additive spectrum. Relative intensities of elemental lines in different samples would also give an estimate of

Figure 3.2 The *K*-spectrum of molybdenum measured using a scintillation detector in conjunction with a 0.15mm collimator and a LiF_{220} crystal to give high resolution. The upper curve (a) shows the electron excited spectrum from a molybdenum target X-ray tube operated at a potential of 60kV. Note that the characteristic lines are superimposed on a continuous background which rises sharply from the Duane-Hunt limit at $10^\circ\ 2\theta$. (The limit is here rather less defined than in other cases since, at 10° 2θ, much of the radiation passes over the analysing crystal without being diffracted.) Note also the very large Compton peaks (Co) obtained when boric acid is used to scatter the incident radiation (compare figure 3.16). The lower curve (b) shows the spectrum from pure molybdenum but excited by primary radiation from a tungsten X-ray tube operated at 60kV. Note the absence of the continuous background and the very obvious 'half order' peaks produced by diffraction from the LiF_{220} crystal.

concentrations. He published photographs of the spectrum of brass showing that it was made up of lines from copper and zinc, but his claim, that the method's 'advantage over ordinary spectroscopic methods lies in the simplicity of the spectrum and the impossibility of one substance masking the radiation of another', has since proved to be somewhat over-optimistic!

Before considering practical techniques for elemental analysis, two background aspects of the method require discussion, these being the excitation of the spectrum and its analysis.

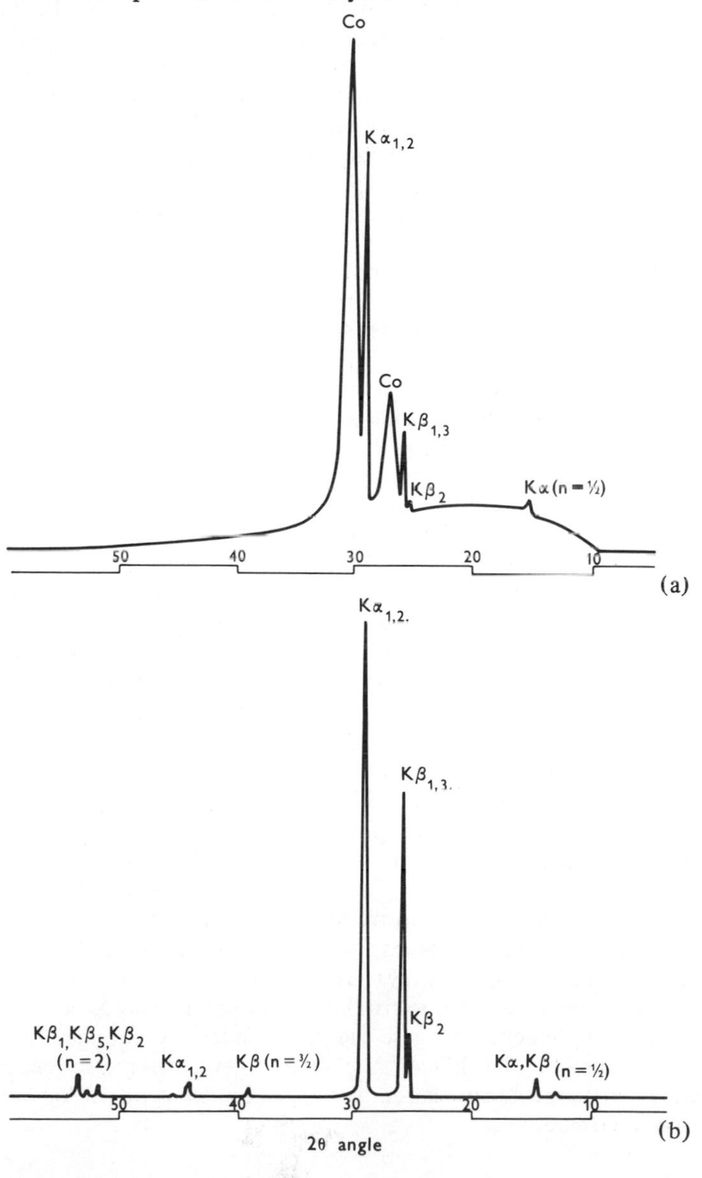

3.2.1 Excitation of X-radiation

The energy to eject electrons can be supplied in two principal ways, by collision with an accelerated electron or by irradiation with high energy photons in the X-ray or the γ-ray regions of the spectrum. A recent paper by Johansson, Akselsson and Johansson [5] describes the use of high energy α-particles, which allows determinations to be made in the parts per billion range for some elements, but the method is not generally applicable.

Electron bombardment is now mainly used to produce X-ray photons from a metal target for use in further excitation, or to produce monochromatic X-rays for diffraction studies (Chapter 7). This is, of course, the basis of the X-ray tube, where accelerated electrons impinge on a water, or air, cooled anode, which is usually of a heavy metal, such as copper, coated with a thin layer of the metal whose spectrum is to be excited. A small proportion, about 1.2%, of the incident energy is converted into the 'characteristic' X-rays of the target material, together with an X-ray continuum, as shown diagrammatically in figure 3.2. The remainder of the energy is lost, mainly as heat.

Historically, it is interesting to note that early X-ray spectroscopists used direct electron excitation of characteristic radiation by coating their samples on to the target of a demountable X-ray tube, in place of the usual metal layer. Thus, von Hevesy and Coster [6] discovered hafnium in 1923 by this method, and Finnish workers in the 1940's were able to determine rare earth elements in granites at the 0.01ppm level, by using multiple film exposures lasting over many hours. Currently, however, only two spectrometers, the Telsec betaprobe and the Hilger & Watts instrument, use direct electron excitation for X-ray fluorescence analysis, and its major application is in the field of light element analysis, atomic number 5 (boron) and above as described by Price [7]. An electron beam is also used in the related technique of electron probe microanalysis (Chapter 11), where it is possible to focus the exciting beam to a spot 0.5-50μm in diameter. This is impossible with an X-ray beam if high intensity is required.

The direct method of excitation, however, is generally inferior to other methods of excitation, especially those based on primary X-ray sources, since, as figure 3.2 shows, the observed spectrum obtained by electron excitation contains not only the lines characteristic of the elements present but also a continuous spectrum, which raises problems of high background noise during the detection stage of the analysis. The spectrum obtained by primary X-ray excitation consists mainly of the characteristic lines of the element, or elements, present in the sample and so is easier to detect and to interpret. Hence, almost all spectrometers employ an X-ray tube in which electrons excite primary X-rays, which are in turn used to produce secondary, characteristic X-rays from the sample to be analysed. Nevertheless, direct electron excitation is still superior to X-ray excitation for the lightest elements, because the gain in intensity obtained outweighs difficulties due to the higher background level, and also because of reductions in the problems caused by the low critical depth of light element radiation (section 3.4.4).

A third mode of excitation has recently come into use, based on radio-isotope sources emitting α-, β-, or γ-rays. These radiations are quite capable of producing fluorescent X-rays from any material, and the small size of the sources makes them potentially of great use in portable X-ray fluorescence units, as described by Bowie [8] and in the review by Rhodes [9]. They suffer from the same disadvantages as direct electron excitation methods, in that they generate a continuous spectrum, and they have the additional drawback that their overall intensities are very low.

Continuous X-ray Spectra

Figure 3.2 shows clearly the continuous spectrum of X-ray wavelengths produced in addition to the characteristic spectrum, when high energy electrons, or very high energy photons, impinge on a material. The continuous spectrum starts abruptly at a wavelength λ_{min}, rises to a maximum at approximately $2\lambda_{min}$, and then progressively decreases in intensity towards longer wavelengths. The minimum, known as the Duane-Hunt limit, decreases with increasing accelerating voltage on the X-ray tube, and represents the point at which the total momentum of an impinging electron is given up in a single collision and converted into a photon of X-radiation. The continuum of lower energies is formed because the majority of electrons will lose varying amounts of energy as they pass through the target material before colliding with an atom and releasing an X-ray photon. Kramers' formula [10] is the most successful approximation to the shape of the continuum, but this does not allow for modification of its shape by such factors as self-absorption by the anode material, or tube window filtration.

The principal properties of the continuum may be summarised as:

(a) the short wavelength limit depends only on the applied tube voltage, is independent of the target material, and is given by the expression

$$\lambda_{min} = \frac{12{,}395}{V} \qquad (3.3)$$

where V is the applied tube voltage in kilovolts and λ_{min} is in Å;

(b) the maximum of the continuum is also independent of the target material;

(c) the total integrated intensity of the continuum is proportional to ZV^2, where Z is the atomic number of the target material, and so increases with increasing atomic number of the target element and markedly with the accelerating voltage.

Characteristic X-ray Spectra

Many modes of interaction are possible between high energy electrons or photons and an atom, but only one will produce a characteristic spectrum. This is the case in which an inner orbital electron of the atom is ejected, and replaced by an outer, higher energy electron to produce a spectrum unique to that element.

In Chapter 1 it is shown that this process requires the exciting photon to have an energy slightly greater than that of the emitted X-rays. Therefore, in direct excitation, provided there are incident electrons with energies sufficient to eject electrons from the target atoms, a characteristic spectrum will be produced. This will, of course, be superimposed on the continuous spectrum. Unlike the continuous spectrum which is produced at any voltage, a minimum voltage is required to produce the characteristic spectrum, and this may be calculated by substituting the wavelength corresponding to the absorption edge of the target element for λ_{min} in equation (3.3). Absorption will be further discussed in section 3.2.2. Tables of absorption edge values are generally available, for example Lonsdale [11], and the ASTM tables [12]. Excitation potentials range from 1.1keV for sodium up to 115.6keV for uranium, in the *K*-series, and from 0.06keV to 21.8keV in the corresponding *L*-series. The higher excitation potentials for the *K*-series lines and the increase towards the heavier elements are both explicable in terms of the modified Bohr model for the atom, since the binding energies of the innermost electrons are greater than those for the outer electrons, and the inner electron levels of a heavy element lie at a lower energy than the corresponding levels for a light element. As a corollary, the wavelengths of both *K*- and *L*-series lines decrease as the atomic weight increases, as shown by Moseley [4].

It would, therefore, appear that the characteristic spectrum from a heavy element target tube might be used to excite the characteristic spectra of lighter elements, in primary X-ray source excitation. This, in general, is not employed, principally because the *K*- or the *L*-series lines for a given sample element are most efficiently excited by incident photons of wavelength approximately 0.2Å less than the appropriate absorption edge value, see figure 3.3. The continuous spectrum from the target, however, contains wavelengths corresponding to the most effective exciting wavelengths for all elements. In addition the integrated intensity of the available wavelengths is also an order of magnitude greater than the intensity of the characteristic spectrum. Hence, in the analysis of multi-element samples using primary X-rays for excitation it is not the characteristic line spectrum of the target element (Cr, W, Mo, Ag) that is most effective in producing the characteristic spectra of the elements in the sample, but rather the continuous spectrum.

In some energy-dispersive spectrometers a series of secondary targets of different elements may be employed, partly to overcome line overlaps, and partly because the use of a single exciting wavelength greatly simplifies the calculation of interelement correction factors (section 3.4.5). These instruments do not require high count rates so that the low intensities can be tolerated.

Choice of Primary X-ray Tube Source

The above discussion would appear to suggest that the best primary source should be a heavy element tube operated at the highest possible voltage, since this would provide the greatest integrated intensity. Unfortunately, the electron backscatter is also greatest for heavy elements and gives rise to higher temperature gradients around the tube window. Because of this, relatively thick beryllium windows are required which do not efficiently

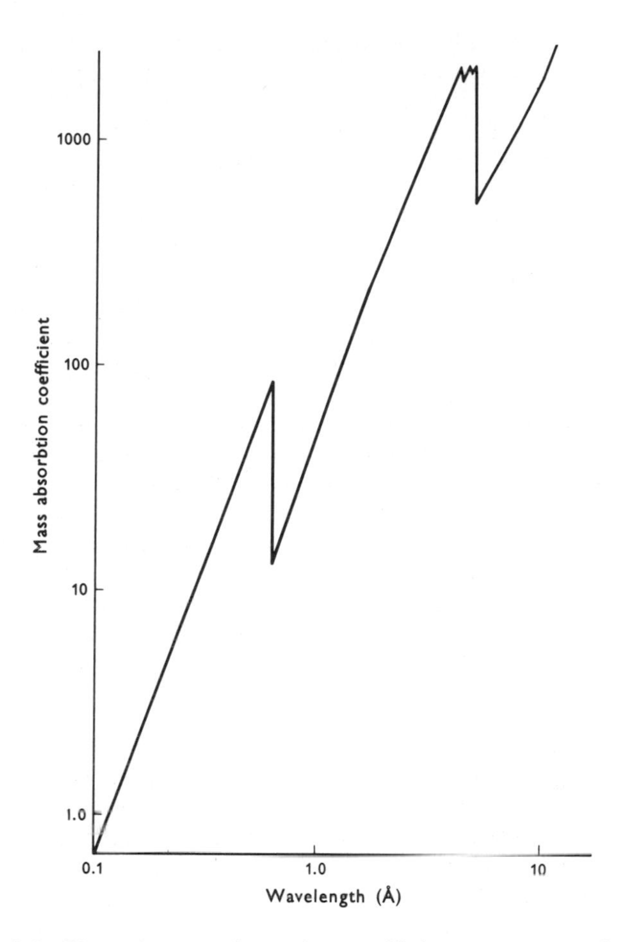

Figure 3.3 Plot of mass absorption coefficient versus wavelength for molybdenum, showing the single K-absorption edge and the three L-edges.

transmit the long wavelength portion of the X-ray continuum, reducing the tube's effectiveness for exciting the lighter elements. Lighter element target tubes do not have the same problems with back-scattered electrons and so can use thinner windows to improve the long wavelength transmission.

Hence, as in much of X-ray fluorescence analysis, a compromise is sought, whereby a heavy metal target is used for general purpose excitation, but a lighter element target is employed for the quantitative analysis of elements below atomic number 24, i.e. chromium. Gold, tungsten and molybdenum target tubes are in common use, with chromium as the preferred target for the lighter elements. Dual anode X-ray tubes, for light and heavy element excitation, are thus not practical unless they could be provided with two sets of windows, about 500μm thick for the heavier target and about 200μm for the lighter.

Tubes containing silver or palladium targets offer a further compromise, since each is relatively efficient for both light and heavy elements, giving about 70% of the intensity of the optimum targets when operated at equivalent power. The effect results from the favourable positions of the *K*- and *L*-series lines for the elements relative to the majority of other characteristic spectra, but, unfortunately, such tubes cannot yet be run at the maximum power of 3kW routinely used for other tubes. This of course reduces the intensity, and hence the excitation efficiency for many elements.

Similar use of the characteristic spectrum of the target element may be made to improve the efficiency of the analysis of specific elements. Thus, the geologically interesting elements Cu(29) through Y(39) are best excited with a molybdenum tube.

Finally, care must be taken in choosing the primary source to ensure that its characteristic spectrum will not interfere with the analysis. Inevitably, some of its radiation will be transmitted to the detector and counted along with the fluorescent radiation of the sample. It will be obvious that the primary characteristic lines from a chromium target tube will interfere with the detection of chromium, but it must also be remembered that interelemental interferences may also occur. Thus, chromium $K\beta$ radiation also interferes with the detection of the manganese $K\alpha$ line, and the *L*-series of tungsten can interfere with the determination of nickel, copper and barium.

3.2.2 Absorption of X-rays by Matter

Initial investigation of the properties of X-rays soon showed that they were attenuated on passing through matter by an amount depending on the wavelength of the X-radiation and the nature and density of the absorbing medium. Solids were found to have a greater absorption than either liquids or gases, but even a gas could absorb or scatter an appreciable number of X-ray photons, the effect increasing with the atomic number of the gas. Hence vacuum or helium atmospheres are used in light element studies.

An equation almost identical with Beer's law relates the emergent intensity, I, to the intensity, I_0, of the incident beam after it has traversed a thickness x of an element of density ρ:

$$I = I_0 \exp(-\mu\rho x) \tag{3.4}$$

where μ is the mass absorption coefficient for the element, and is related to the linear absorption coefficient, μ', by

$$\mu = \mu'/\rho \tag{3.4a}$$

Four processes govern the loss of intensity, representing the different mechanisms by which energy is absorbed from the X-ray beam:

(i) characteristic X-rays are generated, by the methods discussed above, and the wavelengths at which this occurs are dependent on the atomic species present;

(ii) photo-electrons are generated;
(iii) electrons are scattered without loss of energy, the Rayleigh, or coherent, scatter;
(iv) electrons are scattered with slight loss of energy, the Compton, or incoherent, scatter.

The total absorption can thus be expressed as the sum of τ, the photo-electric absorption term, and σ, which represents the mass scattering, terms (iii) and (iv), in the form

$$\mu = \tau + \sigma \tag{3.5}$$

Figure 3.3 shows the variation of μ for molybdenum plotted as a function of the wavelength of the incident radiation. It can be seen that the absorption is greater for longer wavelengths, i.e. lower energy radiation, than for short. Moreover, the curve shows abrupt discontinuities, the absorption edges, which represent the energies required to expel electrons from the various shells in the atom. The highest energy discontinuity, the K-absorption edge, represents the ionisation potential of a 1s electron; the three L-edges ionisation of the 2s and 2p electrons; and so on for the five M-edges, although these can be detected, with some difficulty, only in the heavier elements. Values of μ at different wavelengths have been published by Victoreen [13], Heinrich [14], and Theisen and Vollarth [15], as a result of accurate determinations for many elements and interpolation for the remainder. The last two references contain the current best estimates for the lighter elements.

In practice, the value of μ is controlled by the value of τ, since σ is largely wavelength-independent at an approximate value of 0.04, and so the observed shape of the absorption curve depends on the phenomenon of photo-electron scattering.

Photo-electron Scattering

Photo-electron scattering involves ionisation of electrons from the various levels in the atom, and so includes the initial process leading to the generation of the characteristic line spectrum for the element. The electrons set free generally lose energy by the normal collision mechanisms and are finally recaptured by another atom, the excess energy being dissipated as heat. The photo-electron absorption coefficient, τ, can be expressed as

$$\tau = K\frac{N}{A}Z^4\lambda^3 \tag{3.6}$$

where K is a constant representing the efficiency with which the incident radiation is absorbed, the factor N/A is the number of atoms per gram of the element, and Z and λ have their usual meanings. The value of K will, in turn, depend on the number of electrons that can be ionised by the incident radiation, and so it can be written

$$K = k_K + k_L + k_M + k_N + \ldots \tag{3.6a}$$

where k_K, k_L, . . . , are constants relating to the ionisation of the K-, L-, . . .

electrons. The abrupt changes in μ at the absorption edges thus correspond to changes in K as the energy of the incident wavelength falls below the limit for ionising the 1s, 2s, 2p, . . . , electrons and the k_K, k_L terms drop out of equation (3.6a).

Composite Materials

If absorption occurs in a composite material, the total mass absorption coefficient is given by the sum of each elemental mass absorption coefficient weighted for the mass fraction of the element in the composite, i.e.

$$\mu_{\text{total}} = \mu_A \cdot w_A + \mu_B \cdot w_B + \mu_C \cdot w_C + \ldots \tag{3.7}$$

where w_N is the weight fraction of the element or compound with mass absorption coefficient μ_N. Thus, we may calculate the total mass absorption coefficient for galena, PbS, for zinc $K\alpha$ radiation (λ = 1.437Å), using Heinrich's tables [14], as follows

$$\mu_{PbS} = \mu_{Pb} \cdot w_{Pb} + \mu_S \cdot w_S$$

$$= 194.6 \times \frac{207.2}{239.3} + 73.3 \times \frac{32.1}{239.3}$$

$$= 179.9$$

This value could then be used to calculate the overall mass absorption coefficient in a lead-zinc concentrate, containing, for example, 25% PbS and 75% ZnS by weight, for the Zn $K\alpha$ line.

$$\mu_{\text{mix}} = 179.9 \times 0.25 + 57.0 \times 0.75$$

$$= 87.9$$

Clearly, the total absorption is more easily calculated for a pure compound than for a mixture in which the weight fractions of the compounds are not known exactly, and this problem will receive further attention.

X-ray Scattering

Characteristic radiation from the primary X-ray tube source will be scattered by the sample and give rise to both coherent, Rayleigh, and incoherent, Compton scatter. Rayleigh scatter occurs with no energy loss, and so its wavelength is that of the characteristic of the tube, but Compton scatter involves a certain energy loss and so results in a scattered radiation of wavelength λ_{Compton} which differs from that of the incident radiation, $\lambda_{\text{characteristic}}$, by an amount

$$\lambda_{\text{Compton}} - \lambda_{\text{characteristic}} = 0.0243(1 - \cos\psi) \tag{3.8}$$

Figure 3.4 Photograph of the evacuable crystal chamber of the Philips PW1450 spectrometer. X-rays enter through the variable collimator on the right, are diffracted at one of the five crystals held on the rotatable mount near the centre, and are detected in the flow proportional detector on the left. Short wavelength radiation passes, via a long collimator, to the scintillation detector mounted behind the proportional unit.

where ψ is the angle between the primary beam and the collimator direction, and is normally 90°. The theoretical difference between the Rayleigh and the Compton scatter peaks is thus 0.0243Å, but in practice the Compton peak is diffuse, since the primary beam itself subtends an angle of around 30° at the specimen surface, and lies between 0.020 and 0.028Å from the characteristic peak, as shown in figure 3.16.

3.3 METHODS OF ANALYSING THE FLUORESCENT RADIATION

Once the characteristic radiation of an element or group of elements has been excited, provision must be made to separate the individual wavelengths generated, identify them, and measure their intensities. Two methods are currently available, one based on the older wavelength dispersive techniques and the other on more recently developed energy dispersive methods.

3.3.1 Wavelength Dispersive Systems

The principal parts of a wavelength dispersive X-ray fluorescence system (figure 3.4) are shown diagrammatically in figure 3.5. Basically, these systems are based on the X-ray diffractometer, as described in Chapter 7, section 7.6.3, with the sample acting as the X-ray source and a crystal of accurately known *d*-spacing as the analyser. In most instruments, the fluorescent X-rays are excited from the underside of the sample, thus ensuring simple, close, and reproducible coupling with the primary X-ray

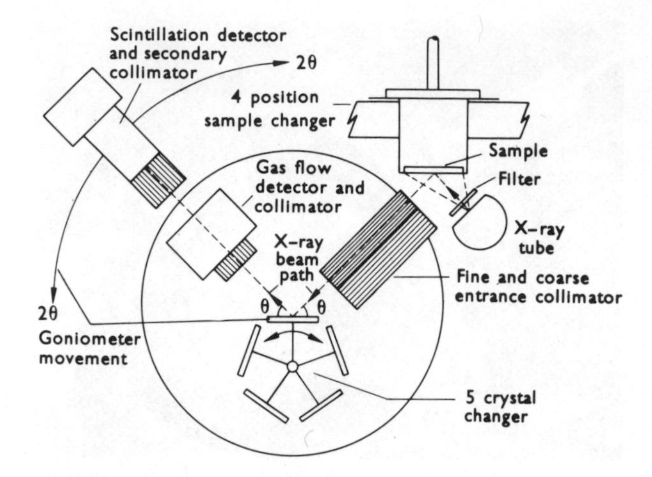

Figure 3.5 Diagrammatic representation of figure 3.4, showing the X-ray paths and the various coupled goniometer movements for a sequential-measuring dispersive spectrometer. (Courtesy of Pye-Unicam Ltd.)

tube source, as well as simplifying the method of loading. The cone of secondary radiation passes through parallel metal foil Soiler slits to provide a parallel beam of X-rays from the greatest possible area of the sample. Modern instruments normally have two interchangeable collimators, one coarse, around 0.45mm spacing, and one fine, around 0.15mm. The former is used for the determination of lighter elements whose fluorescent yields are low but whose interelement wavelength separations are relatively large, and the latter for heavier elements where the shorter wavelengths require higher resolution and the yields are greater. Following collimation, the beam is diffracted at the crystal to produce a spectrum which is scanned by one or more detectors, the resulting signal amplified, sorted electronically to complete the resolution, and displayed either on a chart recorder or digitally. The wavelengths of the spectral lines generated in the sample are related to the diffraction angles, 2θ, by the Bragg equation,

$$n\lambda = 2d \sin\theta \tag{3.9}$$

Where d is the known d-spacing between the planes in the analysing crystal, and θ is the angle of incidence on the crystal (cf. Chapter 7, section 7.5.2).

The Analysing Crystal

The heart of a spectrometer is the crystal which, as shown in Chapter 7, acts in a manner analogous to the grating in an optical spectrograph, to diffract X-rays with different wavelengths and so make it possible to isolate lines of interest. In practice this means that the whole system must have a resolving power of better than 0.3Å for sodium $K\alpha$ radiation, λ = 11.9Å, or 0.005Å for zirconium $K\alpha$, λ = 0.788Å. Moreover, the crystal should diffract as much of the incident radiation as possible into the detector to maintain high sensitivity and accuracy. As in so many cases, these requirements may often conflict with one another and compromises must be sought.

It will be clear that, since a modern spectrometer must be able to analyse elements from fluorine to uranium, covering a wavelength range from 18.3Å to 0.12Å, and since a crystal cannot diffract X-rays with wavelengths greater than twice its *d*-spacing, a range of crystals is needed to cover the wide span of wavelengths with the required degree of resolution. It may be noted that a further limitation is set by the mechanical problems inherent in measuring 2θ angles much in excess of 145-150°. A lower angular limit is also set by the need for the crystal to intercept all of the incident beam of secondary radiation which passes through the primary collimator. This value will vary depending on the size of the primary collimator slit and the crystal length.

Secondly, the crystal must give a good dispersion of the incident radiation, in order to provide the best possible degree of resolution for the separation of adjacent lines. An expression for the dispersing power of a crystal can be obtained by differentiating equation (3.9) to give

$$\frac{d\theta}{d\lambda} = \frac{n}{2d} \times \frac{1}{\cos\theta} \tag{3.10}$$

and hence, for a given diffraction angle, the best resolution will be given by a crystal with the smallest *d*-spacing that can satisfy equation (3.9). The total dispersive power of a spectrometer, however, is not solely that due to the crystal lattice but includes contributions from the collimator, the surface layer of the crystal, and the Gaussian spread of wavelengths generated from each element in the sample. The width of the slit in the primary collimator clearly controls the semi-angle of the cone of radiation impinging on the analysing crystal and this, in conjunction with the fact that the surface of the crystal comprises a large number of small crystallites all in slightly different orientations results in a small spread of angles of incidence for the beam of generated radiation, and hence for the beam of diffracted rays. These result in a curve of width approximately 0.2-0.4° of 2θ for the characteristic radiation from an element, rather than the theoretical narrow band at the 2θ angle. The resolving power can be improved by treating the crystal surface to decrease the angular spread and by using a narrower collimator, but with appreciable losses in intensity in the diffracted beam. Secondary collimation can also help, but again with considerable loss in detectable intensity.

The intensity of the diffracted beam clearly controls the sensitivity of the spectrometer for the element in question, since the greater the fraction of the incident intensity diffracted by the crystal, the smaller the amount of the element required to produce a signal sufficiently large to be distinguished from the background noise. As a corollary, the optimum crystal must produce a minimum of unwanted radiation of wavelength similar to that of the required radiation, to maintain as low a background level as possible and to avoid interferences.

Over thirty different crystals, or different orientations of crystals, are presently available and provide a wide range of $2d$ spacings and reflection efficiencies, although no more than six or seven are needed for the general analysis of all elements. Table 3.1 lists the more commonly used crystals, and the choice for any given application is governed mainly by the value of $2d$, the dispersion, and the reflection efficiency, as outlined above, although such factors as the existence of interferences due to second order reflections from the crystal, crystal fluorescence, etc., can have an important bearing in some applications.

Crystal fluorescence is seen when the radiation from an element in the sample is of the correct wavelength efficiently to produce secondary X-rays from an element in the crystal. Thus, calcium radiation is capable of exciting potassium K lines from a potassium acid phthalate (KAP) crystal and rubidium L lines from rubidium acid phthalate (RbAP), both of which interfere with the determination of magnesium.

Spurious reflections may arise from two sources when natural crystals, such as topaz, are used. High defect concentrations may upset the perfection of the space group symmetry to such an extent that nominally unpermitted reflections may be generated, and appreciably misoriented crystallites may find themselves in the correct orientation to diffract radiation from element B when the bulk of the crystal is correctly positioned for element A, for example the interference between Sn $K\beta$ and Pb $L\beta$ instanced by Jenkins and De Vries [16]. The various cuts of lithium fluoride tend to be used in place of such crystals today, for short and medium wavelength work, but even here interferences are found in the useful LiF_{220} cut, in which the crystal has been cut and polished parallel to a (220) plane, because of the appearance of disallowed odd order reflections corresponding to (110), (330), etc. reflections from the relevant wavelengths, which produce broad peaks at orders 1/2, 3/2, etc. in the observed spectrum [17]. Such spurious lines should be looked for during the initial studies of any new samples for analysis, and of course may be predicted if the elements in the sample are known [18].

Where second order lines overlap the analysis line, and cannot be separated by pulse height selection, it is necessary to use a crystal with roughly the same d-spacing but which does not generate second order reflections. Germanium or silicon in the (111) orientation are generally used. For example, the escape peak for the calcium $K\beta$ second order reflection cannot be separated from the phosphorus $K\alpha$ without using a germanium crystal.

A final difficulty arises from the physical nature of the crystals themselves. The accuracy with which the wavelength of the fluorescent radiation can be measured, or the accuracy with which the spectrometer can be set up to detect a radiation of given wavelength, depends on an accurate knowledge of the d-spacing of the crystal used. Unfortunately, all materials expand on heating, and some more than others, hence the d-spacings change with temperature. This is particularly true of pentaerythritol (PET) and LiF_{220} crystals, where large peak shifts can occur with only moderate

TABLE 3.1

Standard Analysing Crystals for X-ray Spectrometry

Crystal	Chemical Formula	Reflecting Plane		Useful Element Range using 1st Order		Reflection Intensity	Dispersion	Characteristic Radiation	Resistance to X-radiation, Temperature and Vacuum
		(hkl)	$2d$ in Å	K-Spectrum Z	L-Spectrum Z				
Lithium fluoride (cut 100)	LiF	(200)	4.026	19 to 58	$\geqslant 49$	Very good, universally usable crystal	high	–	very good
Lithium fluoride (cut 110)	LiF	(220)	2.84	22 to 70	$\geqslant 58$	Very good (2 to 3 times higher than that of topaz)	very high, therefore favorable for separation of neighb.lines (eg $MnK\alpha$–$CrK\alpha$) or for mixtures of rare earths.	–	very good
Germanium	Ge	(110)	6.53	15 to 46	$\geqslant 40$	Medium No even order extinctions	medium	Ge	very good
Pentaerythritol (PET)	$C(CH_2OH)_4$	(002)	8.750	13 to 41	$\geqslant 36$	Good for elements $Z = 13$ to 20, particularly favorable for Al and Si	medium	–	restricted
Ammonium dihydrogen-phosphate (ADP)	$NH_4H_2PO_4$	(011)	10.648	12 to 37	33 to 92	Medium, favorable for $Mg(Z = 12)$	low	P	good
Potassium hydrogen phthalate (KAP)	$KHC_8H_4O_4$	(100)	26.62	(8) to 24	22 to 62	Good for light and very light elements, particularly for $Na(Z = 11)$ and $Mg(Z = 12)$	very low	K	restricted
Rubidium hydrogen phthalate (RbAP)	$RbHC_8H_4O_4$	(100)	26.12	(8) to 24	22 to 26	As KAP	very low	Rb	restricted
Thallium hydrogen phthalate (TlAP)	$TlHC_8H_4O_4$	(100)	25.76	8 to 24	22–62	Very good for light elements	very low	Tl	restricted
Lead stearate (PbSt) on glass	$[CH_3(CH_2)_{16}CO_2]_2Pb$	–	100	(5) to 12	16 to 34	Good for very light elements, for F ($Z = 9$) better than KAP (with equal beam)	very low	Si (glass)	good (up to 60°C)

changes in temperature. Modern spectrometers are usually temperature controlled to obviate this difficulty, but provided that the crystal remains in the X-ray beam when the instrument is not in use, and so maintains a reasonably constant temperature, little difficulty should be encountered even in older systems.

Choice of Crystal

In general, for short and medium wavelength work, i.e. for the determination of the heavier elements, the various cuts of lithium fluoride are the most efficient. The choice of a particular cut is governed by the required degree of dispersion and by the sensitivity.

The factors to be reconciled may be illustrated by the following example for the analysis of rubidium using the $K\alpha$ line [19]. The LiF_{200} cut has an excellent reflection efficiency, giving a lower limit of detection [equation (3.28)] for 100 seconds counting of 0.51ppm, whereas LiF_{220}, with a superior dispersion but lower reflection efficiency, gives a corresponding detection limit of 0.96ppm. Rubidium is normally found associated with strontium and, since the rubidium $K\alpha$ line (0.9269Å is almost overlapped by the strontium $K\beta$ line (0.8767Å), the poorer sensitivity of the LiF_{220} must be accepted to allow resolution of these peaks.

The elemental range from potassium to aluminium is best covered by PET, whose high reflection efficiency more than offsets the disadvantage of its temperature instability.

The remaining elements, magnesium, sodium, and fluorine, present something of a problem. RbAP has the best reflectivity in this region but it suffers from severe crystal fluorescence which is shared, albeit to a lesser extent, by KAP, ADP, and gypsum. These latter crystals show lower reflection efficiencies, however. The recently developed ammonium acid phthalate [17] does not show fluorescence but it appears to have a very low reflection efficiency. The choice of crystal in this region must depend on the nature of the sample to be analysed and must be made to minimise spurious effects as much as possible, bearing in mind the very low fluorescent yield for these elements*.

Crystal Sources

Certain crystals, such as gypsum or mica, are available from natural sources, but the majority of crystals currently in use are made synthetically. While this move has resulted in an overall improvement in the quality, there are still problems associated with the state of the art in crystal growing and, even with the greatest care during the growth process, large areas of the crystal may become mis-aligned relative to the bulk of the crystal, due, for example, to the occurrence of twinning [17]. This can result in line intensities from separate crystals cut from a single boule varying by as much as 50% from the maximum value. Improvements in the techniques for

* The newly announced thallium acid phthalate (TlAP) crystal [66] would appear to be able to solve many of the difficulties encountered in the analysis of Mg, Na, and F.

growing crystal boules should remove this variability and also increase the overall intensity values.

The crystals, after growth, are cleaved or cut to give a face parallel to the required crystallographic plane and the surface polished as required. The crystal may be used either with a flat surface or with one that has been bent to a curve or curved and ground, and the various geometries have different advantages and disadvantages [16]. The surface is usually lightly abraded after polishing to produce the desired mosaic of slightly mis-aligned crystallites which gives the improved reflection efficiencies, albeit at the expense of some line broadening.

The cost of an analysing crystal is relatively high (ca. £150-£400), but it should be remembered when considering new materials, or improved methods of growth, that 10-50% increases in intensity can be achieved, for example when comparing the use of KAP and RbAP for the analysis of sodium. Similar increases may be obtained with improved detectors, amplifiers, collimators, and X-ray tubes and generators, but with a generally unacceptably higher expenditure.

The Detector

After dispersion, the X-ray photons must be detected and converted into measurable electrical pulses. This is normally achieved using two of the detectors discussed in Chapter 1, section 1.3.3, the gas flow proportional and the scintillation detector. Geiger-Muller counters are not used because pulse height analysis cannot be applied to their output signals, and because of their large dead time losses.

Scintillation detectors are most efficient in the wavelength region up to 3Å, which includes the heavier elements, and for this reason they are normally used in conjunction with a long (~10cm), narrow (~0.15mm) secondary collimator, to provide the high resolution needed at these wavelengths. Gas proportional detectors are useful for wavelengths equal to or greater than 1.5Å and so are more useful for the lighter elements.

Gas flow proportional detectors are provided with extremely thin windows of Mylar, polypropylene, or polycarbonate film, lightly flashed with aluminium and between one half and six microns in thickness. Even so a six micron window will absorb up to 50% of the energy of the aluminium $K\alpha$ radiation. The disadvantage of such films is that the gas filling is able to diffuse from the detector, especially when the spectrometer is operated under high vacuum, and hence a continuous flow of gas is needed. This in turn causes problems as the gas stream inevitably carries impurities, and dirt, which will be deposited on the anode wire causing a deterioration in resolution. The gas is usually a commercially available mixture of 90% argon with 10% methane added as a quenching agent.

A limiting factor in the use of the flow counter is that of dead time [equation (1.11)], partially caused by the time taken to discharge ions reaching the detector cathode. During this period photons incident on the detector are not counted. Moreover, the scaler circuits used to count the pulses after detection have their own dead time, and together these two dead times bring that of the complete system to between one and two micro-seconds. These factors limit the maximum count rates which may be measured to between 50,000 and 150,000 counts per second.

The most modern spectrometers correct for dead time electronically, but in other instruments it is essential, if high count rates are to be measured, for the individual dead times to be known, and corrected for.

A simple method of finding the dead time is to determine the ratio of counts between the $K\beta$ and $K\alpha$ peaks of an element, at a fixed voltage but varying amperage. Such peaks are relatively close in wavelength, and vary in the intensity ratio of $K\beta/K\alpha$ from 1/150 for aluminium to 1/15 for niobium. If the $K\beta$ and $K\alpha$ intensities at each milliamp setting are plotted, a curve results which is linear at the lower count rates, where dead time is minimal, and flattens at the high $K\alpha$ rates. Using the linear region of the curve as the true ratio of counts, R_t may be calculated, and knowing R_m, the dead time value, t in equation (1.11) can be determined.

Both flow proportional and scintillation detectors provide output pulses which contain an approximately Gaussian spread of voltages, arising mainly from the manner in which the detector responds to incoming photons and only indirectly connected with the spread of wavelengths in the incoming beam, and so the resolution of the detector can be defined as the peak width at the half height divided by the voltage of the maximum pulse amplitude, as shown in figure 3.6. Theoretically, the resolving power of a flow proportional detector is 1keV at the wavelength of the iron $K\alpha$ line, and that of the scintillation detector is 2.8keV, these values increasing for lower energy, longer wavelength, X-rays. Since the Fe $K\alpha$ wavelength corresponds to an energy of 6.40keV, the theoretical resolutions for the flow proportional and scintillation detectors are 16% and 45% respectively. Hence the gas flow proportional detector is clearly superior to the scintillation detector. Serious problems can arise, however, due to the deterioration caused by impurities in the gas stream, which can result in a drop in resolution from 16% to 30% (figure 3.6) over a period of time, with attendant difficulties in cases where interfering lines have similar energies and can no longer be separated by the pulse height analyser. Contamination of the anode may also markedly alter the shape of the output pulse, as will an asymmetric anode wire, or faulty flashing of the window [20]. The scintillation detector is more stable but does deteriorate over a period of years, and of course can only be used for the shorter wavelength photons.

The pulse height discriminator is used in spectrometry to pass only those signals with energies between two pre-set limits (figure 3.6), which serves not only to remove background noise but also, more importantly, to remove second or higher order reflections arising from diffraction at the crystal of unwanted wavelengths and other spurious signals. Such reflections may have very similar 2θ values to that of the required element, as shown earlier, and so will be detected along with the required reflection, but their energies will be quite different and so can be removed electronically, knowing the esolution of the detector and the energy of the interfering line. This last can be readily calculated from equation (3.1).

For example, the second order Zr $K\alpha$ ($\lambda = 1.576$Å) overlaps the first order Hf $L\alpha$ ($\lambda = 1.575$Å), but this overlap is removed by pulse height selection.

$$E_{Zr\,K_\alpha} = 15.730\ keV \qquad E_{Hf\,L_\alpha} = 7.870\ \text{keV}$$

Detector resolution is normally measured at Fe $K\alpha$ (6.399 keV) and would ideally be 16% for the flow proportional detector.

$$\text{Resolution Zr } K\alpha = \sqrt{\frac{6.399}{15.730}} \times 16 = 10.2\%$$

$$\text{Hf } L\alpha = \sqrt{\frac{6.399}{7.870}} \times 16 = 14.4\%$$

If the Hf $L\alpha$ is set to peak at 25 volts in the pulse height selector the Zr$K\alpha$ will be at

$$25 \times \frac{15.730}{7.870} = 50 \text{ volts}$$

As the resolution equals the peak width at half height divided by the maximum energy of the pulse, peak widths will be

$$W_{Hf} \qquad 25 \times \frac{14.4}{100} = 3.6 \text{ volts}$$

$$W_{Zr} \qquad 50 \times \frac{10.2}{100} = 5.1 \text{ volts}$$

The shape of the pulse approximates to a Gaussian distribution, and the width at half height corresponds to 2.366σ, whence

$$\sigma_{Hf} = 1.53 \qquad \sigma_{Zr} = 2.16$$

Therefore 99.7% (3σ) of the peak area of Hf $L\alpha$ is between 25 ± 4.59, while the corresponding limits for Zr $K\alpha$ arc 50 ± 6.48. The two peaks are fully resolved and the Zr $2K\alpha$ could be fully resolved if the flow proportional detector were used. If a scintillation detector of resolution 50% for Fe $K\alpha$ was used the pulses would not be resolved, and the determination of IIf is impossible unless a Ge crystal were employed which does not permit second order reflections. Pulse height discrimination further improves the counting statistics of the spectrometer system by removing these high order reflections from background readings also, thus increasing the limits of detection for a given element.

Flow counters also suffer from the phenomenon of pulse shift. This only occurs at high count rates, and with dirty counters [21]. The effect is to move the position (amplitude) of the pulse, often to such an extent that it partially lies outside the pre-set discriminator window, and hence causes an error in the measured intensity.

A further interference associated with these detectors, and not discussed in section 1.3.3, is the phenomenon of the escape peak. Radiation incident on a detector may, in addition to causing ionisation, also excite the characteristic X-radiation of the gas or scintillator crystal, generating an additional pulse with an energy corresponding to the difference between the incident and the characteristic radiation energies. Thus, for a flow proportional detector using argon gas, the relative voltage of the escape peak, V_e is given by

$$\frac{V_e}{V_n} = \frac{E - E_A}{E} \tag{3.11}$$

where V_n, is the voltage corresponding to the incident radiation of energy E, and E_A (2.95keV) is the excitation energy of the argon $K\alpha$ line. The peak will, therefore, always appear at a lower energy than the natural peak, as shown in figure 3.6. Equation (3.11) can be applied to any detector gas, and the magnitude of the effect in each case depends on the fluorescent yield of the gas for the wavelength involved. Thus, the effect will be strongest for elements with atomic numbers just greater than that of the gas involved since their wavelengths lie in the region of maximum absorption by the gas. The effect also increases with the atomic number of the gas, being virtually zero for neon, noticeable for argon, but very large for xenon, in which the escape peak for molybdenum $K\alpha$ is about three times the measured intensity of the natural peak [16]. A similar effect exists in scintillation detectors, and involves excitation of the iodine characteristic lines, but it is only a practical problem when K lines with energies greater than 33keV, that is lanthanum $K\alpha$, occur in the sample.

The existence of escape peaks presents two problems. Firstly their production abstracts energy from the incident beam which may result in a lower measured intensity if the escape peak lies outside the pulse height window. Experiments have shown that this can amount to 5-10% of the observed count rate for the relevant elements when using argon gas. Clearly, the problem will not apply to elements lighter than argon, which cannot excite the escape peaks. For elements heavier than argon it will depend on the fractional separation between the natural and escape peaks, which reduces with increasing atomic number, and the fluorescent yield, which decreases with increasing atomic number. In practice, only the elements between calcium and cobalt are affected, since the escape peak for potassium is very small and can be ignored, and the escape and natural peaks are not resolved for nickel and heavier elements. Similar 'problem ranges' will exist for other gases. Ideally the escape peak should be counted with the natural peak, but care should be taken when the peak lies close to the lower level of the discriminator. Deterioration of the resolution could seriously reduce the count rates, as seen in figure 3.6.

Secondly, problems can arise indirectly through crystal fluorescence effects. Thus, if a gypsum crystal is used in the analysis for sodium, the

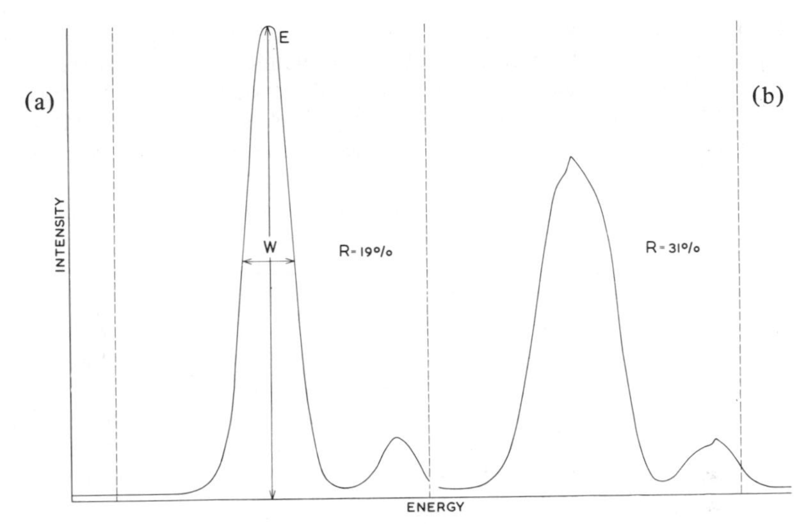

Figure 3.6(a) Energy resolution of a gas flow proportional detector for Fe $K\alpha$ radiation. The resolution is calculated as $(100 \times W/E)$. The small, low energy peak is the iron escape peak with the argon gas of the detector and would not be present with a scintillation detector. (b) Illustration of the progressive deterioration in resolution caused by dirt in the counter gas. Resolution is restored by cleaning the anode wire. Note that the pulse height discrimination limits are normally set at the levels shown, and attention should be paid to the position of the escape peak when the resolution is poor.

calcium $K\alpha$ line may be excited in the process and, whereas this line can easily be resolved and rejected by the pulse height analyser, its escape peak with argon (at 0.74keV) cannot be separated from the sodium $K\alpha$ line (at 1.04keV) and so will cause an interference. This could be solved by changing the detector gas, which is generally impractical and very expensive. It is better to use a different crystal, for example KAP, for the analysis.

It should be noted that the amplitude of the pulse varies with wavelength, and hence a different pulse discriminator setting is required for each peak. In automatic equipment this is performed either by a number of pre-settable windows, or by varying the window position with θ, using a sinusoidal potentiometer to feed a servo-motor system and vary the window position by connecting a linear potentiometer to the θ drive of the goniometer. Such an arrangement varies sinusoidally the degree of amplification of the signal and ensures that the pulses occur at a fixed voltage.

Ultra Light Element Detection

The practical limit for light element detection is fluorine $K\alpha$, using commercial equipment, but it is possible to determine elements with longer wavelengths on modified spectrometers. As discussed in section 3.2.1, the first problem lies in exciting the required radiation, since X-rays of suitable energies are generally absorbed at the tube window. Special tubes have, therefore, been developed which are either windowless or have thin organic film windows and are continuously pumped. A system incorporating a demountable target is available to provide optimum excitation conditions for individual elements in this range [22]. Quantitative analysis of elements down to oxygen is possible with such equipment, but the tubes are difficult to operate on a routine basis and require much maintenance.

A more satisfactory system uses direct electron excitation, which provides much higher intensities while the disadvantage of high backgrounds due to the 'white' radiation has largely disappeared at these long wavelengths. The separation between the white band and the lines of the characteristic spectrum is sufficiently great to allow the two to be resolved without any difficulty. Electron excitation is used in the Telsec Betaprobe [7] and in the electron probe microanalyser (Chapter 11) and can permit analyses to be made of elements down to boron, with detection limits of 200ppm or better.

The very long wavelengths, in the range 20-70Å, generated by the light elements require analysing crystals with correspondingly large *d*-spacings, greater, in fact, than the acid phthalates in common use. The lead stearate pseudo-crystal is made by repeatedly dipping a plate into an organic solution of stearic acid and water containing a lead salt so that it becomes coated with successive layers of oriented stearate molecules attached to heavy metal ions. This gives an effective crystal some hundred layers thick, with a repeat distance of the order of 100Å. Lead stearate is the most popular crystal in use although others, including clinochlore [23], can be used at least for elements down to oxygen.

3.3.2 Wavelength Non-Dispersive Systems

Two additional methods of X-ray fluorescence analysis have been developed over the past few years which do not depend on the use of a crystal analyser to select the required wavelength. One method is based on the different absorption coefficients of elements, and the other on the very high resolutions available with semi-conductor detectors. Both can be used with conventional X-ray tube and portable radioactive sources, and the semi-conductor system can also be used with electron beam excitation.

Filter Spectrometers

These instruments, which have been described by Bowie [8], and others [9], are based on the fact that the *K*-absorption edges for two elements can be found which lie respectively at wavelengths slightly less than and slightly greater than that of the *K* or the *L* line for the element to be analysed. Thus the *K*-edges of zirconium (0.6888Å) and strontium (0.7697Å), or better yttrium (0.7277Å), lie on either side of the molybdenum $K\alpha$ line (0.7107Å). If two sheets of these elements are prepared of such thickness that they absorb the molybdenum $K\beta$ to the

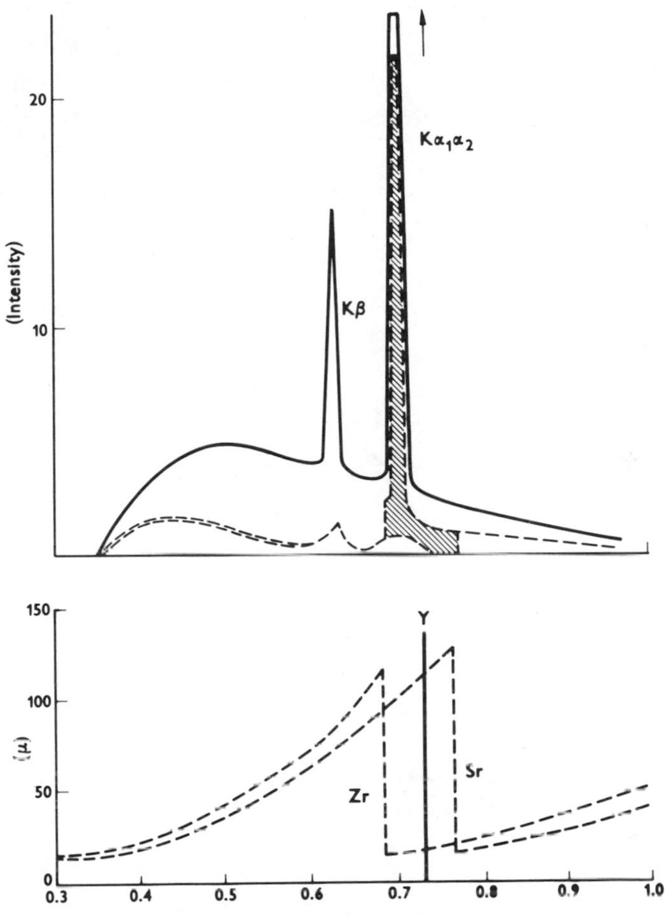

Figure 3.7 Diagrammatic representation of the principle of the absorption edge spectrometer. The edges for Zr and Sr (Y may also be used) lie at wavelengths which bracket the Mo $K\alpha$ emission line. If two intensity readings are taken with the above filters individually in the Mo X-ray beam, the partially absorbed spectra shown by the dashed lines in the upper diagram would be observed, rather than the full spectrum shown by the solid line. The difference between the two readings corresponds to the hatched area, which represents the intensity due to Mo $K\alpha_{1,2}$.

same extent, then their bulk absorption curves will match everywhere except in the region between their *K*-edges, as shown in figure 3.7. It follows that any difference in the intensity of a polychromatic beam of X-rays passed through first one and then the other sheet will be due only to the intensity of the molybdenum $K\alpha$ line, and so the difference reading can be used as a measure of the amount of molybdenum present in the sample studied. Figure 3.7 shows schematically the intensities of a beam containing

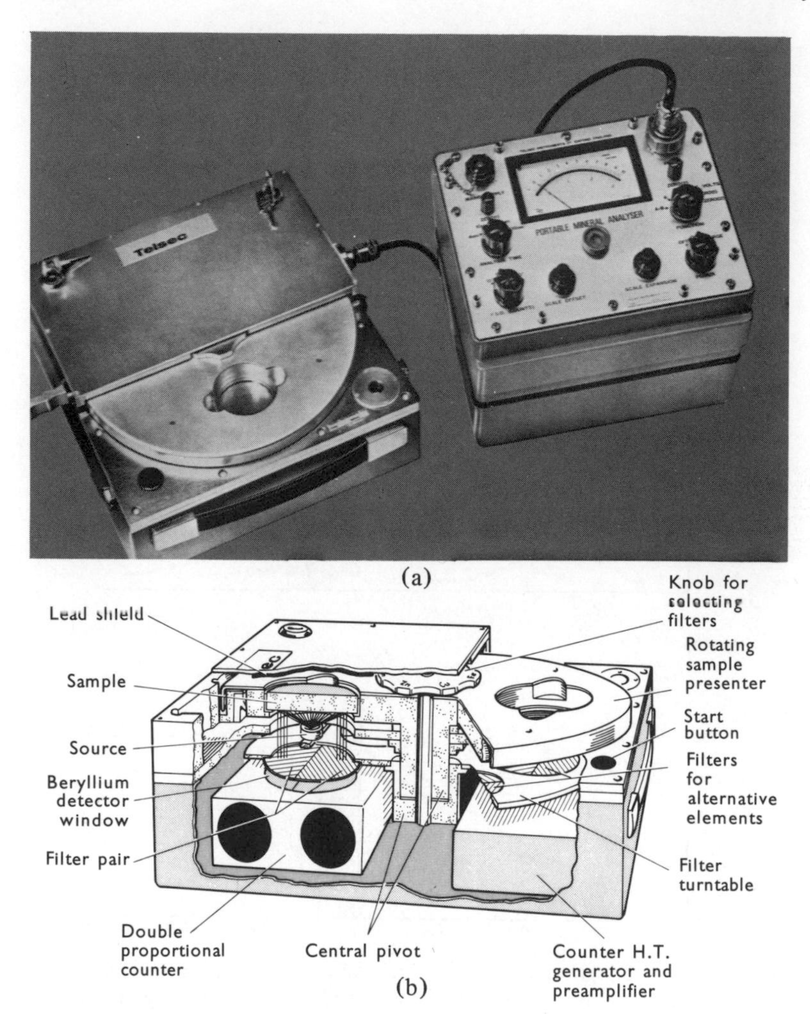

Figure 3.8 (a) The Telsec portable, battery powered, absorption edge spectrometer. The sample holder, filters, and detectors are housed in the plain case on the left of the photograph, with the power supply, amplifier, and read out on the right. (b) Exploded diagram of the portable mineral analyser. The sample is excited by radiation from an isotope source mounted close by, and the characteristic fluorescent X-radiation passes through one pair of filters to a pair of matched proportional counters. (Six pairs of interchangeable filters are normally available on a single rotatable mounting in the housing, for ease of multi-element measurement.) This configuration sends two signals simultaneously to the analysing circuits, which can then generate a difference signal directly and so allow the meter scale to be calibrated directly in terms of the concentration of the element involved, in a given matrix. (Photograph and diagram by courtesy of Telsec Instruments Ltd.)

molybdenum radiation, unfiltered and after passing through the elements of a zirconium-strontium balanced pair filter. Note that it is possible for $K\beta$ and L lines of other elements to fall between the edges and thus be counted. Similar balanced pair filters, or Ross-filters, exist for other elements, and Lonsdale [11] has listed representatives for some common elements. The filter sheets are of the order of 1-10μm thick, either in the form of thin metal foils or, where non-metallic elements are required, as a powder set in a resin. The correct thicknesses are obtained by etching or polishing in the usual way.

A typical instrument is shown in figure 3.8a and diagrammatically in figure 3.8b and includes a radioactive source to provide the primary radiation, the balanced filter pairs in a rotating mount, paired proportional detectors, and the associated electronics. Isotope sources can be used, despite their low intensities, because no collimator or crystal disperser is used and the sample can be placed very close to the source and the detector, thus minimising energy losses. Moreover, the use of an isotope source and a battery to power the electronics makes the instrument easily portable for field use.

The signal from the sample after passing through the balanced filters is measured in two separate proportional detectors, the counts then being electronically subtracted so that the meter can be adjusted to read directly as concentration. Pulse height discrimination may also be applied for further resolution of energies. Less sophisticated instruments may require two separate measurements for the two filters, and may use simpler detectors. At the other end of the scale there may be provision for gas flow proportional detectors, scintillation detectors, sealed low power X-ray tubes, and multiple automatic sample changers [24].

Two types of radioactive source have been employed for excitation, those giving bremsstrahlung radiation analogous to an X-ray tube, for example Pm^{147}, but with a much lower intensity, and sources which are essentially monochromatic. These are normally X-ray emitters, and are usually disc or annular shaped with the back and sides shielded. A recent development is the high energy point, or central source, used in conjunction with a well-shaped target from which secondary X-rays are excited.

If monochromatic radiation is used, the optimum excitation is obtained when the source energy is just greater than that of the absorption edge of the sample element. From this it would appear that the best sensitivity will be obtained with a source that efficiently excites the wanted radiation but, while this is true for the element alone, it may not be true for the element in a mixture. It will be shown, in Chapter 6, that unwanted radiation may simultaneously be excited from other elements present and if this is not absorbed by the filters or eliminated by the pulse height analyser, the resulting high background may depress the sensitivity below the value obtained with a less efficient source which does not excite the extraneous radiation. Thoughtful interpretation of the results obtained in a given case will often indicate the best source to be used in a given situation. For the majority of applications two sources, 30mCi Pu^{238} emitting uranium *L* X-rays and a 20mCi Am^{241} source emitting 60keV γ-rays and neptunium *L* X-rays, will suffice, the former for the lighter elements from calcium to strontium, and the latter for the heavier elements from rubidium to thulium. Below calcium, Fe^{55} which emits manganese *K* X-rays can be used, and for the extreme end of the *K* spectrum, Co^{57} giving 122keV gammas will be needed.

The sample may be presented to the instrument either as a powder or liquid in the usual way, or the instrument may be placed against the surface to be analysed. For quantitative results the sample must be flat and fine grained, and of course all the problems discussed in section 3.4.4 must be considered. These conditions may not apply to some samples of ores or rocks to be analysed, especially in the field or mine. The portability of these instruments does allow their use in field exploration, for example Wollenberg *et al* [25] studied the distribution of niobium, zirconium, and lanthanum plus cerium in southern Greenland, but the low sensitivity and limitations of absorption edge filters means that the results obtained have limited accuracy. In general the need to crush samples to obtain good accuracy is a drawback to field use.

Energy Dispersive Spectrometers

The advent of the lithium-drifted silicon or germanium semiconductor detectors, discussed in Chapter 1, section 1.3.3, with their very high resolution, has opened up a new field of spectral analyses in the X-ray region, especially in terms of qualitative elemental identification. As explained in the earlier section, these detectors respond to all the energy incident upon them, converting the radiation into a series of electrical impulses having voltages corresponding to the energies of the photons in the incident beam, with a final resolving power of between 140 and 200eV. The pulses are then separated and counted in a multichannel analyser to provide a complete spectrum of the radiation generated from the specimen. This is clearly of great use as an aid at the qualitative analysis stage, since the spectrum usually permits the principal, and some of the minor, elements to be identified without much difficulty, although overlap between the $K\alpha$ and $K\beta$ lines of the lighter elements and the *L*- or even the *M*-lines of the heavier elements can give rise to ambiguities. Quantitative estimates of elemental abundances can also be made by obtaining the integrated number of counts under the peak for the required element and applying the normal conversion methods (section 3.4).

The spectrometer itself is shown in figure 3.9, and comprises the detector crystal and its associated pre-amplifier both cooled in liquid nitrogen from the dewar storage vessel, an amplifier and pulse shaper, the analogue-to-digital converter, the multi-channel analyser with up to 4096 channels and a storage CRT display unit. A ratemeter and chart recorder output may also be provided, together with magnetic tape memory, and it is also possible to use small computers in the identification of elements and in their

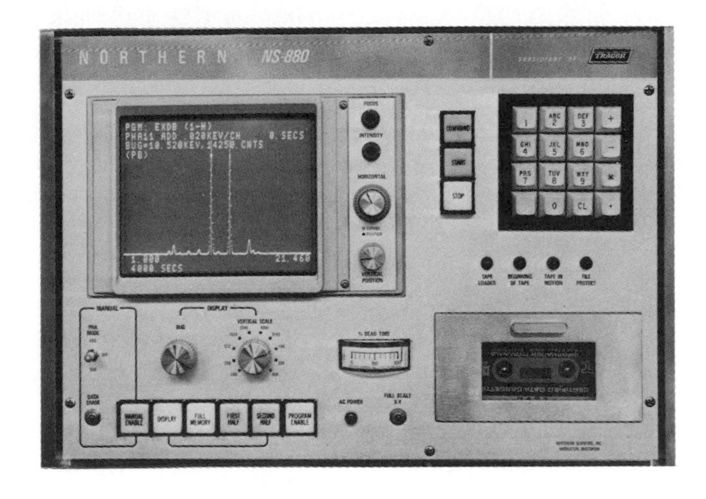

(a)

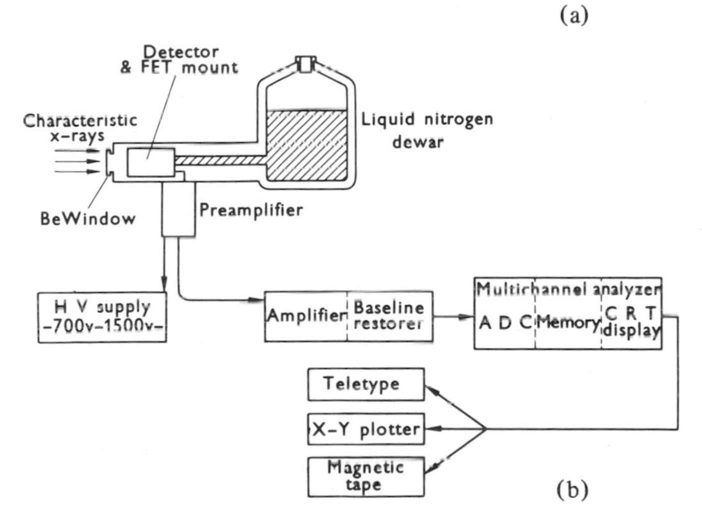

(b)

Figure 3.9 (a) Photograph of the electronic analyser and display section of the Northern energy dispersive X-ray analyser. The cathode ray tube display shows the spectrum stored in the 1024 channels of memory, together with superimposed alphanumeric data from the small built-in computer to indicate the scale (0.020keV per channel), the energy of the line marked by the white dot (10.520keV corresponding to the As $K\alpha$ line), and the number of counts in the marked channel. A cassette system (lower right) permits spectra to be stored for future use, for example for comparison with other spectra directly on the CRT display, and this, together with the built-in computer, allows complete processing of the analytical data to be done on the machine. [By courtesy of Tracor (G.B.) Ltd.] (b) Diagrammatic representation of the energy dispersive spectrometer.

quantitative analysis. The A.S.T.M. publication STP 485 [26] gives full details of the requirements and operation of analysers and detectors.

Normally the analyser is calibrated so that each channel corresponds to a spread of 20eV, 10eV or 5eV depending on the number of memory channels available, and the range of elements to be identified. In simple instruments the element peaks are recognised from energy [27] or wavelength [12] tables, and the intensity measured as counts per channel or from the integrated area of the peak. More sophisticated systems with small computers will display identification of the peaks, figure 3.10, and will perform spectrum stripping on overlapping peaks, and also interelement correction procedures.

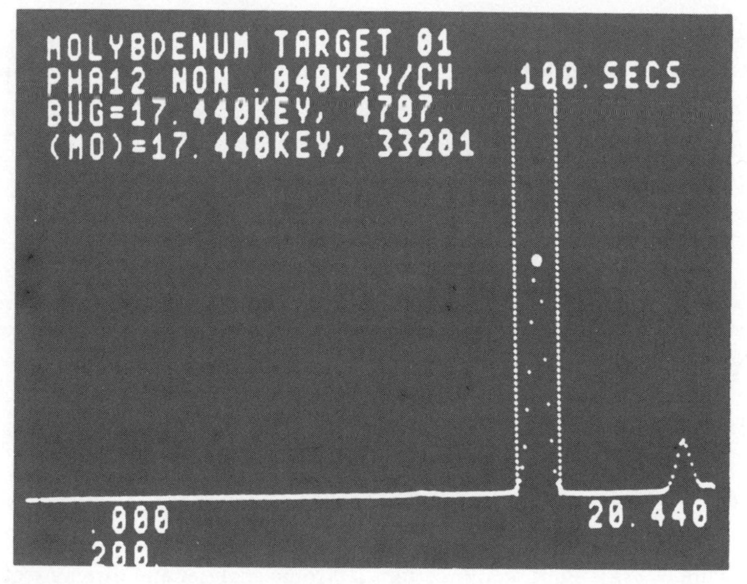

Figure 3.10 One type of display obtained from the energy dispersive analyser shown in figure 3.9 (see also figure 3.11). The spectrum from a molybdenum target excited by an americium^{241} source (60keV). The display presents the energy of the peak maximum found by moving the bright spot (BUG = 17.440keV) and the number of counts (4707) in the corresponding channel, collected in 100 sec. The Mo peak is then identified as having the same energy [(MO) = 17.440keV] and the integrated counts, (33201) over all the channels between the vertical dotted lines (i.e. over the whole Mo peak), displayed.

The geometry of the energy dispersive spectrometer is simple, and the sample can again be placed close to the source, allowing isotope excitation sources to be used in portable units. Greater versatility and sensitivity is, however, provided by the use of specially designed low power X-ray tube sources, in which interchangeable targets may be available to give the required range of exciting wavelengths. The instrument has become popular as an attachment to scanning electron microscopes (Chapter 11), where the low intensity of the X-ray signals generated make this the ideal method of detection. The simplicity of installation, and the ability to detect the complete spectrum are additional advantages.

The energy dispersive spectrometer suffers from two major limitations. Firstly, it can only detect elements heavier than fluorine since, if operated in an air path, the absorption of the signal will seriously reduce sensitivity for the light elements. If used with vacuum or helium paths, the beryllium detector window still attenuates the signal by about 8.4% per micron of window for sodium $K\alpha$. Using special detectors the determination of oxygen and carbon is possible under very high vacuum [28], but as yet not on a routine basis.

Secondly, the multichannel analyser and amplifier circuits impose a major restriction by limiting the count rates to a current maximum of 50,000 per second before serious dead time losses occur. This latter restriction applies to the total counts from the whole incident spectrum and compares with the value of 100,000 counts per second for a scintillation detector operating in a conventional spectrometer and measuring at one fixed wavelength. As a result the simultaneous determination of major and trace elements is difficult.

A further disadvantage is the relatively poor resolution of the lighter element K lines, where overlaps such as that of chromium $K\beta$ on manganese $K\alpha$ necessitates spectrum stripping procedures, and indeed over most of the working range resolution is poorer than the best obtainable with a crystal spectrometer.

Despite these limitations, the energy dispersive instrument is finding many applications, because of its major advantages of speed and ability to display simultaneously all elements within the energy range of the instrument. Commercial applications to process control are discussed in Chapter 6.

A recent development in this field has been the use of programmable absorption edge filters to reduce the intensity of the major elements present and so minimise dead time losses, especially when minor elements are being determined. Such modifications, however, together with the added complications of the data collection and subsequent processing stages, now normally computerised, add to the complexity of the equipment and remove some of the technique's advantages over conventional spectrometers.

3.3.3 Comparison of Systems

The absorption edge spectrometer, as has already been explained, is limited in its accuracy, sensitivity, and in the range of elements that can be analysed. Its virtue lies in its portability and the major applications appear to be in the area of rapid surveys in the field, or for checking samples in the mine, pit, stockpile, etc. It will rarely be used as a quantitative instrument except for very specific tasks such as monitoring a given element in a fairly homogeneous matrix, for example sulphur in fuel oils [24].

The major choice of an analytical system for general qualitative and quantitative use, therefore, lies between the conventional wavelength dispersive and the new semiconductor detector spectrometers. Clearly, of the two the energy dispersive spectrometer is the superior instrument for qualitative identification of all major elements and, when used with an isotope source, is portable provided a supply of liquid nitrogen is available. Ambient temperature Si(Li) detectors are on the market but the resolution of around 9keV makes their use impractical. Even with X-ray tube excitation these instruments are not as expensive as an automatic crystal spectrometer and they will provide a complete analysis in a few minutes. Their sensitivities, however, are low in comparison with crystal spectrometers and the instrumentation is less well advanced. In particular, sample changing units, and any form of automation, are as yet unsophisticated.

A simple crystal spectrometer can be as cheap as the non-dispersive instrument and, with its superior sensitivity, is still able to rival the newer instrument. It is however completely manual and each analysis will involve manually selecting the crystal, collimator, tube power, 2θ angle, and sample. Therefore, although such spectrometers are versatile, the rate at which analyses can be performed is low.

In the more complex crystal spectrometers now available all these functions can be preset and automatically chosen via a series of programs which may be pre-wired or, in the most sophisticated instruments, controlled by a computer. This can also accept the intensity readings from the detector and convert them directly to concentrations. Peak search routines to select the elements for analysis, and the standards with which they are to be compared, are also possible. However the sequential manner in which such spectrometers are forced to operate for multi-element samples again implies fairly long analysis times, and so, when high speeds are required, the simultaneous dispersive spectrometer may be used. In this instrument, which is further discussed in Chapter 6, fixed individual crystal-collimator-detector channels are provided for each wavelength to be measured. The total analysis time is, therefore, that of the longest channel counting time. Automatic sample holders and changers are now available for up to 108 samples sequentially, including a standard or standards which can be scanned at predetermined intervals to check the stability of the system.

At the present state of the art, instrument choice must be governed by many factors, including the available precision, sensitivity, speed of analysis, and elemental range, not to mention the cost of the instrument. Since the prices range from about £1500 for a simple filter spectrometer up to about £50,000 for a fully automatic crystal spectrometer, the type of cost benefit analysis discussed in Chapter 12 must be made carefully before making a commitment to one or other instrument.

3.4 ANALYTICAL PROCEDURES

As the foregoing discussion has indicated, X-ray fluorescence techniques, and particularly those based on crystal spectrometer instruments, currently provide a rapid method of qualitative and quantitative analysis whereby solids and liquids may be investigated non-destructively with a high degree of precision. Clearly, the different types of instrument available can carry out the various measurements associated with the range of analytical problems with varying degrees of efficiency. The three types of spectrometer have developed over many years and are unlikely to show dramatic improvements in performance, although the youngest, energy dispersive, instrument is capable of further development. However, once the X-ray signal from an element has been isolated the problem of converting this to a concentration is similar for each spectrometer.

3.4.1 Qualitative Investigations

Qualitative investigations are concerned with identifying the elements present in a sample, by determining the wavelengths or energies of the characteristic radiations generated from the sample on irradiation. This involves scanning the fluorescent X-ray spectrum for peaks, a task which is ideally suited to the crystal spectrometer or the lithium-drifted detector instrument. Clearly this is difficult, or at best inconvenient, to attempt with a filter spectrometer, because of the need to use a separate filter pair for each element and the limited range of filter pairs available. This last instrument can realistically be used to identify the presence of a limited number of expected elements in a mixture, but it cannot perform a complete spectral characterisation of an unknown material.

The crystal spectrometer performs a sequential scan of intensity versus 2θ angles in order to identify the positions of the peaks in a given pattern, in very much the same way as a powder diffraction pattern is obtained in the analysis of a crystalline mixture (Chapter 7, section 7.6.4). The problems of collimation, uniformity of source intensity, scanning speed, etc., will be similar in the two cases. As discussed in section 3.3.1, a range of crystals and, possibly, different exciting radiations may be required to cover the whole spectrum. This, together with maximum scan speeds of $4^\circ 2\theta$ per minute makes the procedure time consuming. Instruments fitted with lithium-drifted semiconductor detectors can analyse and display the whole spectrum in seconds, or minutes, without changing conditions, although the slightly poorer resolution may hinder identification of peaks and changes in the exciting radiation may again be needed for light and heavy elements. It is likely that such instruments will take over the majority of qualitative analysis in the future.

The data from both techniques comprise sets of peak positions expressed either as 2θ values or channel numbers. Fully comprehensive tables are available for all the commonly used crystals, listing the 2θ values for each element line [12] although for new or little used crystals the Bragg equation [equation (3.9)] must be used. For the energy dispersive system all that is required is a knowledge of the energy calibration, in keV, of the spectrometer.

The first problem in assigning the peaks, particularly with a crystal spectrometer, lies in deciding whether the observed line corresponds to an elemental emission or to a spurious effect. Thus a weak signal may arise from lattice defects in the crystal, from crystal fluorescence, or from escape peak phenomena in the detector. It is usually possible to resolve doubts by comparing scans from different crystals, and possibly from two detectors. Energy dispersive spectrometers do not use crystals for dispersion and so are not prone to this effect, but they are susceptible to the masking of peaks by line overlap, and, if radio-isotope excitation is used, to interfering peaks from the heavy element shielding around the source container.

Once it has been decided that all the peaks correspond to fluorescent X-radiation, it then becomes necessary to decide whether a line corresponds to the K series, the L series, or even the M series of a heavy element. With a crystal spectrometer there is also the problem of second or higher order lines from the crystal planes, as predicted by the Bragg equation. A further point in both types of instrument is the occurrence of lines from the X-ray tube target which may be too weak to figure in many tables but, because of the nature of the target, will appear as relatively intense peaks. Some clues to the probable assignment may be obtained from the expected composition of the sample, but it is usually better to try to extract as much information as possible for each element from the complete spectrum. In this context, the presence of $K\alpha$ and $K\beta$ lines and the various L lines should be checked before finally identifying the element. Figure 3.11 shows this procedure where the computer has identified the major peak as molybdenum $K\alpha$ at 17.44keV, and as a check displayed the positions of the $K\beta$, $L\alpha$ and $L\beta$ energies. The relative intensity of K and L lines is well illustrated. (It is usual to identify as a peak any signal greater than 2σ, where σ is the standard deviation of the background.) As a further guide, wavelength and energy tables list the relative intensities of each series of lines to that of the α line. In the present example (figure 3.11), the relative intensity of the molybdenum $K\alpha$, which is the sum of the unresolved $K\alpha_1$ and $K\alpha_2$ lines, to the $K\beta$ ($K\beta_1 + K\beta_3$) is 150/24.

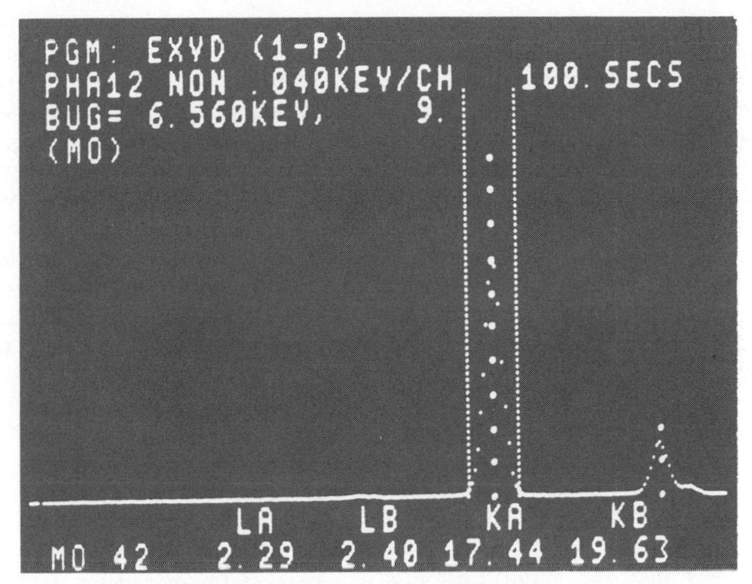

Figure 3.11 Another type of display obtained from the dispersive analyser shown in figure 3.9 (see also figure 3.10). A computer generated spectrum is now displayed to show all the peaks for molybdenum, with their line energies shown along the bottom of the photograph, and this permits positive identification of the upper spectrum as that of the Mo $K\alpha$ and $K\beta$ lines.

3.4.2 Quantitative Analyses

Quantitative estimates of the amounts of the elements present in a sample may also be made using X-ray flourescent measurements. Basically, the method involves measuring the peak intensity for a given line above the local background level and relating this reading to the concentration of the element, using a predetermined calibration. In practice, the technique normally involves determining accurately the position of the peak maximum, measuring the integrated counts at this setting, determining a similar count rate for the local background, and finally subtracting this background from the peak to give the required intensity.

The counting rate may be determined using either a fixed time or a fixed total count mode. In the former, more usual mode, counts are integrated at the peak maximum angle over a preset time, typically between 5 and 100 seconds. Longer times may be needed for elements lighter than potassium, or for those present in very low concentrations. In the latter mode, the time required to accumulate a predetermined number of counts is measured. The choice of method depends on the nature of the analysis being undertaken and each has its advantages and disadvantages, although fixed time is generally the more convenient. The fixed count measurement suffers from two main disadvantages. Firstly, excessively long times may be spent in accumulating data for elements present in very low concentrations. Secondly, as will become clear in the next section, the precision with which these elements are detected will be unnecessarily high. It does possess, however, the advantage of simplifying the correction for system dead time.

It may be noted in passing that the lengths of time over which the peak and the background counts are accumulated may not be equal but may be related to the count rates [equation (3.17)]. Moreover, a ratio method is often used, whereby the count rate for an element in the sample is expressed as a fraction of the count rate for the same element in a known standard. The statistical requirements of the three modes differ and these points, which will be further discussed in the next section, have been well covered by Jenkins and De Vries [29].

3.4.3 Error Corrections

The raw intensity values must be corrected for the various sources of error inherent in the measurement. These include errors due to the composition of the sample and the method of preparation, and also those from the instrument and the counting chain. After these adjustments it may be expected that the count intensity will be directly proportional to the elemental concentration, but exact proportionality is difficult to achieve, principally because of the manner in which X-rays are generated in even a homogeneous sample and the physical state of the sample generally used. In the following sections, the effect of instrumental factors and counting statistics on the accuracy of a determination will be considered first, and then the more important errors introduced by the preparation and physical state of the sample discussed.

Instrumental Errors and Counting Statistics

Instrumental errors, both random and systematic, are common to all devices with mechanical resetting. In the case of a crystal spectrometer, the possible errors are basically those discussed in Chapter 7, section 7.6.4, and relate principally to the setting of the primary collimator, the detector and secondary collimator, the position and height of the sample, the placing and temperature of the crystal, and the adjustment of the primary source. In addition, the sequential nature of the data collection procedure means that variations in the electrical supply can affect the long and short term drift, principally through the stability of the X-ray tube and generator and of the detector electronics. Such sources of errors are readily isolated experimentally and their effects corrected. In modern instruments with well built mountings and good electrical stabilisation circuits, their total contribution to the overall precision error is of the order of 0.2% for measurements made over a time of the order of minutes. This figure can rise to 0.5-1.0% for measurements made over several hours, when long term drift in the instrumentation becomes appreciable [29].

X-rays are generated randomly from a sample, and it can be shown that the number of photons produced in a given time shows a spread of values which closely approximates to a Gaussian distribution. It follows that readings made under identical counting conditions will exhibit a range of values from which it will be possible to calculate a standard deviation (σ) about a mean value. In practice, the relative standard deviation (ϵ), equal to the standard deviation divided by the number of counts accumulated (N), is more important and is expressed as a percentage according to the relationship

$$\epsilon\% = \frac{100}{\sqrt{N}} \tag{3.12}$$

since σ is equal to $\sqrt{N}$.

Thus, in the absence of instrumental error, the precision with which an intensity can be measured depends on the number of counts accumulated, i.e. on the length of time spent on the analysis. It is important to remember that improving the precision by a factor of 2 means quadrupling the number of counts, or the time, taken.

The total random error arising from the combination of the instrumental and the counting errors is given by

$$\epsilon_{\text{total}} = \sqrt{(\epsilon^2)_{\text{counts}} + (\epsilon^2)_{\text{generator}} + (\epsilon^2)_{\text{instrument}}} \tag{3.13}$$

Equations (3.12) and (3.13) can be combined to estimate the maximum number of counts that should be taken, for a given spectrometer, before the instrumental error exceeds the counting error and becomes the limiting factor in the overall precision. For the instrument quoted above, with a short term precision of 0.2%, it will be wasteful to collect more than 10^6 counts in any single measurement, since above this the short term drift contribution will become the limiting factor. A larger drift factor may have to be used if the counting time becomes excessively long, thereby further reducing the precision. When a precision better than that obtainable with a single measurement is needed, multiple measurements can be made. If for a single measurement the precision is $\epsilon\%$, for a set of n replicate measurements the overall precision is given by

$$\epsilon_{\text{overall}} = \epsilon_{\text{total}} \times \frac{1}{\sqrt{n}} \tag{3.14}$$

The above analysis applies to single measurements of intensities, but the observed intensity for an element is usually obtained by measuring the peak intensity (N_p), the local background (N_b), and subtracting the latter from the former (section 3.4.2). The general formula for the relative standard deviation of the observed intensity ($N_p - N_b$) then becomes [29]

$$\epsilon\% = \frac{100\sqrt{N_p + N_b}}{N_p - N_b} \tag{3.15}$$

Jenkins and De Vries [29] have shown how equation (3.15) must be modified for the three most widely used counting regimes, the fixed time, optimal fixed time, and fixed count modes.

In the fixed time mode, both peak and background are counted for the same predetermined length of time. If the times of counting on the peak and on the background are T_p and T_b respectively, the relative standard deviation of the observed intensity is given by

$$\epsilon\%_{\text{F.T.}} = \frac{100\sqrt{2}}{\sqrt{T}} \cdot \frac{\sqrt{R_p + R_b}}{R_p - R_b} \tag{3.16}$$

where R_p and R_b are the count rates for the peak and background, in counts per second, and T is the total counting time, i.e. $T = T_p + T_b$.

In the optimal fixed time mode, the total counting time is divided between the peak counting time and the background counting time in such a manner that the standard deviation of the observed intensity is a minimum. Then, although T is still equal to the sum of T_p and T_b, the ratio of peak to background counting time becomes

$$\frac{T_p}{T_b} = \sqrt{\frac{R_p}{R_b}} \tag{3.17}$$

and

$$\epsilon\%_{\text{F.T.O.}} = \frac{100}{\sqrt{T}} \cdot \frac{1}{\sqrt{R_p} - \sqrt{R_b}} \tag{3.18}$$

Finally, in the fixed count mode, $R_pT_p = R_bT_b = N$ and $T_p + T_b = T$, whereupon the relative standard deviation becomes

$$\epsilon\%_{\text{F.C.}} = \frac{100}{\sqrt{T}} \cdot \frac{\sqrt{R_p + R_b}}{R_p - R_b} \cdot \sqrt{\frac{R_p}{R_b} + \frac{R_b}{R_p}} \tag{3.19}$$

The magnitudes of the relative standard deviations for the three modes will, of course, depend on the peak to background ratio, but it can be shown [29] that for all cases in which $R_p > R_b$, the values increase in the order

$$\epsilon\%_{\text{F.T.O.}} < \epsilon\%_{\text{F.T.}} < \epsilon\%_{\text{F.C.}}$$

The Ratio Method

In many cases the count rate is expressed not in counts per second, as in the 'absolute' methods, but as an arbitrary rate normalised to a standard. It is not necessary to know the composition of this 'standard', only that it should provide a reasonable number of counts for the element to be analysed, and that its composition should not change with time. The measurements of standard and sample are made over a time T_s needed to accumulate N counts on the standard. In such a scheme, the standard is measured at frequent intervals instead of only rarely, as in the absolute method. At first sight it would appear that the precision is reduced by using this method, since the relative standard deviation for the counting procedure is given by

$$\epsilon_x\% = \frac{100}{\sqrt{N_s}}\sqrt{1 + \frac{R_s}{R_x}} \tag{3.20}$$

where the subscript s refers to the standard and x to the sample. For R_s and R_x of the same order of magnitude, equation (3.20) approximates to

$$\epsilon_x\% = \frac{100}{\sqrt{N_s}}\sqrt{2}$$

which is poorer than the absolute deviation by the factor $\sqrt{2}$. However, this is only part of the total error, and in fact the total relative standard deviation can be reduced using the ratio method, since the short term drift value can be used in place of the long term drift, as is required in the absolute mode. Using the ratio method effectively eliminates almost all the instrumental error. The ratio method is not always superior to the absolute method, however, and Jenkins and De Vries [29] have again discussed the relative merits of the two modes, in conjunction with their extensive explanation of counting strategy. Müller [30] also discusses counting statistics in detail.

Sampling Errors

In X-ray fluorescence analysis, in common with all other analytical methods, the majority of errors are introduced by the sample. Initially there is the problem of selecting a representative sample of 0.2g to 10g from the original material submitted for analysis. After this selection, depending on the wavelength used in the determination, only a very small fraction may be analysed. A typical rock powder sample of 7g would produce a pellet 32mm in diameter and 5mm in thickness. The secondary X-radiation for, say, iron $K\alpha$ would come from a depth of around 100μm and an area 30mm in diameter, or around 2% of the sample volume. The size of the primary collimator will further reduce the area of the sample seen, restricting the analysed weight to around 0.1g. This weight will increase for the shorter wavelength radiation, but, of course, reduce considerably for the light elements [equation (3.21)]. When a binding agent is added to the sample, the weight is obviously further reduced. Extreme care must therefore be taken when sampling material for X-ray analysis, and the points discussed in Chapter 12 borne in mind at all times.

In the following sections it will be assumed that a representative sample has been chosen. All problems will therefore be associated with the optimum methods for preparing samples for presentation to the spectrometer, and the factors associated with the manner in which fluorescent X-rays are generated and emitted from the sample.

3.4.4 Sample Preparation

The ideal sample for presentation for X-ray fluorescence analysis should be completely homogeneous and form a disc or layer with a perfectly planar surface. Liquids approximate to this ideal condition, but they do present some problems. The liquid must be contained, and the design of most spectrometers is such that they irradiate and measure the base of the sample. This means that the liquid must be supported, normally by a thin organic film. Purely mechanical problems arise because the film may stretch in the holder and so distort the smooth lower surface. Rupture of the film is also a considerable hazard with the hot, thin, beryllium window of the X-ray tube immediately below the sample. Also measurement of the lighter elements is made difficult, since their low energy radiation is absorbed in the organic film and in the air-path of the spectrometer. This last problem can be minimised by flushing the spectrometer with helium or by using special cups designed to contain the liquid sample within a vacuum environment. In addition, the energy of the incident X-rays can cause heating and evaporation of the liquid phase, which can in turn set up concentration gradients in the body of the liquid. Precipitation and bubble formation are also possible. But despite these problems, liquid samples can be extremely useful, especially for small quantities and 'one off' applications. Carr-Brion shows, in Chapter 6, how convenient solution and suspension sampling can be in control applications.

A restricted sample type which has many of the advantages of a solution without its disadvantages is one made up in a glass, which is, of course, simply a supercooled solution. Dispersion of the sample in a glassy matrix can provide the required homogeneity while the rigidity of the product removes the mechanical problems. Even here, however, the method of manufacture can cause concentration gradients in the material, and surface contamination is a constant hazard. Glasses are widely used today and will be considered in detail in a subsequent section.

Metals

Bulk metals may, at first sight, seem to comprise simple solutions which present few, if any, problems, but this is generally not so in practice. All metals possess a crystalline structure in the sense that different regions of the bulk sample contain different phases and segregation of the elements present may well occur between these phases. If these components occur at different levels in the prepared sample, problems will arise. On the one hand, fluorescent X-radiation derived from layers lying below the surface is absorbed by the over-lying layers and radiation from light elements is more strongly absorbed than that from heavy elements. Sodium $K\alpha$ radiation may therefore be completely absorbed if it originates from a depth of more than 5μm in the sample, whereas Fe $K\alpha$ may be detectable from depths of 100μm. Hence a variable X-ray response will be obtained from a body depending on the nature of the phases lying close to the surface. On the other hand, variation in the response will be obtained depending on whether

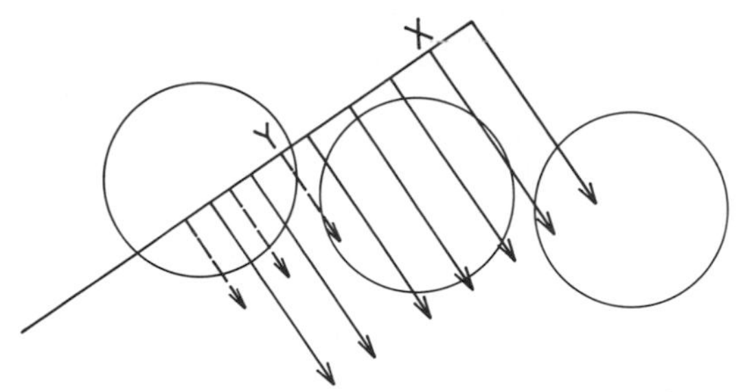

Figure 3.12 Idealised representation of the surface of a granular sample. The lengths of the arrows indicate the critical depths for two different radiations, from which it will be apparent that the rays X and Y will be relatively highly attenuated by the high points of the grains.

the sample constitutes a homogeneous crystalline solution or an inhomogeneous material in which precipitation of one or more components has occurred. Thus cast iron and steel will give a different X-ray response since in the former graphite has precipitated out while in the latter the carbon is more uniformly distributed as cementite.

Surface finish may also introduce errors, since the cutting and grinding procedures normally used to prepare the specimen can result in scoring of the face to be analysed. Secondary X-rays are normally measured at an angle of about 35° to the analytical surface (the take-off angle) and this geometry permits any surface topographical features, such as machining grooves, to give the same effect as concentration gradients. As figure 3.12 shows, the high points can absorb radiation generated in the hollows, the longer wavelengths preferentially, and so give spurious readings. The critical depth formula

$$d = \frac{4.6}{\mu p} \sin \psi \tag{3.21}$$

(where ψ is the take-off angle) gives the maximum depth from which the fluorescent X-radiation from an element can be detected in a given matrix, and can be used to give some idea of the standard of surface finish required to elimate such problems. From this it may be found that polishing is necessary to produce the required finish. While this may work well for uniformly hard materials, care is needed with non-uniform alloys, when the softer component may become smeared over the surface and so mask the harder component, or the polishing medium may become embedded in the surface and again give rise to spurious values.

Powders

In the mineral industries, powders are the most likely form in which the sample, usually a silicate or a sulphide, is available and it is tempting to analyse the powder directly either loose or in a tablet form. In fact, such techniques are widely used but they will rarely give the highest accuracy, although they may be justified by the amount of time saved or where only rough concentrations are needed or in process control applications where the material is in the form of a powder on a conveyor belt (Chapter 6).

Some of the problems encountered with powders are similar to those mentioned above and, in reference to other samples, figure 3.13 illustrates a typical situation diagrammatically. In a rock or other sample comprising several mineral species, any grinding procedure will give a product in which each mineral will show a grain size distribution [31] with the median for each mineral occurring at a different size, depending largely on the hardness of the minerals. As an example of a simple rock, granite will, after grinding for three minutes in a tungsten carbide disc mill, produce feldspar grains with sizes less than 20μm, quartz in the 30-60μm grain size, and very much larger biotite flakes, and prolonged grinding does not improve this situation. Figure 3.13 shows the effect of these size distributions on the physical make-up of a specimen if the powder is prepared for X-ray measurement in a loose and a compacted form. The fine feldspar powder occupies much of the volume adjacent to the sample surface and, in the compacted pellet, the biotites tend to orientate themselves parallel to the tablet face. Both of these factors will produce incorrectly high intensities for the lighter elements and the alkali and alkaline earth metals, because the quartz fraction is buried by the other two components, which are over-represented in the surface layers.

Monomineralic powders will also show a variable X-ray response depending on their grain size. Gunn [32] has shown that there is a marked decrease in intensity as the path length of the analysed radiation approaches the grain diameter. This effect is also present in heterogeneous samples but it

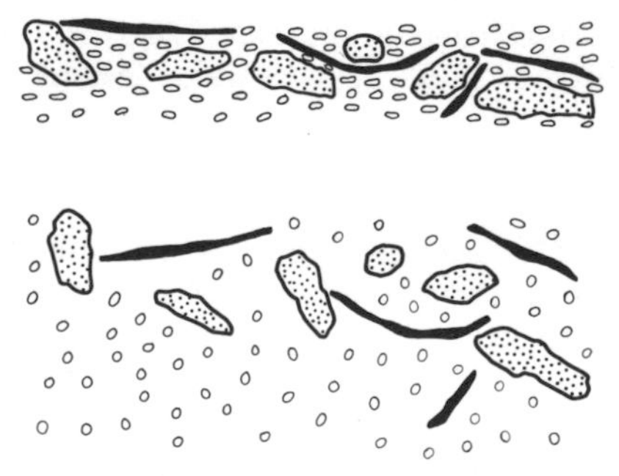

Figure 3.13 The effect of compaction is illustrated for a crushed granite sample, in the upper part of the diagram. After crushing, lower part of diagram, the biotites (black lines) will remain as large plates, quartz (stippled areas) will be partially reduced in size, and the feldspar (open circles) reduced to small grains. When compressed to form a pellet, the elongated particles will be aligned parallel to the surface, and elements present in these minerals will tend to be over-represented in any analysis measurement. Radiation with a small penetration depth, moreover, may indicate anomalously high concentrations of elements present in the fine grained fraction.

is found that compacting the powder under pressures of up to 50 tons per square inch helps to minimise the problem. An extensive theoretical discussion of mineralogical and grain size effects is to be found in Müller [30] and de Jongh [33].

Where powder samples can be tolerated or when, as in the analysis of trace elements, it is necessary to use them, various methods of preparation, summarised by Hutchison [34], can be used. In the author's laboratory, stable and relatively robust pellets are made by mixing 7g of powder with 10-20 drops of a 20% solution of Mowiol* and pressing in a die of 31mm diameter under a pressure of 15 tons. Difficult samples may be pressed in cups formed from cellulose, boric acid, or stearate, or in aluminium cups or lead rings. The last two methods possess the advantage of preventing fracturing of siliceous samples caused by radial expansion of the pellet on extrusion from the die.

Fused Glass Samples

The difficulties encountered in using powder samples have now led the majority of analysts to remove mineralogical effects from the specimen by first converting it to a glass by fusing with a suitable flux and casting into a smooth-surfaced bead. Claisse [35] was the first to apply the technique, using borax as flux, casting the button on a heated, polished metal plate, and annealing. Since then many fluxing agents have been used, including sodium carbonate, potassium bisulphate, and ammonium meta-phosphate, but today the most widely employed is lithium tetraborate, normally used as a eutectic mixture with lithium carbonate, i.e. a mixture of lithium carbonate 44.5% w/w and lithium tetraborate 55.5% w/w prior to fusion. The low melting point of this mixture (810°C) allows the fluxing reaction to proceed efficiently with good internal stirring at 1050°C using a muffle furnace or blast burner. Almost all rocks can be dissolved with this flux, although it is necessary to add an oxidising agent, such as sodium, potassium, or ammonium nitrate, to the system when sulphides or metals are present. Limestone, gypsums, and phosphate rocks do not produce stable discs, but this can be overcome by adding a known amount of silica to the reaction mixture. Other ores and metallic slags, which will not dissolve fully in lithium borates, will dissolve in a sodium borate flux, although the beads have the disadvantage of being hygroscopic.

The other important consideration in preparing a fusion bead is the ratio of flux to sample. Clearly sufficient flux must be used to ensure that the specimen is completely decomposed, but excessive amounts will introduce unnecessary contamination from the inevitable impurities in the flux and these will interfere with the determination of low concentrations of elements in the sample. Generally, flux to sample ratios between 1:1 and 50:1 are used, and a ratio of 5:1 is ideal for most applications, since it provides sufficient flux for decomposition while still allowing sodium and

*Mowiol is available from Hoesch Chemicals, catalogue number M-8-88.

other minor elements to be measured. Extremely high ratios, up to 100:1 may well impede accurate analysis, since a reasonable contamination level of 20ppm of each trace element in the flux then becomes the equivalent of 2% in the sample. In addition even these high dilutions do not completely remove inter-element effects [36]. At the other end of the scale, if low flux to sample ratios are used it is important to check microscopically or by powder X-ray diffraction that complete decomposition and dissolution has been attained, and also that unmixing and devitrification have been avoided.

The actual choice of fluxing mixture and flux to sample ratio must ultimately depend on the nature of the sample to be dissolved, on the elements to be determined, and on the precision with which they are to be measured, and each case must be treated separately to find the best conditions for specimen preparation. Details of fluxes are available from the various chemical manufacturers, in particular the Johnson Matthey Company who produce a wide range of spectrographically pure mixtures in their 'Spectroflux' range*.

The fusion itself is best performed in a crucible made from a platinum-5% gold alloy, which is not wetted by the borate fluxes used and so allows total recovery of the sample-containing liquid. Melting is normally carried out in a muffle furnace, with heating on a blast burner as an alternative, cheaper, method. Induction heating, as described in Chapter 5, section 5.2.1, will provide very rapid fusion, but temperature control is difficult and may cause reproducibility problems where the sample contains a large proportion of volatile elements.

Most samples are completely decomposed and mixed after heating for 10 to 15 minutes in a muffle furnace, but it may be necessary to stir or swirl the crucible during fusion to ensure homogeneous beads. This largely depends on the viscosity of the melt. Repeated fusion and crushing of the bead has been advocated but this is time consuming, and the difficulty can usually be overcome by a change of flux.

The bead is finally cast either on a preheated polished plate in a retaining ring or in a preheated mould. The latter method, which was originally described by Norrish and Hutton [37] and in a modified form by Harvey *et al* [38], is widely used in rock and mineral analysis and is particularly useful where only small sample weights are available, typically in the range 0.1-0.4g. In every method of preparing a disc or pellet it is, of course, essential to ensure that the bead used is always thicker than the critical depth of the most energetic radiation to be measured [equation (3.21)].

Carbon crucibles are apparently a cheap alternative to Pt/Au, and possess the advantage that the bead may be formed simply by allowing the mixture to cool and anneal in the crucible [39]. The bead surface must, however, be polished to remove the inevitable carbon film, especially if the lighter elements are to be determined. The life of these crucibles is also short, about ten beads, so that the overall cost is comparable to that of platinum.

*Johnson Matthey Chemicals Ltd., 74 Hatton Garden, London EC1P 1AE, England.

A major difficulty with all fluxes is the determination of ignition loss from both the sample and the flux mixture. This is particularly important if temperatures above 1100°C are used for the fusion step, when not only are water and carbon dioxide lost but alkali metals and the flux itself may volatilise out of the system. Harvey *et al* [38] have discussed the methods whereby such losses may be determined, but where large amounts of oxidising agents are required, as in the analysis of sulphides, the measurement of the true ignition loss is impossible. The double concentration procedure of Tertian [40] successfully compensates for ignition losses and will be described in section 3.4.6.

3.4.5 Matrix Effects

The previous section discussed the deviations arising from counting statistics, sampling errors, and grain size and shape, but even when all these problems have been eliminated most samples still show deviations from linearity when the observed intensity is plotted against the known concentration of an element. Such effects arise from the manner in which X-rays are generated within the sample.

It has been shown, in section 3.2.1, that the characteristic fluorescent X-radiation for an element, wavelength λ_j, can be excited by X-radiation of wavelength λ, provided that the energy of the exciting radiation is greater than that of the absorption edge for the element. It follows, therefore, that the observed intensity of the fluorescent radiation at the surface of the sample will depend firstly on the efficiency with which the element absorbs the exciting radiation, which depends on the mass absorption coefficient of the element for wavelength λ, and secondly on the extent to which the matrix absorbs both the exciting and the fluorescent radiation. Therefore, for a sample irradiated by a wavelength λ and producing a fluorescent radiation of wavelength λ_j, the excitation efficiency $[E(\lambda\lambda_j)]$ is given by

$$E(\lambda\lambda_j) = \frac{\mu_j(\lambda)}{\Sigma_i C_i \mu_i(\lambda) + A \Sigma_j C_i \mu_i(\lambda_j)} \tag{3.22}$$

where $\mu_j(\lambda)$ is the mass absorption coefficient for the element of interest at the wavelength λ, and C_i and μ_i are the concentrations and mass absorption coefficients at wavelengths λ and λ_j of the other components in the sample. A is a constant for the spectrometer and is related to the angle of incidence (ψ_1) of the exciting beam and the take-off angle (ψ_2) by

$$A = \frac{\sin \psi_1}{\sin \psi_2} \tag{3.23}$$

Equation (3.22) relates to excitation by monochromatic radiation, but in most spectrometers the principal excitation is produced by the 'white' radiation from the tube continuum. In such cases, λ should be replaced by an integrated wavelength taken between the Duane-Hunt limit (λ_{min}) and the absorption edge for the element considered (λ_E). This integral, as shown by Kalman and Heller [41], can be considered to be a weighted average wavelength ($\bar{\lambda}$), and Müller [30] has illustrated the calculation of the wavelength from the expression

$$\int_{\min}^{\lambda_E} \frac{\mu(\lambda)\cdot N(\lambda)d\lambda}{\mu(\lambda)\sin\psi_1 + \mu(\lambda_j)/\sin\psi_2} = \frac{\mu(\bar{\lambda})\cdot N(\bar{\lambda})}{\mu(\bar{\lambda})/\sin\psi_1 + \mu(\lambda_j)/\sin\psi_2}(\lambda_E - \lambda_{\min}) \tag{3.24}$$

where N is the number of photons. The integral is replaced by a rectangle of equal area, with one side the distance between the lower and upper limits of the integral. The height is the intensity at $\bar{\lambda}$.

The ultimately observed intensity due to an element in a mixture, however, also includes a small contribution from the secondary excitation of its fluorescent X-radiation by the characteristic X-rays from the other elements in the sample. The radiation causing this 'secondary enhancement' effect must, of course, be of a shorter wavelength than the element's absorption edge. Calculation of the magnitude of the secondary enhancement contribution is difficult, but the problem has recently been considered by Rasberry and Heinrich [42] and Tertian [43]. Fortunately the effect is small compared to absorption and need only be applied over a limited elemental range.

Direct Derivation of Composition

The preceding discussion has indicated that it should be possible to calculate the concentrations of the elements in a multicomponent sample directly from the observed intensities of the characteristic fluorescent X-rays, provided that the necessary corrections derived from counting statistics, sample variation, and matrix absorption effects are made. Basically, the method involves obtaining intensity data for all the major elements present (oxygen may be determined by stoichiometry and the lighter elements ignored unless they form a significant proportion of the sample) and using the raw data to estimate an approximate composition for the sample. From this it is possible to substitute the various mass absorption coefficients and concentrations into an equation analogous to (3.22) in order to correct the raw intensities for absorption by this first approximation matrix. The new intensities are then used to calculate an improved composition, from which a new matrix correction is calculated, and the process is iterated to convergence. Obviously, this procedure is tedious, although greatly simplified by using an electronic computer, and requires extremely accurate data relating to all the major elements present. To date such methods have been largely developed for use with the electron probe microanalyser, where it is often the only applicable method. A variant of the method, applicable to the X-ray fluorescence analysis of simple mixtures, will be considered in section 3.4.6.

In general, however, empirical methods of determining interelemental effects are employed, although it is considered advisable to use the theoretical, direct method as a check on the results so obtained. Most empirical methods use a calibration technique, whereby the sample is effectively compared with known standards containing the required element

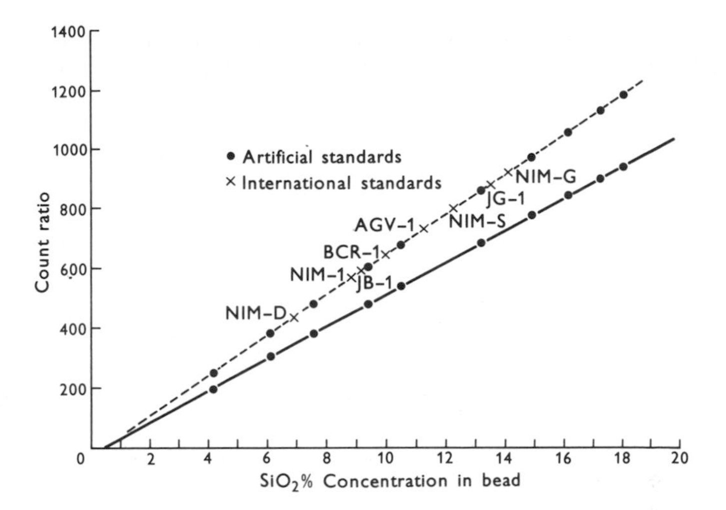

Figure 3.14 A typical calibration curve produced for SiO_2 in a set of international standard samples, fused in a lithium tetraborate/lithium carbonate/lanthanum oxide flux. The excellent linearity over a wide concentration range results from adding the heavy absorber, while the positive intercept on the concentration axis is caused by the presence of a low level of impurities in the flux itself. The dashed curve is the calibration line obtained some months after the original determination, during which time the ratio standard, a fusion bead, had deteriorated. This emphasises how essential it is to have a stable ratio standard.

in different concentrations in a matrix that is closely similar to that of the sample.

Calibration Methods

The simplest method of calibration involves the use of a series of external standards in which the concentration of the required element is varied over a limited range in a matrix whose composition is chosen such that the overall absorption is very close to that of the sample. The exact range of compositions that can be covered by this technique is difficult to specify, since it depends on the element to be measured and on the other elements present in the matrix, as well as on the degree of accuracy required. It is, however, generally accepted that for satisfactory results the overall mass absorption coefficient of the standards should not vary by more than 5% from that of the samples.

In practice, the calibration curve is best constructed using the ratio method (section 3.4.3) where the counts of each calibration standard relative to the ratio standard are plotted against the known elemental compositions of the standards. The ratio standard may itself be one of the calibration standards, or it may be a specially prepared sample. This should give relatively high count rates for all elements to ensure good counting statistics. Unknown samples are treated in the same way as the standards and their concentrations determined by comparing their count ratios to the calibration curve.

Harvey *et al* [38] have pointed out that if a large concentration range is to be covered by such a calibration, residual absorption effects will be found, and the count ratios found for the standards will require an absorption correction which is the reverse of that applied to the unknown samples.

Two points are worth noting. Firstly, since the ratio standard must be used continuously as a reference for all the samples used, it is important both that it should not change its composition with time (figure 3.14) and that it should be mechanically robust, since its loss will entail complete recalibration. Lumb [44] has described a method of making stable standards by setting the material into an araldite disc, and Padfield and Gray [45] have used enamels of compositions similar to the unknowns fused onto copper discs. Secondly, this simple calibration method can cover only limited compositional ranges, although the technique has been successfully used in compensating for mineralogical effects in powder samples [46].

The Heavy Absorber Method

It will be clear that the major limitation to the widespread application of the calibration method arises from the need to maintain the overall absorption coefficient effectively constant and that, if this problem can be overcome, it should be possible to construct wide-range calibration curves for any element in virtually any matrix. Following equation (3.22), this can be accomplished by diluting the sample with a large excess of a low absorption material, so that the absorption coefficient becomes effectively that of the diluent. Alternatively a component with a high absorption coefficient can be added in order to swamp the effect of the other elements present. Both techniques have been used with fusion methods, but the very high flux-to-sample ratios needed in the dilution method, of the order of 100 : 1 and over, make this method less attractive than the heavy absorber, partly because of the increased amount of impurity added with the flux and partly because, as a simple calculation will show, the heavy absorber is more effective in reducing absorption effects. Lanthanum in the oxide form, La_2O_3, is the heavy absorber most commonly employed in silicate analysis, and Norrish and Hutton [37] have discussed its use in some detail.

A typical flux composition is lithium tetraborate 47.0, lithium carbonate 3.67, lanthanum oxide 16.3*, which is then fused and ground and added to the sample in the proportion of five parts of flux to one of sample. This composition, it is claimed [37], (1) improves the peak-to-background ratio for most elements, and (2) improves the relative sensitivity for the lighter

*Available commercially as Johnson Matthey Spectroflux 105.

elements. Lanthanum oxide absorbs all X-radiation, but the relative change in observed intensity is less for the lighter elements than for the heavy elements which, in silicate analysis, is a considerable advantage considering the low fluorescent yield of the light elements. Care should, of course, be taken to avoid interferences from lanthanum peaks, for example the overlap of the La $2L\alpha$ escape peak with the S $K\alpha$ peak, or the possible interference of La $4L\beta_1$ with the background measurement of Mg $K\alpha$.

Calibration curves covering much wider concentration ranges may be prepared using this technique, and figure 3.14 shows a typical plot. Note the excellent linearity. The positive intercept on the intensity axis is the result of impurities in the flux, and must be subtracted from apparent concentrations derived from the curve. An adequate number of blanks must also be measured for each new batch of flux, to determine the possible variation in impurity level. A constant flux-to-sample ratio should, of course, be maintained throughout the work.

It must be remembered that both the dilution and the heavy absorber methods result in a net loss of sensitivity for all the elements present in the sample, and this may preclude their use at sufficiently high concentrations to swamp completely the matrix effects. This applies particularly if minor element analysis is required on the same samples (see section 3.5).

Direct Estimation of Absorption Coefficients
For elements heavier than iron, the secondary absorption of the characteristic fluorescent X-rays by the matrix is more important in determining the observed intensity than is the absorption of the primary beam, and this condition permits the effective overall mass absorption coefficient to be simply calculated or directly measured. Norrish and Chappel [47] have described a method whereby thin, specially prepared samples of varying thicknesses can be used to determine the matrix absorption. These are placed in a holder in front of the scintillation detector of a dispersive spectrometer and the intensity of a beam of the required wavelength measured with, and without the absorbing sample. The absorption coefficient of the sample can then be calculated from a knowledge of the area of the pellet, which is constant for a given die, and the weight of absorber present, using equation (3.4). Absorption coefficients for elements down to calcium may be obtained by diluting the sample with a material of very low absorption coefficient, such as cellulose.

Once the effective overall matrix absorption coefficient has been determined for a given sample type, it is easy to correct the observed intensity for the required element and use the value derived to estimate its concentration by reference to the intensity for the pure component.

Internal Standard Methods
The ratio and calibration methods discussed above utilise a comparison with external standards to determine the concentration of a given element in a sample. Internal standards can also perform this task, albeit somewhat more laboriously, in situations where external standards are difficult to use, for

example where secondary absorption is high, or in trace element analysis, or in 'one-off' situations when the work involved in preparing a calibration curve is not justified. The internal standard used may be either a different element from that being measured, or the element itself.

If a different element is chosen, it is essential that its absorption characteristics should be similar to those of the required element. Cobalt might be used as an internal standard in the determination of iron in the presence of chromium, since both their emission lines lie at wavelengths slightly shorter than that of the chromium *K*-edge. Manganese would be useless since it lies on the opposite side of the chromium edge and will behave in a different way from iron in the presence of varying chromium content. The use of a particular 'different element' internal standard presupposes that the element is not present in the sample.

In the above example, the ratio of the count rates of iron and cobalt is proportional to their concentrations and, knowing the amount of cobalt added and the proportionality constant for the count rates, the amount of iron is simply obtained. The proportionality constant is measured separately by making measurements on known iron-cobalt mixtures.

If several elements are to be determined, the choice of internal standard elements which are not present in the sample and which have lines suitably placed in relation to absorption edges may be difficult, if not impossible. In addition, the weighing and mixing of the internal standards will introduce errors. This last problem may, however, be overcome for a limited number of elements by commercially available fluxes containing internal standards.

In many cases where it is difficult to find a suitable 'other element' internal standard the pure element may be used as its own internal standard. The technique provides an accurate method of overcoming matrix effects, especially in trace element analysis. In practice, an accurately weighed amount of the pure element is carefully mixed with a portion of the sample, making sure that the total concentration of the element is not appreciably raised, and a series of sub-samples prepared by successive dilution of this mixture with the sample itself. The plot of the peak-minus-background intensity against added concentration of the element should be linear, since the total absorption coefficient depends on the major components present. The concentration of the element present in the original sample is found by extrapolating the line to intersect the negative portion of the concentration axis. This method of 'spiking' is useful for elements in concentrations below 2%, but errors may be introduced by the different 'mineralogy' of the added chemical or by the coating of sample grains with the extremely fine powders in which ultra pure standards are supplied.

If samples with different total absorption coefficients are carefully 'spiked' for a given element, curves may be prepared in which the intensities of the scattered portions of the tube characteristic lines or the background may be related to the slope factor, of counts versus concentration in the samples, as shown in figure 3.15. Using this external standard variant, trace element analyses may be made against accurately known synthetic samples, by measuring only the intensity of a suitable scatter line and the peak and

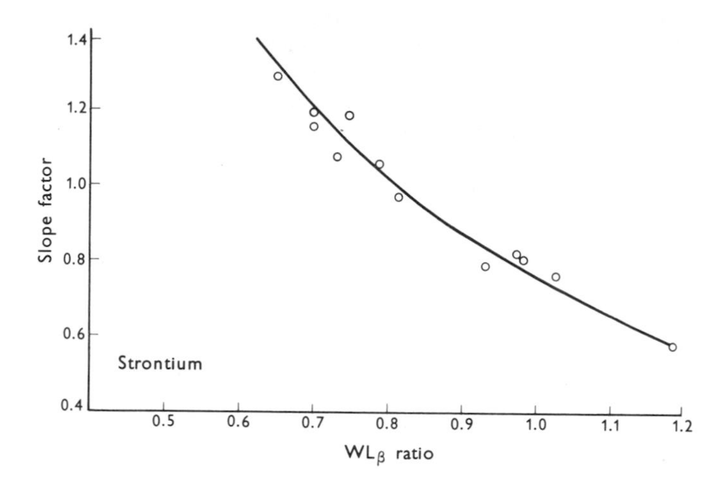

Figure 3.15 Slope factors, expressed as counts per second per ppm, for a range of rocks spiked with strontium are shown plotted against the intensity ratio of the scattered $WL\beta$ line. Such a curve may be used to correct for changes in absorption between standards and unknown samples.

background intensities of the unknown element. The interdependence of the scatter intensity and the slope factor may also be checked against directly determined values of the overall secondary absorption for the matrix in which the low concentration element is being determined, to assess the accuracy of the method.

The effect of primary absorption on the observed intensity of the element may also be determined by this comparison. The intensities of the Rayleigh and the Compton scattered radiation will depend on the density and average atomic number of the sample. It follows, therefore, that the scattered intensity is proportional to $1/\mu_{\text{overall}}$, where μ_{overall} includes both the primary and the secondary scatter and Cullen [48] and Reynolds [49] have demonstrated how measurements of the scattered intensity may be used to estimate the combined effects of primary and secondary absorption. Incoherent scatter should provide a better measure of the absorption, since the coherently scattered radiation is dependent on the nature of the matrix. In practice the Compton scatter is less satisfactory, at least using molybdenum K radiation, as the intensity appears to be affected by the mineralogy of the sample.

3.4.6 Determination of Matrix Effects in Solutions

The problems of grain size and mineralogy which are encountered in X-ray fluorescence analysis have been discussed in the previous section, and the conclusion was drawn that, for the accurate determination of the major elements, solution methods are required. Under such conditions the secondary X-rays are derived from a homogeneous sample in which the elements are evenly distributed, hence the intensity for each element should be proportional to its concentration. As previously discussed, any deviations from proportionality will be caused by changes in the sample's chemistry which will result in a diminution or an enhancement of the observed X-ray intensity for individual elements, because of changes in the overall matrix absorption coefficient. By the use of suitable standards the interactions between elements in a mixture can be studied empirically, especially in fusion methods where artificial mixtures can be made without the problem of different mineralogy between the artificial samples and the material to be analysed. These methods allow all such effects, e.g. primary and secondary absorption and enhancement, to be calculated by simple empirical techniques. Many correction procedures have been published, the better known being those of Sherman [50], Beattie and Brissey [51], Marti [52], and Lachance and Traill [53] which are concentration based and which have been shown by Muller [30] all to reduce to the same form. Lucas-Tooth and Price [54] have derived a method which is based purely on the intensities.

The simplest case to consider is that of a binary system containing two components A and B and is illustrated in figure 3.17. Here the radiation from element A is absorbed by element B, resulting in a change of slope of

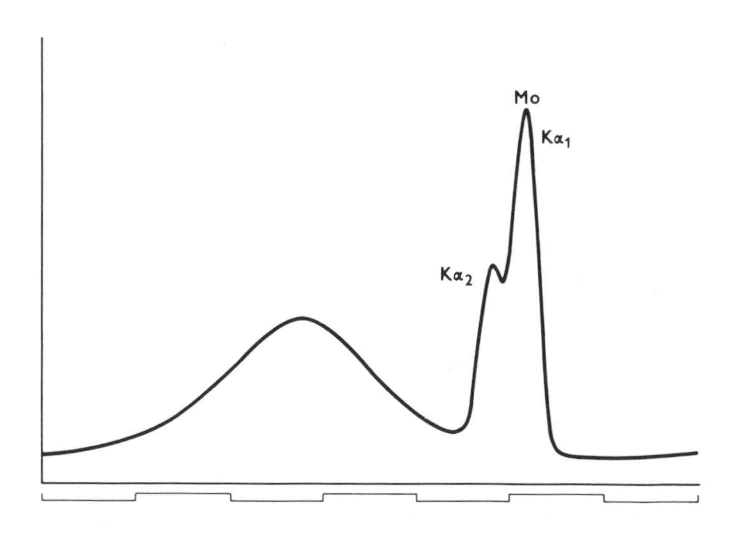

Figure 3.16 The spectrum obtained using a LiF_{220} crystal with a scintillation detector for the radiation from a molybdenum target X-ray tube, scattered by a pure silica disc. The diffuse nature of the Compton scatter peak is caused by the primary beam subtending an angle of approximately 30° at the sample surface [equation (3.8)]. Note the difference in Compton peak height compared with that of boric acid, shown in figure 3.2.

the curve relating the concentration of A to its count rate. The interelement effect arising from the presence of B may be expressed in the form

$$C_A = \frac{R_A}{m_A}\left[1 + \sum_B K_{AB}C_B\right] \tag{3.25}$$

for the element A in concentration C_A and giving a count rate R_A for a slope factor m_A affected by the element B present at a concentration C_B. The factor K_{AB} is the 'influence factor', a constant representing the effect of the presence of B on the observed intensity of the fluorescent radiation from A. Now, measurements are normally made by the ratio method against a separate standard and in this case equation (3.25) converts to the form

$$C_A^x = \frac{R_A^x}{R_A^s} \cdot C_A^s \left[1 + K_{AB}(C_B^x - C_B^s)\right] \tag{3.25a}$$

or, by rearranging the terms,

$$\frac{C_A^x \times R_A^s}{C_A^s \times R_A^x} = 1 + K_{AB}(C_B^x - C_B^s)$$

where the superscripts x and s refer to the sample and the ratio standard respectively, and the slope factor m_A has been eliminated. Therefore, by

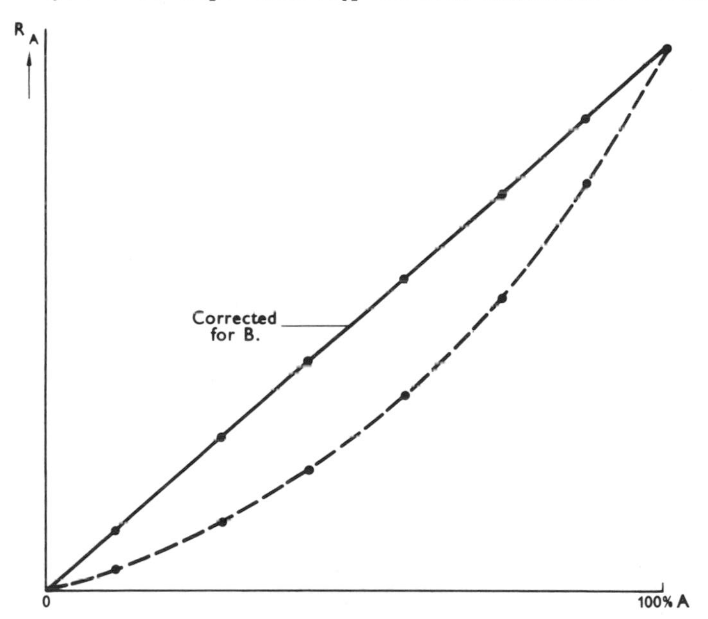

Figure 3.17 If a plot is made of the count rate, R, versus concentration of element A in a series of binary mixtures A-B, the dashed curve results if B absorbs the characteristic radiation of A to any appreciable extent. A correction may be applied to the measured count rate from A following the method shown in figure 3.18.

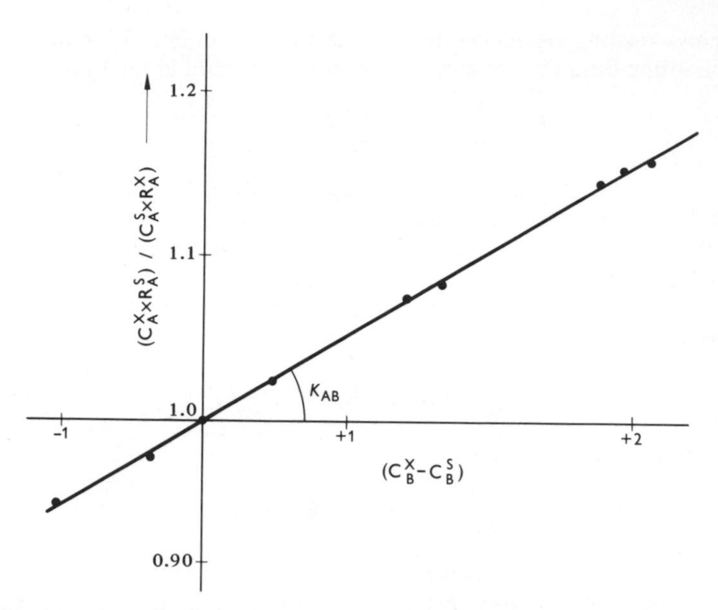

Figure 3.18 This plot, derived from equation (3.25a), allows the determination to be made of the correction coefficient (K_{AB}) for the influence of element B on the intensity of the characteristic radiation from A, and hence on the apparent concentration of A, in a binary system.

plotting $(C_A^x \times R_A^s)/(C_A^s \times R_A^x)$ against $(C_B^x - C_B^s)$, as in figure 3.18, a curve is obtained whose slope is the influence factor K_{AB} for the effect of B on the intensity of X-rays from A.

The effect of introducing a third element into the mixture on the observed intensity of A is shown in figure 3.19, from which it will be clear that it is not possible to evaluate the total correction factor by the above simple graphical method. It is, however, readily shown that the total correction factor is composed of contributions from both B and C, weighted for the concentrations of B and C, and so it becomes necessary to measure influence factors for B on A and for C on A in order to complete the calculation of the total correction factor. Jenkins and Campbell-Whitelaw [55] have discussed the procedures involved in establishing these correction factors.

The correction coefficients for complex mixtures may be expressed as a series of simultaneous equations. Thus, following Lachance and Traill [53], for a three component system we obtain

$$C_A = \frac{I_A^x}{I_A^{100}} (1 + C_B K_{AB} + C_C K_{AC}) \tag{3.26a}$$

$$C_B = \frac{I_B^x}{I_B^{100}} \left(1 + C_C K_{BC} + C_A K_{BA}\right) \tag{3.26b}$$

$$C_C = \frac{I_C^x}{I_C^{100}} \left(1 + C_A K_{CA} + C_B K_{CB}\right) \tag{3.26c}$$

where I_A^{100} and I_A^x, etc, are the intensities of A, B and C in the pure component and the mixture respectively. Note that the intensity from a suitable standard may be substituted for that of the pure component in equations (3.26), with the necessary modifications to their form to allow for this [56]. Müller [30] has discussed regression and other methods in conjunction with this technique at some length, and also described the ways of evaluating the influence factors K_{AB}, etc. Note that the value of K_{AB} is not equal to that of K_{BA} because of the different absorption coefficients in the two cases.

Obviously the determination of influence factors for pairs of elements is a long and laborious process, involving, as it must, accurate determinations of count rates on large numbers of accurately analysed standards or, in fusion methods, of artificial standards, and using equation (3.25a). Once

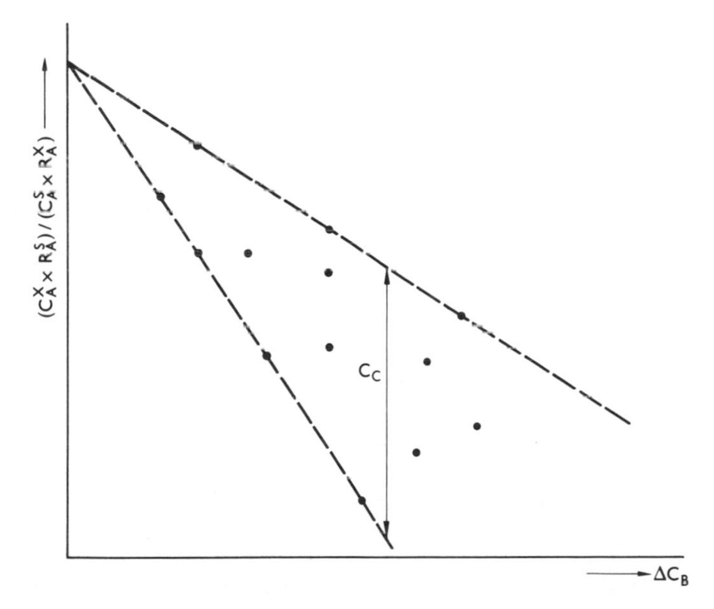

Figure 3.19 Addition of a further element C, which also interferes with the characteristic radiation from A, results in an increased spread in the plot relating the count rate and concentration of A to the concentration of B in a series of similar ternary mixtures.

determined, however, the factors are interchangeable between instruments, particularly if the influence factor is rendered small either by choosing a standard which lies in the middle of the range of compositions to be measured, rather than the pure element, or by adding a heavy absorber to the mixture. The advantage of the latter variation is shown by the large number of geological laboratories who use it in conjunction with the coefficients published by Norrish and Hutton [37].

If the factors are to be derived from a number of analysed samples rather than a set of artificial binary mixtures, there is a danger that incorrect influence factors will be produced by the mathematics of solving a large number of simultaneous equations of the form of equation (3.26). A computer program, 'Alphas' of de Jongh [57], derives theoretical coefficients which at least provide a check on the magnitude of the factors, and at best allow the determination of the correction coefficients with the very minimum of standards.

The principal disadvantage of using this direct technique for all analyses is that all the major components of the sample must be determined before the concentration of any one of them can be calculated. Hence, if only one element is required, much time and effort will be wasted and this may be a considerable disadvantage in many industrial applications where speed is essential. The method has many advantages, principally in giving high precision, when total analyses, for example of a rock, are required.

Tertian [40], following Sherman's improved method [58], has proposed a simple and elegant way of overcoming many of the problems attendant on the mathematical approach to the correction procedure. The sample is mixed with a flux or a solvent or other diluent to give two portions with different sample-to-diluent ratios, the ratio of the two portions having a fixed value to provide an intensity difference of around 2:1. Each sample is then measured and a 'corrected intensity', I', is produced from the intensities of the two portions, thus

$$I' = (n - 1) \frac{I_2 I_1}{I_2 - I_1} \tag{3.27}$$

where I_1 and I_2 are the intensities for the concentrations C_1 and C_2, and $C_2 = nC_1$.

The concentration of an unknown element can thus be obtained by reference to the corresponding value for one standard only, which may be either a well analysed natural sample or an artificial mixture, and precisions of 0.5 to 1.0% are possible [59]. Tertian and Genisca [60] further point out that, provided the intensity ratios of the two portions are close to the optimum value of 2, matrix effects both of absorption and enhancement are adequately corrected. Moreover, the obvious disadvantage in terms of extra work involved in preparing two portions for each sample is offset, at least in the fusion methods, by the automatic correction for ignition loss [59]. It is surprising that this method is not in more general use.

3.5 TRACE ELEMENT ANALYSIS

X-ray fluorescence is particularly useful in the determination of elements present in low concentrations or at the trace level, since the non-destructive nature of the method makes it possible to carry out the analysis without any major preparation of the sample, other than grinding. This prevents the possible loss or contamination introduced by such processes as dissolution and subsequent precipitation.

In much geological and mineralogical work the trace elements of interest are those with high atomic numbers, for which mineralogical effects are greatly reduced owing to the large critical depth compared to the average grain size of a well prepared sample. Undiluted powders may therefore be analysed directly, as even simple dilution is undesirable, not only because it will reduce the intensity from the trace elements and so lower the sensitivity, but also because of the possibility of contamination. It is, for example, quite possible to determine 60ppm of chromium in a bead prepared by the Norrish method (section 3.4.4) but since the flux-to-sample ratio is 5:1, extreme purity of the flux with respect to chromium and other interfering elements is vital.

The major problems encountered in trace element analysis are thus found in contamination and line overlap. It cannot be stressed too strongly that, since the work is concerned with low count rates, a very slight contamination of the sample or overlap from low intensity peaks in the spectrum will cause large errors in the final analysis.

The sample preparation will inevitably tend to introduce some contamination, and the insidious danger at this stage is that the elements introduced may be a minor proportion of the material of the crushing tools. Thus the crushing equipment may add steel, tungsten, silica or alumina depending on the composition of the liners or edges involved, and while a small amount of iron added to a rock powder may not affect the accuracy of the determination of total iron, the large proportion of, for example, chromium found in hard steels may effectively double the concentration of that element in the sample. Typically a tungsten carbide disc mill will, in crushing a quartzite, add up to 2000ppm of tungsten plus an amount of cobalt proportional to that in the carbide (usually about 6%).

Sieves will also add contamination, especially the commonly used brass sieves which give copper and zinc from the mesh and tin and lead from the solder. Liquids used in mineral separation may also introduce unexpected elements.

Bennett discusses the problems of crushing and grinding at some length in Chapter 12.

Contamination may also be introduced by the spectrometer itself, an example of which is given by Leake and Hendry and co-workers [46] who found a 420% increase in the sulphur content of the ratio standard which was continuously in the spectrometer over a period of several months. This was eventually traced to sulphur in the oil backstreaming from the vacuum pump, and was cured by introducing a molecular sieve trap into the vacuum line. Finally, other operations in adjacent laboratories may cause errors, as was found in the author's laboratory where the apparent potassium content of the international peridotite standard PCC-1 increased from 9ppm to 149ppm over a period of 12 hours in the spectrometer, due to feldspar dust in the atmosphere entering the instrument [19].

Problems associated with the counting procedure are more difficult to enumerate than those associated with direct contamination. Certain line overlaps are well known, for example that of the vanadium $K\beta$ with chromium $K\alpha$ or cerium $L\alpha$ with barium $L\beta$, and allowance can be made for these relatively easily, but others may be less obvious. Thus the second order of the tungsten $L\beta_1$ line will interfere with the barium $L\beta$ line, and so may cause difficulties if a tungsten tube is used for excitation. Similarly, the overlap of the tungsten $L\alpha$ Compton peak with the Rayleigh peak of tantalum $L\alpha$ may not at first be expected, but using a tungsten target tube the Compton peak intensity may be significant. Within the detector, escape peak phenomena have been discussed in section 3.3.1 and mentioned in relation to crystal fluorescence and the spurious counts arising from these effects must also be guarded against. Once again, it is relatively easy to recognise first order escape peak interferences, what is more difficult to remember is that second and higher order escape peaks caused by the major elements present may also pose problems. The escape peak arising from the second order of the lanthanum $L\alpha$ will overlap the sulphur $K\alpha$ line, making the analysis of trace sulphur in a fusion bead difficult without using a germanium crystal for dispersion. A more obscure example is the fourth order Cr $K\beta$ escape peak overlap with the A1 $K\alpha$ line, important when using a chromium target to excite the lighter elements. Here again the two peaks cannot be fully resolved using pulse height selection. Similar overlaps with the positions selected for background measurement must also be guarded against, both by a careful study of the wavelength tables in conjunction with the crystal and the detector used, and by scanning a series of samples to check for possible deviations.

A final problem arises from the occurrence of contaminant elements in the X-ray tube, especially iron, manganese, chromium, nickel, copper and zinc,. depending on the nature of the tube. The presence of these contaminants can, fortunately, be detected by scanning either a sample of distilled water, or a pure organic material, or, ideally, polished discs of spectrographically pure silica (Spectrosil)*. It must be remembered, however, that the amounts of contaminants will increase with the ageing of the X-ray tube, and so a continuous check must be kept on their levels. The necessary corrections for these contaminants are made by finding the ratio of the elemental intensity to that of a Compton or Rayleigh line from the tube target element, or to the scattered background at a suitable angle, this ratio being measured with the radiation scattered from a high purity sample such as the silica disc. Provided that no major element absorption edge lies

*Available from Thermal Syndicate Ltd.

between the two measured wavelengths, the ratio of the scattered intensities will be substantially constant for varying matrices. Thus the nett intensity of the scattered tube target line may be measured in an unknown sample, and the calculated concentration of interfering element subtracted from the measured concentration to give a true value [61]. Some spectrometers also employ a programmable filter in the primary beam to remove the interference from both characteristic and contaminant elements in the X-ray tube. The relatively large amounts of copper found in most tubes can be effectively removed by the use of an aluminium filter, while the tungsten L_l line can be used to correct for the small amounts of nickel present in a tungsten target X-ray tube, using the LiF_{220} crystal.

By definition, the concentration and intensity of a trace element are both very low, and so the level of the background measurement becomes of prime importance, as can be seen from the equation for the lower limit of detection of an element,

$$LLD = \frac{3}{\left[\frac{R_p - R_b}{c}\right]} \sqrt{\frac{R_b}{t_b}} \tag{3.28}$$

where c is the element concentration, R_p and R_b are the count rates for the peak and the background respectively, and t_b is the counting time on the background, typically 100 seconds. This equation assumes that a peak is detectable when its count rate exceeds two standard deviations of the background rate.

Great care must be taken in choosing the excitation and instrumental conditions, such as the crystal used for dispersion, the detector, the size of the collimator, and the particular emission line for the element. For heavy elements it is generally better to measure an L-line rather than a K-line, although the intensity of the latter may give a better peak plus background intensity. The lower background in the region of the L spectrum usually results in better detection limits, from equation (3.28). Jenkins and de Vries [29] have considered the statistics of this problem and have concluded that the 'figure of merit' given by

$$(\sqrt{R_p} - \sqrt{R_b})$$

is a better measure of optimum instrumental conditions than is the more commonly used ratio of peak to background, and that for trace elements the optimum is found by maximising the expression $(R_p - R_b)/\sqrt{R_b}$.

It is often worthwhile investigating the combination of a low dispersion crystal (LiF_{200}) with fine collimation, and a high dispersion crystal (LiF_{220}) with coarse collimation. Thus where the L spectrum is measured, a slight reduction in resolution may combine the $L\beta_1$ and $L\beta_4$ to produce a higher intensity than that available from the $L\alpha$. This of course assumes that high resolution is not required to separate interfering lines.

Minimum detection limits are, of course, dependent on the counting time as well as on the element in question, but taking 100 seconds as a reasonable maximum counting time, limits of between 1 and 6ppm are generally possible for all elements heavier than potassium, with the lowest detection limits occurring around atomic number 30 (zinc). The detection limits for elements lower than potassium rise sharply and may reach levels of 100ppm and over for magnesium and sodium [31].

3.6 ACCURACY

The precision with which an elemental concentration may be determined in a single sample has already been discussed with respect to the random and non-random effects inherent in the use of an X-ray spectrometer and its counting chain, and there now remains to be considered the absolute accuracy of the technique. Clearly, the accuracy is partially controlled by the precision with which the measurements can be made. High accuracy implies high precision, although high precision does not necessarily imply high accuracy. The problem of determining accuracy levels is also discussed in Chapter 12.

The overall precision of a method is obviously affected by factors other than those encountered in the spectrometer, which include sample homogeneity and variations in the standards and the instrument used, especially over a long time. Thus Le Maitre and Haukka [62] report changes of between 12% and 40% in the count rates for silicon, aluminium, magnesium, and sodium in lithium tetraborate beads over a period of 200 hours irradiation (see figure 3.14). Table 3.2 illustrates the range of concentrations measured from four samples of the same rock powder used as monitors over a period of four years. From these it is apparent that significant differences in elemental concentrations can be detected between the samples, and that these are greater than the possible errors due to the measuring procedure. The system of running frequent checks using several, rather than one, samples of the same material cannot be too strongly recommended, especially where calibration curves or reference standards are to be used over a long period of time.

The accuracy of any result is, by its very nature, the hardest factor to assess in any analytical method, particularly where trace elements or elements which are difficult to determine accurately by 'conventional' methods are concerned. As Bennett has pointed out in Chapter 12, wet chemical methods tend to be the standard against which instrumental methods such as X-ray fluorescence are judged, but while these are eminently suitable for a large number of major elements, they may be quite inapplicable to trace elements. Fortunately, the spread of instrumental methods has encouraged analysts to produce natural and synthetic standard samples which have been circulated to laboratories throughout the world and for which 'true' compositions are now available, and the type of spread arising from the different techniques applied in different laboratories may be assessed.

TABLE 3.2

	Sample Number															
	700				701				702				703			
	Mean	S.D.	C.V.	n	Mean	S.D.	C.V.	n	Mean	S.D.	C.V.	n	Mean	S.D.	C.V.	n
SiO_2	58.208	.096	.165	5	58.417	.175	.300	14	58.285	.199	.342	14	58.251	.187	.321	9
TiO_2	.935	.0016	.169	5	.936	.006	.684	14	.935	.006	.641	14	.935	.006	.639	9
Al_2O_3	14.666	.065	.443	5	14.718	.081	.549	14	14.707	.148	1.005	14	14.788	.046	.308	9
Fe_2O_3	6.859	.052	.762	5	6.827	.044	.641	11	6.823	.047	.694	12	6.767	.054	.790	7
MnO	.098	.0007	.722	5	.096	.0014	1.47	11	.096	.0012	1.27	13	.096	.0011	1.168	8
MgO	3.901	.057	1.471	5	3.818	.066	1.72	15	3.820	.077	2.01	14	3.837	.082	2.13	11
CaO	6.077	.017	.274	5	6.079	.032	.526	14	6.062	.031	.506	14	6.071	.034	.564	9
Na_2O	3.704	.087	2.361	6	3.636	.099	2.71	16	3.639	.090	2.47	17	3.637	.084	2.31	11
K_2O	2.513	.014	.562	5	2.447	.024	.983	14	2.454	.029	1.20	14	2.421	.038	1.59	9
P_2O_5					.389	.026	6.57	6	.392	.026	6.57	6	.388	.027	6.95	6
S					557	54.2	9.74	8	578	108.8	18.83	8	560	42.6	7.61	8
Cl	654	38.5	5.88	6	654	42.9	6.56	7	655	43.8	6.68	7	675			1
Cr	46.2	1.48	3.21	5	46.4	1.54	3.32	11	46.5	1.33	2.86	13	47.3	1.08	2.28	8
Ni	34.6	1.14	3.20	5	36.2	2.29	6.34	9	36.4	1.98	5.44	13	37.3	1.37	3.67	7
Rb	84.8	2.85	3.37	9	82.5	3.29	3.99	18	82.6	2.91	3.53	18	80.3	1.87	2.33	9
Sr	601	4.56	.76	9	596	7.16	1.20	22	595	6.57	1.10	18	592	5.59	.94	9
Y	29.9	3.37	11.3	8	29.4	1.79	6.07	17	29.6	1.29	4.37	17	29.1	.75	2.57	9
Zr	478	3.91	.82	9	486	7.41	1.53	19	474	7.53	1.59	19	471	9.41	2.00	10
Nb	21.7	1.29	5.94	9	21.2	1.78	8.40	18	21.5	3.44	16.04	18	21.7	2.23	10.29	9
Ba	1136	17.5	1.55	4	1152	47.9	4.16	6	1164	25.2	2.17	6	1181			2
La					109	3.82	3.51	4	109	2.12	1.95	6	108	2.51	2.32	6
Ce	202	5.26	2.60	5	204	7.89	3.87	11	203	7.28	3.58	13	201	10.01	4.98	8
Pb	22.5	2.24	4.98	9	22.8	5.72	25.1	22	20.6	3.79	18.5	18	23.2	4.37	18.9	9
Th	7.8	1.73	22.1	8	8.5	2.56	30.0	18	8.7	2.66	30.4	17	10.0	2.96	29.5	9

Results on four samples taken from a finely ground, homogenised rock powder. These pellets were used to monitor instrumental precision over a period of four years. (Sample 700 was damaged after 18 months).
S.D. = standard deviation, C.V. = coefficient of variation, n = number of measurements.

Values in the upper part of the table, for the major elements, are given as percentages and in the lower, for the minor elements, as p.p.m.

A list of currently available silicate standards has been compiled by Flanagan [63] and samples of more commercial interest are available from such organisations as the U.S. Bureau of Standards or the British Chemical Standards.

It is again worth stressing that such internationally accepted standards should be used as the final check on the accuracy of determinations, rather than an artificial standard made up for use within separate laboratories. Such standards do represent a considerable investment in time and, considering the real cost of a single element determination, money. They should therefore only be used as a final determination of the accuracy of a *proven* method, particularly if the sample is destroyed either by solution or fusion. Routine checks of the instrument, and periodic re-calibration with, for example, new batches of flux should be made with sub-standards made in the individual laboratory.

Table 3.3 illustrates the type of result obtained in such an international comparison, from the results of major element determinations on two international rock standards, granite G-2 and diabase W-1, circulated by the U.S. Geological Survey. The concentrations obtained by the use of short range calibration curves on powder samples, fusion with a heavy absorber (essentially after Norrish), and the double concentration fusion method of Tertian all show very good agreement with the accepted values [64].

TABLE 3.3

	U.S.G.S. G–2				U.S.G.S. W–1				S.S.C. SU–1	
	H.A.	D.D.	P.	R*	H.A.	D.D.	P.	R*	F.	R+
SiO_2	69.29	69.43	70.11	69.11	52.68	52.47	52.50	52.64	34.2	34.5
TiO_2	0.52	0.50	0.49	0.50	1.07	1.07	0.99	1.07		
Al_2O_3	15.42	15.36	15.73	15.40	14.84	14.89	14.90	15.00	9.30	9.3*
Fe_2O_3	2.70	2.77	2.69	2.65	11.08	11.15	11.21	11.09	32.9	32.7
MnO	0.03	0.03	0.03	0.03	0.17	0.17	0.16	0.17		
MgO	0.81	0.82	0.77	0.76	6.60	6.59	6.65	6.62		
CaO	1.96	1.96	1.99	1.94	10.93	11.01	10.56	10.96		
K_2O	4.46	4.48	4.49	4.51	0.65	0.64	0.67	0.64		
Na_2O	4.16	4.13	4.08	4.07	2.12	2.21	2.09	2.15		
P_2O_5	0.13	0.11	0.13	0.14	0.14	0.13	0.15	0.14		
S									12.00	12.04
Ni									1.4	1.3*
Cu									0.8	0.8*

H.A. Fusion with lithium tetraborate/lithium carbonate/lanthanum oxide.
D.D. Double dilution by fusing with lithium tetraborate/lithium carbonate.
P. Compressed powder disc, close range calibration.
R Recommended value, *Flanagan 1973, +Webber 1965.
F. Fusion with sodium tetraborate/lanthanum oxide.

TABLE 3.4

Element	G–1			G–2			GSP–1			AGV–1			BCR–1			W–1		
	U	C	R	U	C	R	U	C	R	U	C	R	U	C	R	U	C	R
Y	13	13	13	9	10	12	23	28	30	16	20	21	30	35	37	19	24	25
Sr	278	252	250	501	485	479	222	233	233	585	660	657	333	331	330	200	199	190
Rb	235	218	220	182	172	168	246	258	254	58	70	67	41	45	47	20	21	21
Th	52	51	50	26	25	24	92	101	104	4	5	6	9	7	6	3	2	2
Pb	62	50	48	43	32	31	64	54	51	35	35	35	15	17	18	9	9	8
Zr	268	207	210	402	320	300	589	523	500	251	223	225	210	194	190	120	108	105

(All sample columns are under the spanning header "Sample".)

Trace element analysis of U.S.G.S. standard rock samples: G–1 and G–2, granites; GSP–1, granodiorite; AGV–1, andesite; BCR–1, basalt; W–1, diabase.

Note: U = uncorrected results based on a calibration from a single spiked sample
C = corrected results using the intensity of the *WL*-line to give an estimate of the absorption coefficient
R = recommended values after Flanagan [64], 1973.

Results for the Canadian Association of Applied Spectroscopy standard Sulphide Ore 1, using a sodium tetraborate flux with lanthanum oxide absorber [65], are also shown and again show good agreement.

The accuracy of trace element analyses is more difficult to estimate and there is still considerable dubiety about the values of many trace elements in the standard samples G-1 and W-1, despite the numerous results published over the past fourteen years. Table 3.4 shows values for some of these elements determined after correcting the observed intensities using scattered tube lines.

Thus, high accuracy, comparable to that obtained by standard chemical methods, is obtainable for major elements provided that the precautions outlined in this chapter are carefully followed. The accuracies of trace element determinations are also, in general, high and again values obtained by X-ray fluorescence agree extremely well with concentrations determined by the more lengthy techniques of neutron activation analysis and isotope dilution.

3.7 SUMMARY

X-ray fluorescence analysis provides a rapid, convenient, and accurate method of determining the elements present in a sample both qualitatively and quantitatively. It is particularly useful for the determination of trace elements as the competing techniques are more lengthy and generally more expensive. Care must be taken in applying the technique quantitatively, but provided that suitable precautions are taken in preparing the sample for analysis and in accumulating a sufficient number of counts for each element, high accuracy can be obtained. Where large numbers of samples are to be analysed, automated X-ray fluorescence analysis is probably the cheapest method available, in view of the speed and small number of personnel involved.

REFERENCES

1. DEPARTMENT OF EMPLOYMENT AND PRODUCTIVITY, 'Code of Practice for the protection of persons exposed to ionising radiations in research and teaching'; H.M.S.O., London, 1968.
2. 'Report of International Union of Crystallography', *Acta Cryst.*, **16**, 324 (1963).
3. R. JENKINS and D. J. HAAS, *X-ray Spectrom.*, **2**, 135 (1973).
4. H. G. J. MOSELEY, *Phil. Mag.*, **27**, 703 (1914).
5. T. B. JOHANSSON, R. AKSELSSON, and S. A. E. JOHANSSON, *Nucl. Instrum. Meth.*, **84**, 141 (1970).
6. G. von HEVESY, 'Chemical Analysis by X-rays and its Applications'; McGraw-Hill, New York, 1932.

7. B. J. PRICE, 'Metals and Materials', 140 (1973).
8. S. H. U. BOWIE, *Mining Mag.*, **118**, 1 (1968).
9. J. R. RHODES, in *A.S.T.M. Special Tech. Publ.* 485, 243 (1970).
10. H. A. KRAMERS, *Phil. Mag.*, **46**, 836 (1923).
11. K. LONSDALE (ed), 'International Tables for X-ray Crystallography', Vol. III; International Union of Crystallography, Kynoch Press, Birmingham, 1962.
12. A.S.T.M., 'X-ray Emission and Absorption Wavelengths and Two-Theta Tables', A.S.T.M. Data Series DS 37A (1970).
13. J. A. VICTOREEN, *J. Appl. Phys.*, **20**, 1141 (1949).
14. K. F. J. HEINRICH, 'The Electron Microprobe'; John Wiley and Sons, New York and London, 1966.
15. R. THEISEN and D. VOLLARTH, 'Tables of X-ray Mass Attenuation Coefficients'; Verlag Stahleisen M.B.H., Düsseldorf, 1967.
16. R. JENKINS and J. L. de VRIES, 'Practical X-ray Spectrometry'; Macmillan, London and New York, 1967.
17. R. JENKINS, *X-ray Spectrom.*, **1**, 23 (1972).
18. B. POST and R. JENKINS, *X-ray Spectrom.*, **1**, 161 (1972).
19. G. L. HENDRY, *X-ray Spectrom.* (In Press).
20. R. JENKINS and P. W. HURLEY, *Can. Spectros.*, **13**, 35 (1968).
21. L. V. SUTFIN and R. E. OGILVIE, in *A.S.T.M. Special Tech. Publ.* 485, 197 (1970).
22. B. L. HENKE, 'Advances in X-ray Analysis'; **5**, 288 (1961).
23. W. L. BAUN and E. W. WHITE, *Anal. Chem.*, **41**, 831 (1969).
24. C. F. GAMAGE, *X-ray Spectrom.*, **1**, 99 (1972).
25. H. WOLLENBERG, H. KUNZENDORF, and J. ROSE-HANSEN, *Econ. Geol.*, **66**, 1048 (1971).
26. A.S.T.M., 'Energy Dispersion X-ray Analysis: X-ray and Electron Probe Analysis', *A.S.T.M. Special Tech. Publ.* 485, (1970).
27. A.S.T.M., 'X-ray Emission Wavelengths and KeV Tables for Non-diffractive Analysis', *A.S.T.M. Report DS46* (1971).
28. R. G. MUSKET and W. BAUER, *J. Appl. Phys.*, **43**, 4786 (1972).
29. R. JENKINS and J. L. de VRIES, 'Practical X-ray Spectrometry'; Macmillan, London and New York, 1967, Chapter 5.
30. R. O. MÜLLER, 'Spectrochemical Analysis by X-ray Fluorescence'; Adam Hilger, London, 1972.
31. G. L. HENDRY, *Proc. 5th. Conf. on X-ray Anal. Methods, Swansea*, 72 (1966).
32. B. GUNN, 'Advances in X-ray Analysis', **4**, 382 (1960).
33. W. K. de JONGH, *Proc. 6th Conf. on X-ray Anal. Methods, Southampton*, 1 (1968).
34. C. HUTCHISON, 'Laboratory Handbook of Petrographic Techniques', Wiley Interscience, 1974.
35. F. CLAISSE, *Quebec Dept. Mines, Prelim. Rept.*, 32 (1956).
36. G. K. CZAMANSKI, J. HOWER, and R. C. MILLARD, *Geochim. et Cosmochim. Acta*, **30**, 745 (1966).
37. K. NORRISH and J. T. HUTTON, *Geochim. et Cosmochim. Acta*, **33**, 431 (1969).
38. P. K. HARVEY, D. M. TAYLOR, R. D. HENDRY, and F. BANCROFT, *X-ray Spectrom.*, **2**, 33 (1973).

39. P. R. HOOPER and L. ATKINS, *Miner Mag.*, **37**, 409 (1969).
40. R. TERTIAN, *Spectrochim. Acta,* **24B**, 447 (1969).
41. Z. H. KALMAN and L. HELLER, *Anal. Chem.*, **34**, 946 (1962).
42. S. D. RASBERRY and F. J. HEINRICH, *Proc. Colloq. Spectroscopicum Internationale XVI, Heidelberg,* **1**, 337 (1971).
43. R. TERTIAN, *X-ray Spectrom.*, **2**, 95 (1973).
44. P. G. LUMB, *Proc. 7th Conf. on X-ray Anal. Methods, Durham,* 27 (1970).
45. T. PADFIELD and A. GRAY, *Bull. FS 35,* Anal. Equipment Dept., Philips, Eindhoven, 1970.
46. B. E. LEAKE, G. L. HENDRY, A. KEMP, A. G. PLANT, P. K. HARVEY, J. R. WILSON, J. S. COATES, J. W. AUCOTT, T. LÜNEL, and R. J. HOWARTH, *Chem. Geol.*, **5**, 7 (1969).
47. K. NORRISH and B. W. CHAPPEL, 'Physical Methods in Determinative Mineralogy', J. Zussman (ed.); Academic Press, London and New York, 1967, Chapter 4.
48. T. J. CULLEN, *Anal. Chem.*, **34**, 812 (1962).
49. R. C. REYNOLDS, *Amer. Min.*, **48**, 1133 (1963).
50. J. SHERMAN, *Spectrochim. Acta,* **7**, 283 (1955).
51. H. J. BEATTIE and R. M. BRISSEY, *Anal. Chem.*, **26**, 980 (1954).
52. W. MARTI, *Spectrochim. Acta,* **18**, 1499 (1962).
53. G. R. LACHANCE and R. J. TRAILLE, *Can. Spectros.*, **11**, 43 (1966).
54. H. J. LUCAS-TOOTH and B. J. PRICE, *Metallurgia,* **64**, 149 (1961).
55. R. JENKINS and A. CAMPBELL-WHITELAW, *Can. Spectros.*, **15**, 32 (1970).
56. W. JOHNSON, *Proc. 4th. Conf. on X-ray Anal. Methods, Sheffield,* 73 (1964).
57. W. K. de JONGH, *X-ray Spectrom.*, **2**, 151 (1973).
58. J. SHERMAN, 'Advances in X-ray Analysis', **1**, 231 (1958).
59. R. TERTIAN, *Spectrochim. Acta,* **27B**, 159 (1972).
60. R. TERTIAN and R. GENISCA, *X-ray Spectrom.*, **1**, 83 (1972).
61. W. K. STEELE, *Chem. Geol.*, **11**, 149 (1973).
62. R. W. LE MAITRE and M. T. HAUKKA, *Geochim. et Cosmochim. Acta,* **37**, 708 (1973).
63. F. J. FLANAGAN, *Geochim. et Cosmochim. Acta,* **34**, 121 (1970).
64. F. J. FLANAGAN, *Geochim. et Cosmochim. Acta,* **37**, 1201 (1973).
65. N. G. WEST, G. L. HENDRY and N. T. BAILEY, *X-ray Spectrom.*, **3**, 78 (1974).
66. R. VIE le SAGE and B. GRUBIS, *X-ray Spectrom.*, **2**, 189 (1973).
67. G. R. WEBBER, *Geochim. et Cosmochim. Acta,* **29**, 229 (1965).

CHAPTER 4

Radiotracers in Minerals Engineering

V. I. Lakshmanan and G. J. Lawson

Department of Minerals Engineering
University of Birmingham
Birmingham B15 2TT
England

4.1 INTRODUCTION

Radioisotopes and general radiochemical techniques find a multitude of uses in the mineral processing and metallurgical industries, for two principal reasons. Both are based on the simplicity of detecting the radiation emitted by radioisotopes and of using the radiation not only to identify the radioisotope involved but also to measure its concentration in the system being monitored. Thus, on the one hand, a radioactive isotope may be introduced into a reacting system and, because the isotope will behave chemically in a manner identical to all the other isotopes of the same element present, it then becomes possible to follow the reaction without chemically interfering with the system in any way. This ability may be crucial, for example in studying, surface reactions such as froth flotation, where the concentrations of the reagents involved are very small; in studies involving interchange between the solid and the gas phase, as in many processes in pyrometallurgy; or in studies of solvent extraction or ion exchange methods. On the other hand, the fact that many non-radioactive isotopes can be converted into a radioactive form, *in situ*, can provide a sensitive, non-destructive method of detecting very small concentrations of these elements.

Radioisotopes have been used in the minerals industries since the early nineteen-fifties, and the intention of this chapter is to introduce the basic principles on which the subject is based and to discuss some applications that have already been made of the technique, to illustrate how the method may be used and perhaps suggest new uses for it.

4.1.1 Radioactive Decay and Growth

An element is characterised by the number of protons contained in the atomic nucleus, but associated with the protons there can be a variable number of neutrons. Atoms with the same number of protons but different numbers of neutrons constitute the 'isotopes' of an element, and radioisotopes are metastable isotopes which can undergo spontaneous change to a stable form. The phenomenon of radioactivity arises from the emission of radiation which occurs as the isotope changes from a high energy to a low energy form. This change may result in the formation of an isotope of a different element, when the process involves emission of an α- or a β-particle, or it may result in the formation of a stable form of the same isotope, when it is accompanied by the emission of a γ-ray photon. One of the most important characteristics of radioactivity is the regularity of the disintegration process, based as it is on a purely random decay probability for each atom in a sample. This process of radioactive decay is independent of the nature of the neighbouring nuclei and is unaffected by such factors as pressure and temperature.

The instantaneous rate of decay of a radioisotope depends on the number of atoms of the radionuclide present at the time, and so follows first order rate law kinetics

$$-dN/dt \propto N \tag{4.1}$$

where dN is the change in the number of atoms in time dt, and N is the number of atoms present. By introducing a proportionality constant, λ, equation (4.1) becomes

$$dN/dt = -\lambda N \tag{4.2}$$

integration of which gives

$$N = N_o \exp(-\lambda t) \tag{4.3}$$

where N_0 is the number of atoms present at time $t = 0$, and N is the number remaining after time t. The constant λ is known as the decay constant. It has dimensions of reciprocal time and is usually expressed in reciprocal seconds.

The characteristic decay rate of a nuclear species is normally quoted as the 'half life', $t_{1/2}$, which is the time taken for one half of the initial number of nuclei to decay. Hence,

$$N_0/2 = N_0 \exp(-\lambda t_{1/2}) \tag{4.4}$$

or

$$t_{1/2} = (\ln 2)/\lambda = 0.693/\lambda \tag{4.5}$$

This relationship permits λ to be measured quite simply, by measuring the half life of the radioisotope.

In practice, it is difficult to measure the number of atoms, N, in a sample, and even the rate of change, dN/dt, is not normally measured. The procedure usually adopted involves measuring a quantity which is proportional to N, using some form of electronic, photographic or other detection system. This quantity is the 'activity', A, and it will be clear that

$$A = C\lambda N = C(-dN/dt)$$

where C, the detection coefficient, depends on the nature and the efficiency of the measuring system and on the geometrical arrangement of sample and detector. The various decay laws can be rewritten in terms of the activity, and so equation (4.3) becomes

$$A = A_0 \exp(-\lambda t) \tag{4.6}$$

and the half life becomes the time for the activity to fall to one half of its initial value.

The independent nature of radioactive decay processes means that, when several unrelated radioisotopes are present together in the same sample, the total observed activity, A, is given by the sum of the separate activities, i.e. by

$$A = A_1 + A_2 + A_3 + \text{etc.}$$

whence, by substituting for the separate activities,

$$A = A_1^0 \exp(-\lambda_1 t) + A_2^0 \exp(-\lambda_2 t) + A_3^0 \exp(-\lambda_3 t) + \text{etc.} \tag{4.7}$$

Growth of Radioactivity

In many cases the daughter atom produced by radioactive decay of a parent radioisotope may not be stable and may in turn decay to a stable nuclear form. Such a condition may be represented as

$$A \xrightarrow{\lambda_A} B \xrightarrow{\lambda_B} C$$

where the radioactive daughter, B, is formed at a constant rate from nucleus A and decays to C, again at a constant rate characterised by the decay constant λ_B. Then,

$$dN_B/dt = \lambda_A N_A - \lambda_B N_B$$

or

$$dN_B/dt + \lambda_B N_B = \lambda_A N_A^0 \exp(-\lambda_A t) \tag{4.8}$$

and integration of equation (4.8) gives

$$N_B = \frac{\lambda_A N_A^0}{(\lambda_B - \lambda_A)} [\exp(-\lambda_A t) - \exp(-\lambda_B t)] + N_B^0 \exp(-\lambda_B t) \tag{4.9}$$

where N_B^0 is the number of B nuclei present at time $t = 0$. It may be noted in passing that a radioisotope can be formed by means other than by the decay of a parent nucleus, as will be discussed in section 4.4.1.

4.1.2 Radiation Emission from Radioactive Material

As mentioned earlier, radioactive decay processes may involve changes in either, or both, the number of neutrons and/or the number of protons in a nucleus. This change can occur by conversion of a neutron into a proton, with the emission of a β-particle, or electron, or the reverse process, when a positron is emitted, thus

$$n \rightarrow p^+ + e^- + \bar{\nu}$$

$$p^+ \rightarrow n + e^+ + \nu$$

The neutrino (ν) or anti-neutrino ($\overline{\nu}$) emitted is not easily detected and will not further concern us. An example of such a process is

$$Pr_{59}^{144} \rightarrow Nd_{60}^{144} + e^-$$

and the neodymium nucleus formed is in its ground state and so is stable. Note that there is no change in the mass number of the isotope, although the atomic number has changed by +1. A variant on this β-decay process involves the capture of an extra-nuclear electron, usually from the *K*-shell, with an attendant change of −1 in the atomic number. Subsequent electronic transitions from higher energy shells to fill the resulting vacancy in the *K*-shell result in the generation of characteristic X-radiations (cf. Chapter 3).

Radioactive decay may also occur with the emission of α-particles, whose identity with helium nuclei is well established. In such a case, the atomic number changes by −2 and the mass number by −4, and the particles are emitted from the nucleus with a wide range of energies, usually between 4 and 9MeV. They lose their energy rapidly, by molecular collisions with the medium through which they pass, and so are of limited diagnostic use, although they are useful in some applications requiring simple detection of radiation only (cf. section 4.6).

In addition to the transformations that change the numbers of protons and neutrons in a nucleus, a further class exists which involves no change in atomic weight or number. Such transformations produce γ-radiation, which is high energy electromagnetic radiation with no charge and zero rest mass, as discussed in Chapter 1. Many cases are known in which the daughter atom produced by, for example, β-decay is itself unstable and exists in an excited state. The excess energy can be given out as a γ-ray photon. Thus, Na^{24} decays by β-decay to an excited form of Mg^{24}, which subsequently decays to its ground state by emitting a γ-ray photon,

$$Na^{24} \rightarrow e^- + Mg^{*24} \rightarrow Mg^{24} + \gamma$$

where Mg^{*24} indicates the excited state of Mg^{24}.

An important point about β-particles and γ-rays is that both correspond to energies which are characteristic of the isotope from which they have been emitted. Hence, identification of the energy of β- or γ-radioactivity permits identification of the radioisotope, or isotopes, present in the sample.

Thus, in the investigation of radioactive decay, there are three important characteristics through which the process may be identified:

(i) the emitted particle or gamma-ray
(ii) the half life of the parent isotope
(iii) the energy involved in the transition

4.1.3 Determination of Half Lives

Half lives in the range from several seconds to several years can usually be determined experimentally by measuring the radioactivity with an appropriate instrument at a number of suitable time intervals. If log A is plotted against t, as in figure 4.1, the result is a straight line of slope λ, provided, of course, that the sample is free of any other radioactive isotopes. The half life can be determined simply by inspecting the plot, and interpolating or extrapolating as necessary.

Frequently a measured radioactive decay does not exhibit a straight line relationship, and this is due to the presence of more than one radioactive component. Under these conditions, however, provided that only a few isotopes with markedly different half lives are present it may be possible to extract the separate half lives from the decay curves alone. Figure 4.2 shows the data for a sample containing two radioisotopes, one having a long, and the other a short, half life. With increasing time the short lived radioisotope will disappear and the final portion of the curve will depend only on the long lived component and becomes a straight line from which the half life of this latter component can be derived. Furthermore, if this line is extrapolated back to zero time, the portion of the counting curve due to the long lived isotope can be subtracted from the total curve, and the difference curve so obtained represents the decay of the short lived isotope.

The half life of a long lived parent radioisotope may be determined in either of two ways:

(i) This method is limited to radioisotopes which have had time to reach equilibrium with short lived decay products. Under these conditions, the fact that the daughter nucleus (B) is in equilibrium with the parent (A) means that the rate of disintegration of both nuclei is the same, i.e.

$$\lambda_A N_A = \lambda_B N_B \tag{4.10}$$

If the ratio of the numbers of atoms of each element can be determined, e.g.

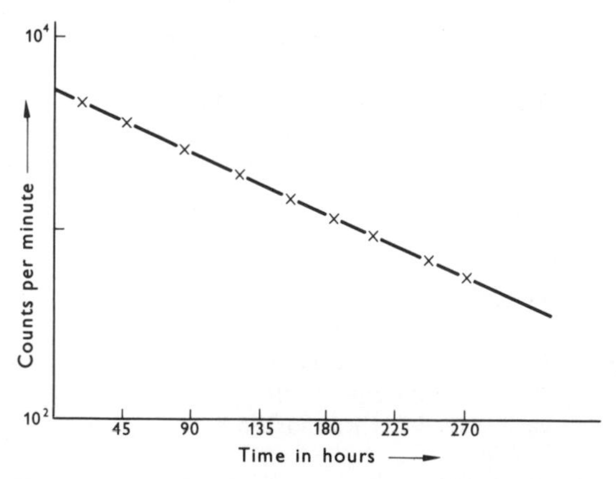

Figure 4.1 Decay curve of a simple radioactive nuclide with a half life of 90 hours.

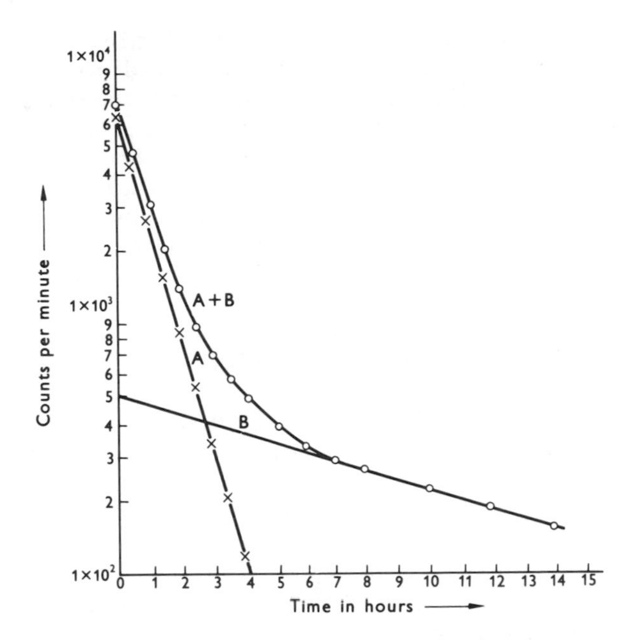

Figure 4.2 Decay curve obtained from a sample containing two radioactive nuclides, one with a short half life of about 1 hour, represented by the line A, and the other of longer half life, about 8 hours, represented by line B.

by chemical analysis, and the half life of one of the radioisotopes is known, then that of the other may be calculated.

(ii) Since, for a radioisotope with a very long life, the activity $A = C\lambda N$ does not change significantly during a measurement, the value of λ can be determined if N is known, from chemical analysis, and the rate of disintegration can be absolutely determined through a knowledge of C. The value of $t_{1/2}$ is obtained from λ by substituting in equation (4.5).

The half lives of very short lived radioisotopes can be determined by the variable delay coincidence technique, described by Friedlander, Kennedy and Miller [1].

4.1.4 Units of Radioactivity

Although units such as the 'rutherford' (1×10^6 counts/sec) have been proposed in the past, the 'curie' has emerged as the generally accepted practical unit. The curie is defined as the quantity of a radionuclide that produces 3.7×10^{10} disintegrations per second. The sub-units millicurie and microcurie are most often used in laboratory work.

The specific activity of a sample of a radioactive substance is the ratio of its activity to its mass (or volume of solution, or number of moles, or any other convenient unit proportional to mass) and gives a relative measurement of the radioactive strength of the sample. In practice, the specific activity is expressed in counts per second per unit mass for a particular measuring procedure. This is a fraction of the absolute activity of the sample, expressed in disintegrations per second per unit mass, the value of this fraction depending on the efficiency of the counting arrangement. Hence, studies of radioactivities are based on comparative methods with strict adherence to the same measurement procedure when following the fate of a tracer in any experiment. Delicate measurements of absolute activities are seldom, if ever, necessary in analytical tracer applications [2].

4.2 DETECTION AND MEASUREMENT OF RADIOACTIVITY

The characteristics of the detectors used in studies of radioactivity have been discussed in Chapter 1, section 1.3.3, together with a consideration of

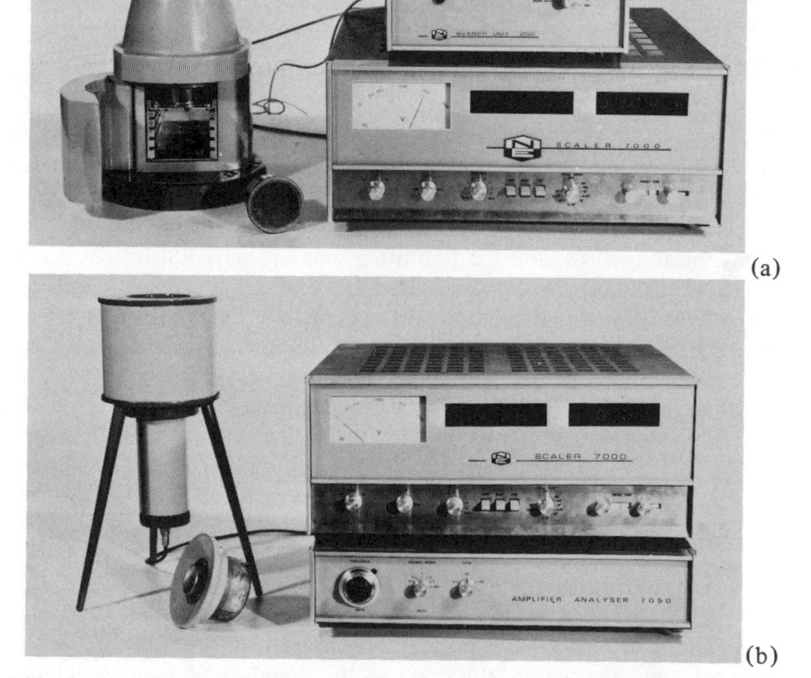

Figure 4.3 Detection equipment for radioactive materials, (a) Lead castle and counting equipment for beta-rays, with a typical Geiger-Muller tube shown demounted. In use, the counter sits under the truncated cone and points downwards into the 'oven' of the lead castle. (b) Lead well, scintillation detector and counting equipment for gamma-rays. Note the pulse height discriminator, the lower unit, labelled 'Amplifier Analyser 7050'.

the modes of interaction of β-particles and γ-rays with their active components. All the types of detector listed there are used, and details of their applications will be given in subsequent sections.

Beta-emitters are handled in a lead-castle counting unit such as that shown in figure 4.3a. The detector is mounted in the upper part of the unit and the sample, lying on a flat metal plate or planchet, is mounted in the lower part so that a maximum area of sample is presented to the detector. The distance between the sample and the detector can be varied, by inserting the planchet into slots at different levels in the counting chamber, and aluminium foils of various thicknesses can also be inserted in the slots, either singly or in groups to give a wide range of total thicknesses of absorber, for energy measurement (section 4.2.2). The whole is surrounded with lead shielding to minimise background levels due to stray radiation either from other radionuclides present or from the cosmic radiation which is always present.

A typical counting unit for gamma-emitters, based on a scintillation detector, is shown in figure 4.3b. The larger cylinder contains an annular lead ring, again to act as a radiation shield, into which are inserted the scintillation crystal and photo-multiplier tube, contained in the smaller cylinder. The crystal faces a well into which the sample is placed, either on a planchet, usually made of aluminium, for solids or in a glass vial for liquids, and this can be covered with a lead plug to further minimise stray radiation, if desired.

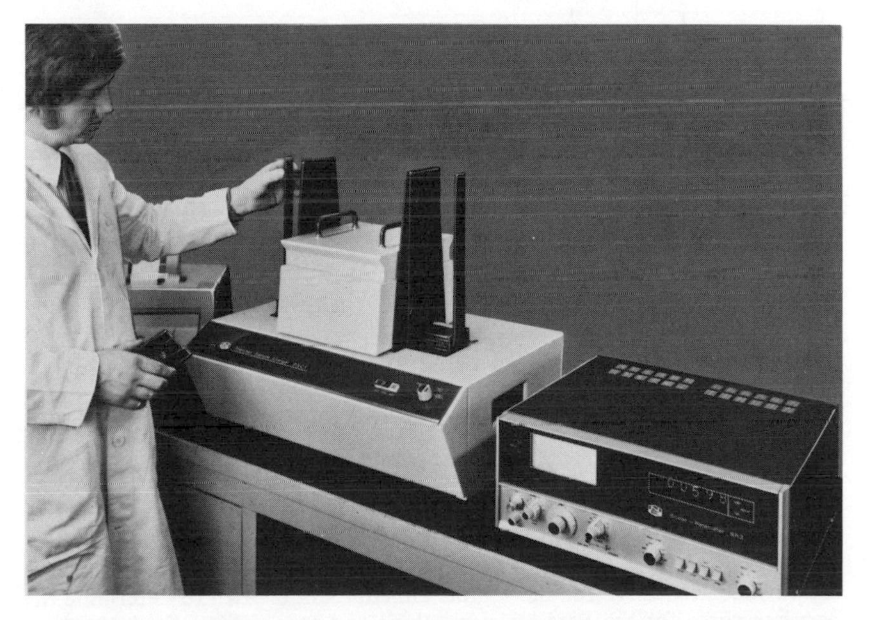

Figure 4.4 Automatic sample changer capable of handling up to 40 samples in sequence. (Courtesy of Nuclear Enterprises Ltd, Edinburgh.)

Both units can handle either solid or liquid samples, but each can handle only one sample at a time. Larger units are now commercially available which can handle large numbers of samples automatically, giving each a predetermined counting regime, and a typical example is shown in figure 4.4.

4.2.1 Characterisation of Radiation

Determination of Beta-ray Energies

Accurate beta-ray energies can be determined using an electron spectrometer, but absorption measurements suffice for many purposes. Basically, the method entails finding the thickness of a metal, commonly aluminium, required to reduce the transmitted intensity of the beta-rays to predetermined fractions of the original intensity. Aluminium foils, of various thicknesses, are placed between the sample and the counter until a

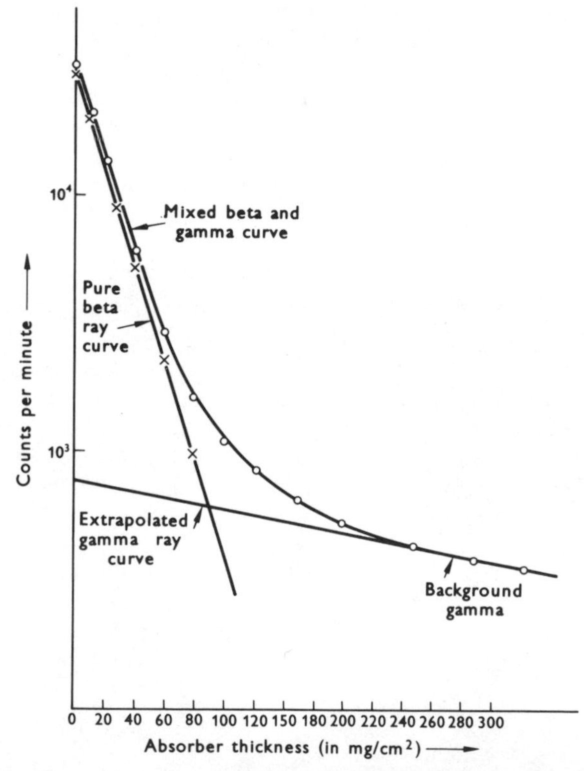

Figure 4.5 Absorption curve for sample emitting both beta and gamma radiation, plotted as log (total counts) against absorber thickness. Note how the extrapolated curve for the gamma emission can be used to determine the accurate curve for the beta emission alone and hence obtain the 'maximum range' and 'half thickness' values.

thickness is reached beyond which no appreciable change in the count rate can be detected. The constant value for the counting rate at this limit may be due to the natural background radiation or it may include a contribution from γ-ray emission from the sample. By plotting the observed data as the logarithm of the apparent activity against the 'density thickness' of the absorber, expressed as milligrams of absorber per square centimetre of absorption path, as shown in figure 4.5, it is possible to determine the minimum absorber thickness necessary to prevent all the beta-rays from reaching the detector, and the thickness required to reduce the number of particles reaching the detector to one half the original value. These thicknesses are the 'maximum range' and 'half thickness' respectively. The value for the maximum range, R, can then be substituted in equation (4.11) to find the maximum beta energy, E, in MeV:

$$R = 0.543E - 0.160 \tag{4.11}$$

This equation holds for beta energies in the range 0.6MeV through 15MeV, and correlation of maximum range with energy for lower energies (<0.6MeV) has been discussed by Friedlander *et al* [1]. Feather [3] has described a method for accurate measurements at lower energies.

Determination of Gamma-ray Energies

Gamma-ray energies may be determined by absorption methods similar to those used for beta-ray energies, although precise measurements can be made only with a spectrometer. Lead filters must be used instead of aluminium, since γ-rays are much more penetrating than are beta rays. It is important to note that, unless the absorption of the γ-radiation occurs entirely by photoelectric processes, the condition that only photons corresponding to the incident energy are measured is not easy to meet experimentally. It is necessary to have either a good detector geometry or a detector which can respond only to a narrow range of energies. In particular, if the detector receives some of the degraded radiation produced in Compton scattering and pair production processes, deviations from the true exponential absorption will occur unless the absorber is thick enough for equilibrium with the secondary radiation to be established. The determination of γ-ray energies by the use of γ-ray spectrometers will be discussed in a later section.

4.3 NUCLEAR REACTIONS

In common with chemical reactions, nuclear reactions can be accompanied by absorption or emission of energy (Q). As implied in the earlier sections of this chapter, processes which proceed with a loss of energy are 'decay' processes, and result in the destruction of a radioisotope and the production of a more stable isotope. Nuclear reactions involving absorption of energy can result in the synthesis of radioisotopes, particularly if the sample is subjected to irradiation by sub-atomic particles – neutrons, protons, dueterons, α-particles, etc.

The probability that a nuclear reaction will occur is expressed as the cross section (σ), the capture cross section for a synthetic process, for the reaction. This has the dimensions of area and the unit is the 'barn' (10^{-24} cm^2). The sub-unit 'millibarn' (10^{-27} cm^2) is in common use.

Among the various factors that govern the probability of a nuclear reaction, the nature of the emitted or the bombarding particle and of the reacting isotope are of major importance. Hence, for example, the probability that reaction will occur in a target isotope bombarded with slow, or thermal, neutrons as projectiles is greater than with fast neutrons, and the emission of γ-rays in a synthesis-decay process involving thermal neutrons is readily accomplished. For the emission of charged particles, however, fast, high energy neutrons are needed, despite their lower capture cross section. The probability of the occurrence of various reactions, and of whether a particular reaction can take place or not, under given conditions, can be calculated from a knowledge of the coulombic and centrifugal values for the reaction [1].

4.3.1 Radiochemical Notation

As with most specialised subjects, radiochemistry possesses its own shorthand notation for writing nuclear reactions. Such reactions, as has been seen earlier in this chapter, can be written exactly as chemical reactions, with the reactants on the left and the products on the right of the equation. Thus, for the conversion of the isotope Re^{185} to the isotope Re^{186} by neutron bombardment, a reaction which also gives rise to a γ-ray photon from the product nucleus, we can write

$$Re^{185} + n \rightleftharpoons Re^{186} + \gamma$$

The same equation can be written in shorthand in the form

$$Re^{185}(n,\gamma)Re^{186}$$

where the reactant, or target, nucleus is written to the left of the bracket and the product nucleus to the right. Within the bracket are written the bombarding particle, in this case a neutron, and the particle ejected from the product nucleus, in this case a γ-ray photon. In like manner, the conversion of Li^6 into tritium when it is bombarded by neutrons can be written

$$Li^6(n,\alpha)H^3$$

4.4 ACTIVATION ANALYSIS

4.4.1 Basic Principles

The method of activation analysis was first employed by Hevesey and Levi [4], then by Seaborg and Levinghood [5], and by Griёnberg [6].

Extensive research into activation analysis techniques has since been pursued in all corners of the world by such workers as Smales in the United Kingdom; Leddicote, Meinke, Morrison, Wainerdi, and Ehmann in the United States of America; Comar in France; Hoste in Belgium; Girardi in Italy; and Haldar in India.

Activation analysis is based on the principle that when a material is irradiated by the nuclear particles produced in a nuclear reactor, or by sealed sources, or by particle accelerators, some of the atoms present in the material will interact with the bombarding particles and be converted into different isotopes of the same element, or isotopes of different elements, depending on the atom undergoing reaction and the nature of the bombarding particle. In many cases the isotopes produced are radioactive. If only one type of radioactivity is induced, or, since more than one radioactivity is normally induced, if each different radioactivity can be distinguished or separated from all the others, the amount of each radioactivity is a measure of the quantity of the parent, non-radioactive isotope present in the sample. Since absolute activities are difficult to measure, it is customary to irradiate a reference standard with the sample and to compare their measured activities after identical treatments. This method, known as comparative activation analysis, has the further advantage that the absolute flux of incident particles and the activation cross section for the reaction in question need not necessarily be known. The mass of the element X in the sample is given by

$$\frac{\text{Mass of X in sample}}{\text{Mass of X in standard}} = \frac{\text{Radiation intensity from X in sample}}{\text{Radiation intensity from X in standard}}$$

Formation of Radioactive Species

When a target material is exposed to neutrons, the nuclear reaction and the product decay may be represented by

$$A \xrightarrow{n} B^* \xrightarrow{\gamma} C$$

The stable isotope A, on bombardment with neutrons, is transmuted into the radionuclide B*, which subsequently decays, with its characteristic decay constant λ_B, to the stable isotope C. The rate of formation of B* will depend on the number (N) of atoms of A present in the target, the activation cross section (σ) of A, and the neutron flux (f), defined as the number of neutrons passing through a square centimetre of area per second. Since B* is disintegrating at a rate $\lambda_B N_B$ at the same time as new B* is being formed, the rate of accumulation of B* will be the difference of the rates of formation and of disintegration, i.e. the overall rate of growth of activity during bombardment is given by

$$dN_B/dt = fN\sigma - \lambda_B N_B \tag{4.12}$$

where N_B is the number of radioactive, B*, atoms present. Integration between $t = 0$ and $t = t$ gives

$$N_B = \frac{fN\sigma[1 - \exp(-\lambda_B t)]}{\lambda_B} \tag{4.13}$$

where t is the time of irradiation.

At the end of the irradiation, the activity A_t, in disintegrations per second, of B* is equal to $\lambda_B N_B$, and so

$$A_t = fN\sigma[1 - \exp(-\lambda_B t)]$$

$$A_t = fN\sigma[1 - \exp(-0.693t/t_{1/2})] \tag{4.14}$$

where $t_{1/2}$ is the half life of the radioactive nuclei. The term in brackets is termed the growth or saturation factor and reaches a maximum value of unity for infinitely long irradiation times, when

$$A_t = A_{t=\infty} = fN\sigma \tag{4.15}$$

So far it has been assumed that the element is mono-isotopic, but in general one must consider the presence of several isotopes, not necessarily of the same element. If the sample contains W g of an atom of atomic weight M, and ϕ is the isotopic abundance of this target atom which gives rise to the radioactivity, substitution for N in equation (4.14) gives

$$A_t = \frac{f(6.02 \times 10^{23} W\phi)\sigma \times 10^{-24}[1 - \exp(-0.693t/t_{1/2})]}{M} \tag{4.16}$$

where the factor 10^{-24} has been introduced to convert σ in barns to cm^2. After the irradiation has stopped, the activity formed will, of course, decay with its characteristic half life, $t_{1/2}$, so that at a time t' after cessation of the irradiation the activity has dropped to

$$A_{t'} = A_t \exp(-0.693t'/t_{1/2})$$

where A_t is given by equation (4.16). Care must be taken to ensure that $t_{1/2}$ and t' are expressed in the same units.

From this relationship it can be seen that high activity (and therefore high sensitivity) is obtained for a given mass if the flux and the activation cross section are high. Also, other things being equal, sensitivity is greater for lower atomic weight elements and for those with high relative abundances of the target isotopes concerned. The half life of the radionuclide does not control the inherent sensitivity of the method, except in so far as it becomes a practical limitation, either for a long lived radionuclide at the time of irradiation or for a short lived radionuclide between the time of irradiation and the time of measurement of the activity. The characteristic radiation emitted by the radionuclide formed also affects the sensitivity of the method, and should be taken into account when assessing its potentialities.

In activation analysis one has a choice from among several incident activating particles. These fall mainly into three categories

(i) Neutrons, which are subdivided into
  (a) lower energy, or thermal, neutrons, up to $\simeq$0.025eV obtained by slowing down (moderating) fast neutrons with such materials as water or paraffinic hydrocarbons.
  (b) higher energy, or fast, neutrons, up to 14MeV, obtained by reactions such as that between deuterons and a tritium target. Neutrons of intermediate energies may also be used in special cases.

(ii) Charged particles, usually of moderate energies, such as protons and alpha-particles.

(iii) High energy photons, for photonuclear activation analysis.

4.4.2 Neutron Activation Analysis

Neutron Sources

Neutrons can be obtained from three principal types of source, giving a low intensity beam from a radioactive source, a medium intensity beam from an accelerator, and a high intensity beam from a nuclear pile.

Low Intensity Isotope Sources

These sealed neutron sources may be divided into three types, of which two involve two-stage reactions of the (γ,n) or (α,n) varieties, and the third involves spontaneous fission.

A mixture of Sb^{124} and Be is a commonly used (γ,n) type neutron source. Sb^{124} emits 1.71 and 2.04MeV γ-rays with a half life of 60 days, while the threshold for the $Be^9(\gamma,n)Be^8$ reaction is 1.7MeV. One curie of Sb^{124} can produce up to 10^4 neutrons/sec$\cdot$cm^2. The neutrons are fast and can be slowed down with graphite or paraffin wax.

The energies of neutrons obtained in (γ,n) reactions usually range up to 4MeV, whereas in spontaneous fission the emitted neutrons have energies of about 2MeV. Table 4.1 summarises data for selected (γ,n) and spontaneous fission sources.

The high flux obtainable from Cf^{252} makes it appear to be the most promising of the sealed neutron sources and, with its light weight, it may prove useful in field prospecting, since a source yielding 10^{11} neutrons/sec could perhaps be prepared from as little as 30mg of this isotope. At present it is possible to produce Cf^{252} from Pu^{239} or Pu^{242}, only at the Savannah River facility in the U.S.A., where a reactor flux of the order of 10^{16} neutrons/sec is available. A typical neutron source based on Cf^{252} is estimated to cost about £5,4000. Strain [7], Zeman [8], Meinke [9], and Kusaka [10] have discussed various aspects of these isotope sources.

TABLE 4.1

Sealed Neutron Sources

Type of Source	$t_{1/2}$	Neutron yield / sec. curie	Neutron Energy
(γ,n) sources			
Po^{210}/Be	138.9days	2.5×10^6	4MeV
Cm^{242}/Be	163.0days	3.0×10^6	4MeV
Th^{228}/Be	1.9yrs	2.0×10^7	4MeV
Cm^{244}/Be	1.8yrs	3.0×10^6	4MeV
Ac^{227}/Be	22.0yrs	2.0×10^7	4MeV
Pu^{238}/Be	86.0yrs	2.0×10^6	4MeV
Am^{241}/Be	458.0yrs	2.0×10^6	4MeV
Am^{241}/Cm^{242}/Be	163.0days	3.0×10^6	4MeV
Pu^{239}/Be	24,000yrs	1.5×10^6	4MeV
Spontaneous fission sources			
Cm^{244}	18.0yrs	2.0×10^5	2MeV
Cf^{252}	2.6yrs	4.0×10^9	2MeV

Small Accelerator Neutron Sources

These sources have advanced rapidly in recent years. The most widely used type of accelerator is the small Cockcroft-Walton machine, which produces 14MeV neutrons by bombarding a tritium target with 150keV deuterons. At a D^+ beam current of 1000μA, a 1 curie/cm^2 tritium target emits, isotropically, about 10^{11}neutrons/sec. At positions where samples of appreciable size (1-10g) can be placed and irradiated without excessively large flux gradients being present, 14MeV neutron fluxes of about 10^9neutrons/sec.cm^2 can be obtained. The safety shielding for such an accelerator requires about 2m thickness of concrete plus 3-5m of clear space, for continuous use at 10^{11}neutrons/sec.

Unfortunately, tritium is evaporated from the target during bombardment, due to the heating effect of the deuteron beam, and only a small fraction of the tritium consumed in the target actually contributes to the generation of neutrons. In the operation of a neutron generator a rather rapid decrease in neutron yield is observed over the first few minutes of operation with a fresh target, followed by an approximately exponential decrease in the yield with a half time of approximately 34 milliampere-hours of target exposure. Several attempts have been made to improve target life. The older oil-diffusion vacuum pump has been largely replaced by the vac-ion pump, to prevent deposition of material from the pump oil on the surface of the target. Heating of the target has been reduced by using refrigeration cooling based on freon instead of water cooling systems, and by using larger, rotating targets with useful lives of 5 or 10 times that of 'normal' targets. Improvements in target life can also be achieved by magnetically deflecting away D_2^+ ions, present in the D^+ beam. These D_2^+ ions cause target heating, as do the D^+ ions, but contribute very little to neutron production. With accelerators which employ a radio-frequency ion source, D_2^+ ions cause little trouble since the beam from such a source is about 90% D^+ and only 10% D_2^+.

The need for frequent target changes, with their attendant risks of contamination to personnel and laboratory, has led to the development of high yield, sealed tube generator systems in which the tritium supply is continuously replenished [11]. These systems operate by accelerating a mixture of deuterium and tritium ions into the target and recirculating the resultant gases through a penning-type ion source. Neutron yields in excess of 10^{11} neutrons/sec have been obtained, with little or no diminution in the yield over several hours of operation [12].

A detailed review of generators and their uses has been given by Strain [13].

Nuclear Reactors

The most intense neutron sources have been provided by neutron chain reactors utilising fission reactions. Fluxes currently available are many orders of magnitude greater than those available from the laboratory neutron sources discussed above, and so the sensitivity of neutron activation analysis may be greatly improved by their use. The neutrons produced in a nuclear pile can be classified as fast neutrons ($>$1MeV), resonance neutrons ($\sim$0.4eV), and thermal neutrons ($<$0.4eV). Although appreciable amounts of fast neutrons are always present in a reactor, pure thermal fluxes can be obtained by carrying out the irradiation in 'thermal columns', which consist of graphite or other moderator material in sufficient quantity to slow the more energetic neutrons to thermal energy levels before interacting with the sample. Since both neutron density and velocity are decreased in the process, thermal fluxes are much lower than the maximum pile flux.

The pool type of reactor is most widely used for neutron activation analysis, and can provide neutron fluxes in the range 10^{12}-10^{14} neutrons/sec.cm^2, depending upon the reactor type, power level, and location of the sample within the pool or the core. One of the newer types of pool reactors is that which employs uranium-zirconium hydride fuel elements, the TRIGA reactor, and this is now widely used. Research reactors and their application to neutron activation analysis have been described by Buchanan [14], Guinn [15], and Meinke [16]. Moreover, improved sensitivity for the detection of elements giving rise to very short lived induced activities can be obtained by taking advantage of the high intensity, short duration pulses of neutrons generated in a pulsed reactor.

4.4.3 Sample Preparation and Irradiation

Prior to the activation of a sample it is necessary to avoid any steps which could result in the contamination of the sample with elements which might also become activated. The precautions to be taken against such contamination are well described in 'Radioactivation Analysis' by Bowen and Gibbons [17]. Contamination by inactive material after irradiation is unimportant, provided that it does not occur in amounts sufficient to affect the chemical yield. The restrictions imposed on neutron activation by self shielding, i.e. the reduction in flux to the inner parts of the sample due to absorption in the outer layers, which depends on the absorption cross section for the reaction, may necessitate the use of small samples and either liquid or solid dilution of the pure standard material to overcome this difficulty. In research reactors, where the operating temperature is low, short irradiations may be performed with samples in polythene containers, provided that liquids are not used. Where long irradiations are necessary, or for small volumes of liquid, it is customary to seal samples and standards in silica ampoules. The samples must be weighed prior to irradiation.

As mentioned earlier, in neutron activation analysis the content of an element in a sample is evaluated with the help of a reference standard. The sample is activated together with the standard containing a known amount of the element in question. The required element may be present in the standard either as the pure element, or as an oxide or a salt. All samples, whether of meteoritic material, stones, or other minerals, are crushed in preparation for subsequent chemical separations. Care should be taken to ensure that the sample and standard are of approximately the same weight, shape, and thickness, so that the analysis may not be affected by such factors as variations in flux and self attenuation during irradiation. If the element occurs as a cation, its carbonate, acetate or chloride is the preferred medium since effects due to activation of these anions are unimportant. Similarly, ammonium salts are preferred for anions. A neutron flux monitor, usually gold, indium, or silver foils whose activity after irradiation under controlled conditions is a measure of the total neutron flux, may be used to measure the integrated neutron flux during exposure. Cobalt wire has also been used for this purpose.

Self shielding effects may become very important if a major component of the substance under study, not necessarily the element being measured, has a high thermal neutron absorption cross section, but this can be minimised by using relatively small samples, and the use of internal standards to correct errors due to this effect has been described by Hoste *et al* [18, 19]. A number of samples and standards of different weights can be irradiated and analysed in order to determine whether self shielding will present difficulties in a particular analysis.

4.4.4 Chemical Purification

After irradiation of sample and standard, one can proceed to the qualitative and quantitative identification of the nuclides present using either a purely instrumental technique (gamma-ray spectrometry) or by including rapid radiochemical separation of the elements of interest before making the

TABLE 4.2

Radiochemical Separation Methods

Method	Minimum Time in Minutes	Selectivity	Dependence on Concentration	Applicability
Distillation	1	Good-excellent	None	Non-metals
Solvent Extraction	1	Moderate-good	None	Most metals
Chromatography	30	Good-excellent	None	Most metals
Precipitation	2	Poor-moderate	Marked	Most elements
Electro-deposition	30	Moderate-good	Slight	Most metals

instrumental measurements. The method followed usually depends on the facilities available in the individual laboratory and the types of nuclides resulting from the irradiation, among other factors. Where interferences are severe and cannot be sufficiently resolved by the proper choice of irradiation and decay times, post-irradiation radiochemical separation methods, using inactive carriers and scavengers, may be used to great advantage. These are, however, time consuming and tend to preclude the effective utilisation of induced activities with short half lives. Meinke [20] has developed very rapid radiochemical separation techniques, which can achieve separations in periods of 5 to 10 minutes in some cases.

The chemical behaviour of a minor constituent may be unpredictable when carrier atoms are absent, since the small weights of irradiated material present in certain situations may not be extracted at the expected stage, or they may be totally or partially absorbed or coprecipitated at an unexpected stage. Consequently, it is usual to add carriers at the earliest possible stage in the separation process to normalise the chemistry of the system. Solutions should be so treated that the carrier and the element in the sample are in the same chemical form, and the addition is made before chemical treatment is begun in the case of solids. Usually 3-100mg of carrier are used, with 10mg the normally recommended amount. An excess of carrier is particularly undesirable if the final measurements involve detection of beta-emissions, due to the need for self absorption and self shielding corrections in such cases. When a carrier of the same element is not available, one of an element with similar chemical properties may be used. Table 4.2 lists the most useful methods for radiochemical separations.

All the methods listed depend on distributing the wanted material between two phases, and each has its advantages and disadvantages. Thus, methods based on distillation require a volatile component, while chromatographic methods, which can be very selective, tend to be time consuming. Solvent extraction is more rapid, and its selectivity can be enhanced by:

a) the use of masking agents such as cyanide, tartrate, EDTA, and fluoride to produce charged unextractable species containing possible contaminants in systems involving the extraction of neutral chelates of the wanted elements;

b) combining several organic layers containing the desired element and back-extracting with an aqueous solution of suitable composition such that the distribution ratio for the wanted element favours its retention in the organic phase, and so little of it is lost, while the ratios for the unwanted contaminants tend to favour their extraction into the aqueous phase.

Precipitation methods are widely applicable, but they can suffer from the disadvantage of adsorption of unwanted activities in the precipitates. Adsorption of unwanted activities is usually avoided by scavenging with gelatinous precipitates which have large surface areas suitable for adsorption, e.g. $Fe(OH)_3$. It should be noted that monovalent ions are not adsorbed effectively by this means. In electro-deposition, the range of voltage over which a cation is deposited usually overlaps the ranges of other cations present, resulting in co-deposition of activities in certain cases, and so the method is not normally used as a primary method of separation, but rather in the final stages of preparation of a pure source, after potential contaminants have been minimised by one of the above methods.

Thus, isolation and detection of a wanted element may require a fast efficient step to isolate the element from the bulk sample with minimum contamination, especially by other radioactive materials, followed by a counting method specific for the nuclide in question. After isolating radiochemically pure specimens from the sample and the standard, the chemical yield of the method can be calculated from the standard by conventional means. The inclusion of this step makes it unnecessary for the radiochemical separation steps to be quantitative since, provided that both sample and standard are treated in an identical manner at every stage, the final measured activity can be corrected for losses in the separation stage by using the measured chemical yield.

Substoichiometric Method of Separation

The majority of radiochemical separation methods used attempt to approach quantitative extraction of the wanted element wherever possible. This newer method, suggested by Ruzicka and Stary [21] and applicable to precipitation and ion exchange methods, although it is best used with solvent extraction techniques, involves using a deficiency of the separating agent rather than the more usual excess. After activation and decomposition, the same amounts of carrier and separating agent are added to both sample and standard. The same chemical yield, with the appropriate fraction of the active isotope, is obtained from both sample and standard and determination of the absolute chemical yield is unnecessary. The greatest advantage of the substoichiometric method lies in the fact that substantially greater selectivity can be achieved than with the use of excess reagent.

Ruzicka *et al* in cooperation with Stary, Ruzicka and Zemann [22], have developed further uses of this method in extraction procedures for activation analysis.

4.4.5 Source Preparation

After completing the separation step, where applicable, the radioactive element, and carrier, must be mounted in a form suitable for the measurement of the radiation emitted from the unstable nuclei. The choice of the compound to be so mounted depends on the element to be measured, but it should be stoichiometric and have low solubility during the final separation steps, which are usually precipitation steps. Bowen and Gibbons [17] have discussed details of the practical problems associated with this stage of the method.

4.4.6 Measurement of Induced Activity

Measurement of the activity induced in an irradiated sample involves counting either the beta-particles or the γ-ray photons emitted during the decay process. The first step usually involves a qualitative study of the most prominent radiations present, in order to decide on the most satisfactory counting technique to use. This usually involves determining the decay schemes of the various isotopes in the sample. Since most of the elements have more than one stable isotope in nature, and different isotopes of the same element may react differently under irradiation, consideration must be given to the activities produced from the different isotopes, their relative isotopic abundances, their activation cross sections, and the half lives of the induced activities, in order to predict the most prominent forms of activity from a given sample exposed to a given bombardment. Although bombardment times may be chosen to favour formation of one particular radioisotope, it will be clear that some residual activity due to other isotopes, or to daughter products of any or all of the radionuclides present, may be detectable and thus may cause interference. Hence, even with a single element present, difficulties can be encountered at the counting stage in resolving the different signals from a number of different radioisotopes in the least favorable cases. Fortunately, radioisotopes with relatively short half lives can usually be allowed to decay to insignificant levels before measurements are started, and there are few problems in distinguishing beta- from gamma-emitters.

Beta Counting

The various methods available for distinguishing between beta-radiations from the same source have been discussed in sections 4.1.3 and 4.2.1. Geiger-Muller tubes with an end window are the most frequently used detectors, and end-window flow-type proportional ionisation counters have also become popular. A detailed discussion of the factors affecting the counting yield, which represents the ratio of the measured counting rate to the disintegration rate, has been given by Friedlander *et al* [1].

Gamma-ray Spectrometry

The gamma-ray spectrometer can be used as a qualitative tool in activation analysis, for the identification of possible trace impurities prior to quantitative analysis or for the confirmation of the purity of a separated species after a radiochemical separation, and such applications depend only on the ability of the spectrometer to identify the photon energies of the incident γ-rays, as explained in section 1.3.3.

The quantitative applications of the instrument depend on its ability to integrate the total number of counts for a given energy band, either electronically or by measuring the heights or areas of photo peaks produced by a scanning method. The numbers so derived from the sample are compared with those obtained, in an identical manner, from the standard, in the usual way, to determine the amount of the wanted element present. Covell [23] has discussed the methods available for evaluation of photo peak areas.

4.4.7 Non-Destructive Neutron Activation Analysis

It is clearly advantageous if determinations can be carried out non-destructively on a sample by direct evaluation of the spectrum of the induced radioactivity. This, however, becomes possible only if individual photo peaks can be easily distinguished, and if the activity of the Compton continuum does not exceed that of the given photo peak.

With the advent of the small neutron generator and semi-conductor detector systems linked to multi-channel pulse height discriminator circuits, it has become apparent that neutron activation analysis could be effectively employed in competition with many other methods of elemental analysis. It has been estimated that half of the elements in the periodic table may be determined at levels of 1mg, or less, with commercially available neutron generators producing 14MeV neutrons [24]. In this purely instrumental method, the greatest emphasis is placed on the resolution of the gamma-ray spectrum from the sample, and it is here that the greatly superior resolution offered by Ge(Li) detectors, compared with that of NaI(Tl) scintillators, is decisive. Figure 4.6 shows that the energy resolution of the most commonly available NaI(Tl) crystals is only about 50keV at 0.7MeV (fwhm), whereas a good Ge(Li) detector will show a resolution of 3–4keV for the same radiation [24]. Thus the Ge(Li) system will often allow resolution of very complex gamma-ray spectra without recourse to a complex computer-based resolution technique.

One of the main disadvantages of the Ge(Li) detector is that it must be maintained at the temperature of liquid nitrogen, and it has not yet been possible to make Ge(Li) detectors which approach currently available NaI(Tl) crystals in size and counting efficiency. Even relatively large detectors with 35–40cm^3 active volumes have a photo peak efficiency only 5% of that of a cylindrical crystal of NaI(Tl), 3in. diameter and 3in. deep, for 1MeV γ-rays, and the difference becomes even greater for higher energy radiation, due to the higher photo-electric efficiency of iodine at higher

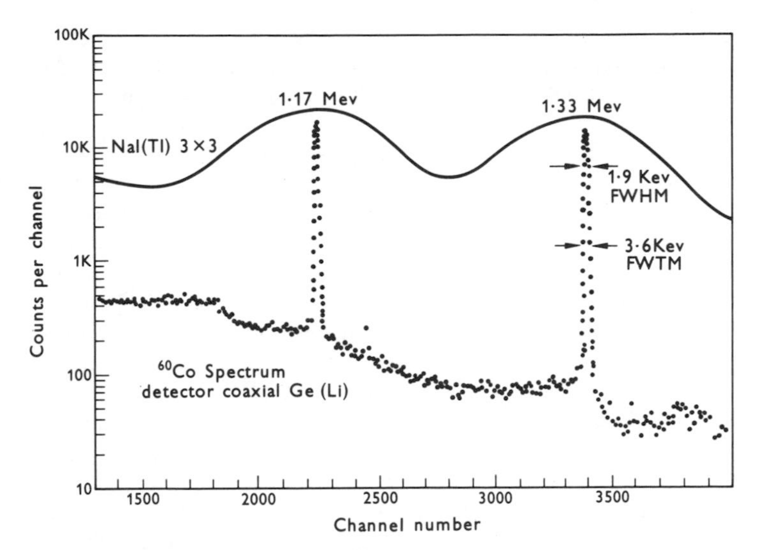

Figure 4.6 Comparison of Co^{60} spectrum obtained using a scintillation counter fitted with a NaI(Tl) crystal, upper curve, and a solid state Ge(Li) detector. Note the much higher resolving power of the Ge(Li) detector, 3.6keV at the half height position. (Courtesy of Nuclear Enterprises Ltd., Edinburgh.)

energies. Furthermore, the more complicated electronic equipment required for use with the Ge(Li) system must be capable of storing a considerable amount of data in order to make optimum use of the excellent resolution characteristics of the system, and this will inevitably increase the cost.

Estimated sensitivities for non-destructive neutron activation analysis using a large volume Ge(Li) detector have been given by Ehmann [24].

Of course, the instrumental method lends itself well to automation and computer-based techniques. The conditions needed for the application of automatic methods include [25]:

(i) that the elements present should have favourable activation cross sections;

(ii) that each element to be analysed must produce one or more gamma-ray emitting radioisotopes on neutron activation;

(iii) that these radioisotopes must have half lives suitably situated between certain upper and lower limits;

(iv) that the nuclear characteristics of each non-radioactive element and of its activation products must be known; and

(v) that the elements to be analysed must be present in sufficient quantities.

Kuykendall and Wainerdi [26], Anders and Beamer [27], Nostrand *et al,* [28], Burrus [29], and Morrison and Sunderman [30] have described applications of automatic systems.

4.4.8 Sensitivity

Sensitivity may be expressed as limit sensitivity, or detectability, denoted by l.s. and given in micrograms, or as limit concentration, denoted by l.c. and given in ppm.

Neutron activation analysis and spark source mass spectrometry have general sensitivities for many elements that are much superior to either atomic absorption or absorption spectrometry. Neutron activation analysis has also the peculiar feature that its greatest sensitivities are realised for some of the rarer elements, where high sensitivities are particularly advantageous, as shown by Winchester and Cataggio [31] who have compared the sensitivities of neutron activation analysis, spark source mass spectrometry, atomic absorption spectrometry, and absorption spectrometry with the abundance of the elements in the earth.

Yule [32] has drawn up a table of sensitivities which can be achieved in practice with neutron activation analysis in the determination of individual elements. The data relate to the elements oxygen through lead, omitting neon, krypton, and xenon, activated using a thermal neutron flux of 4.3×10^{12} neutrons/sec.cm^2 for one hour in a reactor, and give the yield in terms of count rate within the range of the main photo peak for 118 product radioisotopes formed in the reactor. Table 4.3 presents the minimum detectable concentrations, based on these measurements, corresponding to the use of a 10g sample. The data were obtained after electronic differentiation of the particular radionuclide activities by gamma-ray spectrometry, and Jenkins and Smales [33] and Meinke [34] have reported sensitivity limits for experiments under different conditions.

TABLE 4.3

Elemental Detection Limits

Range (ppm)	Elements
$10^{-5}-10^{-4}$	In, I, Dy, Au
$10^{-4}-10^{-3}$	A, V, Br, Cs, Sm, Ho, Er, Hf, Re, Ir
$10^{-3}-10^{-2}$	Na, Al, Sc, Co, Cu, Ga, As, Se, Sr, Rh, Ag, Cd, Sn, Ba, La, Nd, Gd, Yb, Lu, W
$10^{-2}-10^{-1}$	Cl, K, Ti, Zn, Ge, Rb, Zr, Mo, Ru, Pd, Te, Ce, Pr, Tb, Ta, Pt, Hg
$10^{-1}-10^{0}$	F, Mg, Cr, Y, Nb, Tm, Pb
$10^{0}-10^{1}$	Ca, Os
$10^{1}-10^{2}$	Si, S
$10^{2}-10^{3}$	Fe
$>10^{3}$	O

4.4.9 Precision and Accuracy

The precision of activation analysis, when all other variables connected with the irradiation and the separation steps have been eliminated, is limited by statistical variations in the counting rate. The precision will tend to decrease as the amount of element estimated, and the activity present in the final source derived from it, become smaller. Fortunately, statistical variations in the counting rate are rarely of importance except at extreme sensitivities, and even then the effect can be minimised by increasing the counting period, within limits depending mainly on the half life of the radioisotope involved. The accuracy of the method depends ultimately on the availability of standards of accurately known compositions, and these are usually obtainable. On the basis of information currently available there appears to be no reason to accept a coefficient of variation greater than 5%. Detailed discussions of the errors associated with activation analysis are to be found in the books listed in the reference list for the chapter, and the review articles by Mathur and Oldham [35], on 14MeV neutron activation methods, and by Mott and Orange [36], on precision, are especially recommended.

4.4.10 Recent Developments in Related Techniques

Although neutron activation analysis has been extensively developed, activation by other means is being actively studied, and mention may be made of recent advances in the closely related fields of charged particle, photonuclear, and 'prompt gamma' activation analysis. Charged particle activation, as the name implies, involves the use of protons or the nuclei of the lightest elements as projectiles to give daughter nuclei which are isotopes of different chemical species, instead of isotopes of the same species when neutrons are used. Mahony and Markovitz [37] have discussed the use of He^3 charged nuclei, and Pierce [103] has discussed the technique in a survey of latest developments in the field of activation analysis.

In 'prompt gamma' activation analysis, after the capture of a thermal neutron, which is the activating particle most commonly involved in this technique, the resultant compound nucleus promptly decays, usually through several intermediate stages, yielding a spectrum of gamma-rays characteristic of the original nuclide. The advantage of the technique lies in the fact that the initial excited states resulting from neutron capture may have energies of 7MeV or more, so that the spectra generated are often complex and the 'normal' gamma-ray spectrum from multi-component samples is often difficult to analyse, but the 'capture spectrum' can be measured independently of the rest of the spectrum and so provide very useful analytical data. Either NaI(Tl) or Ge(Li) detectors may be used.

Photonuclear activation analysis involves activation by gamma-ray photons instead of by nuclear particles, and the reaction cross section for this process depends on the incident gamma-ray energy. The method is particularly useful for light elements, such as carbon or oxygen, because the superior penetrating power of gamma-rays helps in the analysis of bulk samples where neutrons would be absorbed. High energy gamma-ray photons are required to obtain a reaction cross section adequate for high sensitivity determinations, and these are normally produced by allowing a beam of accelerated electrons to strike a heavy metal target and produce a 'bremsstrahlung' effect. Provided that the photon energy of the radiation is controlled, interferences can be avoided, and techniques have been developed to achieve this.

4.5 ISOTOPE DILUTION ANALYSIS

Basically, the method of isotope dilution analysis involves adding a known amount of a radioisotope of known specific activity to a known amount of the non-radioactive sample under study, isolating the wanted element in a known form radiochemically, and measuring the specific activity of the product. Comparison with the specific activity of the original additive gives the dilution factor due to the inactive material in the sample, and from this it is possible to calculate the amount of inactive material present in the sample. The method has the advantage that irradiation of the sample is not required, but it does need a supply of the required radioisotope. If co-operation with a reactor, or other irradiation facility, is possible, the method can make radiotracer techniques available to virtually any laboratory without the expense of the radiation installation. Three main variants are currently in common use.

4.5.1 Direct Isotope Dilution

In this variant, a known amount of a solution of the radioactive isotope is added to a known amount of a solution of the unknown, then excess of a reagent is added to precipitate the desired element in a known form. From the weight of the precipitate and the measured activity, the specific activity of the precipitate can be obtained and compared with that of the original standard. The extent of dilution is a measure of the amount of inactive isotope in the sample. Sensitivities in the range 10^{-9} through 10^{-12} g can thus be obtained. The separated compound must be pure, but need not be recovered quantitatively, provided that sufficient material is recovered for determination of the specific activity.

The mathematics underlying the method are quite simple. Let W_x be the unknown weight of the element or compound in the sample, and W_o be the known weight of tracer, of specific activity S_o, added to the sample. Assuming that complete mixing has occurred, the observed specific activity of the isolated material, S_f, is given by

$$S_f = \frac{W_o S_o}{(W_x + W_o)} \tag{4.17}$$

whence

$$W_x = W_o\,[(S_o/S_f) - 1] \tag{4.18}$$

If the specific activity of the tracer is very high, so that the weight of tracer added is negligible compared to that of the inactive material, i.e. $W_o \ll W_x$, equation (4.18) can be simplified by replacing the specific activities by the measured activities, A_o and A_f, of the standard and of the recovered material respectively, and replacing the sum of the weights of the active and the inactive isotopes by the weight recovered from solution, W_f. Equation (4.18) then reduces to

$$W_x = W_f(A_o/A_f) \tag{4.19}$$

4.5.2 Substoichiometric Isotope Dilution Method

The direct dilution method suffers from the drawback that it is necessary to know the weights of various compounds, particularly the weight of the isolated pure compound. Because the sensitivity of mass-determination is very much poorer than the sensitivity of radioactive counting, it becomes difficult to measure specific activities accurately using sub-microgram quantities of the final product, in some cases. The use of substoichiometric isotope dilution methods can help to overcome this difficulty. A known aliquot of tracer solution is added to a solution of the sample, and both mixed solution and the standard tracer solution are treated with a small aliquot of a reagent to isolate the required compound. The amount of this latter aliquot is chosen to give incomplete precipitation, or separation, in both cases, and so the same absolute amount of material is extracted from both solutions, but in one case the active isotope will be diluted by the inactive in the ratio in which they are present in solution. Comparison of the activities of the two samples provides a measure of the amount of the unknown in the sample, by

$$W_x = W_o[(A_o/A_f) - 1] \tag{4.20}$$

since the masses of material from both sample and standard are the same. The principal advantage lies in the absence of a weighing step, and higher sensitivities and greater accuracy can be achieved by this means. Furthermore, solvent extraction methods can be used to provide a superior separation free from the problems associated with precipitation techniques.

4.5.3 Isotope Exchange Reactions

In an early exchange experiment in 1920, Hevesy [44] demonstrated the rapid interchange of lead atoms between $Pb(NO_3)_2$ and $PbCl_2$ in aqueous medium, by using thorium B (Pb^{212}). This pioneering experiment opened up an important field of chemical investigation, and many exchange systems have since been examined, more particularly since the advent of artificial radioactivity. The method is one of the important techniques by which various labelled molecules can be obtained, but its main use tends to be for the study of reaction kinetics and mechanisms, rather than as an analytical tool. Friedlander and Kennedy [1] have discussed these methods at some length.

4.5.4 Radioactive Reagents

When it is possible to obtain, in a radioactively labelled form, a reagent which reacts quantitatively with a wanted element to form a stable compound, a further variant on the dilution method is possible. This variant is especially useful for trace constituents.

Briefly, the method involves treatment of a weighed amount of the sample with the radioactive reagent to convert the trace constituent quantitatively to a labelled derivative. Excess reagent is removed and a much larger portion of the derivative, inactive in this case, is added and the mixture treated to ensure complete mixing of the active and inactive forms. The derivative is finally isolated and purified to a constant specific activity, and this value, usually quoted in microcuries per millimole, is compared with that of the original reagent in order to find the amount of active derivative present in the final sample, and hence the amount of the wanted element present in the original sample. If M_o is the number of millimoles of active derivative associated with M_x millimoles of inactive derivative, and S_o and S_f are the specific activities of the pure reagent and of the recovered derivative respectively, then, since $M_x \gg M_o$,

$$M_o = M_x(S_f/S_o) \tag{4.21}$$

4.5.5 Sources of Error

It will be clear that the principal sources of error in all the dilution methods are those connected with incomplete mixing or homogenisation of solids or solutions containing the active and inactive isotopes of the wanted element. All the variants quoted depend entirely on isolating a representative sample in which the ratio of active to inactive species is the same as that in the mixed state. Normally this means in solution. Thus it is imperative that the isotope be in the same chemical form in both sample and diluent. To take an extreme case, it would be impossible to homogenise a solution in which chromium was present as the chromate anion with one in which the chromium existed as the Cr^{3+} cation, but it would be equally difficult to homogenise solutions in which the same element existed as cations in two different oxidation states, e.g. Mn^{2+} and Mn^{4+}. Suitable treatment must, therefore, be given to convert all of the wanted element to the same state before final isolation and purification, for example digestion in hot dilute acid or complete oxidation or reduction of multivalent cations after addition of the tracer solution.

Attention must also be paid to radiochemical purity, and tracers should contain only one active constituent wherever possible since, if an inert diluent is added to a tracer containing an active impurity, the specific activity of the tracer will be reduced while the activity of the impurity may be left at an unacceptably high level. A very similar situation may be produced by radioactive daughter atoms arising from the decay processes.

Finally, the very small weights involved in some determinations may generate problems in mass determination, especially when a large dilution factor has been used, but, although special techniques may be needed in some situations, these problems are no greater than those encountered in other microanalytical work.

4.6 AUTORADIOGRAPHY

The detection of ionising radiations by their interaction with a photographic plate was the original method whereby Bequerel [45] discovered the phenomenon of radioactivity. Each particle passing through a photographic plate produces an ion-track in the emulsion which, in turn, causes darkening of the plate on subsequent development. The density and appearance of this darkening is determined by the specific ionisation involved, and α-particles are most easily detected since their tracks appear as thick black lines when viewed under the microscope at about 500x magnification. Beta-particles give a more diffuse blackening and are less easily recognised, while γ-rays cause only a general, low intensity blackening of the entire plate with no characteristic features at all, and special methods are needed for their identification and measurement. Early work in this field has been reviewed by Shapiro [46] and more recent reviews have been given by Webb [47], Yagoda [48], Beiser [49], Norris and Woodruff [50], and Goldschmidt-Clermont [51].

Basically, the experimental technique is very simple, involving placing a photographic film in contact with the radioactive sample so that localised radiation from concentrations of radioactive materials in the sample can cause similarly localised darkening in the film after development. The resulting picture is a representation of the distribution of radioactivity in the sample. The method has very obvious direct applications in the studies of uranium- or thorium-bearing minerals, but it may also be applied to cases in which the radioactivity has been induced artificially. Few problems are encountered in studies at the qualitative level, but more care must be taken if quantitative results are required. In particular, the sensitivity of the method for γ-rays must be improved, and this can be done by interposing a lead foil, about 0.1mm thick, between the source and the plate. The γ-rays absorbed by the lead generate secondary electrons which interact with the emulsion and enhance the blackening of the plate.

The technique of radiography can be made quantitative by two methods. Firstly, if the density of blackening is small and the tracks are clearly defined, the number of particles emitted may be counted visually and the lengths of the separate tracks measured. The number of tracks gives an estimate of the amount of radioactive material present, and the track lengths, which depend on the range of the emitted particles and hence on their energies, can give an indication of the nature of the radioactive nuclei causing the tracks. This method is especially applicable to α-emitters, because the energy of emission of an α-particle is characteristic of the emitting nucleus, albeit not so characteristic as that of β-particles or γ-rays, as discussed in section 4.1.2, and the experimental conditions of autoradiography are reasonably well suited to estimating these energies. Secondly, if sufficiently high activity levels are used to cause appreciable blackening over the whole film, the optical densities of the various regions of the film can be measured with a microdensitometer. Such techniques are better suited to beta emitters, and it has been found that 10^7 disintegrations are normally required, but this varies with the amount of β-emitter present.

The choice of film emulsion for a particular problem depends on (a) the sensitivity of the emulsion to different radiations, since some films are sensitive to β-particles and not to γ-radiation, and vice versa; (b) the grain size of the film; (c) the grain density; (d) the emulsion thickness; (e) the background radiation level. The choice of film also usually involves a compromise between sensitivity and resolution, since for high sensitivity one should use a coarse grained thick emulsion film, whereas the highest resolution is given by a fine grained, thin emulsion film. The background radiation level should, of course, be kept as low as possible.

Film for radiography is available commercially in various forms and of various types for maximum sensitivity, best resolution, etc. The film may be supplied mounted on glass plates, on cellulose acetate sheets or rolls, as an emulsion which may be stripped from its support, or as a gel which can be melted at 40°C and applied directly to the specimen surface and so conform to the topography of the surface. Emulsions can also be obtained with a thin gelatin coating to protect against scratches and abrasions.

In all radiographic studies, it is important that the best possible contact be made between film and surface, and that the maximum sample-film distance be kept as small as possible in order that the resulting autoradiographs may show as much detail as possible at the highest resolution obtainable. The aim is to restrict the observed radiation to radiation which is incident normally on the film, and to minimise the amount of oblique radiation. Techniques for improving sample-emulsion contact have been reviewed by Fitzgerald *et al* [52] and by Gross *et al* [53].

Optimum exposure times depend on such factors as the half life and the specific activity of the active nuclei, the sensitivity of the emulsion, and the nature and optical density of the required autoradiographic image. It is usual to take several exposures to emphasise different features and so obtain the maximum information from the sample. Processing of the exposed film requires some care, and Bowie [54,55] has discussed the techniques available. General discussions on points of technique will be found in the reviews cited [47–51].

4.7 FISSION TRACK METHOD

This method, developed by Price and Walker [56], has been extensively used to study the distribution of uranium in mineral samples, and it has recently been extended to thorium-bearing minerals. Fission of the uranium in a polished sample is induced by neutron irradiation, and the activity accompanying the fission can be followed by studying the tracks of the fission fragments in the prepared surface. The tracks are developed by etching, since preferential attack occurs along the line of ionisation and radiation damage caused by the fragments. Currently the major difficulty lies in this etching step, both because precise etching conditions are known for only a few rock-forming minerals, and because the effectiveness of attack, in a given substance, depends on a host of factors, including the choice of etchant (HF, NaOH, etc.); its concentration; the temperature of attack; the crystallographic orientation of the attacked surface; and the history of the surface, especially the concentration of crystalline defects, grain boundaries, and, most particularly, the presence of impurity atoms, which can play a dominant role in determining the geometry of attack. Preferential attack along a particle track, however, is a general phenomenon, and careful experimentation can provide excellent track development, although it has not yet proved easily adaptable to the determination of the aggregate uranium content of a rock, or similar problems.

Fleischer and co-workers [57] and Fisher and Bostrum [58] modified the method, and overcame some of the difficulties, by placing their samples between plates of mica or plastic before irradiation and developing the tracks in the plate rather than in the mineral surface. The plastic 'Lexan' (supplied by the General Electric Co., Massachusetts, U.S.A.) has been found to be particularly suitable for this purpose, and Fleischer *et al* [57] have successfully applied the method to the minerals separated from the Vaca Muerta mesosiderite. Irradiation of polished samples cut from the meteorite, with thermal neutrons in a nuclear reactor, followed by measurement of track length distributions, showed that some of the tracks resulted from the spontaneous fission of uranium impurities. These authors have also reported the detection of uranium concentrations ranging from 4000ppm in a single zircon grain to less than 10^{-3} ppm in hypersthene and anorthite. Murali, Parekh and Sankardas [59] have determined the uranium content of a whole rock sample by combining the fission track method with the technique of standard addition, i.e. by adding known amounts of the element to be studied to the sample [as in isotope dilution analysis (section 4.5)].

4.8 APPLICATIONS OF RADIOTRACER METHODS

In considering analysis in connection with a problem involving minerals it is important that the method selected should be capable of giving the required information in a minimum time, and with the minimal cost and difficulty. The general factors governing the choice of technique are discussed in Chapters 1 and 12, but there are certain special features connected with the use of radiotracers in geochemistry or in any of the other multitudinous facets of the minerals industries. Thus, in considering a method for trace element analysis in minerals, four points that must be borne in mind are [60–64]

(i) how sensitive is the method under consideration?

(ii) what are the relative merits of a destructive and a non-destructive method?

(iii) can the method perform simultaneous multi-element analysis?

(iv) can the technique be adapted for field operations?

Activation analysis would, in many cases, appear to offer an extremely attractive solution to this problem because of its unique ability to measure the concentration of an element in a reasonably large sample, certainly a sample larger than that used for spark source mass spectrometry, with a minimum of sample preparation and at a competitive price.

Today, the uses of radioactive tracers in the various branches of geology and of the mineral processing, metallurgical, ceramic, and coal industries have become so diverse and widespread that it is not possible to describe them with any degree of completeness. Instead, this section will be devoted to indicating how they may be applied, using typical examples from the aforementioned fields.

4.8.1 Activation Analysis

Neutron activation analysis is widely used as a tool for laboratory-scale analysis, especially for the determination of minute traces of elements, but also as a major analysis facility. A completely automated system operating in the Bay Laboratories of General Telephones, Long Island, U.S.A., can provide a complete analysis within minutes of a weighed aliquot of the ground sample being placed in the sample holder. A small computer controls the whole procedure, from injection of the sample into the irradiation chamber to the printing out of the results as percentages, including automatic control of the irradiation time in the accelerator and the various counting times. The technique has also been applied to batch processing of material, and recently Pierce [65] has described its application to continuous process control in the examination of moving sample streams. With the low level counting techniques now available, the small amount of activity induced in a process stream presents no hazard to either operator or consumer.

Geochemistry

Tolmie and Thompson [66] have recently described a portable neutron activation analysis instrument, based on an Sb:Be neutron source with pneumatic sample transport to the irradiation zone, a pair of scintillation counters in a 4-inch-thick lead shield, a Ge(Li) detector, and a small computer. The equipment is compact and reasonably easily transported to, and operated at, a base camp to give rapid ore assays, rock identifications, and the ability to explore a region based on immediately available trace element contours. On the debit side, the capital cost of such a system is in the region of $90,000, and it would be necessary to arrange for a supply of liquid nitrogen at all times. This application of neutron activation analysis and high resolution gamma-ray spectrometry to the measurement of the distribution of elements in the earth's crust has markedly increased in importance over the past few years [67] and the use of Ge(Li) detectors in the determination of trace elements in rocks is excellently illustrated by the work of Gordon and his group [68], who were able to determine up to 23 elements instrumentally in a wide variety of igneous rocks.

Activation analysis of different minerals occurring in nature has been applied extensively, either as a non-destructive method using purely instrumental techniques or in conjunction with more conventional chemical separation processes. Among the elements for which details of schemes of analysis have been given are O, Na, Mg, Al, A^{40}, K, Sc, Ti, V, Mn, Ni, Cu, Zn, Ga, Ge, As, Se, Br, Rb, Sr, Y, Zr, Nb, Mo, Pd, Ag, Cd, In, Sn, Sb, Hf, Ta, W, Re, Os, Pt, Au, Tl, Bi, Th, and U, either as single-element or multi-element analyses. Rakovic [69] has provided an excellent summary of recent work in this field. In addition, Loveberg *et al* [70] have described the use of field gamma-ray spectrometry in the exploration of uranium and thorium deposits in South Greenland, and Czubek [71] and Tittle [72] have recently reviewed geophysical applications of neutron methods in field prospecting, in borehole logging, and in the laboratory. Finally, Pradzynski [73] has reported the use of neutron and photonuclear activation analysis methods for the determination of copper in copper ores and flotation products.

Metallurgy

In the manufacture of steel, oxygen is added to the molten iron to oxidise out the impurities present. Unfortunately, oxygen is itself an impurity in steel and so there is a limit beyond which further additions are disadvantageous. Oxygen should be added until the total impurity level reaches a minimum, and then its addition should be stopped. Obviously, this technique requires very rapid analysis of the oxygen content of the steel if the oxidation process is to be stopped at the optimum point. Activation analysis possesses the required speed, and can be so incorporated into an automatic control loop that this becomes possible. The method is based on the reaction

$$O^{16}(n,p)N^{16} \xrightarrow[7.4\ \text{sec}]{\beta} O^{16}$$

and the intensity of the 6–7MeV electrons emitted is a measure of the quantity of oxygen present. Two reactions may interfere,

$$N^{15}(n,p)C^{15}, \qquad t_{1/2} = 2\ \text{sec}, \qquad E = 5\text{MeV}$$

$$F^{19}(n,\alpha)N^{16} \rightarrow O^{16}$$

but N^{15} constitutes a negligible 0.36% of natural nitrogen and the fluorine content of steel is usually low. The existence of both possible reactions does, however, mean that checks must be run at regular intervals. The results obtained by this method appear to be as reliable as those from other methods, such as vacuum fusion, and, in a case where speed is essential, activation analysis certainly gives the fastest reliable results. The time required, in fact, is less than that required to prepare the sample for most other methods. Methods of analysis of iron and its alloys for C, O, Si, P, S, V, Cr, Mn, Co, Cu, Ga, As, Nb, Sn, Ta, W, and U have also been summarised by Rakovic [69], who also quotes extensive references to analysis of various metals and their alloys for trace elements.

Non-Metallics

Gaudin [74] has advocated the use of neutron activation analysis for monitoring the separation of various minerals, but he has noted that each separate ore body would present a distinctive problem, since the activities induced in a given mineral species would vary markedly depending on its original provenance and on the varying amounts and types of impurity present. Recently, however, Martin [75] has applied conventional neutron activation analysis and prompt gamma techniques to the relatively simple coal system, for the rapid and routine analysis of its carbon and mineral contents (principally Al, Si, and O), using a neutron generator as source. In the cement industry, Poovey and Covault [76] have recommended a method for the determination of calcium in portland and bituminous concretes, and Liebermann *et al* [77] and Hamaguchi *et al* [78] have described methods for multi-element determinations in concretes and silicates. Goode and his co-workers [79] have developed methods for determining up to twenty-five elements in milligram samples of glass using thermal neutron activation analysis and taking advantage of an automated radiochemical separation by a solvent extraction procedure.

4.8.2 Radiotracer Techniques

Radioisotopes find many other uses apart from those allied to elemental analysis. The fact that a radioisotope is chemically identical with its non-radioactive isotopic forms means that the radiotracer can be introduced into a reacting system without affecting the chemistry of the system in any way. The radioisotope is simply one of the atoms of the element which can take part in all the chemical reactions of the element, but which is 'labelled' in such a way that it can be identified and traced by non-interfering physical means. Moreover, the sensitivity and ease with which they can be traced is a distinct advantage, since it means that only small amounts need be added to the system. The property has found particular use in systems involving self-diffusion or sintering studies. There are, however, a few limitations to their use, such as the availability of a suitable tracer with a half life and a specific activity lying within a reasonable working range, and emitting radiations that can be detected under the experimental conditions of the study.

Metallurgy
The use of radiotracers in the iron and steel industries has shed new light on many old problems [80], including investigations of

(i) the thermodynamics and kinetics of metal-slag reactions, including desulphurisation and dephosphorisation processes

(ii) the physicochemical properties of slags, for example the determination of relative mobility in slags.

(iii) the hydrodynamics of molten steel

(iv) the melting of scrap

(v) the movement of the charge material in a blast furnace

(vi) crystallisation and elemental segregation in ingots.

Thus, radiotracers may be introduced into a granule of coke, or ore, or lime, in studying the movement of the burden materials [81], and the 'tagged' granules are then followed through the furnace either by monitoring the radioactivity transmitted through the furnace shell at different levels in the stack, or by monitoring the radioactivity of the pig iron or of the slag produced over a period of time. The latter method, involving as it does simply adding the radioisotope to the feed line at known intervals before the furnace is to be tapped and measuring the activity of the metal and the slag produced, is the easier to apply, since detectors need not be mounted over or beside the furnace, but the data obtained relate only to average rates of movement in the furnace and give no details about differences in rates of travel in the different regions of the blast furnace. For these more precise data, the former method must be used, despite the difficulties inherent in measuring the very highly attenuated signals transmitted through the furnace wall. Rates of travel are measured by recording the intervals between times of maximum intensity registered on rings of detectors mounted at different levels, or by monitoring intensity changes from one level to another as a function of time. The radiotracers may be added to the feed-stock as P^{32}, in the form of Fe_3P, or Fe^{59}, as the oxide; these are usually added to the feed to the sinter strand before agglomeration and the resulting sintered mass is broken into 2–3 inch pieces and distributed through the furnace charge. Coke lumps may be tagged with Fe^{59} and Ce^{141} isotopes, by saturating them with solutions of salts of these isotopes, and artificially irradiated burden can also be used.

The movement of the gases in the blast furnace has been investigated by introducing a chemically inert and radioactive gas, usually radon, with the tuyere blast and monitoring its transit time and distribution in the stack [82]. This method has shown that the transit time of the gas in the centre of the furnace is longer than that of the gas near the stack wall by a factor of 1.5–2.

One of the major problems associated with studies of dephosphorisation of iron, or of the distribution of sulphur and phosphorus in iron slags, lies in the very low concentrations of the elements involved. The addition of sulphur or phosphorus to give levels at which accurate analytical data, using conventional methods, can be obtained is unacceptable, since such a move would upset the overall stoichiometry of the system. The easy and accurate detection of such isotopes as Fe^{59}, P^{32}, and S^{35} make these eminently suitable for use in such studies, since the system involved can be studied under precisely the conditions obtaining in practice. Hence, Winkler and Chapman [83] have studied the kinetics of the transfer of phosphorus from the metal to the slag phase using P^{32}, a group of Soviet workers [81] have used S^{35} and P^{32} to study the distribution of sulphur and phosphorus between iron slags of relatively simple compositions, and Kulikov and Zhokhovitsky [84] have used Fe^{59} to study the rate-determining steps governing the transfer of materials between slag and metal phases. A large body of information on the kinetics of exchange between phases in thermodynamic equilibrium is thus being built up, based on data obtained through radiotracer methods.

Other areas of research in which radiotracers have been applied with great success include studies of the melting of scrap, by introducing iron or cobalt isotopes into the charge and correlating the change in activity of samples taken from the open hearth, as a function of time, with the melting rate [81]. Effects due to changing the order of charging the lime, the ore, and the pig iron have been studied by labelling the lime with Ca^{45}, the ore with P^{32}, and the iron with Co^{60}. The behaviour of all three isotopes can, of course, be monitored simultaneously by electronic discrimination of the gamma-ray spectrum, and this ability again has obvious advantages for correlating the overall behaviour of the charge. The conditions of slag formation in the open hearth furnace have been studied by an isotope dilution method, using Ca, Fe, and P isotopes, and this application has again demonstrated the advantage of the method, not only in determining very small amounts of a constituent in the system, but also the advantages of being able to determine a component without having to separate it quantitatively from the other constituents of the system [81]. In the Azovstal works, U.S.S.R., this method has been used to determine the effects of special slags, oxygen addition, and the briquetting of the charge materials. Radioactive tracers have also been used to study sources of contamination in steel due to non-metallic inclusions [85], using for example Ca^{45}, and Fodor and Varga [86] have used P^{32} for the rapid determination of the phosphorus content in steels during the manufacturing process.

Applications in other metal extraction processes include the use of Au^{198} and Ag^{110} to study the movement of the molten metal in a copper verberatory furnace, reported by Saito [87], and isotope dilution methods have been used to monitor the amount of liquid aluminium present in a furnace at any time without hindering production [88]. Considerable use has been made of Na^{24} in mixing and flow studies in the cement manufacturing industry [89,90].

Surface Studies

Reactions at a solid surface in contact with a solution containing ions of different kinds can be studied in two principal ways, indirectly and directly. The indirect method involves measuring the concentrations of the ions of interest in the solution as a function of time and of distance from the surface, and analysis of such data can give information about the amounts of ions adsorbed by the surface, provided that the concentrations and the changes in concentration are large enough to permit the necessary analyses to be performed. The direct method involves determination of the actual amount of each ion adsorbed onto the surface, by measurements at the surface. For studying the sorption properties of complex solid surfaces, and for obtaining information about the sorption on different phases present in the surface, direct methods must be used. Since only very small amounts of material are actually sorbed, the sensitivity of the analytical method used must be very high, especially in the direct method, and radiotracers are again eminently suitable for such investigations. Gaudin *et al* [91–93], Bogdanov *et al* [94], and Plaksin [95–97] have reported successful applications of radiotracers to studies of flotation, using the reliable labelled reagents now available. Chang and Gaudin [93] have studied the adsorption of labelled *n*-dodecylamine and lauric acid on quartz, using the standard method of determining the specific activity of the sorbate and then its distribution between the sorbant mineral and the aqueous solution.

The most advanced method for studying the mechanism of interaction between flotation reagents and minerals is microautoradiography, which permits very direct study of the distribution of the reagents on the surfaces of individual particles as well as between the different types of particle present in the mineral pulp. Plaksin [98] has used the method to study variations in the distribution of xanthogenates labelled with S^{35} on the surfaces of sulphide mineral particles as a function of changes in the hydrocarbon radicals in the xanthogenate molecules, solution pH, oxygen concentration, and reagent concentration. He has also made a study of the flotation of fluorite using C^{14}-labelled reagents, which has made possible the quantitative evaluation of the significance of the oxygen concentration in the solution as a factor determining the distribution of the collector over the surfaces of particles in the flotation product. The use of microautoradiography in conjunction with radiometry has provided means for studying the joint action of collecting and frothing agents, and has led to proposals for improvements in the recovery of such minerals as pyrrhotite, which is difficult to float. Tridecylamine labelled with C^{14} has similarly been used to study the interaction of a cationic collector with huebnerite and wolframite [97].

Other Fields

Radiotracers today find many applications in fields of somewhat more peripheral interest to the minerals engineer. Bowie [54] has reviewed the use of autoradiography in the wider field of geological research, many examples of which are of very direct interest to minerals engineers, of course. Radiotracer techniques are being increasingly used in research and development, and some applications are now at the production plant level. Radiotracers find a ready use in diffusion studies in metals, alloys, and ceramic or refractory materials, where their chemical identity with the system being studied, coupled with their ease of detection, makes them immensely valuable [99–101] . Finally, mention may be made of their use in studies of refractory wear [81], of pipeline wear [102], and of their recent use in production control [80].

Looking to the future, it would appear that the importance of radioisotopes will increase in all branches of the minerals industries, both at the research and at the production levels. Improvements in detection systems will provide improved means of analysis, and, possibly more significantly, may provide means whereby automatic plant control systems may become commercially viable. The use of labelled reagents in froth flotation is only one of the possible points of application of radiotracers that could lead to efficient automation of a complex process.

ACKNOWLEDGEMENTS

The authors wish to express their thanks to Professor Stacey G. Ward, Head of the Department of Minerals Engineering, University of Birmingham, for his help and constant encouragement.

REFERENCES

1. G. FRIEDLANDER, J. W. KENNEDY and J. M. MILLER, 'Nuclear and Radiochemistry' (John Wiley and Sons, Inc., New York and London, 1966).
2. M. B. A. CRESPI, *Pure Appl. Chem.*, **26**, 259 (1971).
3. N. FEATHER, *Proc. Camb. Phil. Soc.*, **34**, 599 (1938).
4. G. von HEVESEY and H. LEVI, *Kgl. Danske Videnskab Mat. fys. Medd.*, **14**, 5 (1936).
5. G. T. SEABORG and J. J. LIVINGHOOD, *J. Am. Chem. Soc.*, **60**, 1784 (1938).
6. A. A. GRIENBERG, *Achievements in Chem.*, **9**, 771 (1940).
7. J. L. STRAIN, *IAEA. Conf. Radiochem. Methods of Anal.*, Salzburg, SM 55/48 (1965).
8. A. ZEMANN, *Sbornik Ref. Sem. Aktiv. Anal. Modrg.* (Czechoslovakia), 14 (1963).
9. W. W. MEINKE, *Anal. Chem.*, **30**, 686 (1958).
10. Y. KUSAKA, *Zeits. Anal. Chem.*, **172**, 199 (1960).
11. A. J. MOSES, 'Nuclear Techniques in Analytical Chemistry' (Pergamon Press, Oxford, 1964).
12. O. REIFENSCHWEILER, *Proc. Int. Conf. on Modern Trends in Activation Anal.*, Gaithersburg, Maryland. paper 20 (1968).
13. J. E. STRAIN, *Prog. in Nucl. Energy, Ser IX*, **4**, 137 (1965).
14. J. D. BUCHANAN, *Atomparxis*, **8**, 272 (1962).
15. V. P. GUINN, *Proc. IAEA Seminar Prod. and Use Short Lived Radioisotopes from Reactors, Vienna*, **2**, 3 (1962).

16. W. W. MEINKE, *Anal. Chem.*, **31**, 792 (1959).
17. H. J. M. BOWEN and D. GIBBONS, 'Radioactivation Analysis' (Oxford, Clarendon Press, Oxford, 1963).
18. G. LELIAERT, J. HOSTE and Z. EECKAUT, *Nature*, **182**, 600 (1958).
19. J. HOSTE, F. BOUTEN and F. ADAMS, *Nucleonics*, **19**(3), 118 (1961).
20. W. W. MEINKE, *Proc. Int. Conf. on Modern Trends in Activation Anal.*, College Station, Texas, paper 36, (1961).
21. J. RUZICKA and J. STARY, *Talanta*, **10**, 287 (1963).
22. J. STARY, J. RUZICKA and A. ZEMANN, *Anal. Chim. Acta*, **29**, 103 (1963).
23. D. F. COVELL, *Anal. Chem.*, **31**, 1785 (1959).
24. W. D. EHMANN, *Fortschr. Chem. Forsch.*, **14**, 49 (1970).
25. W. E. KUYKENDALL Jr, R. E. WAINERDI and ASSOCIATES, *Proc. Use Radioisotopes Phys. Sci. and Ind.*, Copenhagen, 233 (1960).
26. W. E. KUYKENDALL Jr. and R. E. WAINERDI, *Trans. Amer. Nucl. Soc.*, **3**, 95 (1960).
27. O. U. ANDERS and W. H. BEAMER, *Anal. Chem.*, **33**, 226 (1961).
28. J. W. NOSTRAND Jr, A. J. FAVELE, H. WINTON and M. D. D'ABOSTINO, *Trans. Amer. Nucl. Soc.*, **3**, 412 (1960).
29. W. R. BURRUS, *Trans. Amer. Nucl. Soc.*, **6**, 173 (1963).
30. D. L. MORRISON and D. N. SUNDERMAN, *Trans. Amer. Nucl. Soc.*, **6**, 174 (1963).
31. J. W. WINCHESTER and J. A. CATAGGIO, *Proc. Use Nucl. Tech. Prospecting and Development Min. Resources, IAEA*, Buenos Aires, 435 (1968).
32. H. P. YULE, *Anal. Chem.*, **37**, 37 (1965).
33. E. N. JENKINS and A. A. SMALES, *Quart. Rev. Chem. Soc.*, **10**, 83 (1956).
34. A. K. DE and W. W. MEINKE, *Anal. Chem.*, **30**, 1474 (1958).
35. S. C. MATHUR and G. OLDHAM, *Nucl. Energy*, 136 (1967).
36. W. E. MOTT and J. M. ORANGE, *Anal. Chem.*, **37**, 1338 (1965).
37. J. D. MAHONY and S. S. MARKOVITZ, *Anal. Chem.*, **33**, 329 (1961).
38. H. R. LUKENS, J. W. OTVOS and C. D. WAGNER, *Int. J. Appl. Radiation Isotopes*, **11**, 30 (1961).
39. J. W. OTVOS, V. P. GUINN, H. R. LUKENS and C. D. WAGNER, *Nucl. Instr. Methods*, **11**, 187 (1961).
40. R. C. GREENWOOD and J. REED, *Proc. Int. Conf. on Modern Trends in Activation Anal.*, College Station, Texas, paper (1961).
41. T. B. PIERCE, P. F. PECK and W. M. HENRY, *Nature*, **204**, 571 (1964).
42. T. B. PIERCE, P. F. PECK and W. M. HENRY, *Analyst*, **90**, 339 (1965).
43. J. W. HAFFNER and D. OESTREICH, *Trans. Amer. Nucl. Soc.*, **5**, 290 (1962).
44. G. von HEVESEY and F. A. PANETH, 'Radioactivity' (Oxford, Clarendon Press, Oxford, 1938).
45. H. BEQUEREL, *Compt. Rend.*, **122**, 1086 (1896).
46. M. M. SHAPIRO, *Rev. Mod. Phys.*, **13**, 58 (1941).
47. J. H. WEBB, *Phys. Rev.*, **74**, 511 (1948).

48. H. YAGODA, 'Radioactive Measurements with Nuclear Emulsions' (John Wiley and Sons, Inc., New York and London, 1949).
49. A. BIESER, *Mod. Phys.*, **24**, 273 (1952).
50. W. P. NORRIS and L. A. WOODRUFF, *Ann. Rev. Nucl. Sci.*, **5**, 297 (1955).
51. Y. GOLDSCHMIDT-CLERMONT, *Ann. Rev. Nucl. Sci.*, **3**, 141 (1953).
52. P. J. FITZGERALD, E. B. SIMMEL, J. WENSTEIN and C. MARTIN, *Lab. Invest.*, **2**, 181 (1953).
53. J. GROSS, R. BOGOROCH, N. J. NADLER and C. P. LEBLOND, *Amer. J. Roentg. Radium Therapy,* **65**, 420 (1951).
54. S. H. U. BOWIE, 'Physical Methods in Determinative Mineralogy', J. Zussmann (ed) (Academic Press, London and New York, 1967), Chapter 12.
55. S. H. U. BOWIE, *Bull. Geol. Surv. Gt. Brit.*, **3**, 58 (1951).
56. P. B. PRICE and R. M. WALKER, *Appl. Phys. Lett.*, **2**, 23 (1963).
57. R. F. FLEISCHER, C. W. NAESER, P. B. PRICE, R. M. WALKER and U. B. MARWIN, *Science,* **148**, 629 (1965).
58. D. E. FISHER and K. BOSTRUM, *Nature,* **224**, 64 (1969).
59. V. MURALI, P. P. PAREKH and M. SANKARDAS, *Anal. Chim. Acta,* **50**, 71 (1970).
60. J. W. WINCHESTER, 'Radioactivation Analysis in Inorganic Geochemistry', *Prog. in Inorg. Chem.*, Vol. 2, F. A. Cotton (ed), (Interscience, New York, 1960).
61. G. H. MORRISON (ed), 'Trace Analysis, Physical Methods' (Interscience, New York, 1960).
62. M. PINTA, 'Detection and Determination of Trace Elements' (Daniel Davey, New York, 1966).
63. W. W. MEINKE and B. F. SCRIBNER (editors), 'Trace Characterization, Chemical and Physical', U.S.N.B.S. Monograph 10D, Washington, D.C. (1967).
64. N. D. CHERONIS (ed), 'Submicrogram Experimentation' (Interscience, New York, 1961).
65. T. B. PIERCE, *Sec. Symp. Rec. Devel. Neutron Activation Anal.*, Cambridge, U.K., 3 (1971).
66. R. W. TOLMIE and L. J. THOMPSON, *Proc. Symp. Use Nucl. Tech. Prospecting and Development of Min. Resources, IAEA,* Buenos Aires, 504 (1968).
67. R. E. WAINERDI, E. A. UKEN, G. G. SANTOS and H. P. YULE, *Ibid,* 533.
68. G. E. GORDON, K. RANDLE, G. G. GOLE, J. B. CORLISS, M. H. BEESON and S. S. OXLEY, *Geochim. Cosmochim. Acta,* **32**, 369 (1968).
69. M. RAKOVIC, 'Activation Analysis' (Iliffe, London, 1970).
70. L. LOVBERG, H. KUNZENDORF and J. HANSEN, *Proc. Symp. Use Nucl. Tech. Prospecting and Development of Min. Resources, IAEA,* Buenos Aires, 197 (1968).
71. J. H. CZUBEK, *Ibid,* 3.
72. C. W. TITTLE, *Proc. Symp. Radioisotope Instr. Ind. Geophys., IAEA,* Warsaw, 3 (1968).
73. A. PRADZYNSKI, *Proc. Symp. Use Nucl. Tech. Prospecting and Development Min. Resources, IAEA,* Buenos Aires, 451 (1968).

74. A. M. GAUDIN, *Trans. I. M. M.*, **62**, 29 (1952).
75. T. C. MARTIN, *Trans. Amer. Nucl. Soc.*, **6**, 181 (1963).
76. C. POOVEY and D. Q. COVAULT, *Georgia Inst. Tech., Eng. Exptl. Station, Atlanta, Publ.* SRO-47 (1961).
77. R. LIEBERMAN, C. W. TOWNLEY, C. T. BROWN, J. E. HOWES Jr, R. E. EWING and D. N. SUNDERMAN, *Battelle Memorial Inst., Columbus, Ohio, Publ.* BMI-1505 (1961).
78. H. HAMAGUCHI, R. KURODA, T. SHIMIZU, R. SUGISITA, I. TSUKAHARA and R. YAMAMOTA, *Nippon Genshirhoku Gekkaishi,* **3**, 800 (1961).
79. G. C. GOODE, C. W. BAKER and N. M. BROOKE, *Analyst,* **94**, 728 (1969).
80. A. V. TOPCHIEV, I. T. ALADIEV and P. S. SAVITSKI, *Proc. Sec. Int. Conf. Peaceful Uses of Atomic Energy, Geneva,* **19**, 61 (1958).
81. A. B. TESMAN, *Metal Prog.,* 115 (1959).
82. M. SERIZAWA, *Tetsu to Hagane,* **43**, 938 (1957).
83. T. WINKLER and J. CHAPMAN, *Amer. Inst. Min. Eng. Tech. Publ.* 1987 (1946).
84. I. S. KULIKOV and A. A. ZHOKHOVITSKY, *Prod. and Treat. of Steel,* Vol. XXXII of Collected Works of Stalin Inst. Moscow, State Metalle Publ. Houses 54 (1954).
85. I. N. PLAKSIN, *Annals Acad. Sci. U.S.S.R., Dept. Tech. Sci.,* **1**, 109 (1955).
86. J. FODOR and C. VARGA, *Proc. Sec. Int. Conf. Peaceful Uses Atomic Energy,* **19**, 231 (1958).
87. T. SAITO, *Proc. Sec. Int. Conf. Peaceful Uses Atomic Energy,* **19**, 201 (1958).
88. M. S. BELTSKY and V. P. MATCHOVETZ, *Tzvetnce Metalle, xx,* (1955).
89. U. A. LUOTA, *Nag. Symp. Instr. Ind. Sparamn., Teck. foren. Fin. forhandl.,* 5 (1957).
90. U. BEEN and E. SAELAND, *Proc. First Int. Conf. Peaceful Uses Atomic Energy,* **15**, 170 (1955).
91. A. M. GAUDIN, H. R. SPEDDEN and M. P. CORRIVEAU, *Min. Eng.,* **3**, 780 (1951).
92. A. M. GAUDIN and C. S. CHANG, *Trans. Amer. Inst. Min. Met. Engrs.,* **193**, 193 (1952).
93. A. M. GAUDIN, D. W. FUERSTENAU and M. M. TURKANUS, *Min. Eng.,* **9**, 65 (1957).
94. O. S. BOGDANOV, B. Y. HAINMAN, M. A. YANIS and A. K. PODNEK, *Proc. Int. Conf. Radioisotopes Sci. Res., Paris,* P/72 (1957).
95. I. N. PLAKSIN, *Proc. Sec. Int. Cong. Surf. Act.,* Butterworths, London, 1957.
96. I. N. PLAKSIN, R. Sh. SHAFEEV and S. P. ZAITSEVA, *Dokl. Akad. Nauk S.E. S. R.,* **108**, 905 (1956).
97. I. N. PLAKSIN, *Proc. Conf. Use Radioisotopes Phys. Sci. and Ind., IAEA, Copenhagen,* **1**, 483 (1960).
98. I. N. PLAKSIN, *Proc. Sec. Int. Conf. Peaceful Uses Atomic Energy,* **19**, 249 (1958).
99. Y. P. GUPTA and T. B. KING, *Trans. Met. Soc. AIME,* **238**, 1701 (1967).

100. P. G. SHEWMAN and F. N. RHINES, *J. Metals,* **6**, 1021 (1954).
101. B. BROOK, A. ZAVYALOV and G. KAPIRIN, *Proc. Sec. Int. Conf. Peaceful Uses Atomic Energy,* **19**, 219 (1958).
102. H. L. TAVERES, P. E. AUN, A. A. MAESTRINI and M. T. MAGALHAES, *Proc. Third Int. Conf. Peaceful Uses Atomic Energy,* **15**, 341 (1965).
103. T. B. PIERCE, *Selected Ann. Rev. Anal. Sci.,* **1**, 133 (1971).

CHAPTER 5

Elemental Analysis Using Mass Spectographic Techniques

G. D. Nicholls and M. Wood

Department of Geology
University of Manchester
Manchester M13 9PL
England

5.1 INTRODUCTION

The use of mass spectrographic procedures for determining the elemental compositions of rocks, minerals, and allied materials, rather than for investigation of isotopic inter-relationships, is one of the newer developments in analytical geochemistry. Dempster [1] clearly recognised the possibility of utilising spark source mass spectrography for elemental analysis over thirty years ago, but the technique has been applied to the analysis of geological materials only during the last decade [2-5]. The particular features of this technique which have commended it to geochemists are

(i) the very wide range of elements that can be detected and, if required, determined quantitatively in a single analytical run,

(ii) the inability to 'miss' any element present in the analysed material in an amount greater than the limit of detection for that element, and

(iii) the high sensitivity of the method for almost all elements.

Theoretically all elements could be determined by this technique, but limitations imposed by instrument geometry normally restrict the recorded mass range to x–$30x$. The range most commonly chosen is 8–240, thereby including Th and U among the elements detected but excluding H, He, and Li.

This method of analysis differs from most other techniques of physical analysis of rocks and minerals in that it measures numbers of charged particles rather than the intensity of some form of electromagnetic radiation. Basically, the method involves incorporation of the sample to be studied into a pair of electrodes; excitation of a portion of the sample by sparking between the electrodes; dispersion of the resulting ion beam into separate component beams, each consisting of particles with the same mass/charge ratio; and finally measurement of the numbers of particles in the individual dispersed beams.

Since rocks and minerals are generally non-conductors or, at best, semi-conductors, they must be mixed with a conducting matrix before fabrication into electrodes. Graphite is commonly used for this purpose, thus eliminating carbon from the list of detectable elements, but powdered metals can be used instead if the detection and determination of carbon is of particular interest, or if the presence of carbon would interfere with the detection of other elements. Moreover, since spasmodic sparking can occur to the electrode holder, the metal used for the construction of this part of the instrument, usually tantalum, must again be excluded from the detectable list. In quantitative work, it is customary to use an internal standard and the element chosen for this role is then, necessarily, also eliminated.

For reasons arising from the experimental conditions under which determinations are made, the amounts of Si and O in silicate materials are too great to be determined in the same analytical run as trace elements, and it is usual for several other elements (the so-called 'majors') to be eliminated for the same reason. Moreover, under normal operating conditions the detection of a few elements, such as Sc or Au, presents problems due to interference effects and, although these problems can be overcome by the adoption of various correction procedures, the detection and determination of these elements is made more involved thereby.

Nevertheless, data on the elemental content of approximately 70 elements can be obtained in the course of a single analytical run on a given sample, and this is probably the most distinctive and the most useful feature of the technique. It cannot be stressed too strongly that, with the widely used sensitive plate method of recording or with electromagnetic peak scanning, it is virtually impossible to miss an element, whether its presence was suspected or not. Most analytical methods require a decision on the part of the analyst about the elements to be sought in the material under analysis, and in the majority of analytical problems this is no real disadvantage, but instances do arise in which time and effort are wasted looking for the wrong elements, or not looking for the right elements. Chemical causes of mal-performance of industrial raw materials, or failure of a product batch to meet specifications due to the presence of a trace chemical impurity, are illustrative cases. The clear indication of the presence of an unsuspected element given by spark source mass spectrography is then a very real advantage of the technique.

The full value of the comprehensive elemental coverage in discriminant geological studies has yet to be explored, but it is already clear that the full elemental population of a rock or ore formation is almost as distinctive as a fingerprint. Location of the source of an ore of uncertain provenance and determination of the likely location of rock contamination in imported industrial raw materials are but two of the problems made much more amenable to solution by the use of spark source mass spectrography for analysis of the suspect material. The method is very sensitive and, in many cases, it is unnecessary to use the technique to its maximum sensitivity. Elemental contents down to 0.001ppm(atomic) can be determined if desired and this feature, together with the wide range of determinable elements, permits studies to be made of the distribution of various elements whose geochemistry is still imperfectly known.

5.2 OUTLINE OF THE TECHNIQUE

As shown in the above section, there are three principal stages in the analysis of a mineral or other geological or ceramic material by spark source mass spectrography, namely

(i) preparing the sample,

(ii) sparking the sample and recording the spectrum of dispersed ion beams produced, and

(iii) measuring and interpreting the ion beam spectrum qualitatively and quantitatively.

5.2.1 Sample Preparation

Sample preparation affects the quality of the analytical data produced by this technique more than it does for most other techniques used in the chemical analysis of rocks and minerals. It constitutes the factor which effectively controls the overall analytical precision attainable using the technique. For a determination of an elemental content, in a submitted sample, of the order of 300ppm(atomic), the amount of electrode consumed under the excitation conditions commonly employed is only about 0.25×10^{-7}g, and if the particle diameter of the powder from which the electrodes are fabricated is about 2μm, this corresponds to around 2500 particles. The proportion of matrix, usually graphite, particles to sample particles in this number should, ideally, be the same as in the whole electrode, which may contain 10^{10}-10^{11} particles, and it is clear that good mixing of the powders from which the electrodes are manufactured is essential. When graphite is used as an electrode matrix, approximately 60%, or 1500, of the particles in the powder consumed are graphite. The remaining 1000 particles are of the sample plus any internal standard-bearing compound that may have been added. For the internal standard to have any analytical value, the proportion of internal standard-bearing particles to sample particles in this group of 1000 particles must be very close to the overall ratio of standard to sample particles in the bulk powder from which the electrodes were fabricated. Moreover, if the sample itself is multimineralic, for example if it is a common type of igneous rock such as granite, the proportions of the different mineral phases present in the 1000 particles consumed must again be the same as their proportions in the total sample submitted for analysis.

It will be clear from this that the very small amounts of electrode, and therefore sample, consumed during the sparking process result in the precision of sample preparation, particularly mixing and homogenisation processes, dominating the overall precision of spark source mass spectrography. Pioneer applications of the technique to the analysis of geological materials led to suspicions that the technique lacked the accuracy and precision necessary for high quality geochemical investigations, but it was subsequently shown [5] that, if careful attention was paid to electrode preparation, a precision of better than ±5% could be attained, with an accuracy within the limits of this precision. However, it remains true that informed selection of appropriate sample preparation procedures is vital if a desired analytical performance is to be achieved using this method. It may be fairly considered a disadvantage of this technique that it consumes such a small amount of sample, though there are cases where this feature is most definitely an advantage, as, for example, in the analysis of the full trace element population of individual glass spheres from lunar dust.

General Methods of Electrode Charge Mixing

Before proceeding further with this discussion, it will be useful to recall the general outlines of sample preparation for physical methods of analysis, and for this purpose it will be assumed that the sample is received in the analytical laboratory in the form of a powder with a grain diameter of approximately 0.1mm. This material must be well mixed before a sub-sample of approximately 1g is extracted. This first sub-sample is crushed to a much finer grain size and to it is added the internal standard-bearing material, already crushed to the same grain size as the sample. The resulting powder is again well mixed, and finally mixed with an equal weight of electrode matrix material to give the electrode charge mix. It cannot be too strongly emphasised that good mixing is imperative at all stages of this procedure.

In spark source mass spectrography about 0.25g of this mix is used for the fabrication of a pair of electrodes for a single analytical run. However, the mixing must, ideally, be sufficiently good to ensure that the very small amount of electrode actually consumed in the analysis (0.25×10^{-7}g for analysis at the 300ppm level or approximately 1×10^{-6}g for the 10ppm(atomic) level) is truly representative of the whole electrode charge mix. In a very real sense this small amount of sample actually consumed in the analysis can be regarded as a second sub-sample. When dry mixing methods alone are used in the preparation of the electrode charge mix, very fine grinding of the sample, the internal standard-bearing material, and the electrode matrix material is also vital, to ensure optimum sampling during the mixing and sub-sampling operations. Figure 5.1 shows the effect of grain size on the probable sampling precision for the ratio of matrix particles to other particles for second sub-samples of varying amounts. In addition, figure 5.1 shows the effect of grain size on the probable sampling precision for ratios of internal standard-bearing particles to sample particles, in the quantity of second sub-sample from the electrode charge mix that would give a measurable response from the internal standard, i.e. a sub-sample of 0.25×10^{-7}g containing 300ppm internal standard or a sub-sample of 1×10^{-6}g containing 10ppm.

Curves expressing the relationship between grain size and probable sampling precision, which also take into account sample homogenisation, would further depend on element location in the various particles. For an element uniquely located in one minor mineral component of a multimineralic powder, the curve would be similar to those which refer to the introduction of the internal standard, in figure 5.1, i.e. curves (d) through (f). For an element present in trace amount in a major mineral component (50% by volume) in such a powder, the curve would be closer to those referring to the matrix powder homogenisation, curves (a) through (c). Since instrumental precision on mass spectrographs commonly in use is about 2-3%, it is abundantly clear that, unless all powders are ground to particle sizes in the range 2-5μm when mixed dry in the preparation of the electrode charge, good overall analytical precision cannot be expected, and certainly the ultimate precision of the instrument cannot be achieved. Nicholls and his associates [5] have reported that, when conventional methods of dry mixing of electrode charges were used, overall precision was only ±20-25% and even when slurry mixing was used, to facilitate mixing by minimising the electrostatic attraction between particles, the precision improved only to ±12-20%.

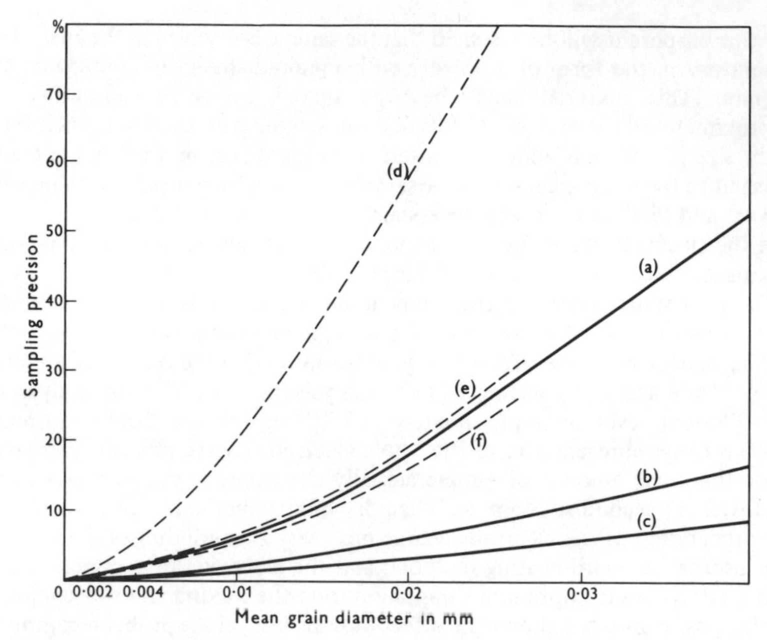

Figure 5.1 Effect of grain size of electrode charge mix powder on probable sampling precision.

(a) Sampling precision for ratio of matrix to other particles in electrode charge mix for 0.25×10^{-7} g sub-sample actually consumed in the analysis.

(b) Ditto for 0.25×10^{-6}g sub-sample.

(c) Ditto for 1.0×10^{-6}g sub-sample.

(d) Sampling precision for ratio of internal standard-bearing particles to sample particles in electrode charge mix for 0.25×10^{-7}g sub-sample with internal standard content of 300ppm.

(e) Ditto for 0.25×10^{-6} g sub-sample (using ion beam chopper to increase amount of sample consumed by a factor of 10).

(f) Ditto for 1.0×10^{-6}g sub-sample (using ion beam chopper to increase amount of sample consumed by a factor of 40).

Therefore, where the requirements of an analytical programme make an overall precision of ±25% acceptable, fine grinding and careful dry mixing of electrode charge mixes may be satisfactory. The internal standard material should be introduced in such amount as to produce a concentration of the internal standard element, expressed as ppm(atomic), in the electrode charge mix within the expected range of elemental contents that are to be determined.

An Improved Method of Electrode Charge Mixing

Two suggestions have been advanced for improving analytical precision by reducing sub-sampling imprecision. In one of these the problem posed by the very small amount of electrode consumed in the course of an analysis is overcome by the instrumental modification of ion beam chopping, which is more fully discussed in section 5.3.1 and which permits only a fraction of the incident ion beam to pass to the recording device, thus directly increasing the amount of electrode consumed for an analysis to a given elemental level. This approach involves no modification of the sample preparation techniques. Theoretically, precision can be improved by a factor of 4-5, but the time required for the analysis is thereby increased by a factor of 16-25.

The second suggestion recognises that inhomogeneity in the sample and in the distribution of the internal standard contributes more to sub-sampling imprecision than does imperfect distribution of the matrix and the other particles in the electrode charge mix. Indeed, when an internal standard is used, imprecision in the ratio of matrix particles to other particles in the electrode mix has little or no effect on the overall analytical precision. The solution adopted in this case seeks to achieve complete homogenisation of the sample-internal standard mix before final grinding, by fusing this mix to a glass.

Basically, the finely ground sample is mixed with a suitable flux containing the internal standard element in known amount, sealed in an air-tight platinum capsule, and heated to a temperature of 1200°C. For rocks and similar materials, Nicholls [6] has suggested the following composition for a suitable flux,

SiO_2	44.5% by weight
Al_2O_3	15.1%
Na_2CO_3	28.8%
K_2CO_3	2.2%
Fe_2O_3	9.4%

and to this the internal standard is added by the method of successive dilution. Rhenium is the most commonly used standard, added as potassium perrhenate.

Such a flux contains none of the elements likely to be sought in the analysis (since all will normally be present as 'major' components), melts at a temperature below 1000°C, to give a low viscosity liquid which is miscible with molten silicates corresponding to rock compositions in all proportions, and liberates small bubbles of CO_2 gas on heating which, by rising through the liquid in the sealed capsule, act as a built-in stirring device and promote mixing. Furthermore, the glass resulting from rapid chilling of the fused silicates has a pumice-like nature, which facilitates crushing and grinding to a powder. Despite all these advantages over dry mixing, however, it is necessary to conduct three successive fusions of a sample to produce a truly homogeneous glass.

In tests of this method conducted by Nicholls and his associates [5], equal amounts of finely ground sample and Re-bearing flux were mixed for 15 minutes in a mixing mill and the resulting powder was tightly packed into ¾ inch by ¾ inch platinum envelopes prepared from platinum foil,

Figure 5.2 Platinum foil capsule containing fusion charge being heated in the loop of a radio-frequency heater, the inner diameter of which is 4.5cm.

0.001 inch thick. The envelopes were sealed, care being taken to exclude air, and the sealed envelopes, or capsules, heated in a radio-frequency heater for 1 minute. Figure 5.2 shows a platinum capsule in the loop of the radio-frequency heater. They were then rapidly chilled by immersion in liquid nitrogen and the resulting glass was removed from the envelope and ground to a fine powder. Addition of 10% by weight of sodium carbonate to the fine powder restored the 'built-in stirrer', and this mixture was packed into a new envelope and the fusion process repeated. Grinding and refusion were repeated, so that the sample received three successive fusions and the finely powdered glass from the third stage was mixed with the requisite amount of graphite to prepare the electrode charge mix. Sampling precision was found to be ±4%, using this method, and virtually independent of the size of the sub-sample actually consumed in the course of the analysis.

Grinding to the very fine grain sizes required in the dry mixing methods is not necessary in the fusion technique. The whole triple-fusion process takes about 30 minutes, which compares favourably with the time required for the careful grinding and mixing necessary with dry mixing to yield a sampling precision of only ±20%. Contamination of the charge with platinum may, of course, occur, and this probability eliminates platinum from the list of detectable elements when the fusion method is used for sample preparation. Nevertheless, the method is strongly recommended for the most precise analytical work.

Electrode Preparation

Whatever the method used for preparation of the electrode mix, the powder must now be compressed into solid electrodes for insertion into the electrode holders of the instrument. Where there is no shortage of sample, and therefore of electrode charge mix, approximately 0.25g of the powder is compressed into a bar, about 20mm long and of 4 square millimeters cross sectional area, in an electrode-forming die (figure 5.3) under a pressure of 10,000-15,000psi. Some operators favour thin electrodes and use only 0.1g of powder to produce electrodes less than 1mm thick and 2mm wide, and there is no doubt that instrumental performance can be improved by using such thin electrodes. Unfortunately they tend to be fragile and require skilful operation of the electrode controls. The bar of compressed electrode charge mix is finally broken into two approximately equal parts, which are mounted in the electrode holders so that the spark is struck between the broken surfaces.

An alternative method of electrode preparation has been introduced by Brown and Vossen [7] to utilise more fully the capability of the technique to analyse very small samples. It is often called the 'tipped electrode technique' and utilizes the equipment shown in figure 5.4. The die is formed from a cylindrical slug of polyethylene into which is drilled a 2mm-diameter hole, about 10mm long, terminating in a cone. The electrode charge mix, or the sample mix, or the sample with no admixture at all, is introduced into this hole, tamped down with a teflon rod, and the remainder of the hole filled with graphite, or whatever electrode matrix material is being used. After further tamping of the powder, the slug is placed in a moulding die and subjected to a pressure of about 130,000psi. The polyethylene tends to flow under these conditions, causing near isostatic pressures to be applied to the enclosed powder, which compacts it into a rod. On releasing the pressure, the polyethylene has sufficient elasticity to regain its original

Figure 5.3 Seidl electrode-forming die used in the preparation of bar electrodes. The scale marker at the centre of the figure is divided into 1 cm sections. In the upper part of the figure, above the scale marker, the die is dismantled to show its component parts. *Top left,* hardened steel base plate; *top centre,* hardened steel confining ring with central hole of conical, not cylindrical form; *top right,* hardened steel extraction ring; *bottom,* two halves of slotted cylinder showing locating lugs and sockets, on either side of the upper slide plate of the die, a bar electrode, and the lower plate of the die.

To use the die, the two halves of the slotted cylinder are fitted together and seated in the conical central hole of the confining ring. The lower die plate is placed in the base of the central slot and the structure 'tightened' by compression of the conical cylinder into the confining ring while the latter is located on the extraction ring. The electrode charge mix powder is poured into the central slot and the upper slide plate inserted into the slot. This plate is then forced down the slot under pressure while the base of the slotted cylinder rests on the base plate, the extraction ring having been removed. After compression, the extraction ring is again interposed between the confining ring and the base plate. Pressure on the upper slide plate then pushes the lower die plate and the bar electrode out of the slot onto the base plate.

In the lower part of the figure, the die is seen after use with, to the left, the slotted conical cylinder still in the confining ring with the upper slide plate of the die still in the slot and, to the right, the extraction ring still located in the base plate. Between the two parts of the die are the lower die plate and bar electrode, which have been pushed out of the slot.

Figure 5.4 Equipment used in the preparation of tipped electrodes. The moulding die consists of four hardened steel parts, shown in the top half of the figure. *Top,* external casing; *left centre,* base plate with anvil; *middle centre,* confining ring, drilled with a cylindrical hole which is recessed at the base to receive the anvil of the base plate snugly; *right centre,* plunger.

To use, the confining ring is firmly located on the base plate and the forming die (a polyethylene slug) introduced into the central cavity of the confining ring. The external casing is then fitted over the assembly and the plunger put into position with its narrow cylindrical end located in the upper part of the central circular cavity of the confining ring above the slug. Pressure is applied to the wide (outer) end of the plunger. The forming die, a cylindrical slug of polyethylene, is shown to the right below the confining ring. The cylindrical hole drilled in this slug has been filled with graphite to make it more visible. *Bottom,* a tipped electrode and to the left the plastic filler block used to introduce the powder into the slug, with a conical depression on the upper surface leading to a 2mm hole against which the slug can be located. The scale marker is divided into 1 cm sections.

shape, thus retracting slightly from the rod and facilitating removal of the tipped electrode. Experience with these tipped electrodes in the authors' laboratory has been satisfactory and they appear to be less likely to suffer minor surface disintegration than the more conventional bar electrodes. They are also less susceptible to surface contamination during electrode fabrication than the bar electrodes. There seems to be no good reason why this method of electrode fabrication could not be used for all samples, although its real value undoubtedly lies in the possibility of handling micro-samples, such as lunar spherules, which cannot be handled by any other technique.

5.2.2. Sparking and Recording Procedures

The second stage in the technique involves the use of the mass spectrograph itself. In Britain most of the instruments currently used for spark source mass spectrography belong to the A.E.I Scientific Apparatus Ltd., M.S.7 series, and in the following account the description of the instrument refers to this type, although other instruments commercially available are basically similar.

There are three essential units in any instrument used for this technique, namely

(i) the excitation source area,
(ii) the dispersion unit, and
(iii) the recording device.

The mass spectrograph is shown diagrammatically in figure 5.5, with the three units referred to above delineated, and a typical commercial instrument is shown in figure 5.6. When in use, all parts of the instrument are under high vacuum, but the three units can be isolated from one another by valves, and these afford convenient points for subdividing the instrument for the purpose of this description.

Excitation of the Beam

The excitation source area (figure 5.7) is a chamber in which the ionisation of the sample takes place. It can be sealed off from the rest of the instrument by a source isolation valve and is provided with independent pumping facilities, as shown in figure 5.5. This permits the source area to be raised to atmospheric pressure for changing electrodes without loss of vacuum in the remainder of the instrument. Access to the source area is by removal of a thick, round lead glass disc, clearly visible in figures 5.6 and 5.7, which, when the source area is under vacuum, is held in place by air pressure forcing it against a Viton gasket. After the electrodes have been clamped into place in the electrode holders, the lead glass disc and gasket are put into position and the source area pumped down to a pressure which should not exceed 5×10^{-6} torr and should, preferably, be about 0.5×10^{-6} torr. By making the source area as small as possible, commensurate with ease of handling the electrodes, and by the use of oil diffusion pumps

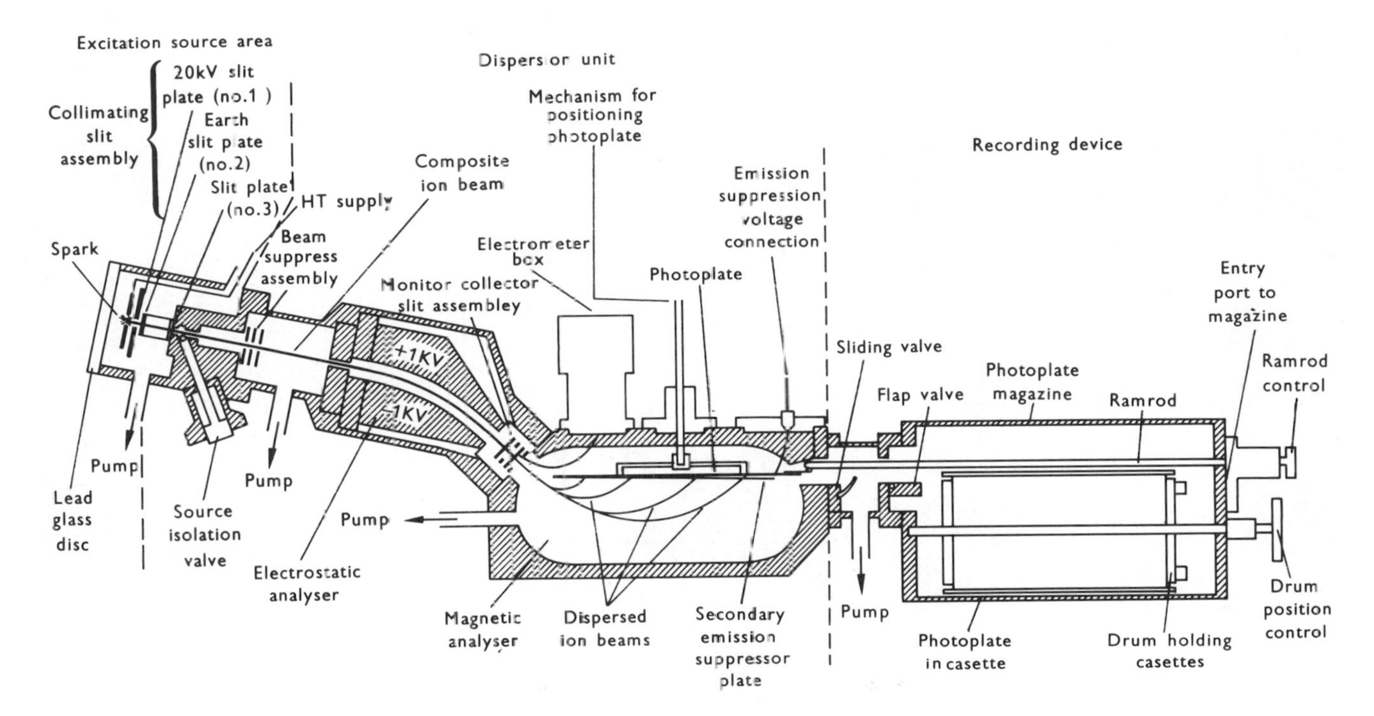

Figure 5.5 Diagram to illustrate the principal features of an A.E.I. Scientific Apparatus Ltd. M.S.7 spark source mass spectrograph using sensitised plate recording. The diagram is discussed in detail in the text.

Figure 5.6 Spark source mass spectrograph (A.E.I. Ltd., M.S.7) in the authors' laboratory, viewed from the control panel end. The excitation source area is directly above the pad of data sheets on the table and is closed against the atmosphere by the thick circular lead glass plate. Electrode manipulation controls are visible to the right of the plate, and the electrode vibrator of the 'Autospark' attachment is to the left. The two units on the bottom right of the photograph are the control units for the 'Autospark' (section 5.3.2) and the ion beam chopper (section 5.3.1).

cooled by liquid nitrogen and backed by rotaries, the pump-down time can be kept as low as 5-10 minutes. Ionisation of the sample is achieved by passing a high-frequency (500kH) spark discharge between the two electrodes.

The high voltage discharge with a large instantaneous current followed by a relatively long 'off' period produces high local temperatures in the electrode tips with consequent evaporation of the electrode material, including both matrix and sample. The resulting plasma consists of a range of ionic species and atoms which are ionised in the discharge. One disadvantage of this type of discharge is that it produces a wide spread of energy levels among the ionic species, as well as considerable variation in the

Figure 5.7 Excitation source area of a spark source mass spectrograph as seen by the operator in the normal working position, but without the tantalum spark guard to show the electrodes more clearly through the lead glass plate. The bar electrodes are visible at the centre of the area, clamped in the electrode holders, which are mounted at the ends of glass insulators. The Viton gasket seal appears clearly as a black ring sandwiched between the metal flange and the lead glass plate. The external diameter of the metal flange is 16cm.

nature of the ionic species generated. Ideally, all the atoms of a given element will be present in the discharge as singly charged monatomic ions for ease in subsequent analysis but in reality multiply charged ions of the elemental atom, i.e. M^{2+} or M^{3+}, or singly charged multiatomic ions, e.g. M_2^+

or M_3^+, will be produced simultaneously, and these can cause confusion and interfere with the detection of other elements, in the subsequent interpretation stages of the analysis.

The aim, therefore, is to produce an ionic population in which singly charged monoatomic ions predominate, and the sparking conditions can be varied, within limits, to suit a particular type of sample. The spark pulse length and the pulse repetition rate can be varied, as also can the spark voltage, though once they have been fixed for a particular type of sample they must remain unchanged throughout the analysis. Inter-element relative sensitivities are markedly affected by changes in the sparking parameters and it is prudent to adopt a standard set of sparking conditions, changing them only when the nature of the rock demands it. For the majority of rocks and minerals a spark voltage of 25kV, a pulse length of 200 microseconds, and a pulse repetition rate of 300 per second have been found to yield the best results.

Too high a spark voltage leads to over-production of multiply charged monoatomic ions as well as overheating of the electrodes, although, of course, the intensity of the ion beam produced is increased. Too low a spark voltage leads to a weak beam and a higher proportion of singly charged multiatomic species in the ion population. Spark pulse rate and pulse repetition rate together determine the ratio of 'live' time to 'off' time during the sparking. If this is too low, a weak beam is produced and the analysis time is unduly extended. If it is too high (greater than 1:10 for most rock and mineral samples) overheating of the electrodes occurs with consequent selective volatilisation of certain elements and some thermal ionisation, especially of those elements having low first ionisation potentials. There is evidence of selective volatilisation in analytical runs on basaltic rock samples conducted with a spark pulse rate of 100 microseconds and a pulse repetition rate of 1000 per second, i.e. a 'live' to 'off' ratio of 1:9. Considerations such as these limit the extent to which analysis time can be decreased by more energetic excitation of the sample.

It is important to remember that the spark also generates low energy X-rays from the sample, and the lead glass sealing disc is designed to protect the operator from this radiation hazard. Alternative closing devices should be used only after careful testing to ensure operator safety.

The charged particles produced in the spark excitation are accelerated towards a collimating slit assembly, see figure 5.5, by a large negative potential, normally 20kV, applied between the number 1 slit and the number 2 slit, which is at earth potential. This slit assembly draws the positively charged particles into a composite ion beam, containing a representative sample of all the ionic species and energy levels generated in the spark. The thickness of the emergent beam is determined by the aperture width of the final, number 3, slit of the assembly, and should not be greater than 0.5μm. A slit width of 0.2μm has been used for applications where instrumental resolution was more important than analysis time. Obviously, the absolute number of ions transmitted into the analyser of the instrument decreases as the aperture width of number 3 slit is reduced. After transmission through number 3 slit, the composite ion beam passes out of the source area into the dispersion unit.

Dispersion of the Beam

The mass spectrographs used for elemental analysis are of the double-focussing type, with two components in the dispersion unit, namely the electrostatic analyser and the magnetic analyser.

The electrostatic analyser is simply an energy level selector, and the need for it arises from the wide spread of energy levels among the ions produced by the radio-frequency spark excitation. It comprises two arc plates of length $\pi/4\sqrt{2}$ radians, of which one is held at +1kV and the other at −1kV, as shown in figure 5.5. Only ionic species for which the energy, expressed as mv^2/e (where m is the isotope weight, v is the velocity, and e the ionic charge), lies within a very limited range can follow the path between the arc plates and reach the exit without striking either plate and being trapped thereon. For two singly charged particles of masses m_1 and m_2 passing through the exit slit,

$$m_1v_1^2 = m_2v_2^2 \tag{5.1}$$

since their kinetic energies are virtually identical, whence

$$v_1/v_2 = (m_2/m_1)^{½} \tag{5.2}$$

On leaving the electrostatic analyser stage, the 'energy selected' composite ion beam passes through a monitor collector slit assembly, as shown in figure 5.5, which is designed to intercept 50% of the ion beam at this point. Since no mass separation has yet taken place, the intercepted 50% will be truly representative of the remainder of the ion beam, which passes on into the magnetic analyser. Before discussing the function of the monitor collector slit assembly it is convenient to follow that 50% of the ion beam which enters the magnetic analyser.

This analyser is a chamber lying within a strong magnetic field, normally 16 kilogauss, produced by two large magnetic coils. Within the magnetic field, whose lines of force are perpendicular to the path of the particles as they enter the chamber, the ionic species follow curved paths, in fact arcs of circles with radii r, governed by the relationship

$$kr = mv/eB \tag{5.3}$$

where k is a constant, m, v, and e have the same significance as before, and B is the flux density of the magnetic field. For two singly charged particles of masses m_1 and m_2 the ratio of the radii of their paths in the magnetic analyser is given by

$$r_1/r_2 = m_1v_1/m_2v_2 \tag{5.4}$$

and substituting for the condition imposed by the electrostatic analyser, from equation (5.2),

$$r_1/r_2 = (m_1/m_2)^{1/2} \tag{5.5}$$

Particles of larger mass, therefore, follow paths of greater radii of curvature and a mass separation takes place with a dispersion of the composite, but 'energy selected', ion beam into a spectrum of beams, each composed of particles of a particular m/e value. Note that a singly charged particle of mass m is equivalent to a doubly charged particle of mass $2m$, or a triply charged particle of mass $3m$, and so on. These beams, are brought to focus at the exit of the magnetic analyser, where the beam spectrum is recorded.

Certain features of this spectrum are of interest. Low mass particles will follow strongly curved paths and be brought to focus at the near, or electrostatic analyser, end of the magnetic analyser, while high mass particles will follow less strongly curved paths to the far end of the analyser. The discrete beams at the focal plane will be arranged in a regular sequence of increasing m/e value, but the spacing between adjacent beams will progressively decrease as their m/e values increase. If, for example, the radius of curvature is r for $m/e = 16$, it will be $2r$ for $m/e = 64$, $3r$ for $m/e = 144$, and $4r$ for $m/e = 256$.

It should be noted that particle path deflection and possible energy transfer during and after the electrostatic analysis stage, by collision of charged particles with residual gas molecules in the two analyser chambers, is avoided, as far as possible, by maintaining a vacuum of 10^{-7} to 10^{-8} torr in the dispersion unit, again by the use of liquid nitrogen cooled oil diffusion pumps backed by rotaries.

Returning now to the 50% of the 'energy selected' composite beam intercepted at the monitor collector slit assembly, this is used to measure the number of ions entering the magnetic analyser during a given period of recording or 'exposure'. The amount of charge collecting on the plates of this assembly during the analysis, due to the impact of the charged particles thereupon, is measured by an electrometer valve circuit, shown as the electrometer box in figure 5.5, and, after appropriate amplification, is displayed on a monitor meter. The charge collected by the monitor slit assembly is a constant fraction, ideally 50%, of the charge carried by the ions leaving the electrostatic analyser, and is, therefore, related by a constant factor to the amount of charge carried by the ions entering the magnetic analyser. Variations in beam intensity during the course of sparking can be observed on the monitor meter. It will be clear that, even with careful manipulation of the electrode controls to ensure that the spark gap is held as constant as is practicable, the ion beam produced by the spark source excitation is variable and so it would be impracticable to determine the duration of recording periods on a time basis. Instead, the duration is determined by the total amount of charge collected on the monitor slit assembly in the course of the recording. The monitor circuit feeds a second electronic circuit which integrates the amount of charge collected and displays this integrated amount on an integrator meter. Recording is continued until a pre-determined amount of charge has been collected. With the sensitised plate method of recording, vid. inf., this is about 150 nanocoulombs for a sensitivity level of 0.01ppm (atomic). If all the charged particles in the composite ion beam were singly charged, i.e. each carried the elementary charge 1.601×10^{-19} coulomb, the amount of charge collected would clearly be a direct measure of the number of particles entering the magnetic analyser during the recording period. This is one of the several reasons why any significant proportion of multiply charged ions in the 'energy selected' beam is undesirable.

Since any interruption in the spark discharge can lead to inhomogeneities in the beam produced, recording periods are terminated not by cutting off the spark but by deflecting the composite beam away from the monitor slit assembly. This is achieved by applying 1300 volts to one of a pair of beam-suppressing half-plates located near the entry of the composite ion beam into the dispersion unit and shown as the beam suppress assembly in figure 5.5. Further recording is then initiated simply by removing this voltage. Only at the end of a complete analysis is the spark discharge switched off.

Since trapping of unwanted ions and charged particles on surfaces within the dispersion unit plays such an important role in the selection of ions to be recorded, it will be clear that these surfaces gradually become contaminated with unwanted material, and so it is necessary to clean these surfaces periodically by baking out under vacuum. This constitutes part of the general routine servicing of the instrument, and will be further discussed in section 5.2.4.

Recording the Dispersed Beams

When spark source mass spectrography was first applied to the analysis of rocks, minerals, and allied materials, the recording device adopted was a sensitised plate, and 10 inch by 2 inch Ilford Q2 plates have been widely used. Since that time, electrical detection methods have been introduced, and these will be discussed in section 5.3.3, under 'Further Developments', but the plate method of recording is still widely used, and much can be learned about the problems of measurement and interpretation of the recorded spectra from a consideration of the use of this method of recording.

Until they are required for recording, the plates are housed in cassettes located in the photoplate magazine, see figure 5.5, which can be isolated from the dispersion unit by two valves, a sliding valve and a flap valve. Eight cassettes, each containing one plate, can be accommodated in the magazine at any one time around the circumference of a drum, which can be rotated without breaking the vacuum in the instrument. After closing the sliding and the flap valves, the magazine can be vented to the atmosphere to permit insertion or extraction of cassettes through the entry port provided, without affecting the high vacuum in the dispersion unit. The magazine is provided with independent pumping facilities to bring it to the required pressure of 10^{-6} atmosphere before opening the valves between it and the dispersion unit. Transport of the plates between the magazine and the recording position is achieved using a ramrod, which is normally almost completely withdrawn from the magazine, although it is shown delivering a plate to the recording position, in figure 5.5. Lugs are provided on the plateholders onto which the ramrod can engage to carry the plateholder, with its plate, from the magazine, through the open valves, into the dispersion unit. Having placed the plate with its sensitised emulsion side next to the exit slit of the dispersion unit, the ramrod is disengaged and withdrawn so that the two interconnecting valves can be closed before analysis is begun. At the end of an analysis, the plate is returned to the magazine by reversing the above procedure, and, with the ramrod fully withdrawn, the cassette drum can be rotated to bring the next plate into the loading position. Thus eight analyses can be conducted without breaking the vacuum in the magazine.

The 'exit slit' of the magnetic analyser is 2.5-3.0mm wide, and 15 successive exposures of a plate to the dispersed ions beams can be made by racking the plate across this exit slit in 3mm steps in a direction perpendicular to the plane of figure 5.5. Generally these successive exposures are made for progressively longer times so that a range of exposure densities may be obtained to aid the estimation of the different elemental species present. The exposure times are controlled by using the beam suppress assembly, as described earlier.

The exit slit, in practice, is a longitudinal slit cut in a metal plate, which also acts as the secondary emission suppressor plate of figure 5.5. The secondary ions emitted from the sensitised recording plate by the action of the incident ion beams are trapped by a small potential applied to this plate. Secondary ion dispersal is minimized, i.e. the spectrograph is focused, by empirical determination of the potential required to produce lines on the processed recording plate of maximum height to width ratio.

After processing the sensitised plates, according to the manufacturer's instructions, line images develop corresponding to the incident dispersed ion beams. The position of the beam on the plate is a measure of its m/e value and hence can be used to identify the charged species, and the intensity of the line is a measure of the amount of the species present. The appearance of a typical processed plate is shown in figure 5.8. Franzen, Maurer and Schuy [8] have shown that the precision of recording using an Ilford Q2 plate should be in the range ±1.5-2.0%.

5.2.3 Interpretation and Measurement of the Ion Beam Spectrum

A variety of particles contribute to the lines appearing on the processed sensitised plates, as may be expected from the discussion in the previous section. The chief contributors are, of course, singly charged isotopes of the various elements present in the electrodes. Unfortunately, as has been shown in Chapter 4, elements are not generally monoisotopic and so some lines can be produced by singly charged isotopes from more than one element. Thus, $^{58}Ni^+$ and $^{58}Fe^+$ both contribute to the line for $m/e = 58$, and there are many other cases where the dispersed beams are not isotopically uniform,

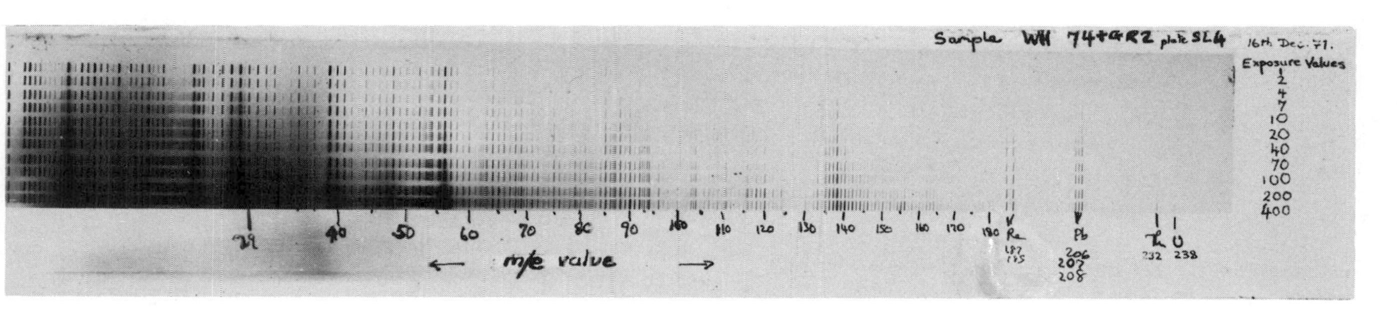

Figure 5.8 Typical processed Ilford Q2 plate showing the ion beam spectrum recorded from a sample of deep sea basalt. The lines at m/e values 185 and 187 are due to the isotopes of rhenium, which was added as the internal standard. Note that the Ba lines at 134–138 inclusive are strong, so $^{138}Ba^{++}$ might be expected to contribute to the line at $m/e = 69$ as well as $^{69}Ga^{+}$. Further examination of the spectrum over the range 67–69 shows the existence of lines at $m/e = 67\frac{1}{2}$ and $68\frac{1}{2}$ which must be due to $^{135}Ba^{++}$ and $^{137}Ba^{++}$, so interference in the line at $m/e = 69$ is proved, see section 5.2.3. Note also that doubly charged lead, between 103 and 104, and rhenium, at $92\frac{1}{2}$ and $93\frac{1}{2}$, are also visible.

even considering only the singly charged species. On the other hand, consideration of the probable relative concentrations of different isotopes in nature, and hence of their relative contributions to the ions recorded at a given *m/e* ratio, may lead to the conclusion that no serious error will be introduced if the line is attributed to the dominant contributory isotope. Thus, the line for $m/e = 138$ is not strictly uniform, since contributions can be made by the isotopes $^{138}Ba^+$, $^{138}La^+$, and $^{138}Ce^+$. The isotopic abundances of the species quoted, however, are, for ^{138}Ba, 71.50%, for ^{138}La, 0.089%, and for ^{138}Ce, 0.258%. Moreover, barium is generally much more abundant than either lanthanum or cerium, and so the line at $m/e = 138$ can safely be attributed to ^{138}Ba.

In addition, even with the most careful control of sparking parameters, some doubly, and triply, charged ions of certain isotopes are produced in the source and may appear at positions corresponding to $m/2$ or $m/3$, where m is the mass number of the isotope. Both $^{28}Si^+$ and $^{56}Fe^{++}$ may contribute to the $m/e = 28$ line. Fortunately, evidence of second-order, or doubly charged, contributions to lines can often be deduced by considering the total isotopic population of the element producing the second order contribution. The principal isotopes of barium are ^{138}Ba and ^{137}Ba, with abundances of 71.5% and 11.4% respectively. It is possible that $^{138}Ba^{++}$ might contribute to the $m/e = 69$ line, in addition to $^{69}Ga^+$ to which this line is usually attributed. If the barium contribution was significant, however, a weak line should be present at $m/e = 68\frac{1}{2}$, representing $^{137}Ba^{++}$. Since the sparking conditions demand that the same percentages of ^{138}Ba and ^{137}Ba be produced as the doubly charged ion, the intensity of the $^{137}Ba^{++}$-line will be about one-seventh of the contribution of $^{138}Ba^{++}$ to the $m/e = 69$ line, and so the latter contribution can be allowed for in interpreting the 69 line. Note that the non-integral nature of m/e for the $^{137}Ba^{++}$ line makes it relatively easy to identify, despite its low intensity. Further evidence of second order contributions can come from slight mass defects in the isotopes involved. Deviations of the actual mass of the isotope from its mass number, which is the total number of nucleons present in the nucleus, may result in such mixed order lines appearing as doublets which can be resolved if care is taken over the focusing of the spectrograph (cf. section 5.2.2). Equation (5.5) shows that this resolution is most likely to be successful for smaller m/e values. Thus, if $^{62}Ni^{++}$, actual mass 61.928, is produced in any quantity together with $^{31}P^+$, actual mass 30.974, a doublet will appear at the $m/e = 31$ position, the higher component ($m/e = 30.97$) being due to $^{31}P^+$, and the lower component ($m/e = 30.96$) being due to $^{62}Ni^{++}$.

Generally, multiply charged monatomic ions give less trouble than polyatomic species, for two main reasons. In the first place, as seen above, second and higher order contributions often appear at positions for which m/e is not an integer, and where, therefore, no first order contribution could appear. Secondly, it has been shown [6] that the tendency to increase second and third order contributions increases as the content of the isotope in the electrode increases, so that second and higher order contributions would be expected to be most serious from the major elements in the sample analysed. The largest mass number represented among the isotopes of the major elements of most rocks and minerals is 58. Excluding carbon as the probable matrix element, the only non-major elements, i.e. elements that one would be analysing for, with isotope mass numbers below 29 are fluorine (19), nitrogen (14 and 15), boron (10 and 11), beryllium (9), and lithium (6 and 7). Comparatively poor resolution, or bad focusing, would suffice to resolve first and second order emission at these very low *m/e* values.

Resolution of beams due to singly charged polyatomic ions from those due to singly charged monatomic ions of almost the same mass, on the other hand, is frequently virtually impossible. Singly charged polyatomic ions are produced by certain elements, and carbon is particularly prone to do this when added to the electrode in the form of graphite. Thus, $^{12}C^+$, $^{12}C_2^+$, $^{12}C^{13}C^+$, $^{12}C_3^+$ $^{12}C_2{}^{13}C^+$, and so on, are all produced simultaneously during sparking of graphite electrodes, and the regular series of lines at *m/e* values of 12, 24, 36, 48, 60, etc., is a feature of spectra recorded from graphite-bearing electrodes. On the one hand this can be useful for marking the *m/e* scale across the plate, but on the other it can lead to interference effects at the lower end of the *m/e* scale where the contribution from the polyatomic species is greatest. A line at $m/e = 24$ cannot be attributed solely to $^{24}Mg^+$ if carbon is present in the electrodes since a considerable contribution will have been made by $^{12}C_2^+$, but, with reasonably good instrumental focusing, $^{60}Ni^+$ ($m/e = 59.93$) should be distinguishable from $^{12}C_5^+$ ($m/e = 60.00$). At high *m/e* values, where only very good focusing would separate the beams, and particularly above $m/e = 144$, the contributions from polyatomic carbon ions from electrodes other than pure graphite is so small as to be negligible, fortunately. Note should be taken that carbide ions, such as SiC^+, may also be formed and can lead to line interference. $^{28}Si^{12}C^+$ and $^{40}Ca^+$ would both contribute to the line at $m/e = 40$, and most instruments would be unable to resolve these into a doublet. If polyatomic carbon ions, or carbide ions, are thought likely to prove a serious interfering hazard, some other conducting powder than graphite can be used as the electrode matrix.

Oxide ions present greater problems in the analysis of silicates, and particularly silicon and carbon oxides can interfere badly in certain cases. Thus, most instruments would fail to resolve the $m/e = 45$ line into a doublet representing the contributions from $^{29}Si^{16}O^+$ and $^{45}Sc^+$. Since scandium has only one commonly occurring isotope, a somewhat complex correction procedure, to be discussed later in this section, would have to be used to detect and determine Sc. The major anion element in the sample often contributes to polyatomic ion formation and 'mixed polyatomic' ion production is markedly affected by the major elemental composition of the sample. For example, it has already been shown that $^{60}Ni^+$ may be resolved in the presence of $^{12}C_5^+$, but $^{12}C\,^{16}O_3^+$ ($m/e = 59.985$) would almost certainly interfere if the analysed sample contained a carbonate, and the 60 line would then be suspect. In other systems, the high fluorine contents of fluorspar and fluorspar-bearing rocks result in the production of such polyatomic ions as $^{40}Ca^{19}F^+$, which interferes with $^{59}Co^+$, and $^{88}Sr^{19}F^+$, which interferes with $^{107}Ag^+$. Similarly, high chlorine contents, such as occur in the evaporated residues of natural waters, result in the production of $^{40}Ca\,^{35}Cl^+$ which interferes with $^{75}As^+$, although a fairly simple correction procedure can be used to overcome this difficulty.

It will be clear, therefore, that interpretation of a recorded spectrum requires some care, and a knowledge of the major elemental composition of the sample under analysis. Fortunately, many elements have more than one naturally occurring isotope and if the line produced by singly charged ions of one isotope cannot be used due to interference under the analytical conditions, that produced by singly charged ions of another isotope can often be used. If this solution does not apply, either because the second line is not sufficiently intense, or is itself prone to interference, or because there is no alternative isotope, some other solution, such as changing the electrode matrix, can often be found. Spectral complications due to the above mentioned interferences must not be over-emphasised, however, and it should be noted that those encountered using spark source mass spectrography are much less than those encountered in, for example, optical emission spectrography.

Quantitative Measurement

In quantitative work, lines are selected for measurement on the basis of freedom from interference and of sensitivity and, clearly, details of the manner of selection will vary according to the nature of the material being analysed. It is worth noting, in passing, that lines due to the less abundant isotopes give less sensitivity for the element, for a given exposure, and that, for a single isotope element, a sensitivity of 0.001ppm(atomic) can be obtained with a collection corresponding to 1500 nanocoulombs at the monitor collector slit assembly. Hence the limit of sensitivity for an element for which one is forced to use an isotope with a relative abundance of $n\%$, is raised to

$$0.001 \times \frac{100}{n} \text{ ppm(atomic)}$$

for a collection of 1500 nanocoulombs.

In determining the amount of an element present, the optical densities of the selected lines are measured for several successive exposures, recorded on the same plate. An automatic recording microdensitometer may be used for this, and the instrument made by the Joyce-Loebl Company is a typical example of its kind. When the mass spectrograph has been well focused, the peak heights on the microdensitometer trace are an adequate representation of the line densities. For each line a plot, called the ‘characteristic curve’, is then made of optical density against the logarithm of the exposure, expressed as nanocoulombs.

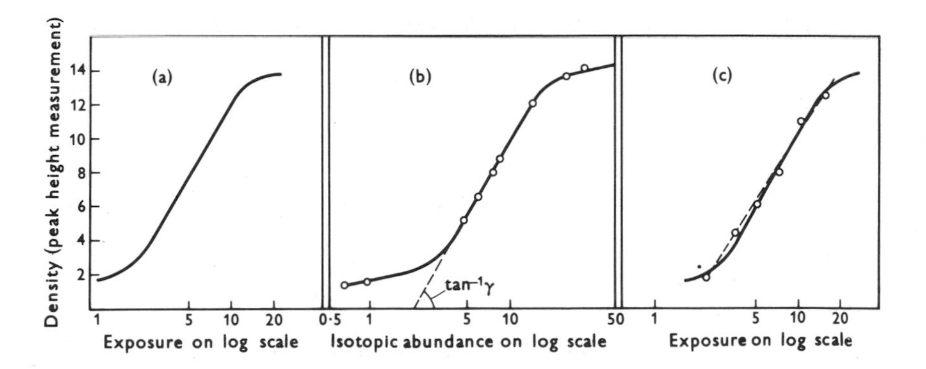

Figure 5.9 Characteristic curves for Ilford Q2 emulsion.
(a) Ideal form of the curve.
(b) Characteristic curve obtained by plotting measurements of line densities for different isotopes of tin in a single exposure against log (isotope abundance), and linear extrapolation to give γ.
(c) Multi-exposure plot of scattered density measurements. The dashed line indicates the probable 'best straight line' that would be drawn if no attention was paid to the shape of the characteristic curve, and the full line shows the better location of the curve.

It is in the plotting of these characteristic curves that any heterogeneity in the distribution of the matrix material in the electrodes shows up most clearly. The form of a typical characteristic curve for Ilford Q2 emulsion should be that shown in figure 5.9(a), but if the electrode matrix has been poorly mixed with the sample plus internal standard mix during preparation of the final electrode mix, the points on the curve will scatter about the line and give rise to uncertainties about the true position of the curve. Fortunately a method exists to allow for such uncertainties. If tin is present in the sample in an amount greater than 100 times its limit of sensitivity for a given exposure, the various isotopes present will have produced a number of lines of different optical densities. By plotting the densities of these lines against the logarithm of the relative abundances of the isotopes producing them, a characteristic curve for the plate emulsion can be obtained from a single exposure. This curve usually shows much less scatter than curves based on density measurements obtained from several successive exposures, as shown in figure 5.9(b). The slope of the single exposure characteristic curve can then be used to draw the best line for the multi-exposure plots, as in figure 5.9(c).

To determine the actual amounts of the elements present in the sample, a value of the optical density lying on the straight line portion of the curve is chosen and the exposures in nanocoulombs required to produce this density for each element is read from the characteristic curves for the different elements. These exposure values are substituted in the equation

$$C_e = C_s \times \frac{\mathrm{Exp}_s}{\mathrm{Exp}_e} \times \frac{I_s}{I_e} \times \frac{1}{R_e} \qquad (5.6)$$

where, C_e and C_s are the concentrations of the unknown element E and of the internal standard element S, both in ppm(atomic), in the electrode analysed;

Exp_e and Exp_s are the exposures in nanocoulombs required to give a line of the chosen density for chosen isotopes of the unknown element E and of the standard element S respectively;

I_e and I_s are the isotopic abundances of the chosen isotopes of E and S respectively; and

R_e is the relative sensitivity factor relating the sensitivity of the total recording procedure for the line used of element E to the corresponding sensitivity for the line used of element S.

The internal standard can be one already present in the analysed sample in previously measured amount or, as is more usual, it is one added in known amount during preparation of the electrode charge mix. In either case, C_s is known, and I_e and I_s can be obtained from published tables of isotopic abundances. The value of R_e for any combination of element and standard must be determined experimentally by sparking standards of known elemental contents, specifically prepared for the purpose, under sparking conditions identical to those used for the analysis.

Since these R-factors must be known as accurately as possible, the fusion technique of electrode preparation should always be used for these standard mixes. Replicate determinations of R-factors are also desirable, to give best mean values. It should be emphasised that a change in the sparking parameters always causes a change in the values of the R-factors, and is the reason why sparking parameters must not be changed during the course of a single analysis. The earlier practice of using more energetic excitation to provide the longer exposures in an analysis is wholly indefensible for this reason.

Correction Procedure for Interference

It is occasionally necessary to use lines subject to possible interference for quantitative measurement, and it is then necessary to introduce some correction for this effect. Reference has already been made to the difficulty of determining scandium in silicates and a possible solution to this problem will illustrate the operation of a correction procedure. The method utilises the presence of the isotopes ^{29}Si, ^{30}Si, ^{46}Ti, ^{47}Ti in their known isotopic abundances, and the electrodes sparked should contain a matrix other than graphite. Silver powder may be used.

If $^{29}Si^{16}O^+$ is contributing to the line at $m/e = 45$ in addition to $^{45}Sc^+$, it is likely that the line at $m/e = 46$ will contain a contribution from $^{30}Si^{16}O^+$ in addition to the contribution from $^{46}Ti^+$. The line at $m/e = 47$, however, is likely to be due only to $^{47}Ti^+$. Characteristic plots are prepared for the 45, 46, and 47 lines using the normal method of successive exposures. Then, for a given exposure, the optical densities of lines 46 and 47 are determined. Now, from figure 5.9(b) it is clear that the optical densities of lines produced by two isotopes of the same element in the same exposure are related by the expression

$$D_a - D_b = \gamma(\log I_a - \log I_b) \tag{5.7}$$

where D_a and D_b are the densities of the lines produced by isotopes a and b respectively, I_a and I_b are again their isotopic abundances, and γ is the gradient of the straight line portion of the characteristic curve. Since a characteristic curve can be plotted for any line, whether of mixed origin or not, γ can be determined from the characteristic plots drawn. The density of the $^{47}Ti^+$ line can be determined. The isotopic abundances of ^{46}Ti and ^{47}Ti are 7.99% and 7.42% respectively, and clearly the optical density attributable to $^{46}Ti^+$ is readily obtained by substitution in equation (5.7). If this theoretical density is less than the observed density of the 46 line, the additional density is due to $^{30}Si^{16}O^+$ and interference will occur in the 45 line because of $^{29}Si^{16}O^+$, but if the values are equal, or very nearly equal, no serious interference will occur. In the former case, the ratio of the contributions of $^{46}Ti^+$ and $^{30}Si^{16}O^+$ to the 46 line can be calculated, and it is now necessary to determine the densities of lines 45 and 46 again at the same exposure but not necessarily the same exposure as the one used for lines 46 and 47. The density attributable to $^{30}Si^{16}O^+$ in the newly measured 46 line can be calculated from the previously measured ratio of contributions to this line and the isotopic abundances of ^{29}Si and ^{30}Si, which are 4.68% and 3.09% respectively. Substitution in equation (5.7) shows that

$$D_{^{29}Si^{16}O^+} - D_{^{30}Si^{16}O^+} = \gamma(\log 4.68 - \log 3.09)$$

and from this the contribution of $^{29}Si^{16}O^+$ to the 45 line can be calculated. Any additional density in this line, provided that carbon is absent from the electrodes, is then due to $^{45}Sc^+$, and so the ratio of the contributions of $^{45}Sc^+$ and $^{29}Si^{16}O^+$ to any intensity of the 45 line can be established. With this knowledge a new characteristic curve can be drawn using the density measurements on line 45, corrected to correspond to the contribution from $^{45}Sc^+$ only. This curve can be used as if it had been obtained from a line not subject to interference. The scandium correction is a particularly complex one, involving, as it does, a two-stage process. Most correction procedures for other lines are only one-stage, and so are easier to apply.

5.2.4 Routine Maintenance

It is hardly necessary to emphasise that care must be taken to avoid contamination at all stages of an analysis. But it is worth pointing out that contamination can arise inside the mass spectrograph itself from ions trapped on the various internal surfaces during previous analyses. It is prudent to change and clean the electrode holders and the number 1 slit of the source area after each analysis, and the whole source area, including number 2 and number 3 slits, should be cleaned about once a week. Periodically, about once a month under normal working conditions and with an average programme of work, the inside surfaces of the dispersion unit must be cleaned as well by 'baking' the instrument. Provision is made for heating the whole of the source area and the dispersion unit to temperatures of up to 250°C, by the installation of heating units, to volatilise any deposits on the inside surfaces. The diffusion pumps are brought into action, by filling their cold traps with liquid nitrogen at least 30 minutes before the heating units are switched off, to suck the contaminants out while the internal surfaces are still too hot to permit redeposition of the volatilised materials. To be effective, 'baking' should be continued for at least 24 hours, and in most laboratories the instrument is 'baked' over a week-end. After 'baking', the instrument should be allowed to cool for at least 12 hours before any mechanical or other operations are conducted on it.

5.3 FURTHER DEVELOPMENTS

The early work on the analysis of geological and allied materials by spark source mass spectrography has been followed by the introduction of several important modifications of the technique, which will constitute the subject matter of this section. The objective of these developments has been an improvement in the precision, accuracy, and speed of analysis, although attention has also been paid to extending the range of elemental contents determinable by this method.

5.3.1 Ion Beam Chopping

Once the significance of electrode heterogeneity on analytical precision had been appreciated, attention began to be paid to means of increasing the amount of electrode consumed for a given total number of ions detected by the recording device. Ion beam chopping has been adopted by a number of laboratories to achieve this end.

In essence, an ion beam chopper is a modified beam suppress assembly mounted at the exit from the source area, and which comprises two plates similar to those in the beam suppress unit but of larger area. A pulsed voltage, of −200 volts, with very short rise and fall times is applied to the ungrounded plate to permit the composite beam to enter the dispersion unit only during the 'off' period of the pulse. The pulsing conditions are chosen so that 5 microsecond pulses of ions are fed to the dispersion unit at a pre-determined pulse frequency, which can be varied between 500 and 110,000 pulses per second. Jackson, Whitehead and Vossen [9] have given details of such a device and have shown that the precision can be improved from ±15% to ±5% by applying ion beam chopping to the analysis of calcium and phosphorus in titanium dioxide pigments. If the time, T, between successive ion beam pulses passing the chopper is less than the spark pulse length, the amount of electrode consumed for a given exposure is increased by a factor equal to T divided by the ion beam length, t. For a pulse frequency of 10,000/sec, $T/t = (100 \times 10^{-6} - 5 \times 10^{-6})/5 \times 10^{-6} = 95/5 = 19$, i.e. 19 times as much electrode material would be consumed for a given exposure than if ion beam chopping had not been used. Figure 5.1 shows the considerable improvement in precision to be expected by increasing the amount of electrode consumed by a factor of 10 for a given elemental content. If T is increased beyond the spark pulse length, however, this simple relationship between T/t and the amount of electrode consumed is no longer valid, unless the chopping frequency is reduced below the spark pulse repetition rate, which is undesirable [9]. A limit is placed on the factorial increase in the amount of electrode consumed, therefore, by the ratio of the spark pulse length to the ion beam pulse length. With the usual spark pulse length of 200 microseconds, recommended for rocks and minerals, a chopping frequency of 5000/sec, and an ion beam pulse length of 5 microseconds, the amount of sample consumed should be increased by a factor of 40 for a given exposure, and this, theoretically, should lead to an improvement in 'sampling precision' by a factor of $\sqrt{40}$, i.e. 6.3, if the electrodes are reasonably well mixed. Thus, a sampling precision of ±25%, such as can be obtained by dry mixing methods of electrode preparation, could be reduced to ±4%.

On the other hand, the analysis time is extended by the use of ion beam chopping but it will be shown that, while the increase in time may be unacceptable in many cases, there are many occasions when the extended time is very advantageous and can lead to an extension of mass spectrography to higher elemental concentrations. Thus, collection of 10 nanocoulombs of charge, which yields measurable lines for elemental contents of approximately 0.2ppm(atomic), typically requires about 2 minutes if no ion beam chopping is used. With ion beam chopping, using the conditions given in the previous paragraph, this would be extended to 80 minutes which is about the limit of acceptability and, for longer exposures, the increase in analysis time would be unacceptable. An exposure of 0.1 nanocoulombs, however, for analysis at the 2–5ppm(atomic) level, takes only slightly more than 1 sec if the recommended sparking conditions for silicate analyses are employed, and calls for considerable operator dexterity. With ion beam chopping, this time period can be extended to 48 sec, a much more easily handled time span, and ion beam chopping can be used to extend exposures of as low as 0.005 nanocoulomb to time periods of 2-3 sec. This is a very real advantage, since it permits retention of the same sparking conditions for very short exposures, which can be carried out with beam chopping, and for longer exposures, which can be performed without beam chopping. The range of elemental contents that could be handled by mass spectrography was restricted at the higher end both by short analysis times before the introduction of ion beam chopping and by the realisation that R-factors are markedly affected by changes in the sparking parameters and that the previously acceptable method of extending analysis times, by decreasing the spark pulse length and the spark repetition rate, could not be accepted. Where a large number of elements of widely differing content levels are sought, it is both perfectly feasible and highly desirable that short exposures be made with the ion beam chopper in use, and that longer exposures be made without this facility in use. Fusion-homogenised electrodes should be used for such studies.

5.3.2 Autospark Control of Electrodes

It was early realised that control of the spark gap and of the electrode positions played an important role in determining the instrumental precision obtained during an analysis. Variations in the positioning of the electrodes relative to the slit assembly and in the spark gap affect not only the intensity of the composite ion beam which is drawn towards the dispersion unit, but also the proportions of the singly charged monoatomic, multiply charged monatomic, and singly charged polyatomic ions in the beam. Careful manipulation of the electrode controls is essential if maximum instrumental precision is to be obtained. In effect, operator skill and experience determine the precision obtained when manual control of the electrodes is adopted.

To reduce the need for this operator skill in obtaining maximum performance from the instrument, the M.S.7 manufacturers, A.E.I. Scientific Apparatus Ltd., have introduced a device known as 'Autospark' which is really an auto-control on the spark. In place of one of the manipulators used for manual electrode control, a vibrator is introduced which vibrates one electrode at mains frequency relative to the other, stationary electrode. A sensing coil is used to monitor the spark discharge and provides a signal proportional to the discharge emission. If this deviates from the maximum value, due to the spark gap being too large or too small, a d.c. voltage, superimposed on the a.c. vibrator circuit, decreases or increases the electrode gap to correct the loss in signal. The cycle may be set to maintain emission within 5% of its peak value, which is the equivalent of controlling the electrode gap manually to within ±0.00635mm. In effect, the 'Autospark' is an operator with better than human skills. Its function is to achieve maximum instrumental precision, unlike ion beam chopping which improves sampling precision. These two devices are, therefore, complementary to one another in improving overall precision.

5.3.3 Electrical Detection Methods

Electrical detection methods of recording have been introduced to spark source mass spectrography with the aim of improving precision and reducing analysis time, although their chief value lies in the latter field since the sensitised plate method of recording has good precision. Bingham and Powers [10] positioned an electron multiplier at the principal radius of the dispersion chamber to collect individual ionic species, thus turning the spectrograph into a spectrometer. Two principal methods of electrical scanning are in common use. In one, known as electromagnetic peak scanning, the current to the magnet coils in the magnetic dispersion unit is varied to alter the field strength in this unit, and in the other, known as voltage peak switching, the accelerating and electrostatic analyser voltages are changed, thus altering the energies of the ions entering the magnetic dispersion unit.

In the peak scanning mode, ions of different mass are swept successively over the collector by initiating an exponential scan of the magnet current, which procedure ensures that equal times are spent on each peak. In peak switching, the relevant voltages are changed incrementally to bring one ion beam, or peak, of interest after another onto the collector position. Measurements for quantitative analyses are obtained by integrating the collector current from each peak while a pre-determined charge is collected by the monitor slit assembly. Bingham and Powers [10] found that the m/e range 8–250 could be scanned in 3 minutes to a detection level of 0.5ppm (atomic) with a precision of ±35%. With a slower scan, of the order of 30 minutes, they found a detection level of 0.1ppm(atomic) with a precision of ±21%. The procedure gives rapid surveys of the whole spectrum at a tolerable degree of precision.

Much better precision can be obtained using the switching mode. For a total monitor collector slit assembly charge collection of 1 nanocoulomb, requiring about 45 sec under the sparking conditions used, a precision of ±5% was obtained at the 1ppm(atomic) level, or ±1-2% at the 50ppm (atomic) level. With this performance, 40 elements could be determined accurately in 30 minutes. The performance depends, of course, on counting statistics and must deteriorate as the number of particles counted falls. Improved precision at lower elemental content levels would require longer periods on each peak and the collection of greater amounts of charge by the monitor collector slit assembly. To obtain a precision of ±5% at the 0.01ppm (atomic) level would require approximately 75 minutes to be spent on the peak, but even this compares very favourably with the time required for analysis using the sensitised plate method of recording. Electrical detection is at least six times faster than the sensitised plate method of recording, when all factors are concerned, since it eliminates all the operations associated with the use of plates, such as plate storage, loading, development, interpretation and line measurement, as well as the subsequent plotting of characteristic curves and the calculations. Among the disadvantages of electrical detection methods are initial cost, servicing, and maintenance requirements, but these are minor when compared with the problems of maintaining the required degree of quality control on sensitised plates, plate deterioration during storage, and the effort and time actually spent on plate interpretation and measurement. There is little doubt that electrical detection methods will supersede sensitised plate recording and that the time required for spark source mass spectrographic analysis be greatly reduced from its present length.

5.3.4 Developments in Analytical Philosophy

The danger of interference by second and third order emission was recognised during the early work in spark source mass spectrography. With the demonstration that variation in the ratio of first to second order emission with elemental content can occur within the 10–100ppm (atomic) range of certain elements, e.g. barium, has come the realisation that, for accurate work, account must be taken of this variation in determining the contents of the elements producing the second, or higher, order emission [6]. When the production of doubly and triply charged ions from an isotope being used for analysis is significant, the second and third order lines must be measured and plotted to give values of Exp_{e^+}, $Exp_{e^{++}}$, and $Exp_{e^{+++}}$, where Exp has the significance defined for equation (5.6). Equation (5.6) must then be re-written

$$C_e = C_s \times \frac{I_s}{I_e} \times \frac{1}{R_e} \times \frac{\dfrac{1}{Exp_{e^+}} + \dfrac{1}{Exp_{e^{++}}} + \dfrac{1}{Exp_{e^{+++}}}}{\dfrac{1}{Exp_s}} \tag{5.8}$$

This need may be illustrated by a study of the geochemical standard 'W.1'. Without taking second and third order emission into account, the measured value of 145ppm for the barium content did not agree well with Fleischer's value of 180ppm [11], which is the recommended value for this standard. When the second and third order emissions for barium were taken into account, the barium content was returned as 175ppm [6], which agrees well with the recommended value within the limits of the analytical precision.

Second order emission is displayed most strongly by the alkaline earth metals, to a lesser degree by the rare earths and transition metals, and hardly at all by the alkali metals. If account is not taken of it, analytical results can be seriously inaccurate, on the low side, despite the precision of the analytical method appearing satisfactory, as shown in the above example. If the isotope of the internal standard element used also produces doubly and triply charged ions, the term $1/Exp_s$ in equation (5.8) must similarly be expanded to the form $(1/Exp_{s^+} + 1/Exp_{s^{++}} + 1/Exp_{s^{+++}})$, otherwise determinations of all elemental contents will be too high. It is, of course, notoriously difficult to demonstrate absolute accuracy for any analytical method, and the best that can be done is to demonstrate, by the analysis of internationally accepted standards, that the method produces results in agreement with the accepted compositions of these standards. Properly conducted spark source mass spectrography appears to be accurate within the limits of the analytical precision, and of our knowledge of the standard compositions.

5.4 THE ISOTOPE DILUTION METHOD OF SAMPLE PREPARATION

Another method of sample preparation, which eliminates the problem of interference to a large degree although it reduces the number of elements that can be determined in a single analytical run, involves chemical treatment of the sample before analysis, followed by addition of a tracer isotope to the treated sample. This is known as the isotope dilution method and bears many similarities to the isotope dilution method discussed in Chapter 4. Details vary from one analytical problem to another, but the broad outlines of the procedure are as follows.

A weighed amount of the sample is treated chemically to bring it into solution and a weighed amount, usually an aliquot of a stock solution, of a separated isotope of the element to be determined is added to the solution and the whole well mixed. More than one isotope can be added simultaneously if several elements are to be determined. The element, or elements, sought are then extracted from the solution, for example by the use of an ion exchange column, and liquid containing the required elements is placed on the filament of a thermal ionisation source for the mass spectrometer. Liquid derived from the sample after the same treatment, but without tracer addition, is also placed on a second filament and both are gently evaporated to dryness to leave a solid residue on the filament. The isotopic compositions of the elements in the normal and the enriched samples are determined using electrical detection recording. The elemental contents in the sample can be calculated from

$$\begin{pmatrix}\text{weight of}\\ \text{element}\\ \text{in sample}\end{pmatrix} = \begin{pmatrix}\text{weight in}\\ \text{tracer}\\ \text{added}\end{pmatrix} \times \frac{I_{e_2}(T)}{I_{e_2}(S)} \times \frac{S_{A.W.}}{T_{A.W.}} \times \frac{M_{e_1/e_2} - T_{e_1/e_2}}{S_{e_2/e_2} - M_{e_2/e_2}} \tag{5.9}$$

where $I_{e_2}(T)$ and $I_{e_2}(S)$ are the isotopic abundances of isotope e_2 in the tracer and the normal sample; $T_{A.W.}$ and $S_{A.W.}$ are the atomic weights of the element in the tracer and the normal sample; T_{e_1/e_2}, S_{e_1/e_2}, and M_{e_1/e_2} are the isotopic ratios of the two isotopes e_1 and e_2 in the tracer, the normal sample, and the enriched sample respectively. Single-focusing mass spectrometers are often used in this method of analysis, although a thermal ionisation source can be fitted to a double-focusing mass spectrometer used for spark source work as an interchangeable alternative to the spark source.

This method of analysis has been in use for longer than spark source mass spectrography and has been much used in Rb-Sr geochronology. More recently, it has been applied to the analysis of Rb, Cs, Sr, Ba, and some of the rare earths in lunar material [12, 13]. Fewer elements can be determined in a single analytical run by this method, although better accuracy and precision is claimed for it than has hitherto been achieved in spark source work.

5.5 RAPID SURVEY WORK

Hitherto the emphasis in spark source mass spectrography has been on precise and accurate analysis at high sensitivity levels, but the technique can, like many other techniques, be applied to less demanding analytical problems. In the search for ores and minerals by geochemical prospecting procedures, an analytical method with a precision of ±50%, an accuracy of the same order, and a sensitivity of 1ppm(atomic) would often be acceptable. If this method would, at the same time, disclose the presence of elements possibly not suspected in the analysed material, there would be an incentive to use it.

For spark source mass spectrography to meet such specifications, few of the procedures described earlier in this chapter are necessary. Much time could then be saved in the analytical programme. The ground sample could be hand mixed with graphite and the electrodes sparked for not more than 5 minutes. Visual inspection of the plates, leading to identification of the exposure in which a line of interest was just visible, would permit estimation of the elemental contents within the specified limits. If samples are handled in batches, 25-30 minutes per sample should more than suffice and, since data on 15-20 elements plus notes on the occurrence or absence of others can be obtained in each run, the average time of under 2 minutes per elemental determination compares well with other analytical methods applied to such problems.

The fact that most investigators who have used spark source mass spectrography have concentrated on achieving maximum precision, accuracy, and sensitivity should not obscure other potential fields of application of the technique in the general field of rock and mineral analysis.

REFERENCES

1. A. J. DEMPSTER, *Nature,* **135**, 542 (1935); *Rev. Sci. Instr.,* **7**, 46 (1936).
2. R. BROWN and W. A. WOLSTENHOLME, *Nature,* **201**, 598 (1964).
3. S. R. TAYLOR, *Geochim. Cosmochim. Acta,* **29**, 1243 (1965).
4. S. R. TAYLOR, *Nature,* **205**, 34 (1965).
5. G. D. NICHOLLS, A. L. GRAHAM, E. WILLIAMS, and M. WOOD, *Anal. Chem.,* **39**, 584 (1967).
6. G. D. NICHOLLS, *Proc. Brit. Ceram. Soc.,* April issue, 85 (1970).
7. R. BROWN and P. G. T. VOSSEN, Proc. Eighth M.S.7 Users Conference (A.E.I. Scientific Apparatus Ltd., Manchester, 1970), p.101.

8. J. FRANZEN, K. H. MAURER, and K. D. SCHUY, *Z. Naturforschung,* **21A,** 37 (1966).
9. P. F. S. JACKSON, J. WHITEHEAD, and P. G. T. VOSSEN, *Anal. Chem.,* **39,** 1737 (1967).
10. R. A. BINGHAM and P. POWERS, *A.E.I. Tech. Publ.* No. T.P.26 (1969).
11. M. FLEISCHER, *Geochim. Cosmochim. Acta,* **33,** 65 (1969).
12. P. W. GAST, N. J. HUBBARD, and H. WEISMANN, *Geochim. Cosmochim. Acta,* **34** (Suppl. 1), 1143 (1970).
13. J. A. PHILPOTTS and C. C. SCHNETZLER, *Geochim. Cosmochim. Acta,* **34** (Suppl. 1), 1471 (1970).

CHAPTER 6

X-ray Techniques for Process Control in the Mineral Industry

K.G. Carr-Brion

Warren Spring Laboratory
Gunnelswood Road
Stevenage, Hertfordshire SG1 2BX
England

6.1 INTRODUCTION

In any mineral processing plant, automatic control demands information on the behaviour of the system from sensors measuring the required parameters. In most processes the concentrations of certain elements or crystalline materials must be known at a number of points in the plant, and these can be obtained by X-ray fluorescence and X-ray diffraction analysis respectively. The concentrations can be monitored by either on-line or on-stream analysis, where an on-line analyser automatically takes a discrete sample, prepares it, delivers it to an instrument which carries out the required measurement, and then subsequently communicates the data to the control computer, while an on-stream analyser has all or part of the process stream continually flowing through it and it carries out its measurement also continuously, the output being available when required. X-ray methods are well suited to on-stream as well as to on-line analysis, having high selectivity, large sample throughput, the ability to examine liquids, slurries and powders in an 'as-received' form, and true digital outputs.

The points in a mineral processing plant where determinations of the composition are most often of value are:

(1) The mined material before blending and grinding
(2) The ground material before processing
(3) The concentrated ore
(4) The tailing material

Unfortunately, the material at point (1) is commonly in the form of large lumps and rocks and is not suitable for on-line or on-stream analysis, particularly with X-rays which, as shown in Chapter 3, can penetrate at most only the outer 1mm of the material. In some mines the use of a portable X-ray fluorescence analyser (Chapter 3, section 3.3.2) has proved of value in ensuring good blending to achieve a uniform feed to the mill, but the associated sampling problems are immense. One possibility, if the rocks can be separated on a good-bad basis on the basis of their surface mineral contents, is to use an X-ray fluorescence sensing device in conjunction with an automatic sorting machine. At points (2), (3) and (4) the material is almost always present as a slurry, and occasionally as a dry powder, and is suitable for on-stream X-ray analytical control.

6.2 SAMPLING AND CALIBRATION

The value of any process control analysis depends primarily on the material examined being representative of the whole process stream at that time, and secondly on the accurate analysis of the material by the instrument. The more rapidly the analysis can be made the better, of course.

Ensuring and proving that the sample examined is representative of the process stream presents real difficulties, especially in the mineral industries where wide ranges of particle density and particle size make segregation prevalent. With an on-line instrument looking at samples withdrawn from the process stream, the problem is at its worst, as shown by the sample to sample variations often found associated with such measurements [1]. Typically, a small portion of a large stream passing, perhaps, thousands of gallons per minute has to be removed, dried, possibly ground and pelleted, and presented to the X-ray analyser which may actually analyse the top 0.1mm of a 3cm-diameter disc. If an on-stream analyser is used the sample examined is several orders of magnitude bigger, but even if the instrument is mounted on the whole process stream itself, it is impossible for it to examine more than a small portion of the material passing because of the limited depth from which characteristic X-rays can be detected due to absorption. In the majority of cases, into the bargain, mounting on the main process stream is not feasible, and on-stream analysis of powders or slurries normally involves splitting a representative portion of the material from the process stream before presenting it to the instrument and subsequently permitting it to rejoin the main process stream. The analysis must, therefore, be made on 'as-received' material with no special preparation.

Most slurry samplers use a constant-head tank, fed from the process stream, with the underflow passing to the analysing station and the overflow by-passing it. The volume of this head tank is kept as small as possible to ensure high turbulence and hence good mixing and efficient sampling. A design, which is capable of operating with a wide range of pipe sizes, uses concentric pipes with the sample stream coming from a cone at the base of the larger pipe, as shown in figure 6.1. The performance of such a sampling device can be tested under recirculating loop conditions by varying the solid composition, the particle size, and the flow rate, and checking that the solids in the overflow and the sample streams remain constant in composition, although there may be a significant variation in slurry density even though representative sampling of the solid is achieved.

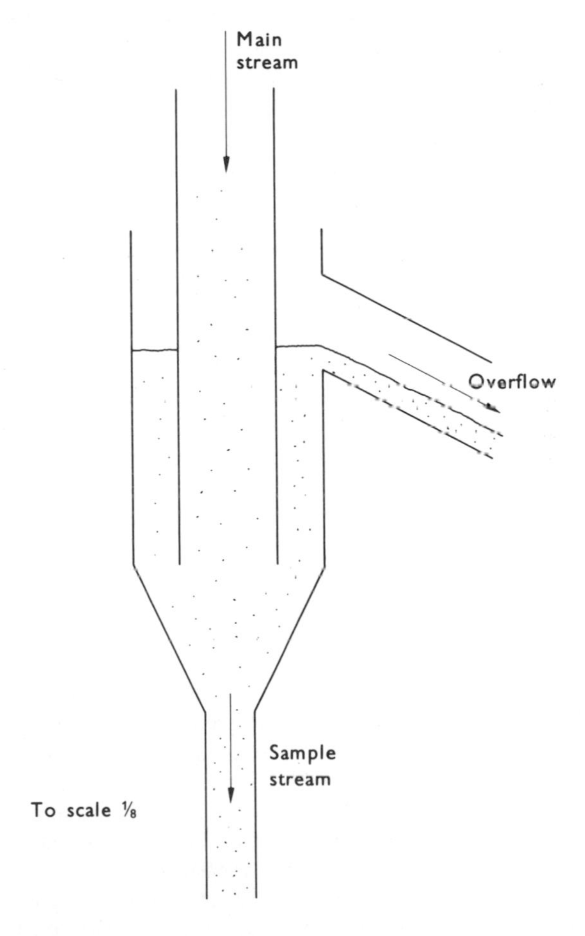

Figure 6.1 Scale drawing of a constant head slurry sampler.

An Archimedean screw, which continuously removes a portion of the stream, is often used to sample a powder travelling in a conduit, although care must be taken to position it in an appropriate position if segregation is present to any marked extent. An alternative system shoots a small cross section of the whole powder stream at intervals into the analytical system, using a blade. This is especially useful for material travelling on belts and should show little dependence on segregation across the stream, although it would be of less value if rapid fluctuations with time were occurring in the composition of the powder.

But even when a truly representative sample has been taken, it is still necessary to ensure that the portion of the primary sample actually seen by the instrument is itself truly representative of the sample in its turn. As shown above, in the case of a slurry this is achieved by having a high turbulence in the sample cell to maintain the constant mixing. Powders are more difficult to handle, and good mixing before the production of the smooth surface required for X-ray analysis is all that can be done in general.

Calibration

No X-ray analyser can function without calibration, as has been shown in Chapter 3. With an on-line analyser, this is carried out by taking measurements on discs of known composition and arriving at an accurate mathematical relationship between X-ray intensity and elemental concentration. The accurate mathematical relationship is, of course, imperative for the application of computer-based control systems. The difficulties inherent in deriving such a relationship have been noted in Chapter 3, and particularly the way in which changes in mineralogy can affect the calibration accuracy, if automatic fusion of the sample is not employed, although fine grinding can reduce such effects.

Calibration of an on-stream analyser is more difficult. The usual practice involves setting up a recirculating slurry loop, taking care to minimise possible segregation within the loop, and examining a series of typical analysed samples, using a kilogram or more of each. The mathematical relationship between observed intensity for a given element and its concentration in the recirculating load is again derived. The composition of the load can be checked by taking samples at a point immediately after the sample cell, using total diversion of the slurry stream. Mineralogical and particle size effects are often greater in slurries than in powders, and care should be taken to ensure that the expected range of each factor be included in the samples used, and that adequate compensation be included in the derived mathematical model. When installed on-stream, correlation between sample analyses and on-stream analyses is often poor, which almost always reflects on the efficiency of the technique for taking small samples.

The relationship between the measured X-ray intensities and the concentrations is often complex and an on-line programmed calculating machine or computer is essential if the required data are to be obtained. If the analyser is being used in conjunction with a control computer with sufficient spare capacity, this unit may be used to carry out the necessary model calculations in a real-time mode. If the main computer is incapable of performing this task, a relatively inexpensive mini-computer is the best alternative. This computer will not only carry out the required calculations, but it can also be programmed to make periodic checks on the performance of the analyser and up-date the calibration data, as necessary, and it will warn of instrumental breakdown.

6.3 INSTRUMENTATION FOR PROCESS CONTROL XRF ANALYSIS

On-line X-ray spectrometers are generally standard automatic laboratory analysers fitted with a device to take, prepare, and load samples, and output the analysis data either directly or via a slaved mini-computer to the control computer. They are commercially available both in Europe and North America and, to date, have been mainly used in the cement industry where, with a dry powder feed, sample preparation by automatic grinding and pelleting, or even fusion, may be essential to achieve the required accuracy. They have been less used in the mineral industry, where the widespread use of slurries has lead to the adoption of on-stream analysers.

Three different types of on-stream X-ray fluorescence analysers are in current use. The first has a high power X-ray tube and a set of crystal monochromators, each set to monitor a selected characteristic X-ray wavelength, as described in Chapter 3, section 3.3.1. In a typical commercial unit, the X-ray tube, monochromators, and detectors are mounted in a rigid casting which can traverse a line of slurry streams flowing through specially designed slurry presenters [2], or the unit may sit on top of a smooth band of moving powder, as shown in figure 6.2.

The second type, which is illustrated in figure 6.3, uses a sealed radioisotope source emitting X- or γ-rays to excite the characteristic X-rays, which are then selected by paired absorption edge X-ray filters backed up by pulse height analysis circuitry in conjunction with an energy sensitive detector, usually of the scintillation type (cf. Chapter 1, section 1.3.3 and Chapter 3, section 3.3.1). Crystal monochromators cannot be used with these sources because of their inefficiency and because of the very low output of these sources compared with a high-power X-ray tube. The methods of energy, and hence wavelength, selection adopted in these simple analysers give poorer limits of detection and of selectivity than can be achieved with the first type of system, but the instruments are much less expensive, are compact and rugged, and are capable of remote operation on or adjacent to the process stream without the need for the high tension supplies which the X-ray tube source demands. Other advantages of this system will be discussed at more length in section 6.4. One special variation on this analyser, which is now undergoing tests, is the in-stream head, where the analyser is actually immersed in the slurry stream and measures the composition and the density of the slurry flowing around it.

Figure 6.2 On-stream X-ray fluorescence analyser for powders, using a high power X-ray tube source and six crystal dispersive monochromators for the simultaneous measurement of up to six elemental concentrations.

The third type uses either a sealed radioisotope source or a low-power X-ray tube together with a semiconductor X-ray detector [4] and acts as a non-dispersive analyser (Chapter 3, section 3.3.2). As explained in the section quoted, this detector converts X-rays incident upon it into electrical pulses whose voltage is very precisely related to the quantum energy of the X-ray.

If the pulses are sorted electronically into their respective voltages and the number of pulses with the same voltage is plotted against voltage, an X-ray intensity spectrum plotted against wavelength results which allows the simultaneous determination of from one to a dozen or more elements. The limits of detection and selectivity of the system lie between those of the high-power X-ray tube source and the simple radioisotope source instruments. The detector must be operated and maintained at liquid nitrogen temperatures, and automatic methods of providing the coolant are desirable in a full-scale plant installation.

All these instruments, except the in-stream unit, require a slurry presenter or sample cell for use with slurries. These have a very thin X-ray-transparent window separating the slurry from the X-ray system and the high turbulence necessary to ensure that the instrument sees a representative sample can lead to rapid wear of the 10μm to 100μm thick window. The art of designing

Figure 6.3 Simpler on-stream X-ray fluorescence analyser for tin-containing slurries using radioisotope sources. The upper unit is fitted with a Pd filter to absorb the background radiation only while the lower has a matched Ag filter to absorb both background and Sn *K* radiation. The difference between the two signals is a measure of the tin concentration.

good slurry presenters, therefore, lies in achieving a balance between high turbulence and window wear, for the slurry to be examined. Further complications can arise from the deposition of such material as calcite or a flotation reagent on the window, thereby making it opaque. This problem can be overcome by increasing the turbulent flow adjacent to the window to a point where the material is eroded as fast as it is deposited and the minimum useful window life of several hundred hours is maintained.

The thinness of the windows makes them liable to rupture, and in the event of an unexpected failure due to the effect of foreign material or a flaw in the window itself, means must be provided to ensure that the instrument and its surroundings are not flooded with slurry. These may take the form of a double window with a moisture-sensitive electrode between the components, coupled to a shut-off valve, or it may be accomplished by operating the cell at reduced pressure so that a rupture results merely in air bleeding into the cell instead of liquid flowing out. The thin window is also quite easily distorted by the motion of the slurry past it, but it should remain in a constant position relative to the X-ray system. To achieve this constancy, the window is normally held under tension, but even then it requires a relatively constant pressure in the cell and this is conveniently provided by the constant-head tank used for the sample splitting. The position of the window is much more critical when using crystal monochromators (permitted movement ±50μm) than with either of the other two systems (permitted movement ±250μm).

With the on-stream analysis of powders, a flat surface having a similar constant position relative to the X-ray system has to be produced continuously. This can be done by means of a suitably designed feed unit with a shoe at a constant height above the conveyor belt, or by passing the powder over an X-ray transparent window. The bulk density of the powder surface will affect the observed intensity of the X-rays generated therefrom and so it must be held relatively constant or else measured continuously and compensated for in the subsequent calculations.

The performance of on-line analysers is generally equal to that of the best laboratory instruments, and they are capable of determining all elements from fluorine upwards with high precision. On-stream analysers using high-powered X-ray tubes, or low-power sources in conjunction with a semiconductor detector, can determine all elements from about aluminium upwards with limits of detection adequate for nearly all requirements of the mineral industry. Table 6.1 gives typical values for a group of common elements. The simpler radioisotope sensors are limited to elements from about phosphorus upwards (chlorine in slurries) but in many cases have limits of detection which are too poor to enable them to be used to determine the concentration of elements such as copper in the tailing streams from low grade ores, as can be seen from the data presented in table 6.1. All are

TABLE 6.1

		Limit of Detection*	
Element Being Determined	Matrix	High-Power X-ray Tube Analyser	Simple Radio-Isotope X-ray Analyser
Phosphorus	Iron ores	0.02	Not detected
Potassium	Clays	0.001	0.08
Titanium	Clays	0.0004	0.01
Iron	Sand	0.0005	0.003
Copper	Ore slurry	0.0004	0.005
Zinc	Ore slurry	0.001	0.005
Molybdenum	Ore slurry	0.003	0.006
Tin	Ore slurry	0.003	0.005
Barium	Ore slurry	0.005	0.010
Lead	Ore slurry	0.001	0.005

*Per cent element giving signal equivalent to 2 sigma background intensity, time 100 sec. Limits of determination (±10 per cent) are ten times greater.

DATE	TIME	–STREAM–	CONTENT: :TIN PERCENT	LB/ TON	SOLIDS P.CENT	IRON P.CENT
20/01/71.	0416	SLIME TAIL	.23	5.3	6.1	5.8
20/01/71.	0423	SAND TAIL	.19	4.3	16.3	3.2
20/01/71.	0430	SLIME TAIL	.24	5.5	5.8	6.2
20/01/71.	0437	SAND TAIL	.19	4.4	16.5	3.2
20/01/71.	0445	SLIME TAIL	.24	5.5	5.9	6.0
20/01/71.	0452	SAND TAIL	.18	4.1	16.6	3.2
20/01/71.	0459	SLIME TAIL	.21	4.8	5.8	5.8
20/01/71.	0506	SAND TAIL	.18	4.1	16.8	3.1
20/01/71.	0513	SLIME TAIL	.23	5.2	5.8	5.8
20/01/71.	0520	SAND TAIL	.18	4.1	16.6	3.2
20/01/71.	0527	SLIME TAIL	.20	4.5	5.8	5.9
20/01/71.	0535	SAND TAIL	.17	3.9	16.6	3.1

Figure 6.4 Computer Output for on-stream tin analyser shown in 12.3 with up-date cycle and analysis of two streams.

capable of an instrumental precision of 0.1 per cent, although 0.5-1.0 per cent is more typical of the simpler systems. The output from a prototype radioisotope analyser used to determine low concentrations of tin in slurries is shown in figure 6.4. This performance is typical of such an analyser on a tailing stream.

6.3.1 Basic Theory for On-Stream XRF Analysis

On-stream X-ray fluorescence analysis differs from on-line and laboratory analysis in that the sample preparation steps described in Chapter 3, section 3.4.4 cannot normally be applied. Hence in cases where an internal standard, grinding, or fusion are used in the laboratory to obtain an accurate analysis by overcoming matrix or heterogeneity effects, other means must be found for reducing these effects if accurate on-stream analyses are to be obtained. In slurry and fluidised powder analysis, the effects of heterogeneity, commonly classified as particle size and mineralogical effects, are accentuated due to differences in the X-ray absorption coefficients of the particles and the suspending medium. All these effects require adequate quantitative understanding if they are to be reduced or allowance made for them, and this understanding has markedly increased during the last six or so years [5].

Correction for inter-element or matrix absorption effects with a high power or semiconductor system can be made by measuring the characteristic X-ray intensities from the elements causing the absorption effects. The true concentrations can then be arrived at by combining all these intensities into a series of simultaneous equations, involving the elemental concentrations and their absorption coefficients, and solving by iteration [6]. With the simpler type of analyser employing filters, this is not always possible and compensation may have to be made by using another X-ray intensity, such as radiation from the source which has been backscattered from or transmitted through the sample, as a calibrant. For example, in the on-stream determination of tin in ores and mill products, the iron content of the slurry affects the relationship between the tin X-ray intensity and the tin concentration. With the first two types of instrument the iron X-ray intensity can be simultaneously measured and used to compensate for the interference. With a simple analyser, measurement of the iron X-ray intensity would require a second analyser since, because of the difference in the quantum energies of the iron and the tin X-rays, different sources and detectors would have to be used. The required compensation can, however, be achieved by measuring the intensity of the primary X-ray transmitted from the source, used for the tin determination, through the slurry and using this figure in a simple equation, discussed in Chapter 3, section 3.2.2, relating it and the tin X-ray intensity to the tin concentration.

The characteristic X-ray intensity from a slurry or powder stream also depends on the size of the particles present and on the composition of the particles actually containing the wanted element. These are both examples of heterogeneity effects. With particle sizes below a rather imprecise limit, depending on the X-ray absorbing power of the particles and of the suspending fluid, little dependence on particle size or composition is observed. For many elements this is well below the range of sizes normally encountered in the mineral industry, and, indeed for lighter elements such as silicon and aluminium, it is beyond the range of standard grinding methods. Above this limit an increasing dependence of intensity on particle size and composition is found. At very coarse particle sizes the rate of change of intensity with particle size again decreases, but deposition of particles, especially on the cell windows, and segregation then introduce additional errors. Both the magnitude of the change in the intensity of the characteristic radiation and the particle size range over which the change is most marked vary with the energies both of the primary exciting X-ray beam and of the characteristic fluorescent X-ray involved [7]. As the quantum energy of either or both X-rays increases, so the size-sensitive range moves to larger size values, while the magnitude of the change tends to decrease.

Obviously, for a given characteristic fluorescent X-ray, it is impossible to reduce the size effect or to shift it to a size range outside the working particle size range by altering the energy of this X-ray, although a more energetic characteristic wavelength from the same element could be used. Thus, lead *K* radiation could be used instead of the more usual lead *L* radiation.

The energy of the primary exciting X-ray beam can, however, be increased to obtain the desired reduction and shift in particle size effects, although, as has been shown in Chapter 1, section 1.2.2, this leads to a reduced efficiency of excitation of the characteristic X-rays, and hence lower intensities and reduced limits of detection. The required change in the energy is obtained, when using an X-ray tube source, by increasing the voltage across the tube to increase the range of the 'continuum' radiation to higher energies (Chapter 3, section 3.2.1) and simultaneously using a filter in the primary beam to remove the lower energy X-rays from it. Note that increasing the tube voltage alone causes only a limited reduction in the heterogeneity effects since it does nothing to remove the lower energy primary X-rays which are the principal cause of the size effects. As an example, the use of a silver primary filter markedly reduces the particle size effects when determining tin in slurries, the silver filter removing the lower energy primary X-rays which efficiently excite tin. Although the absolute fluorescent intensity is much reduced when using this filter, the relative characteristic to background intensity ratio is increased, due to the absorption of the 'white' background X-rays, so that limits of detection remain unaltered. Radioisotope X-ray sources used for fluorescence analysis generally give approximately monoenergetic primary radiation, so that a source can be chosen that excites the required characteristic X-radiation inefficiently and so reduces heterogeneity effects. Thus, excitation of copper by a Pu^{238} source, which emits uranium *L* X-radiation, gives much larger heterogeneity effects than excitation by the 60keV γ-rays from Am^{241}, although in this case the limits of detection are poorer with the higher energy primary γ-ray because of a lower characteristic to background intensity ratio.

6.3.2 Applications of On-Stream XRF Analysis

On-stream X-ray fluorescence analysers using high power X-ray tubes have been in use in the mineral industry since 1962-63. It is estimated that there are more than 20 installations in North America alone with a somewhat smaller number in the rest of the world. One of the earliest successful installations, at Lake Dufault Mine [8], switches slurry streams sequentially through a single sample cell and measures copper, zinc, and iron on seven streams. More typical are those mines, such as Kidd Creek, Kristineborg, and Mattagami Lake [9], where the spectrometer head traverses up to 30 streams successively. The increased mechanical complexity of such systems is offset by the elimination of cross contamination, ease of standardisation, and down time on only one stream if a cell window needs to be replaced. The information on concentrations throughout the plant is used to control automatically such factors as flotation reagent addition and to provide operators with performance data such as metallurgical balances. A typical installation may cost in excess of $100,000, and the purchase of the earlier units can only be regarded as economic acts of faith on the part of the mine managements concerned. However, a steadily increasing amount of information [9, 10] indicates that the economic return from such an installation is sufficient to return the capital cost within 6-12 months simply from the increased efficiency of operation.

Processes currently employing on-stream analysers in the mineral industry include those concentrating aluminium, cobalt, nickel, copper, zinc, molybdenum, silver, and lead ores. Other elements such as sulphur, iron, and arsenic are also monitored in some processes for control and matrix correction purposes.

Simple radioisotope on-stream X-ray fluorescence analysers using balanced filters have been shown to give an acceptable analytical performance for applications in iron, copper, zinc, molybdenum, tin, barium, and lead concentration plants. Actual applications, however, have so far been limited, with the exception of pilot plant work and relatively short trials at mines using prototype equipment. The only applications have been the determination of potassium, zinc, copper, tin, and lead on selected process streams. The failure of such methods to be more widely adopted in the mineral industry is probably associated with a number of factors, including a distrust of radioisotopes from a safety point of view, inadequate limits of detection for tailings from some low-grade ores, e.g. copper, and the uncertain size of the market, which has made instrument manufacturers wary of devoting too much effort to this field. Present indications are that their acceptance will become more widespread in the next few years.

Semiconductor detector systems using radioisotopes or X-ray tube sources are too new a development to be much beyond the pilot plant and field-trial stages. They have been used with success to determine iron, zinc, copper, arsenic, and rare earths on-stream, but their only known application to control has been in the determination of uranium. Fully industrialised systems with automatic cooling are in the process of development, and these should have a wide application in the mineral industry in the near future.

6.4 INSTRUMENTATION FOR PROCESS CONTROL X-RAY DIFFRACTOMETRY

As shown in Chapter 7, X-ray diffractometry permits the detection, identification, and determination of the concentration of crystalline phases, such as minerals. It is conceivable that an on-line diffractometer could be made by automating a laboratory instrument, but no such analyser is known to be used in the mineral industry, mainly because of the difficulties inherent in obtaining quantitative results from the small samples required by the instrument. On-stream analysers look at very much larger samples and so the determination of crystalline phases on-stream seems a much more viable proposition, especially in the mineral industry.

The technique possesses certain similarities with X-ray fluorescence, in that a beam of X-rays from a high-powered tube is directed at a smooth surface, produced by a former-scraper for powders or an X-ray-transparent window for slurries, and a detector or series of detectors is used to measure the X-rays diffracted through angles characteristic of the crystalline phases present (Chapter 7, section 7.6.4). Diffraction patterns from mineral mixtures are complex and this may mean that a relatively large number of intensities have to be measured. In a conventional X-ray diffractometer, the sample is irradiated by a beam of monochromatic X-rays while a single detector scans the diffracted intensity over the required angular range, by rotating both sample and detector about a common axis, and sequentially records the reflexions. Recording a whole pattern may take an hour or more to give the required accuracy. This may be speeded up by using so-called focusing geometry, whereby several detectors can measure diffracted intensities at different diffraction angles simultaneously. There is, however, a lower limit to the angular resolution of such an instrument, since the physical size of the detectors puts a very definite constraint on how closely together these may be mounted. Moreover if, due to temperature changes or a small shift in the sample position, the effective diffraction angle changes by even a small amount, this would be recorded as a change in the concentration, since the diffraction peak would no longer be centred on the detector, with obvious unfortunate results, especially if this occurred for only one of the components being measured.

A newly developed on-stream X-ray diffractometer [11] overcomes these limitations. It uses a polychromatic X-ray source and a single semiconductor X-ray detector to detect and measure all the diffracted intensities from the sample simultaneously.

The method is a variant on the energy dispersive X-ray fluorescence method, but utilising the diffracted radiation and not the fluorescent. The Bragg equation [Chapter 7, equation (7.3)] can be written in the form

$$\frac{NK}{E} = 2d\sin\theta \tag{6.1}$$

where N is a whole number, usually unity; K is a constant; and E is the quantum energy of the incident X-rays. It will be clear that the term (NK/E) corresponds to the monochromatic wavelength in equation (7.3). Normally, the experimental conditions for diffractometry are chosen such that E (orλ) is kept constant, by using the characteristic radiation from the X-ray source, and θ is varied to scan all possible values of d. With the new instrument, however, θ is kept constant and the values of d scanned by making use of the energy sorting property of the semiconductor detector, to measure at what energies, and hence wavelengths, X-rays are being diffracted for the given value of θ. The required polychromatic X-rays are provided by using the continuous portion of the output from a high power X-ray tube. The energy sorting is carried out electronically and the measuring system is able to detect and adjust to a peak shift.

Thus, instead of either measuring a large portion of the pattern sequentially, or at selected wavelengths simultaneously, this instrument measures a large portion of the pattern simultaneously, thereby increasing the amount of information available to the control engineer by a very considerable amount. The instrument, with its radiation shields lifted, and a typical diffraction pattern are shown in figures 6.5 and 6.6 respectively.

Figure 6.5 On-stream X-ray diffractometer for powders, using a high power X-ray tube source with a semiconductor detector. Note the curved shoe in contact with the powder stream and the liquid nitrogen tank, on the right, for cooling the detector head.

Note how both fluorescence and diffraction data can be collected simultaneously, to give information relating both to the elemental and to the phase composition of the stream. For optimum performance the instrument requires a curved sample surface and this is obtained, for powders, by the use of a curved former in contact with a preshaped stream of powder on a belt or, for slurries, by deforming the window by reducing the pressure in the cell.

The performance of the on-stream diffractometer is such that limits of detection of between 0.1 and 1.0 per cent of a phase are possible in favourable cases, although these limits will be higher in more complex mixtures. Precision should be better than 1 per cent relative, at higher concentrations. It is not possible to give general rules about the limits of detection for any particular mineral, except that highly crystalline minerals, such as calcite or quartz, tend to give stronger diffraction patterns, and so are more easily detected, than less crystalline materials, such as clays.

6.4.1 Basic Theory for On-Stream X-ray Diffractometry

The theory of X-ray diffractometry will be covered in Chapter 7, and most of the additional points involved with on-stream slurries and powders can be explained by theory similar to that used for X-ray fluorescence. There is, however, one complicating factor in X-ray diffractometry which is especially applicable to on-stream diffractometry, and this is preferred orientation (Chapter 7, section 7.6.5). As shown in that section, the term describes the tendency for crystals of certain minerals, especially those with platy habits, to become aligned in a preferred direction with respect to the sample surface and so destroy the random orientations required for accuracy. In practice, the effect results in certain *d*-spacings giving higher diffracted intensities and others lower than would correspond to the actual concentration of the phase in the sample. This can be largely overcome in a slurry by ensuring high turbulence in the flow cell, and reduced in powders by ensuring good mixing and reproducible preshaping before forming the

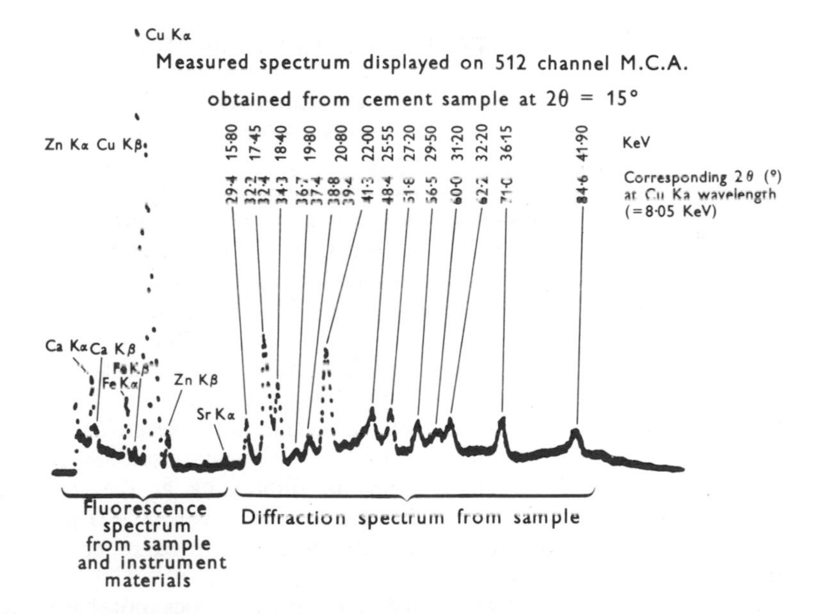

Figure 6.6 Output from an on-stream X-ray diffractometer fitted with a semiconductor detector. The section of the pattern on the right gives information derived from the diffracted X-rays, and hence relating to the crystalline phases present, while the part to the left refers to the fluorescent radiation, and hence to the elemental composition of the powder.

final sample surface. It will be difficult to eliminate the effect completely and, in general, quantitative accuracy with powders will depend on the reproducibility of the residual preferred orientation in the sample surface. One simplifying factor in X-ray diffraction, as compared with X-ray fluorescence analysis is that, since the diffracted intensity from a particular mineral is being measured, mineralogical or particle composition effects are less commonly detected, although they can still be present due to mineral intergrowths in single grains.

6.4.2 Application of On-Stream X-ray Diffractometry

No commercial on-stream X-ray diffractometer has yet been installed in the mineral industry, so application data are not available. Its poorer limits of detection, when compared with X-ray fluorescence, means that it would be of limited value in monitoring the concentration of values in the majority of streams in plants processing low grade ores. Possible applications include the determination of oxidised and sulphide ore ratios for flotation control, a task involving the distinction between two crystalline phases which cannot be performed by X-ray fluorescence, and the monitoring of gangue composition for grinding and flotation control.

6.5 X-RAY SCATTERING OR TRANSMISSION

The intensity of X-rays scattered from, or transmitted by, a sample stream, such as a slurry, can provide useful information for process control. Preferential absorption of X-rays of a given energy by a specific element can be used to measure the concentration of that element. As shown in Chapter 3, figure 3.3, the absorption coefficient for an element changes with the energy of the incident X-radiation, decreasing as the X-ray energy increases until a limiting value is reached, the absorption edge, when the absorption coefficient increases very sharply. The energy of this absorption edge is characteristic of the element in question. If we now monitor the intensities of two sets of X-rays, one with an energy just above the relevant absorption edge and the other just below, it is possible, in principle, to determine the concentration of the element by comparing their intensities after passing through a known thickness of the sample. The X-rays used are usually obtained from targets containing elements fluorescing at the required energies and excited by an X-ray tube or a radioisotope source. Thus, the *K*-series X-rays of zirconium and niobium would be used to determine strontium. Limits of detection with this method are much poorer than those obtained by X-ray fluorescence, and the method does not have the flexibility of XRF, and no applications are known at present in the mineral industry.

In heterogeneous materials, both the scattered and the transmitted X-ray intensities depend on the particle size as well as on the overall chemical composition of the sample. This effect can be used to measure mean particle size in slurries [12], and the method has been applied on the laboratory scale, but as yet no applications exist in the industry.

6.6 FUTURE DEVELOPMENTS

The next few years should see the general adoption of on-stream X-ray analysis spread outside North America. If instrument manufacturers produce suitable analysers, it would be reasonable to expect that semiconductor detector systems using low-power X-ray tubes will take the major share of the market, because of their flexibility and relative insensitivity to changes in the position of the sample surface. The simpler filter systems should establish their place in smaller mines or in those where multi-element, multi-stream analysis is not essential, provided that the cost differential between these and their more powerful, but more expensive, competitors is adequate. X-ray diffractometry should become a standard technique for on-stream analysis, the information it is able to give showing up new ways in which control can be applied.

Technical advances are likely to be made in the improvement of semiconductor detectors, especially with respect to count-rate capability and insensitivity to vibration, and in the field of sample preparation where in-stream analysers and windowless sample presenters will increase reliability and allow greater flexibility in both range of elements and materials which the methods can handle.

REFERENCES

1. G. J. SUNDKVIST, F. O. LUNDGREN and L. J. LIDSTRÖM, *Anal. Chem.*, **36**, 2091 (1964).
2. A. HEWETT-EMMETT, *Acta Imeko,* 361 (1967).
3. K. G. CARR-BRION, *Trans. Inst. Min. Met.* C, **76**, 94 (1967).
4. S. E. BRAMWELL and K. G. CARR-BRION, *Inst. Practice,* **24**, 324 (1970).
5. P. F. BERRY, T. FUTURA and J. R. RHODES, *Advances in X-ray Analysis,* **12**, 612 (1969).
6. J. J. LUCAS TOOTH and B. J. PRICE, *Metallurgia,* **54**, 149 (1961).
7. K. G. CARR-BRION, *Analyst,* **94**, 177 (1969).
8. C. L. LEWIS, W. H. A. TIMM, and A. J. WILLIAMS, Proc. 3rd IFAC/IFIP Conf. 1971, Preprint XIV-2.
9. K. V. KONIGSMANN and Staff, *Can. Min. J.*, **90**(6), 62 (1969).
10. E. J. KROKRISKIA, *Mining Engineering,* **22**, No.10, 99 (1970).
11. ANON, *Mining Magazine,* **125**, No.1, 43 (1971).
12. K. G. CARR-BRION, Proc. 9th Comm. Min. Met. Cong., London, 5 (1969).

CHAPTER 7

X-ray Diffraction

A.W. Nicol

Department of Minerals Engineering
University of Birmingham
Birmingham B15 2TT
England

7.1 INTRODUCTION

It was in 1912, at the University of Munich, that Friedrich and Knipping, following a suggestion from Max von Laue, produced the first X-ray diffraction pattern by irradiating a single crystal of ZnS with a beam of X-rays and photographing the resulting diffracted beams on a photographic plate placed behind the crystal [1, 2]. Analysis of these 'Laue' pictures proved difficult, however, and Bragg [3, 4], in 1913, introduced a simpler relationship which, as the 'Bragg equation', lies at the basis of most of our current X-ray diffraction studies.

Together these techniques represented major advances in our ability to study and characterise materials, particularly crystalline solids, and now X-ray diffraction techniques provide fast, convenient, and simple methods not only for identifying crystalline phases at room temperatures and pressures, but also for characterising materials at high and low temperatures and at high pressures, and as an aid to studying solid state reactions in appropriate circumstances. Such facilities can obviously be of the greatest benefit in studies of minerals, mineral products, ceramics, and refractory and clay materials, both at the research level and in the processing plant. This chapter will discuss the basic theory of X-ray diffraction and its application to mineralogical and related problems, with especial reference to the identification of phases by powder diffraction methods. Studies on single crystals are not of such general usefulness as studies on powders for identification purposes, but it is sometimes advantageous to obtain single crystal data and a short section is included to summarise relevant procedures.

There is now quite a large literature dealing with X-ray crystallography generally, and a bibliography of recommended texts is appended to the chapter. Of these, the books by Bunn and Buerger are probably best for the basic theory, although Klug and Alexander give very good descriptions of equipment and techniques and Peiser, Rooksby and Wilson's volume gives an excellent survey of the powder method in all its aspects. Henry, Lipson and Wooster give a wide-ranging discussion of the interpretation of all kinds of X-ray diffraction data, and particular mention may be made to the chapter on X-ray diffraction by Zussman, in the book which he edited, with its strong emphasis on mineralogical applications. Readers are advised to consult these references for further details or for the solution of special problems beyond the scope of this chapter.

7.2 UNITS

The unit in most common use throughout this chapter will be the angstrom unit, Å, which is 10^{-10}m or 10^{-1}nm. This is the generally accepted unit for X-ray measurements, but readers are reminded that another, older unit exists, the X-unit or the kX-unit ($1\text{kX} = 10^3$ X-units). This latter unit, which was based on the supposed sizes of the spacings in calcite [5], was in common use in the early days of X-ray diffraction until Bearden [6] showed that the basis for the X-unit was slightly in error and that the kX-unit was not exactly equal to the angstrom unit. In fact, wavelengths and other measurements quoted in kX-units should be multiplied by 1.002076 to convert them to absolute metric or angstrom units. The conversion need only be made when working to high accuracies, but readers are warned to check the units quoted, particularly in references dated before about 1950. No current work should be quoted in kX-units.

7.3 SAFETY PRECAUTIONS

Safety precautions to be taken when working with high energy radiations have been discussed in Chapter 1, section 1.5. Readers are reminded that X-rays are dangerous, and that they should pay particular attention to the earlier section on safety and to the publications from the Department of Employment and Productivity [7] and the International Union of Crystallography [8].

7.4 NATURE AND PRODUCTION OF X-RAYS

The nature and general methods of production of X-rays have been discussed in Chapter 3, and they have been well reviewed by Klug and Alexander [9] and by Norrish and Chappell [10]. Today there is a wide range of generating equipment available commercially, sufficient to cover all but the most specialised applications. Almost all modern X-ray generators utilise hot cathode, sealed source vacuum tubes operating at power loadings up to 2kW, depending on the nature of the target material in the tube, although special tubes, often with rotating anodes, with ratings up to 7kW have recently become available. Such tubes may be gas-filled, but otherwise gas-filled tubes are rarely used, due to the difficulties inherent in their operation, albeit that they possess the advantage of interchangeability of targets. Sealed tubes are convenient to use, since they require no vacuum equipment, and they provide the stability of output intensity demanded by modern techniques, factors which outweigh the disadvantage of a single target element.

The most commonly used radiation is Cu $K\alpha$ radiation, generated from a sealed tube operating at 40kV and 16-20mA. This radiation can be used for most crystallographic applications in conjunction with most materials. The major difficulties encountered are with materials rich in iron, since the fluorescent Fe K radiation excited by the Cu $K\alpha$ radiation can give rise to unacceptably high backgrounds on films and to swamping of the counter in diffractometry. It is possible to use Co $K\alpha$ radiation with iron-containing samples and avoid the fluorescence problem, but this radiation suffers from two disadvantages. Firstly, the beam intensity is lower than for copper and exposures are considerably longer, and secondly the wavelength of cobalt radiation is longer than that of copper, making it impossible to compare films taken with the two types of radiation by direct superposition. The fluorescence problem may be minimised in film work with copper radiation either by positioning a filter between the sample and the film or by stripping the emulsion from the side of the film nearer to the sample with a hypochlorite solution, since this is the side of the film more affected by the fluorescent scatter [11]. It can be completely eliminated in diffractometry by using an energy-sensitive counter with pulse height discrimination (cf. section 1.3.3) to remove entirely any contribution from the fluorescent radiation.

Molybdenum radiation is often used in single crystal studies when there is a need to collect accurate intensities relatively free from absorption effects, for crystal structure determinations.

7.4.1 X-ray Generators

The choice of X-ray generator for any application obviously depends on the precise nature of the application, but it may be appropriate to discuss some of the basic considerations which may apply generally. The first consideration must be of the level at which studies are to be made. If it is merely intended that routine, qualitative identification studies are to be made, then it is probably quite sufficient to buy a relatively inexpensive instrument with the minimum of controls, and giving a stability level of about 1% without feedback. It is probably worth paying for a smoothed and rectified set since the savings in the recurrent cost of X-ray tube inserts run in this mode, compared with the rate of replacement for an unrectified set, will be appreciable over the long term. On the other hand, if there is a need for more sophisticated studies, and certainly for the application of quantitative methods, then it will be necessary to buy a generator with full feed-back circuitry giving a stability level of, at worst, 0.05%. Such a set is roughly twice the price of the simpler generator, and it goes without saying that a proper cost-effectiveness study must be made before buying such an instrument. It is worth noting that inclusion of a simple voltage regulator in the input line to a smoothed and rectified generator can improve its stability to around 0.1-0.5%, at a much lower cost than feed-back circuitry, and this is probably sufficient for the great majority of X-ray applications where no more than reasonable accuracy is required.

7.4.2 Provision of Monochromatic X-rays

In Chapter 3 it has been shown that the output spectrum from an X-ray tube includes both the continuous 'white' radiation and the characteristic 'lines' of the target element. The line spectrum of, for example, a copper target tube comprises the $K\alpha_1/K\alpha_2$ doublet and the $K\beta$ line, with the $K\alpha_1$ and $K\alpha_2$ wavelengths very close together, and appreciably longer than the $K\beta$ wavelength, as shown in table 7.1. The majority of X-ray diffraction studies require reasonably monochromatic radiation, and it is clear that the $K\alpha$ lines of a given target element could be used, provided that the white radiation and the $K\beta$ line could be removed. Two methods exist for this, by filtering the primary radiation or by using a single crystal monochromator.

All elements absorb X-rays, and their absorption coefficients vary with the wavelength of the incident radiation, as shown in figure 3.3. Fortunately, the K-absorption edge of the element of atomic number $(N-1)$ lies between the energy levels corresponding to the $K\alpha$ and the $K\beta$ lines for the element of atomic number N. Thus, if the primary beam from a copper target is passed through a thin nickel foil, the white radiation and the copper $K\beta$ radiation will be absorbed to a much greater extent than will be the $K\alpha$ radiation, and the emergent beam will, therefore, be relatively strong in the last component and may be used as an effectively monochromatic beam. It is important to remember, however, that the intensities of the $K\beta$ line and of the white component are not zero, and contributions due to diffraction of these radiations may occur. In particular, very strong powder X-ray reflexions may give rise to weak β-peaks, indistinguishable from the α-pattern at first sight, and Zussmann [11] has pointed out that similar spurious broad diffraction maxima may arise through diffraction of the white component since, for copper, this has a broad peak centered at about

TABLE 7.1

K-Emission Series for Selected Elements; Wavelengths, Filter Materials and Intensity Ratios

Element	$K\alpha_1$	$K\alpha_2$	$K\alpha$‡	$K\beta_1$	$I\beta_1$: $I\alpha_1$†	$I\alpha_2$: $I\alpha_1$	Filter	Thickness (mm)*
Ag	0.55936	0.56378	0.56083	0.49701	0.290	0.499	Pd	0.062
							Rh	0.062
Mo	0.70926	0.71354	0.71069	0.63225	0.279	0.499	Zr	0.081
Cu	1.54051	1.54433	1.54178	1.30217	0.200	0.497	Ni	0.015
Ni	1.65784	1.66169	1.65912	1.50010	0.187	0.495	Co	0.013
Co	1.78892	1.79278	1.79021	1.62075	0.160	0.497	Fe	0.012
Fe	1.93597	1.93991	1.93728	1.75653	0.167	0.500	Mn	0.011
Cr	2.28962	2.29351	2.29092	2.08480	0.179	0.515	V	0.011

*The thickness listed is that required to give an integrated intensity ratio of 1/1000 for $I\beta_1/I\alpha_1$.

†This is the intensity ratio at the surface of the target.

‡$K\alpha$ is the weighted mean wavelength (p. 276).

(After Lonsdale [8]).

0.5Å. Note that there is no way in which the $K\alpha$ doublet can be separated by filters, and the extent to which $K\alpha_1$ and $K\alpha_2$ reflexions are resolved depends on the diffraction angle and on the resolving power of the apparatus used. Table 7.1 lists the wavelengths of some common target materials, the thickness of metal foil required to reduce the intensity of the emergent β-radiation to a predetermined fraction of the α-intensity, and the intensity ratios of the different characteristic lines.

More precise monochromatisation of the X-ray beam can be accomplished by reflecting the primary beam from a suitable single crystal. This technique utilises the diffraction properties of a crystal (section 7.5.2), whereby, for radiation incident on the crystal at some angle θ, only radiation lying within a very narrow wavelength range is reflected, also at the angle θ. Hence it is possible to separate the $K\alpha$ radiation completely from the white and the $K\beta$ contributions, but the price paid is a very marked reduction in the intensity of the emergent beam. Both curved and plane crystal monochromators are available, the former being favoured because of the higher beam intensities provided. Curved crystal monochromators have the further advantage of focusing the beam, and so find important uses in conjunction with focusing cameras (section 7.6.3). Crystal monochromated radiation is also used for high accuracy work on single crystals, especially in the collection of intensity data for structure determinations, but otherwise the disadvantages, in terms of cost and low beam intensities, rule out its use for routine measurements. Brindley [12] and Lonsdale [13] have reviewed suitable materials and designs for crystal monochromators, and recent improvements in technique have made graphite an important monochromator material [14].

7.5 BASIC X-RAY DIFFRACTION THEORY

7.5.1 Laue Diffraction

The analysis of X-ray diffraction patterns obtained by the Laue method, i.e. by irradiating a static single crystal with unfiltered X-rays and photographing the diffracted radiation on a static film, is basically an extension of the theory of diffraction of visible light by a line grating. Figure 7.1 shows the elements of such a one-dimensional grating at a distance a_0 apart with a beam of radiation, with wavelength λ, incident upon it at an angle ϕ and a diffracted beam at an angle ψ. For such a grating, constructive interference occurs when

$$n\lambda = a_0(\cos\phi - \cos\psi) \tag{7.1}$$

In a single crystal, however, the elements of the effective grating repeat in three dimensions, and the conditions for constructive interference become correspondingly more complex. Laue [2, 15] developed the mathematical relationships relating diffracted beam direction to the crystal symmetry and the crystal's orientation relative to the incident X-ray beam,

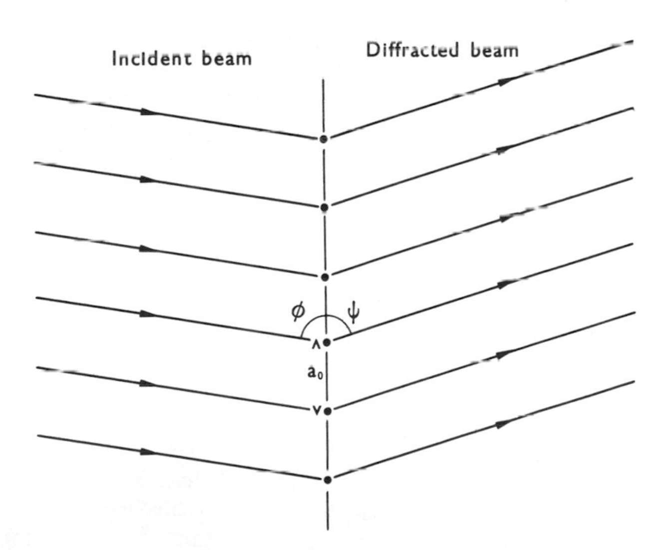

Figure 7.1 Diagrammatic representation of diffraction of a beam of electromagnetic radiation by a row of equally spaced points.

and showed that, for any crystal, the three Laue equations' which must be simultaneously satisfied for constructive scattering, may be written

$$m\lambda = a_0(\cos\phi_a - \cos\psi_a) \quad (7.2a)$$

$$n\lambda = b_0(\cos\phi_b - \cos\psi_b) \quad (7.2b)$$

$$p\lambda = c_0(\cos\phi_c - \cos\psi_c) \quad (7.2c)$$

where λ is again the wavelength of the incident X-rays which produce the diffracted beam; m, n, and p are integers; a_0, b_0, and c_0 are the repeat distances along the principal axes of the unit cell; and ϕ_a, ϕ_b, ϕ_c and ψ_a, ψ_b, ψ_c are the angles of incidence and diffraction respectively relative to the above axes.

Thus the solution of the Laue conditions involves three equations in four unknowns, namely the X-ray wavelength and the three angles of incidence and, in general, these possess no unique solution. This makes detailed interpretation of Laue patterns very difficult for, in the absence of a knowledge of the angles of incidence of the X-ray beam relative to the crystallographic axes of the crystal, it is impossible to assign λ values to the various reflexions obtained. Laue pictures do, however, contain important information about the symmetry of the crystal under study, and so may be used in deriving unit cells and space groups in single crystal studies, and they are particularly useful in orienting crystals for special purposes, for example in setting a crystal prior to cutting parallel to a given crystallographic plane, or for finding the orientation of the crystallographic axes relative to a chosen direction in the crystal. The interpretation of Laue photographs is well covered by Henry, Lipson and Wooster [16]. The paucity of the information which can easily be obtained from Laue photographs, at least in comparison with other methods, has led to this technique being relatively little used, but recent developments have led to a resurgence of interest, and this will be further discussed in section 7.10.

7.5.2 The Bragg Equation

In 1913, Bragg [3, 4] realised that one of the solutions of the Laue equations, for the case of a two-dimensional array of regularly spaced scatterers, was that the plane so defined should act as a mirror plane. Part of the radiation incident on the array should be reflected in a plane perpendicular to the plane of the array, and containing the incident beam, at an angle equal to the angle of incidence. He next showed that a three-dimensional crystal could be considered as a stack of parallel, equidistantly spaced, identical plane arrays and that diffraction from such an array could be explained by assuming that a small fraction of the X-ray energy incident on any plane was reflected while the remainder passed through to repeat the process with the next and subsequent planes. Consideration of the geometry of reflexion from such a stack of mirrors shows (figure 7.2) that, for constructive interference between the beams diffracted from adjacent planes, the path lengths between the individual beams must be integral numbers of wavelengths. The required condition is satisfied if

$$n\lambda = 2d\sin\theta \qquad (7.3)$$

where λ is the wavelength of the monochromatic X-rays used, d is the distance between adjacent planes, and θ is the angle of incidence of the X-ray beam on the plane, normally called the 'Bragg angle'. Note that the angle is measured between the incident beam and the plane, not between the beam and the normal to the plane, as in light optics.

This remarkably simple relationship, the 'Bragg equation', has dominated X-ray diffraction work to the present day. The equation is, of course, much simpler than the Laue equations, involving as it does only two independent unknowns, the X-ray wavelength and the angle of incidence, both of which are relatively easy to define. Normally the conditions under which diffraction is carried out are chosen such that the value of λ is known and θ is varied, usually by rotating the crystal in a beam of monochromatic X-rays, but recent developments in the detectors for electronic counting systems have made practicable the configuration in which θ is held constant and λ is varied by electronic scanning of the energy spectrum of the X-rays diffracted from a crystalline sample. An application of this method is discussed in Chapter 6.

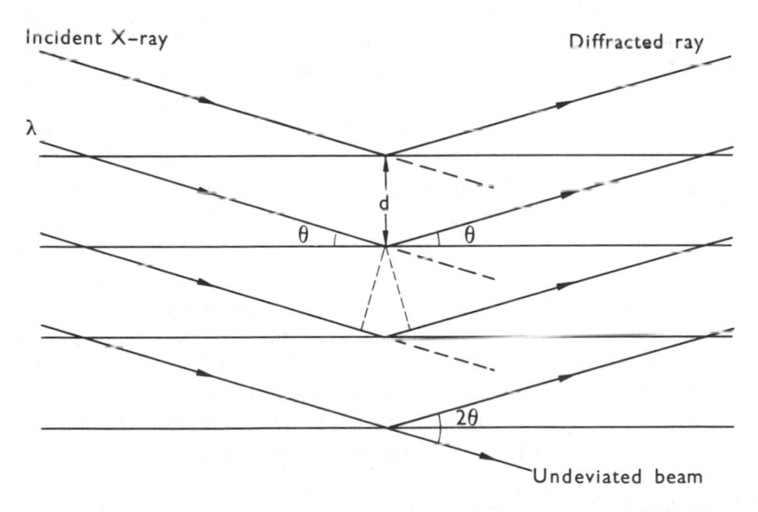

Figure 7.2 Construction to illustrate the derivation of the Bragg equation for diffraction of a beam of X-rays by a crystalline lattice, considered as a stack of equally spaced mirror planes.

7.5.3 The Ewald Sphere

The above methods of analysing X-ray diffraction data are based on the shape and size of the direct cell of the crystalline species under study and are adequate for most simple work, especially work involving powders and their identification. Single crystal data, however, are often more easily handled in terms of the reciprocal lattice and the Ewald sphere, concepts introduced in 1921 by P. Ewald [17]. Both represent geometrical constructions which provide convenient ways of considering the process of diffraction by a crystalline lattice, but neither bears any direct relationship to a real object. The reciprocal lattice, which is an extension of the Miller index notation for crystallographic planes, is constructed by erecting vectors, from an arbitrary origin within the crystal, perpendicular to the various crystallographic planes in the crystal lattice, of lengths ρ given by

$$\rho_{hkl} = k/d_{hkl} \tag{7.4}$$

where k is a positive constant, usually unity but sometimes the X-ray wavelength being used, and d_{hkl} is the inter-planar spacing for the set of planes with Miller indices (hkl). Ewald [17] and Buerger [18] have shown that the ends of these vectors form a regular lattice of points in 'reciprocal space' and each point in reciprocal space, corresponding to the end of a vector normal to a plane in the crystal in real space, corresponds to that plane in reciprocal space. Moreover, if the plane moves in real space, the reciprocal point moves correspondingly in reciprocal space. Figure 7.3 shows diagrammatically how part of a reciprocal lattice can be developed using the above construction. The Ewald sphere is an imaginary sphere, also in reciprocal space, of arbitrarily unit diameter.

Ewald has further demonstrated that, if the origin of the reciprocal lattice of a crystal under study is considered to lie on the surface of the Ewald sphere and an X-ray beam is considered incident on it across a diameter of the sphere, a diffracted beam is generated whenever a reciprocal lattice point encounters the surface of the Ewald sphere, the imaginary encounter being considered to have occurred in reciprocal space. The direction of the diffracted beam from the crystal in real space lies parallel to the radius direction from the centre of the sphere to the point of coincidence in reciprocal space. If the crystal is rotated, new lattice points are brought into coincidence with the surface of the Ewald sphere, to produce the complete diffraction pattern for the crystal. Bernal [19] and Evans and Peiser [20] have expanded this geometrical picture of diffraction and have devised sets of charts [21] to simplify the interpretation of single crystal and rotation Weissenberg photographs. Buerger [18], Henry, Lipson and Wooster [22], and Bunn [23] have reviewed the theory of the reciprocal lattice and of the Ewald sphere in some detail.

7.5.4 The Mechanism of Diffraction

According to the simple wave theory of physical optics, the interference between a set of point scatterers and a plane wave front results in the

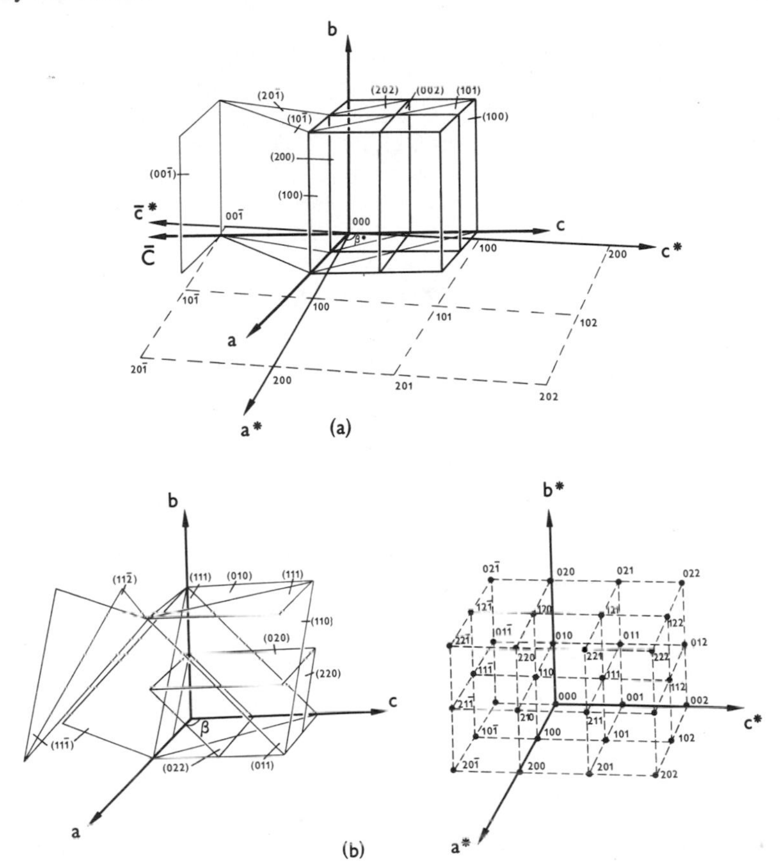

Figure 7.3 Development of the reciprocal lattice as a series of vectors perpendicular to the set of crystallographic planes with lengths inversely proportional to the various interplanar spacings.

formation of a series of spherical wavefronts, each set centred on a point scatterer, and the new wavefronts corresponding to constructive interference are represented by the common tangents that can be drawn across the fronts from adjacent scatterers. This picture can suffice for much routine X-ray diffraction work, but it is better to understand the principles underlying the interaction of an X-ray photon with the atoms in a crystalline solid in order to utilise fully the data obtainable from X-ray diffraction patterns.

X-radiation is an electromagnetic radiation and so it has associated with it a sinusoidally varying electric field. When an X-ray photon interacts with an atom it can cause electrons in that atom to undergo what amount to forced oscillations, at energy levels above the ground state, but without being lost from the atom. If no other energy loss processes intervene, the atom will return to its ground state by re-radiating a photon of the same energy as the incident one, and hence of the same wavelength. Such 'elastically scattered' photons contribute to that part of the diffracted radiation which generates the diffraction pattern. In certain cases, however, an excited electron may lose part of its energy, particularly by recoil processes involving a change in its momentum, before re-radiating the photon. The resulting photon is now of lower energy, and so of longer wavelength, than the incident one, and contributes to the 'inelastic' or 'Compton scatter', named for A. H. Compton [24] who first measured and explained the effect. The Compton scatter accounts for part of the background scatter in an X-ray photograph but, although it contains data useful for certain specialised studies, it is generally ignored. Both types of scattering account for some of the observed X-ray absorption in a material, a topic which has been discussed at some length in Chapter 3. The precise manner in which the atom interacts with the X-ray photon, especially with regard to the elastic scattering, is controlled by the atomic number of the element, as shown in Chapter 3, and by the Bragg angle of the diffracted radiation, so that the 'atomic scattering factor' for any atom is not constant but varies for different reflexions.

7.5.5 The Intensity of Reflexions

The above analyses, of Laue, Bragg, and Ewald all lead us to conclude that the X-ray diffraction pattern obtained from a crystal depends on the shape and size of the basic unit cell of the crystal. But the tacit assumption has been made, in every case, of an array of point scatterers with no reference to the actual atoms present or their spatial relationships one to another. When the analysis is extended to include the mechanism whereby atoms scatter X-rays, it becomes clear that different atoms scatter X-rays to different extents, and the X-rays scattered from different atoms in the same unit cell will, in general, not be in phase with one another, although X-rays scattered from corresponding atoms in different unit cells will be in phase. Hence the actual intensity of any reflexion will depend on the arrangement of atoms within the unit cell, and also on the Miller Indices of the lattice plane giving rise to the reflexion, since these together effectively specify the Bragg angle for the reflexion. If the jth atom in the unit cell has fractional co-ordinates x_j, y_j, z_j in terms of the three unit cell edges, it can be shown that, for the most general case, for the (hkl) reflexion,

$$I(hkl) = p \times F^2(hkl) \times T \times A \times G \qquad (7.5)$$

where

$$F(hkl) = (A^2 + B^2)^{1/2} \qquad (7.6)$$

and

$$A(hkl) = \sum_j f_j \cos 2\pi(hx_j + ky_j + lz_j) \qquad (7.7)$$

$$B(hkl) = \sum_j f_j \sin 2\pi(hx_j + ky_j + lz_j) \qquad (7.8)$$

In equation (7.5), $I(hkl)$ is the observed intensity; p is the multiplicity factor, or the number of planes contributing to the reflexion; T is a temperature correction factor depending on the degree of thermal motion among the atoms; A is the absorption factor for the material for the X-ray wavelength used; and G is a geometrical factor which depends on the Bragg angle. In equation (7.6), $F(hkl)$ is the structure factor for the reflexion, and f_j is the atomic scattering factor for the jth atom for the (hkl) reflexion angle, in equations (7.7) and (7.8). Lonsdale [13, 25] has tabulated the various forms of equations (7.6) through (7.8) for the 230 space groups, and the values of f_j for different atoms in different valence states as a function of the Bragg angle. Readers wishing to pursue this topic further should consult the books by Buerger [26, 27] and Ramachandran [28].

Thus the intensities of different reflexions in a diffraction pattern will vary in a non-regular manner, and it is important to note that, whereas the relative geometrical positions of the diffracted beams, as observed in the resulting diffractogram, provide information about the shape and size of the unit cell, the relative intensities of the reflexions provide information about the spatial relationships between the atoms in the unit cell. Both types of information must be gathered for identification studies.

7.6 METHODS OF OBTAINING X-RAY SPECTRA

X-ray diffraction studies can be made either on single crystals or on powdered crystalline materials. Single crystals give spot patterns which contain much information about the shape and size of the unit cell and about the crystalline structure, but which are somewhat difficult to obtain and so are not widely used for simple identification. Powders, on the other hand, give line patterns which are relatively easily obtained in a reproducible manner, and are particularly suitable for identification work. As indicated earlier, the main emphasis in the succeeding sections will be on powder diffractograms.

7.6.1 X-ray Diffraction by Powders

Many materials, among them minerals, ores, etc., exist in nature as powders – i.e. as masses of tiny crystals which can take up completely random orientations in space relative to one another. Of course, any crystalline material can be converted to this state by careful grinding. The basic sample for powder X-ray diffraction, therefore, is a small quantity of the material, generally 1-2mg but up to 100mg for a diffractometer sample, which has been crushed to -300mesh (<53μm), but which has not been overground until it contains an appreciable amount of extremely fine material. This latter gives rise to problems of line-broadening in diffraction patterns, leading to loss of resolution and inaccuracies in estimating line positions. Obviously, care must be taken initially to ensure that the primary sample taken for crushing is representative of the bulk material, and that the ground material is properly homogenised by dry or wet mixing so that the final sample is again a true sample. The considerations discussed in Chapter 12 will apply here.

If a cylinder of the powder, of radius about 0.1-0.5mm, is prepared and a fine beam of X-rays directed at it, the very large number of discrete, randomly oriented crystallites present ensure that it is statistically almost certain that there will be at least one crystallite in the correct reflecting position for each of the possible crystallographic planes in the crystal. In general, of course, many crystals will be in the correct θ-angle orientation for each plane, and if the cylinder is rotated about an axis perpendicular to the X-ray beam, many more crystallites will be brought into the reflecting position, and these crystallites will be randomly oriented around the beam direction. Figure 7.4 shows two possible reflecting positions for the same reflexion in two separate crystallites, and it will be clear that the locus of all the beams for this reflexion is a cone of semi-angle 2θ centred about the X-ray beam. Note that the semi-angle of the cone is 2θ, from the quasi-mirror reflection geometry of the diffracted beam, and it is this angle that is actually measured in practice, and hence is often quoted, rather than the Bragg angle, θ, itself.

Thus the problem in X-ray powder diffraction studies is one of detecting cones of diffracted radiation, and this can be achieved in one of two major ways, either by using film methods or by using electronic counting techniques. In both cases, the aim is to measure both the 2θ angle and the relative intensity of each diffracted beam from the sample.

7.6.2 Film Methods

In its simplest and commonest form, the powder film camera consists of a metal cylinder, of internal diameter between 5cm and 20cm, supporting a strip of X-ray sensitive film, and surrounding a rod-shaped sample of

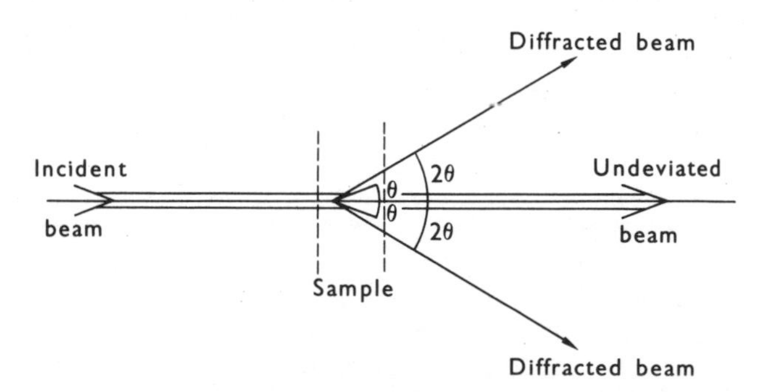

Figure 7.4 Diffraction of a beam of X-rays by two crystallites in a powder specimen with the same (hkl) plane in two diffracting positions oriented symmetrically about the incoming beam.

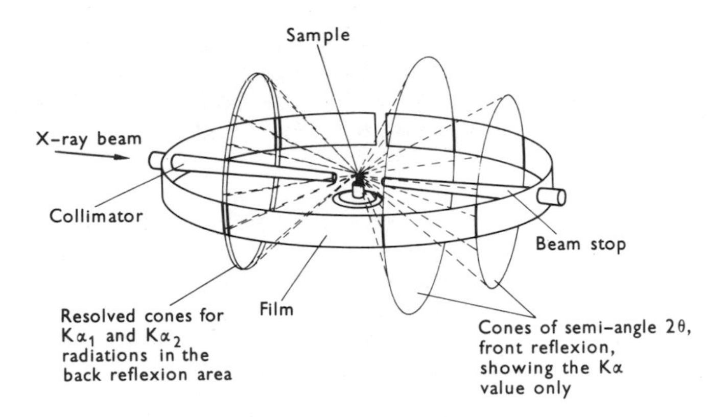

Figure 7.5 Schematic representation of the cones of diffracted radiation generated from a powder specimen and their mode of interaction with a cylindrical film mounted co-axially with the specimen to give a Debye-Scherrer pattern. The film is shown in the Straumanis-Ievins mounting.

powder, about 0.1-0.5mm in diameter. The cylinder is provided with an inlet collimator system for the incident X-ray beam and a back-stop to trap the undeviated beam after passage through the sample, as shown in figure 7.5. Three principal methods of film-loading have been described, and their merits discussed by Nelson [29] and by Henry, Lipson and Wooster [30]. Of these, two involve passage of the X-ray beam through the film once, either as the incident beam (the van Arkel mounting [31]) or as the residual undeviated beam (the Bradley-Jay mounting [32]), and sharp knife edges are used to cast shadows on the film to provide fiducial marks against which the diffraction line positions can be measured. In the third, the Staumanis-Ievins mounting [33], both the incident and the undeviated parts of the beam pass through holes in the film, the centres of which are thus 180° apart, and no fiducial marks are provided. This arrangement gives a self-calibrating picture, provided that lines in the back-reflection region can be measured, since it is possible to measure the actual distance on the film between the centres of the holes, from the mid-points of pairs of front- and back-reflexion lines, and applying the necessary correction factor to all the measured data to correct for deviation of this value from 180°. Figure 7.5 shows diagrammatically how such a typical Debye-Scherrer powder pattern is generated.

Today, probably the most commonly used powder camera is that manufactured by the Philips Company, based on the Straumanis-Ievins configuration. The camera is made in two sizes, the smaller having a diameter of 57.3mm and the larger of 114.6mm. The smaller camera is useful when results are required quickly without a high degree of resolution, often the case in routine work, whereas the larger instrument is better suited for accurate work and for situations where good resolution of the diffraction lines is important. The cameras are shown in figure 7.6(a), and a typical film from the 114.6mm camera is reproduced in figure 7.6(b).

Sample Preparation

The rod-shaped specimen required for this type of camera may be made by several methods. In each case, however, it is important that a fine grained sample, less than 300 mesh size, be used in order that the number of grain orientations present be a maximum and hence that the powder arcs observed on the film be evenly blackened. The use of larger grain sizes can result in 'spotty' arcs due to the limited number of grain orientations represented.

A very common method of specimen preparation, and one that is particularly useful for more heavily absorbing samples, involves diluting the sample with a gum, such as gum accacia, in the ratio 4 sample to 1 gum, moistening the mixture, and rolling a ball of the resulting thick paste first between the fingers and then between two glass slides to form a thin, straight cylinder about 0.3mm in diameter. Alternatively, the paste may be allowed to harden as a sphere, and Hildebrand [34] has shown that this method may produce a specimen with less preferred orientation among the particles than does the above method. Brown *et al* [35] have also described a method whereby the paste, made from the sample plus fish glue or collodion, is extruded from a hypodermic needle. For many purposes, however, it may prove sufficient to coat the specimen onto a thin glass fibre using vacuum grease or shellac as the adhering medium. Finally, lead-free

Figure 7.6 (a) Typical cameras for powder diffraction. That on the left has a diameter of 11.46mm, that on the right of 57.3mm. (Photograph by courtesy of Philips, Eindhoven.)

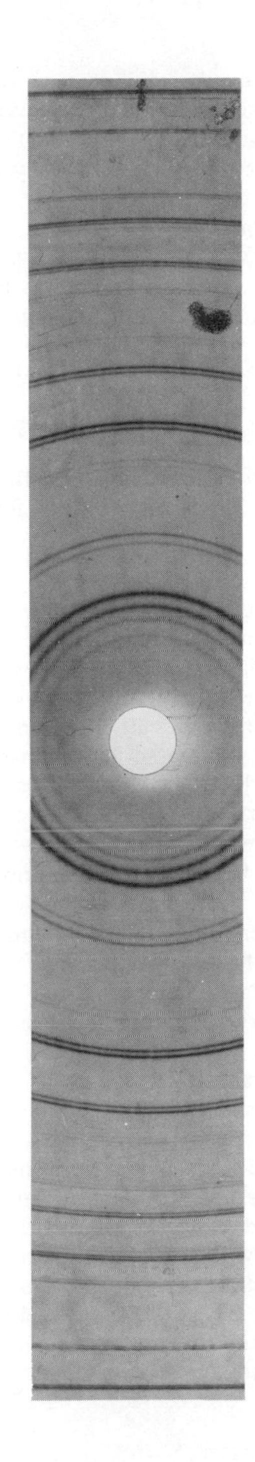

Figure 7.6(b) Powder diffraction photograph of a magnesium aluminate spinel taken with Cu $K\alpha$ radiation in a 114.6mm diameter camera. The upper strip shows the part of the film around the backstop, the low 2θ region, and the lower the collimator or back-reflexion region at high 2θ values. Note how the $K\alpha_1$ and $K\alpha_2$ lines are resolved at high 2θ angles.

borosilicate glass tubes (Lindemann tubes) are available commercially in various sizes and are particularly useful for substances that might be affected by additives, and are imperative for substances which are unstable in air. These tubes may be used for any sample, of course.

Film Measurement

In figure 7.5, if we consider a section through the equatorial plane of the powder camera, it can be shown that, if the distance between the corresponding arcs on either side of the X-ray beam is $2S$, then

$$S/R(\text{in millimeters}) = 2\theta(\text{in radians})$$

i.e. (7.9)

$$57.3 \times kS/R = 2\theta(\text{in degrees})$$

where k is a factor depending on the film shrinkage. But, in the larger of the Philips' cameras, R = 57.3mm and so, for this camera,

$$kS(\text{in millimeters}) = 2\theta(\text{in degrees}),$$

which is quite a convenient arrangement.

Hence interpretation of a powder film involves measuring the distances between the various pairs of arcs, which can be done in various ways depending on the degree of accuracy required. Moderate precision, to within ±1%, for routine identification or comparison work, can be achieved with a good steel rule calibrated in millimeters, and transparent plastic rules calibrated directly in angstroms are also available commercially. With the latter it is important to use the correct scale both for the camera radius and for the X-ray wavelength, and care should be taken to allow for film shrinkage either by marking the positions of fiducial points on the scale or by arranging for the scale to be symmetrical with respect to pairs of lines [36]. Better precision can be obtained with an instrument such as the Hilger and Watts Ltd. instrument, which comprises a long glass plate inscribed with a millimeter scale and a metal cursor fitted with a glass plate on which are scribed fiducial lines and a vernier scale. With this it is possible to estimate line positions to within ±0.01mm. Travelling microscopes may be used, but care is needed in the choice of magnification range and graticule [36], and Sears and Turner [37] have described a modification whereby the film is moved relative to the microscope head. Line positions can be measured to ±0.01mm with these instruments. Recently, projection instruments have been introduced which present an enlarged picture of the film on a ground glass screen and measurement is carried out by moving the line to coincide with a fiducial mark on the glass. Such instruments have the advantage of reducing operator fatigue [38].

Whatever the means of measurement, however, care must be taken in the illumination of the film since even a quite small rise in its temperature can result in its shrinking, possibly unevenly if the heating is localised, with consequent loss of accuracy. Peiser *et al* [36] recommend the use of fluorescent lamps, since they are generally cooler than tungsten filament lamps and spread the heat more evenly.

Conversion of the measured distances into d-spacings, the form in which powder diffraction data are usually handled, may be carried out by converting the millimeter values into 2θ values in degrees, using equation (5.9), and substituting for θ in the Bragg equation. This is tedious, however, and it is more convenient to use tables for the direct conversion of the 2θ value to the d-value in angstroms [39]. Such tables are easily generated using a computer, and little programming skill is required to produce neatly printed listings of d-values at any convenient interval in 2θ, an interval of $0.01°$ being convenient for most purposes, for any wavelength or set of wavelengths.

Estimation of Intensities

Line intensities may be estimated either by eye, for routine work, or using an optical microdensitometer. Estimation by eye is usually made against a relative scale in which the strongest reflexion is assigned strong (s) or very strong (vs), and the gradation goes through medium strong (ms), medium (m), medium weak (mw), weak (w), and very weak (vw). Alternatively a decimal scale, 1-10 by units or 10-100 by tens, may be used. Both scales are logarithmic in nature, very subjective, depending strongly on the person making the estimation, and different workers may grade the same set of lines noticeably differently. An optical microdensitometer is essential for accurate work, e.g. for quantitative studies, and provides intensities on a linear scale. The split-beam instrument made by Joyce-Loebl Ltd. [40] is very useful for this type of work, and the Optronics Company have produced an instrument which can scan a powder film and provide a digital output of intensity along the film [41].

7.6.3 Focusing Cameras

The Debye-Scherrer camera is a useful instrument, both for routine and for many research applications, but it suffers from the drawback, when used with metal-foil filtered X-rays, that the $K\alpha_1/K\alpha_2$ doublet causes line broadening at intermediate 2θ values and line doubling at high 2θ values [cf. figure 7.6(b)], and the resolution at low 2θ values tends to be poor. Focusing cameras can overcome these difficulties, at least to a certain extent.

Several designs of focusing camera have been proposed, and have been discussed by Brindley [12], and while the earlier instruments were based on reflexion from a curved sample using a simple slit system to admit the X-ray beam to the camera, later instruments have used curved crystal monochromators to enhance the focusing effect of the camera. The most widely used design is that proposed originally by Guinier [42] and modified by de Wolff [43]. Figure 7.7 shows that a curved crystal monochromator throws a converging beam of X-rays through a powder sample which is in the shape of a curved plate and lies on the surface of a cylinder shared by

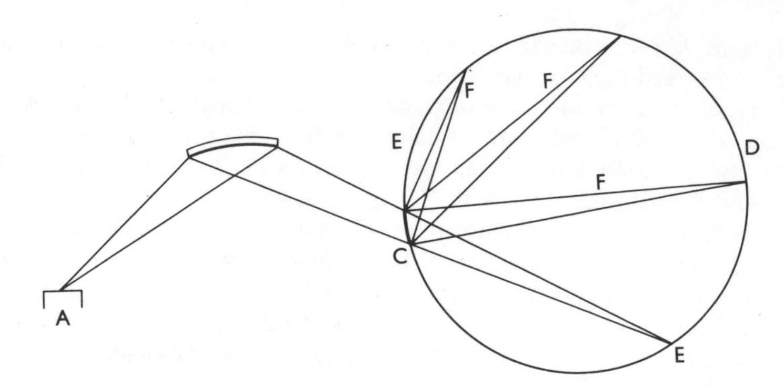

Figure 7.7 Ray diagram for a Guinier-de Wolff powder diffraction camera, showing A – the X-ray source; B – the curved crystal monochromator; C – the specimen smeared onto a curved plate; D – the film on a cylindrical support continuous with the specimen support; E – knife edges indicating $\theta = 0^\circ$ and the angular range of the camera; F – the diffracted beams coming to focus at the plane of the film. (After a diagram by Enraf-Nonius, Delft.)

the photographic film. A knife edge intercepts the X-ray beam to indicate $\theta = 0^\circ$, and a second knife edge delimits the angular range of the camera. De Wolff [43] has pointed out that, since the $K\alpha_1$ and the $K\alpha_2$ separations occur in opposite directions in the two diffraction stages, at the monochromator and at the sample, their final separation at the plane of the film is almost nil over quite a large range of θ, usually chosen to cover the range $\theta = 0^\circ$ through $\theta = 30^\circ$. The resulting very sharp reflexions make the camera useful for studies on clay minerals, because of its consequent high resolving power, and provision is made, in the instrument marketed by Enraf-Nonius Ltd., for up to four samples to be loaded simultaneously and the whole unit is enclosed in a gas-tight casing to allow photographs to be taken under different atmospheres, including vacuum.

The disadvantages of the camera include difficulties in aligning the monochromator, although it is claimed that present cameras are easy to align, and the necessity for a stable base with a firm fixing relative to the X-ray beam but, for many purposes, these are far outweighed by the advantages of precision, detection of low-angle reflexions, sensitivity for weak reflexions, and resolution of closely-spaced lines.

7.6.4 Electronic Counter Diffractometry

The intensity of the radiation diffracted from a powder specimen as a function of θ, or of 2θ in practice, can also be monitored by using a Geiger-Muller or a similar counter, with the appropriate electronics, and displaying its output on a chart recorder while the counter moves along a path corresponding to the position of the film in a camera. Modern diffractometers are based on the para-focusing Bragg-Brentano geom-

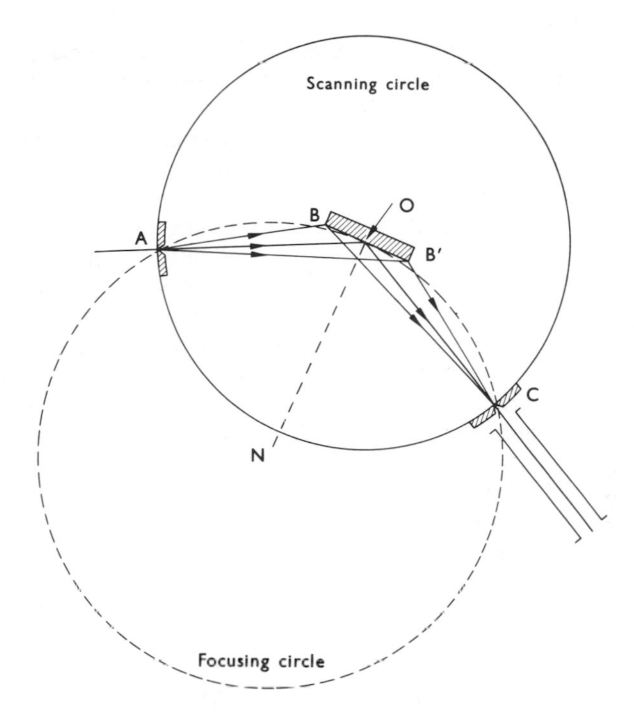

Figure 7.8 The Bragg-Brentano configuration for a modern X-ray diffractometer. A – the X-ray source; BB′ – the plane of the sample; C – the exit slit for the diffracted beam, with the detector mounted behind it. ON is the line of symmetry about which the X-ray source and the detector should move in order to satisfy the ideal conditions of diffraction. (From ref. 51, by kind permission of Chapman and Hall Ltd.)

etry [44], first introduced by Lindemann and Trost [45] and Friedman [46], although other systems have been proposed [47-49].

In the Bragg-Brentano configuration, as figure 7.8 shows, the source of the X-rays, the sample, and the receiving slit of the counter all lie on the circumference of the 'focusing circle', with the X-ray source and the receiving slit equidistant from the sample. It will be clear that a powder sample placed at S, with its surface curved to conform to the circumference of the focusing circle, would ideally focus a divergent beam of X-rays from X onto the point R. Furthermore, if the sample is kept still, it will be possible to scan all Bragg angles by moving the X-ray source and the detector head symmetrically, with respect to the sample, along the locus defined by the 'scanning circle'. Now the radius of the focusing circle is given by

$$r_f = \frac{r_i}{2\sin\theta} \tag{7.10}$$

where r_i is the distance from the X-ray source to the sample and θ is the Bragg angle, and so the curvature of the sample should alter with θ to maintain the ideal reflecting conditions.

Clearly it is very difficult to solve all the mechanical problems involved in the complex interacting movements implied in the above, and so approximations have been made in practice. In the first place, the sample surface is made flat and tangential to the focusing circle. This results in a very slight defocusing at R which is insignificant under almost all conditions of use. In the second place, the X-ray source is normally kept still and the symmetrical motion of the X-ray source and the detector head is simulated by rotating the sample at an angular velocity ω about an axis lying in its surface, perpendicular to the plane of the focusing circle, and using a 2:1 gear train to rotate the detector at a velocity 2ω in the same sense about the same axis. Since the X-ray source, specimen surface, and detector are mechanically set in the same plane at $\theta = 0°$, the detector will always be at the correct 2θ angle when the sample is at the Bragg angle θ relative to the X-ray beam.

Figure 7.9 shows diagrammatically the salient parts of a modern diffractometer. X-rays from a line source at F, viewed at an angle of about 6°, are allowed to diverge through a pre-determined angle, in the plane of the focusing circle, by the 'divergence slit' D and impinge on the flat sample S. Beams diffracted from the sample pass through a 'scatter slit' at M, a 'receiving slit' at G, to the counter at X. The scatter slit is designed to limit the angle of view of the receiving slit/counter assembly to radiation from the sample only, and so it largely defines the resolving power of the instrument as well as cutting down the background from stray radiation. Sets of Soller slits [50] are provided between the X-ray source and the

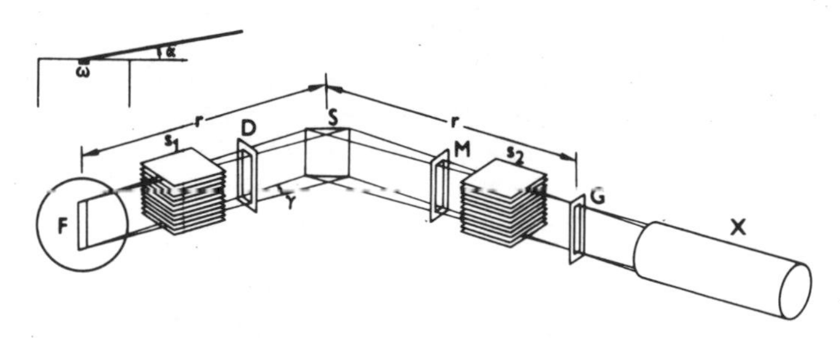

Figure 7.9 Diagrammatic representation of the principal parts of a modern electronic counter diffractometer, showing F – the X-ray source; s_1 and s_2 – the sets of Soller slits on the incident and diffracted beams; D – the divergence slit; S – the specimen surface; M – the scatter slit; G – the receiving slit; X – the detector. The small inset on the upper left illustrates the 'take-off angle' (α) for viewing the source area in the X-ray tube. The angle α is normally set to 6–7°. (After Klug and Alexander, by kind permission of John Wiley and Sons, Inc.)

divergence slit and between the receiving and the scatter slits. These are thin, parallel metal foils placed close together (separations of the order of 0.6mm are usual) to limit the divergence of the various beams in the plane perpendicular to the focusing circle and so permit the use of extended X-ray line sources, with their relatively high intensities, without loss of resolution or line shape. With reference to figure 7.8, the source, F, may be either the line focus of the X-ray tube or a monochromator crystal, and the point of focus of the diffracted beams, G, lies at the receiving slit position. High intensities are obtained with the X-ray focus at F, the configuration used in most diffractometers, and the arrangement with the monochromator focus at F is normally used only when measurements must be made at very high 2θ angles and the physical bulk of the X-ray tube would prevent the detector head being swung round to the required position [51]. Since the distance from the focus is necessarily increased, appreciably lower intensities result. A focusing monochromator may be placed to act on either the incident or the diffracted beam, and the focus of the monochromator lies either at F or at G [51].

Many brands of powder diffractometer are now available commercially, almost all based on the Bragg-Brentano configuration with moving sample and counter, although one instrument has been built based on the idealised movements of the X-ray source and the detector head [52]. The diffractometer may be constructed with the focusing circle either horizontal or vertical, as shown in figure 7.10, and each mounting has its advantages and disadvantages. Mechanically, the horizontally mounted instrument is superior, because a stronger support arm can be provided for the detector head to keep the slit system aligned accurately relative to the sample at all times. The varying gravitational moment acting on the support arm of the detector, in the vertical mounting, due to the weight of the assembly tends to twist the slits out of alignment to different extents depending on the orientation of the arm. If the diffractometer is aligned correctly at $2\theta = 0°$, the change in orientation of the detector head that can occur at higher 2θ values will result not only in a reduction in the apparent intensity of the diffracted radiation, due to reduced transmission through the slits, but also to a displacement in the absolute line positions, due to bending in the whole support arm. Such problems are very much minimised in the horizontal instrument. On the other hand, the need for a vertically mounted powder sample in the horizontal instrument presents its own problems in persuading a relatively loose powder to remain in the vertically mounted sample holder during the loading operation. Hence, the vertical circle diffractometer may be the more useful where the need is for an instrument to carry out large number of routine analyses with reasonably good accuracy, but the horizontal instrument must be the choice when the highest accuracy is required. In both cases it is advisable to mount the tube rigidly onto the diffractometer itself.

X-ray powder diffractometry requires an X-ray generator with a stability better than 0.1%, since different regions of the same powder pattern are

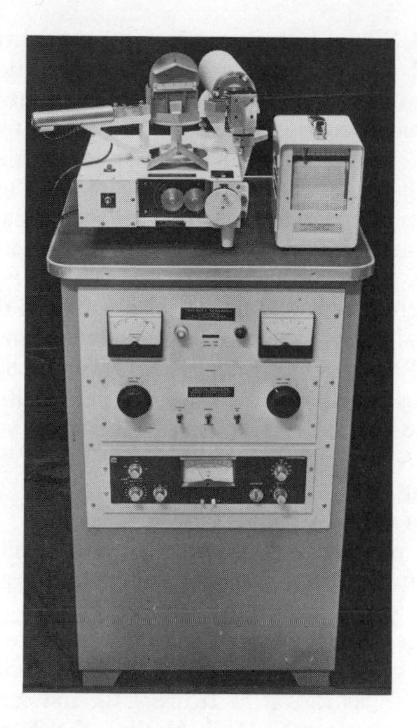

Figure 7.10 Illustration of a commercial diffractometer, with the focussing circle lying in the horizontal plane and the X-ray source made an integral part of the goniometer mounting. A close-up view of this mounting may be seen in Figure 7.15. (By courtesy of Tem-Pres Research, Inc.)

being measured at times instead of simultaneously as in film work. The line focus is used, and is usually viewed at an angle of 5-7°. Fixed slits giving precisely defined divergence, receiving, and scatter angles are commonly provided, but infinitely variable slit units are available on some instruments. Narrow slits will clearly give the best resolution, but they also give low intensities. The best compromise for much routine work is a divergence slit passing a 1-2° angle and a receiving slit of 0.05°-0.1° aperture with a scatter slit width of 5mm, but narrower slits must be used for very low angle work, e.g. in clay studies, when large *d*-values with small differences in 2θ must be determined. Fortunately, the resolving power of the instrument improves at higher 2θ angles, and so the general decrease in intensity encountered at higher angles can be compensated for by using wider slits without loss in overall resolution. It is usual to use a 4° divergence slit and a 0.2° receiving slit for work above $2\theta = 70°$.

For much routine work a simple Geiger-Muller counter with a ratemeter feeding signals to a strip chart recorder suffices to display the diffracted intensity as a function of diffraction angle. Simple removal of the

β-radiation is accomplished with a beta-filter either in the incident or the diffracted beam, the latter having certain advantages in that it can reduce noise due to air scatter from around the sample. Such systems, however, tend to have poor signal-to-noise ratios and an improved system will use either a gas proportional or a scintillation counter in conjunction with a pulse height discriminator. By setting the 'window' of the discriminator to pass only pulses of the desired magnitude, normally corresponding to the $K\alpha$-radiation from the X-ray tube, not only is the overall signal-to-noise ratio greatly improved but the electronic discrimination can be used to filter out unwanted signals arising from the white and the $K\beta$-components of the tube output and from any fluorescent radiation generated from the sample, or from metal parts of the diffractometer itself. The resulting trace will, therefore, be much sharper and a distinct advantage will be gained in that it will be possible to study, for example, iron-containing samples with copper radiation without problems from the iron fluorescence. A detailed discussion of detectors has been given by Arndt [51].

Since the pattern is recorded on the strip chart by scanning the required angular range while the chart recorder runs at a predetermined paper speed, it will be clear that both the goniometer and the recorder speeds must be stable. This throws great importance onto the transmission systems, and particularly onto the recorder, since the gear train of the goniometer is the maker's responsibility. Any good gear-driven recorder may be used, provided the motor speed is sufficiently constant, but belt-driven recorders, or recorders in which the chart speed is changed by altering the voltage to the motor, are not to be recommended.

Nature and Form of Sample

A powder of -350 mesh size (<45μm) can be used for diffractometry, although Klug and Alexander [53] point out that the restricted range of motion of the sample demands a more finely ground specimen, passing 400 mesh, to give results comparable in accuracy with Debye-Scherrer methods. The simplest way of forming the plane specimen required is to pack the powder into a depression in a glass or metal plate and smooth the surface with a microscope slide drawn across it. Some powders will remain in the cavity without any further treatment but some need to be mixed with a binding fluid to prevent their falling out, especially if a horizontally mounted diffractometer is used. With the majority of minerals, moistening with acetone suffices.

Alternatively, the material may be mixed with a little acetone and the resulting slurry spread on the surface of a microscope slide and allowed to dry. The latter procedure is particularly useful if only a small amount of powder is available, or if high accuracy is not required. Both methods can result in preferred orientation in the surface of the specimen, and McCreery [54] has described a method of filling the cavity from the reverse side of the holder which almost completely eliminates problems due to this effect (cf. section 7.6.4).

Setting Up and Recording the Pattern

Before use, the X-ray source, specimen surface, and detector system must be set co-planar at $\theta = 2\theta = 0°$. This is normally accomplished using the attachments provided by the manufacturer, and it is important that care be taken to ensure that the settings are made as accurately as possible and, once made, that they are not altered in any way during the routine use of the instrument. In particular, the instrument must not be moved relative to the X-ray source, and it is here that instruments in which the X-ray tube mounting is integral with the goniometer base, as shown in figure 7.10, are at an advantage over the more conventional mounting with the diffractometer simply set next to the X-ray tube.

The diffraction pattern may be recorded either by scanning from low to high 2θ angles, or from high to low, and a laboratory should standardise on one direction for all its work since

a) patterns from different samples may be compared directly by superposition, and

b) although the data collected by scans in opposite directions are basically the same, slight differences in peak position may occur due to mechanical backlash in the gear-train of the diffractometer, slight eccentricities in the spindle bearing of the diffractometer, pen response speed on the chart recorder, and several other factors.

Both the angular scanning speed on the goniometer and the paper speed on the chart recorder can normally be varied. The combination of angular scan speed and paper speed chosen obviously depends on the requirements of the study in hand, and Klug and Alexander [55] have specified suitable combinations of these factors with the divergence and receiving slit sizes for a range of conditions. In general, a scanning speed of 1° or 2° per minute with a divergence slit of 2° and a receiving slit of 0.05-0.2° will suffice for almost all routine identification work. For high resolution, the slits must be narrower and the scanning speed slower, and for measurement of relative integrated peak intensities the divergence slits should be slightly narrower, and the scanning speed be slower. Klug and Alexander [55] advise chart speeds of no more than 30 inches per hour at a scanning speed of ½° per minute but higher for higher scanning speeds or for accurate studies of line profiles.

The scan may be made over any angular range, but the range 70° through 5° of 2θ is commonly used for minerals, since most of the important information required for characterisation can be obtained in this range. In special cases, more restricted angular ranges may be used to look for specific materials having easily recognisable patterns, and accurate determinations of cell parameters for the higher symmetry classes may require data collected at higher 2θ values. In practice the diffractometer is set to the starting 2θ angle and the chart to a starting position on the recorder. Both are switched on simultaneously and, since sample and detector rotate at a constant angular velocity and the chart recorder runs at a constant speed, the output of the detector is displayed on the chart in such a manner that the

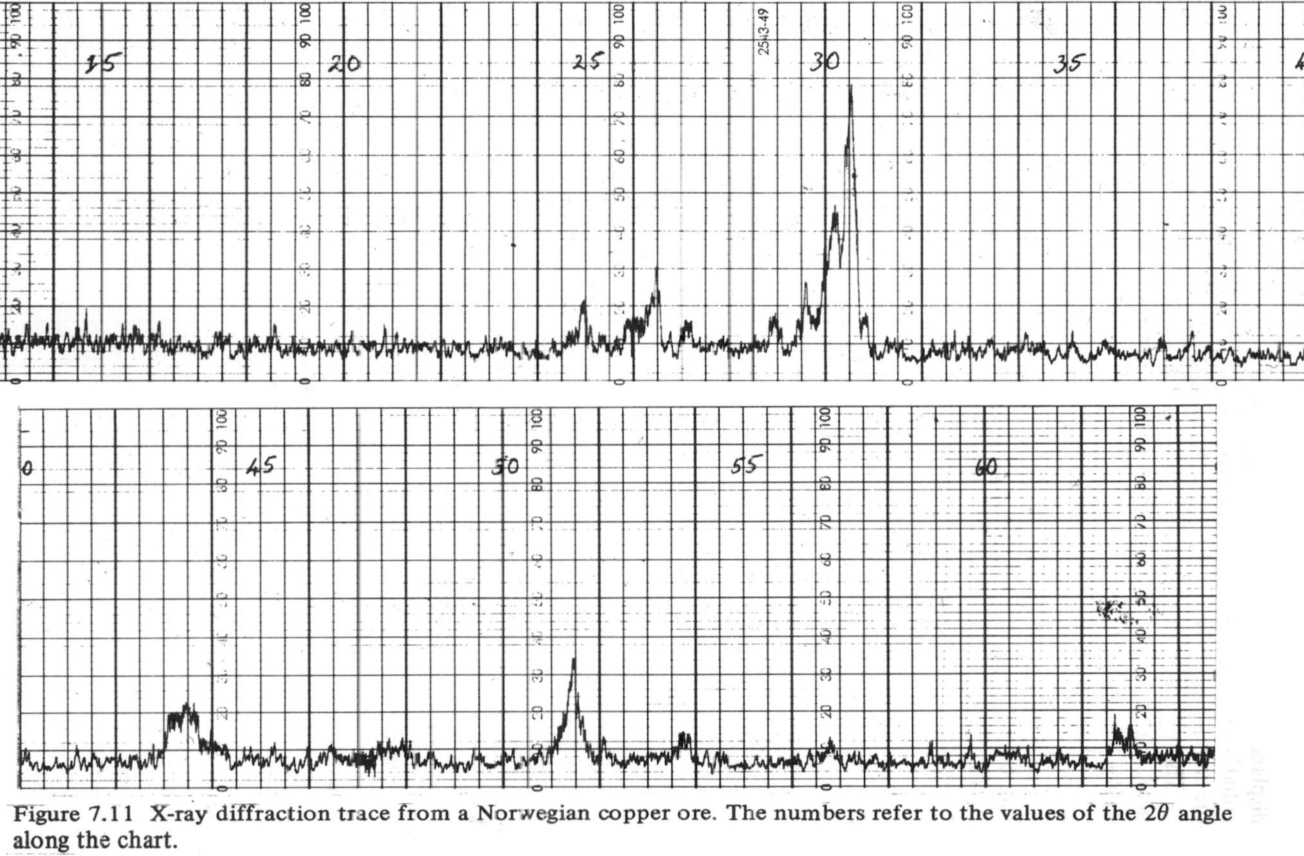

Figure 7.11 X-ray diffraction trace from a Norwegian copper ore. The numbers refer to the values of the 2θ angle along the chart.

displacement across the chart represents the intensity of the diffracted radiation at the 2θ angle represented by the distance along the chart from the starting position, as shown in figure 7.11.

The peak positions on the chart may be measured at any point on the peak, but usually at the centre of the half height position for an unresolved α_1/α_2 doublet, in which case the weighted mean wavelength

$$\lambda = \tfrac{2}{3}(2\lambda\alpha_1 + \lambda\alpha_2)$$

can be used. The high resolving power of the diffractometer, however, means that the α_1/α_2 doublet can be resolved at 2θ values as low as 40°, at slow scanning speeds, and it is advisable to measure 2θ values for both wavelengths in such a case, although the weighted mean value can be used to give a reasonably correct *d*-spacing in routine work when high accuracy is not required. The use of the peak apex to define peak position is not recommended.

Instrument Calibration

Since it is difficult to define accurately the positions for both $2\theta = 0°$ and $2\theta = 180°$, the diffractometer should be calibrated using a standard material of accurately known *d*-spacings. This can be done with an 'external' standard, which is a material with accurately known *d*-spacings and preferably with an absorption coefficient similar to that of the type of sample being studied and which is run as any other sample and the calculated and observed 2θ values compared to give a correction curve for the instrument. The criterion used to define the standard peak positions must, of course, be the same as that used in all other determinations [56]. Alternatively, an 'internal' standard may be used, mixed with the normal sample, and the known lines so produced used to calibrate the instrument for each separate run. The standard chosen must have no lines which overlap with those of the sample, and so two runs must normally be made, one without and one with the standard. The external standard method suffices for most routine work, but the internal standard is required for most accurate work. Wilson [57] and Parrish [58] have discussed the determination of peak positions and the corrections which should be applied.

7.6.5 Error Sources in X-ray Diffraction Methods

The mechanisms involved in the production and detection of diffracted X-rays give rise to various inescapable sources of error in both film and diffractometric methods. Fortunately these errors tend to be systematic in nature and there are only a few sources, associated almost entirely with the manufacture of the instruments used, which are not amenable to mathematical treatment. Edmunds, Lipson and Steeple [59] have discussed the determination of accurate lattice parameters in some detail.

Differentiation of the Bragg equation (7.3) with respect to the inter-planar spacing, the wavelength, and the Bragg angle gives

$$d(\lambda) = 2\sin\theta . d(d) + 2d \cos\theta d(\theta)$$

whence (7.11a)

$$\frac{d(d)}{d} = \frac{d(\lambda)}{\lambda} - \cot\theta \, d(\theta)$$

from which it is clear that the most accurate data may be obtained at Bragg angles approaching 90°, since $\cot\theta$ tends to zero as θ tends to 90°, assuming that the error in λ may be ignored. From this, Kettman [60] has shown that errors may best be eliminated by calculating a set of d-values for a zone and extrapolating to $\theta = 90°$. The problem lies in deciding on the best extrapolation function for a given method. In the following analysis uncertainties in the wavelength used are ignored, since these are of the order of 0.001-0.01% of the value of λ. Note that, for many routine analysis procedures, the as-gathered data are sufficiently accurate and do not need to be corrected.

Systematic Errors

The systematic errors encountered in X-ray diffraction studies arise from such sources as the crystalline imperfection of the sample, absorption of the X-ray beam by the sample, missetting of the sample in the recording instrument, divergence of the X-ray beam, and film shrinkage in the camera or variations in recording speed in the diffractometer. Of these, the effect of film shrinkage can be eliminated quite easily, as shown in section 7.6.2, although application of a correction factor based on the separation of fiducial points on the film, to values derived from different regions of the film, does assume that all parts of the film shrink evenly. Fortunately, the available evidence indicates that this assumption is justified in the majority of cases.

Errors due to the perfection and the size of the separate crystallites in the powder sample contribute to the width of the powder line, rather than to its displacement. The effect of crystallite size is not noticeable above submicron sizes, since the intensity of radiation scattered at small angles away from the Bragg angle falls off very rapidly with angle for larger crystals. The extent of the scatter is, however, inversely proportional to the thickness, t, of the diffracting medium, according to

$$\delta(2\theta) \sim \frac{\lambda}{t\cos\theta} \qquad (7.11b)$$

where $\delta(2\theta)$ is the half-breadth of the reflexion, after Scherrer [61]. For t = 1mm, $\delta(2\theta) = 0.001'$, but for $t = 1\mu$m, $\delta(2\theta) = 0.2°$, and this is easily measured. This topic will be further discussed in section 7.8.2. Line broadening and displacement can also arise through stacking disorder or the presence of mixed layering, which occurs in many clays, where the different layers possess closely similar d-spacings giving rise to overlapping reflexions, or in strained or chemically inhomogeneous crystals, and the effect may be very noticeable indeed. Finally, the mosaic nature of the small crystallites comprising the separate particles in the powder can give rise to broadening of the reflexions due to the angular mismatch between the crystallites. This last effect tends to be slight, however, and broad reflexions usually indicate either a very small crystallite size or a poorly crystallised product.

Absorption by the sample can give rise to very important errors, especially in film methods using large or appreciably absorbing samples, since the total absorption for X-rays is a function of the product of the absorption coefficient for the material and the pathlength of the X-ray beam through the material. Figure 7.12a shows the paths of a number of X-ray beams through a cylindrical sample and figure 7.12b shows possible intensity profiles across the resulting diffracted beam for different total absorptivities. It will be clear that only in the case of a weakly absorbing sample will the position of maximum intensity, which is the point normally used to estimate line position, coincide with the centre of the line. In all other cases the measured value of 2θ will be high, and so d will be low, and, in the limiting case of a very highly absorbing sample, the reflexion may appear as a pair of thin, closely spaced lines lying one on either side of the correct line position. Detailed consideration of the geometry of diffraction, however, shows that this absorption effect also decreases as θ tends to 90°, or as 2θ tends to 180°, since the path lengths for different rays become almost equal, and so more accurate estimates of line position can be made

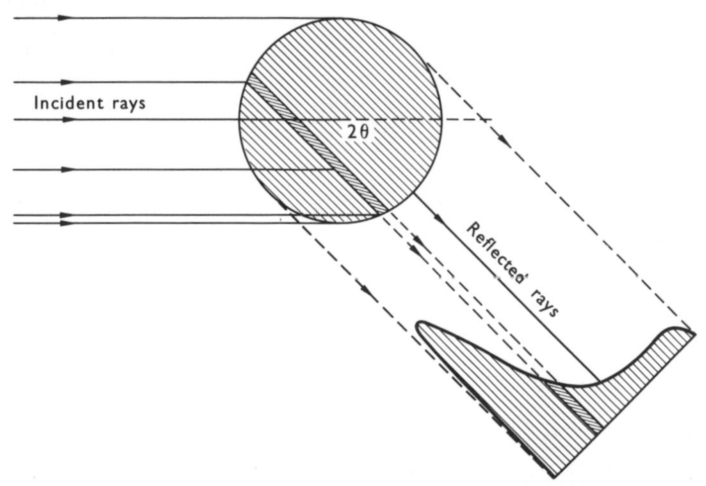

Figure 7.12 (a) Calculated intensity profile for a beam of X-rays diffracted from a sample with an absorptivity (μr) of 2.0 at 2θ angle of 45.0°. Note how the position of maximum intensity has become displaced from the centre of the diffracted beam and that the 2θ angle measured using this position as the line position will give a high angular reading, and hence a low d_{hkl} value.

for high angle reflexions. Also, absorption doubling is absent in high angle reflexions, and so a pattern showing closely spaced pairs of lines at low but not at high angles must always be suspect as a case of excessive sample absorption. Absorption may be reduced by using a smaller sample with the same X-radiation, or, since absorption coefficients depend on the wavelength of the incident X-rays as well as on the elements present in the

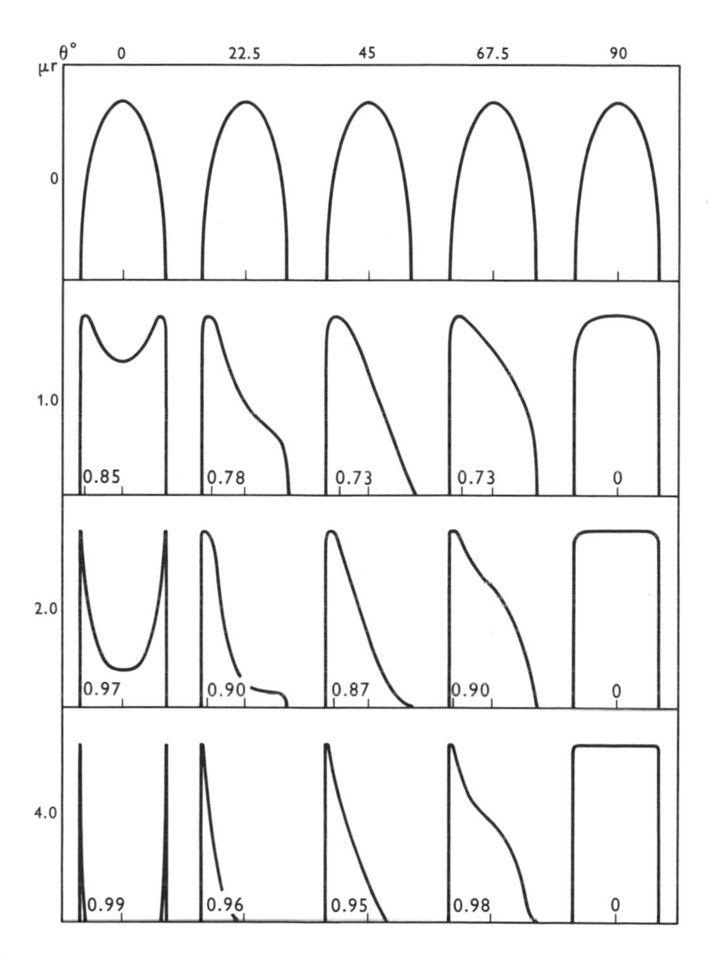

Figure 7.12 (b) Calculated intensity profiles for a range of sample absorptivities over a range of θ angles. Note (i) how the position of maximum intensity, and hence the observed line position, always approaches the centre of the diffracted beam as θ approaches 90°, explaining why high angle reflexions are used to derive accurate cell parameter values, and (ii) how doubling of lines at the low angle end of the pattern may occur for materials with very high absorption values. (Reproduced from ref. 36, by kind permission of Chapman and Hall Ltd.)

sample, the effect may be minimised by using a different radiation for which the absorption coefficient is low and comparing the results with those for which the absorption is high. Dilution with a weakly absorbing, non-crystalline material, as discussed in section 7.6.2 under sample preparation, is often effective, and the use of an internal standard for correction will be discussed in the next sub-section. The error is proportional to $\cos^2\theta/\sin\theta$ or $\cos^2\theta/\theta$ and is amenable to mathematical treatment.

Absorption in the powder diffractometer affects the observed line positions in the opposite sense to that in the film camera, since optimum line positions are obtained with thin, highly absorbing samples. For such samples, the penetration of the beam into the sample to cause reflexions from planes not lying at the level of the main axis is minimised. The error is proportional to $\sin 2\theta/4\mu R$, which has a minimum at $2\theta = 180°$.

Errors can also arise from missetting the sample in the camera or diffractometer. Displacement of the sample in a direction perpendicular to the beam, in a camera, has virtually no effect on the line positions, but displacement along the direction of the beam results in high or low values of 2θ depending on whether the sample is displaced towards or away from the X ray source. The error in this case is proportional to $\cos^2\theta$ and vanishes as θ tends to 90°. A similar effect is observed if the plane of the sample is displaced away from the plane of the axis of rotation, in the plane of the focusing circle, in the diffractometer. Appreciable errors in 2θ can result, observed values being erroneously high if the displacement is towards the centre of the focusing circle, and low if it is away from the centre. The error is proportional to $\cos\theta \cot\theta$ and again extrapolates to zero at high angles.

Divergence of the beam results in errors similar to those produced by absorption in the camera, and the effect is not significant in diffractometry. The extrapolation function is again either $\cos^2\theta/\sin\theta$ or $\cos^2\theta/\theta$.

Several extrapolation functions have been proposed to eliminate these systematic errors, but probably the most widely used one is that derived by Nelson and Riley [62] in determining the cell parameters of cubic materials, for use with Debye-Scherrer camera data only. They calculated a_0 values from indexed d-values at various 2θ positions, plotted them against the function,

$$\tfrac{1}{2}\left(\frac{\cos^2\theta}{\sin\theta} + \frac{\cos^2\theta}{\theta}\right)$$

and obtained remarkably straight line plots, especially when high angle data were used. McMaster [63] has shown how the extrapolation function can be applied to any pattern, to correct the observed d-values at any 2θ position, provided that the absorption coefficient of the sample is known. The method is claimed to give accuracies of better than 3 parts in 10^5, and this is at least as good as is needed for any but the most accurate work.

Further systematic errors may arise from mechanical faults in the construction of the equipment used and random errors through subjective mistakes in estimating line positions. Thus, the cassette of the film camera may not be truly cylindrical or the gear train of either the goniometer or the chart recorder of the diffractometer may not run evenly. In both cases such errors can be detected and allowance made for them by constructing calibration curves, by measuring the observed *d*-spacings for a standard reference material and comparing with the *d*-values supplied. Strain-free spinel, silica or silicon samples are often used, and these and other suitable reference materials may be obtained from the National Physical Laboratory, in Great Britain, or the National Bureau of Standards in Washington, D.C. Faults due to subjective error are the responsibility of the operator!

It should be noted that missetting of the indicated zero position of the 2θ drum on the diffractometer will produce a constant error in 2θ which can be detected only by drawing a calibration curve, and missetting the counter head relative to the diffracted beam direction, as shown earlier, produces complex errors in 2θ, which vary with 2θ, and these can again be eliminated only by a calibration technique. Note that the last condition also results in a serious loss of intensity which can make it difficult to detect peak positions and intensities accurately.

Preferred Orientation

As was shown in section 7.6.1, successful X-ray powder diffractometry depends on obtaining a mounted specimen in which all the crystallites have taken up truly random orientations relative to the incident X-ray beam. In certain substances, particularly materials such as clays with a well developed platy habit, this randomness is difficult to obtain, since the separate particles in the sample tend to orientate themselves so that their basal planes are all parallel, and this is particularly so in samples prepared by the usual method for the diffractometer. The result is that the intensities of some reflexions are accentuated relative to the intensities of others, and the pattern of relative intensities is disrupted, simply because certain crystallographic planes are never brought into the diffracting position. In the case of a clay mineral, the pattern may show strong (00*l*) reflexions only, with virtually no indication of general (*hkl*) lines at all. Such patterns are very useful tools in diagnostic work on clays.

The effect may be eliminated, or at least minimised, by careful specimen preparation, as described by McCreery [54] (cf. section 7.6.3). Strum and Lodding [64] have recently discussed the problem of preferred orientation and have shown that a more accurate correction may be made by using a completely oriented specimen and correcting the observed intensities using a hemispherical normal distribution function. They claim that this method escapes uncertainties in the degree of remanent orientation in specimens prepared by non-orienting methods, but their technique does involve careful setting up of the specimen, and careful control of the angular movements involved in the recording procedure.

7.6.6 Comparison of Film and Counter Methods

Both film and counter methods possess advantages and disadvantages, and the choice of method will depend, to a large extent, on the requirements of the study. Basically, the film camera will provide rather good diffraction data somewhat slowly whereas the counter diffractometer will produce reasonably good data quickly. Angular resolution is better in the diffractometer than in the Debye-Scherrer camera, with the Guinier-de Wolff focusing camera intermediate between the two, absorption is less of a problem, and line profiles and line intensities are obtained directly and do not have to be subsequently measured, as in the case of film. On the other hand, the film camera can handle smaller samples than the diffractometer, the whole pattern, including the back reflexion region where intensity is generally low on a diffractometer, can be obtained simultaneously, and broad, weak reflexions are more easily identified on film.

In the main, for routine identification studies, the diffractometer is probably the better instrument, despite its considerably higher cost. Its greater flexibility makes it generally much faster to use than a film camera, in particular very fast data collection is possible when specific substances are being sought. In favourable conditions, it may be feasible to scan only a very limited 2θ range at a fast scanning speed to check the presence or absence of a component giving rise to specific reflexions within the region. On the other hand, film data probably still provide more accurate spacing data and should be the preferred choice in high accuracy studies.

7.6.7 Ancillary Methods

Apart from the studies made at room temperature, X-ray diffraction finds an important role in determining the effects of temperature and pressure changes on crystalline materials, and particularly in studying materials actually at elevated temperatures and pressures.

High Temperature Methods

Many high temperature powder cameras have been described [65, 66] but possibly the best known is the Unicam high temperature camera designed by Goldschmidt [67]. The camera is shown in figure 7.13, and employs two hemispherical, platinum-rhodium wire wound furnaces to surround the specimen, with thermocouple detection and control of temperature, and rhodium-plated radiation shields to prevent excess heat from reaching the film, which is held in a modified Bradley-Jay mounting [32]. Temperatures up to 450°C can be achieved with good temperature distribution and stability. The central unit surrounding the specimen can be evacuated or filled with any chosen gas, to prevent oxidation or volatilisation of the specimen. Buerger and Chesley [68] have designed a camera based on a Straumanis-Ievins mounting using a very small heater comprising a 3/16 inch cylindrical coil of chromel wire which can be plugged into the camera as required. No provision is made for atmospheric control, and the camera is intended for use only up to 600°C, using known transition points for

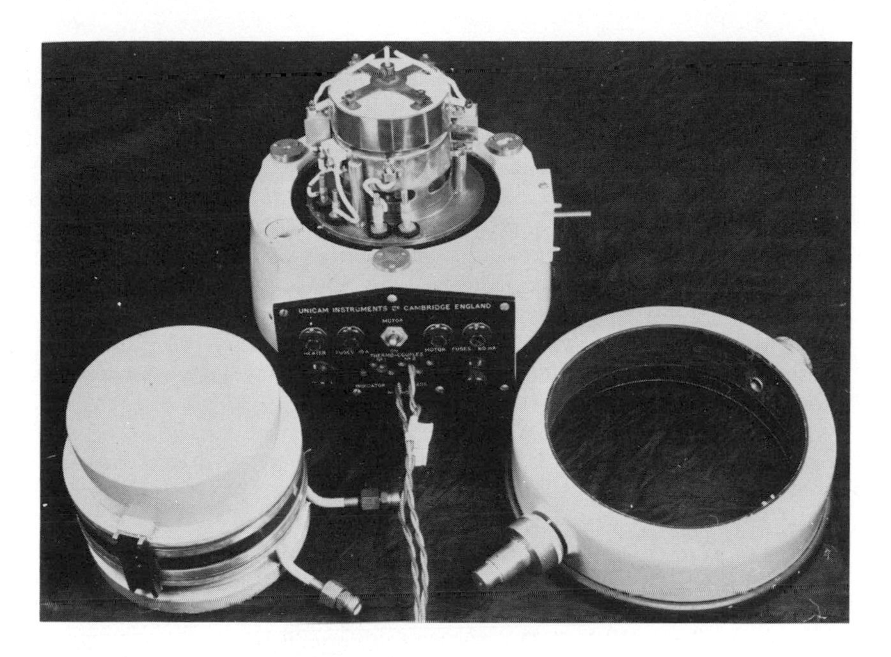

Figure 7.13 The Goldschmidt high-temperature powder camera. (Photograph by courtesy of Mr. D. M. Evans and Mr. G. Titmus.)

temperature calibration since the low heat capacity make it difficult to use a thermocouple for control.

Recently, the Enraf-Nonius Company have marketed the Guinier-Lenné camera [69], which is a Guinier-de Wolff camera modified for high temperature work, following Lenné's original design [70]. The specimen is surrounded by two cylindrical platinum wire wound furnaces, thermocouple controlled, and the film is held in a cassette which can move in a direction perpendicular to the plane of the diffracted rays. A motor drives the film slowly past the exit slit for the diffracted beams as the temperature of the furnace rises, and a continuous record of the diffraction pattern from room temperature to around 1200°C can thereby be obtained. The camera is shown in figure 7.14a and a typical film in figure 7.14b. Various heating programs can be used, and the camera appears to have a great potential for use in studying processes occurring at different temperatures and comparing their relative importances under different heating conditions.

High temperature film methods are excellent for studying phases at high temperatures, either for identification or for determination of physical properties, such as temperature coefficients of thermal expansion, etc., but the slow response of films and the relatively long exposures needed can mean that transient phase changes may be missed, and kinetic rate studies are difficult, if not impossible, to make. The very much faster response of counter diffractometers make them well suited for such studies and several designs of furnace are now available commercially to fit almost all diffractometers. Particular mention may be made of the double-hemispherically wire wound furnaces made by Tem-Pres Research Inc. and by Materials Research Company of New Jersey, which provide easily controlled, even, and well stabilised heating over the whole sample at temperatures up to 1850° and 2250°C respectively. Both units permit the use of controlled atmospheres, and the Materials Research instrument can also be used under vacuum. Figure 7.15 shows the Tem-Pres furnace mounted on a diffractometer. Strip furnaces have been built, especially for use with the Philips diffractometer, but these can suffer from difficulties arising from the inevitable temperature gradients generated in a sample heated from only one side.

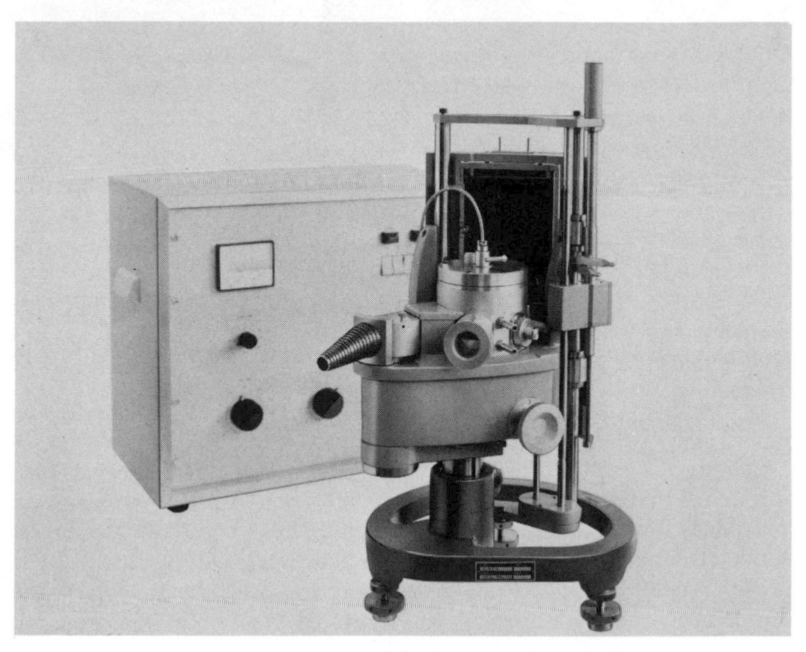

(a)

Figure 7.14 (a) The Guinier-Lenné high temperature camera. (b) The decomposition of magnesium perchlorate followed by high temperature X-ray diffraction photography. A typical film from the Guinier-Lenné X-ray diffraction camera showing the changes in pattern which occur as $MgClO_4$ is heated from 20° through 800° C. Note the changes in structure that occur at 170°, 230°, 240°, 330°, 530° and 770° C, and particularly how the continuous recording technique shows up the transient intermediate formed at 230–240° C, a phase which would, all too likely, have been missed by conventional methods of study. (Photographs by courtesy of Mr. D. Schoeneveld, Enraf-Nonius, Delft.)

High Pressure Methods

The ability to carry out X-ray diffraction studies at pressures as high as hundreds of kilobars became feasible about 1950, but the major development in this field has occurred since 1960. Such studies are obviously important in the investigation of the probable behavior of minerals at great depths, but they are also significant in cases where a high pressure modification cannot be quenched in and studied at room pressure. In practice, the apparatus used, although basically simple, must be a compromise between a strong, rigid support system to generate and contain the pressure, and a small, open environment to allow the incident X-ray beam into the specimen and the diffracted rays out. Equipment is now available which can provide pressures up to 500 kilobars at room temperature or lower pressures at temperatures up to 1000°C.

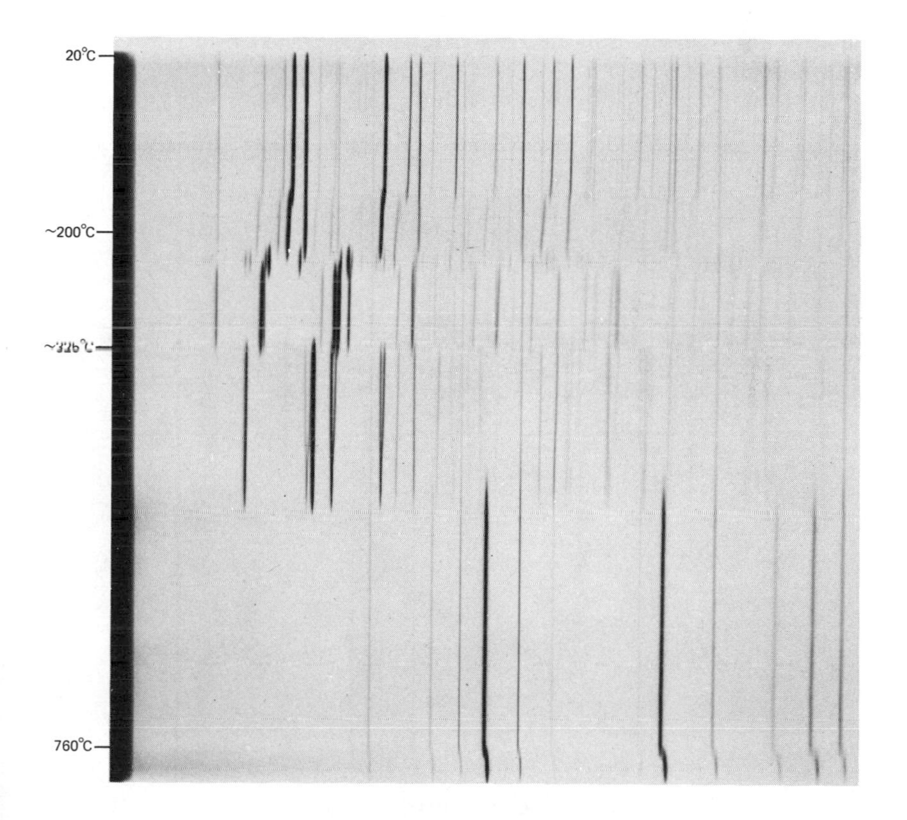

(b)

McWhan [71] has summarised the various designs available. These are based either on uniaxially opposed or tetrahedrally opposed anvils, although recent developments have permitted the use of small, pre-pressurised units which can fit into existing diffraction equipment in some cases.

Thus, Jamieson and Lawson [72] have described a cell in which the pressure is generated within an amorphous boron disc compressed between tungsten carbide anvils. The X-ray beams lie in the plane of the disc, i.e. perpendicular to the pressure axis, and the authors describe its use within a modified Guinier camera. An alternative design by Piermarini and Weir [73] uses two diamond anvils with the X-ray beam directed parallel to the pressure axis, i.e. passing through the supports and the anvils themselves. It is claimed that this configuration has the advantage of allowing optical measurements to be made on the sample under pressure, but the design suffers from the disadvantage that the diamond anvils must be carefully aligned and from difficulties in loading and handling the small sample. The boron disc unit is easy to build and to use. Pressures of up to 150-200 kilobars are available using tungsten carbide anvils, and up to 300 kilobars with diamond.

Such simple uniaxial cameras suffer from the usual problem of the

Figure 7.15 A furnace for high-temperature X-ray diffractometry mounted on a horizontal X-ray diffraction goniometer stage. Note the integral tube mount, which ensures the stability of the X-ray source position relative to the sample and the slit system of the instrument. (Photograph by courtesy of Tem-Pres Research, Inc.)

pressure gradient generated across the anvil faces, the pressure being highest in the centre and decreasing towards the periphery. Calibration using materials of known pressure dependence is unreliable, since the method depends on the assumption that calibrant and sample are at the same pressure and Jamieson and Olinger [74] have shown that this is not generally true. The effects of differences in the bulk moduli and the Poisson's ratio of the two materials can be such that the true pressure on the standard can differ by as much as 50% from the indicated pressure.

To achieve higher pressure, and more approximately hydrostatic conditions, Drickamer and co-workers [75] have built a cell based on an opposed anvil design coupled with a split-girdle configuration to contain the pressure, and LiH windows to transmit the X-rays. This instrument requires careful assembly, but pressures up to 500 kilobars can be obtained at room temperature. This basic design has been adapted by Freud and Sclar [76] to include boron-sheathed graphite rods in the girdle to provide resistance heating and give pressures up to 100 kilobars at 1000°C. Weir, Piermarini and Block[77] have successfully applied this design to the study of single crystals at high pressures.

Larger samples can be handled under consitions which approximate even more closely to hydrostatic conditions in a design described by Barnett and Hall [78] which uses four tetrahedrally opposed anvils to contain the sample. The maximum pressure obtainable in such a configuration is lower than with opposed anvil devices, but the increase in sample size and the improved homogeneity in the pressure profile across the sample outweigh this for many purposes. Barnett, Pack and Hall [79] have described a further modification allowing internal heating to give temperatures up to 1000°C, but the complete apparatus, in both cases, is very bulky and expensive since it requires that the necessary scanning diffraction equipment be built round the press, rather than having the press built into the diffractometer.

More recently, attempts have been made to develop small scale pressure cells to fit into existing cameras, and Kasper [80], Vereschagin [81], Meyer [82] and Davis and Adams [83] have all described useful designs. These tend to be restricted to a maximum of 30 kilobars, and Meyer's design can include a heater to give temperatures up to about 900°C.

The restricted X-ray paths available in all the designs makes imperative the use of Mo $K\alpha$ radiation, with its low dispersion angles. This restricts studies to forward reflexions only and does reduce the accuracy with which d-spacings and cell parameters can be determined. Freud and Mori [84], however, have used non-dispersive, energy scanning methods to determine d-spacings at a constant 2θ value. They claim that this system gives better accuracy than does the more conventional dispersive method, since calibration with nuclear X-ray and γ-ray sources is quite easy to give routinely accuracies of 0.1%, the speed of measurement is increased, the signal-to-noise ratio is improved, and absorption in the pressure medium and sample is reduced because the higher energy portion of the white spectrum is being used. Moreover, the use of a constant 2θ angle simplifies the die manufacture, since only thin grooves need be provided instead of the broad fan slots which reduce the strength of the system.

7.7 INTERPRETATION OF X-RAY POWDER DATA

It was shown, in section 7.5, that the observed *d*-spacings in a powder pattern were associated with the shape and size of the unit cell of the material producing the pattern, while the intensities of the reflexions were associated with the nature and dispositions of the atoms comprising the material within the unit cell. It follows, therefore, that the pattern of reflexions obtained from a crystalline material can be used to characterise the material or, alternatively, that the pattern is unique for that crystalline modification and so can be used as a 'fingerprint' to identify the compound, since all materials, with a few specialised exceptions, possess different structural parameters and so give rise to different patterns.

Direct use of powder pattern data alone to characterise a substance is usually difficult, since they normally need to be combined with data derived from single crystal studies for completeness, and so powder data are usually used for identification purposes. Such studies are mainly qualitative in nature, being confined to indicating which crystalline phases are present in the sample and giving a very rough estimate of the major and minor components in a mixture, but they can be applied quantitatively to give reasonably accurate estimates of the relative amounts of each phase present, or to give information about, for example, the degree of inverse nature in a spinel or to identify the chemical composition in a solid solution series, provided that the data, and especially the intensities, are collected carefully.

7.7.1 Simple Identification Procedures

Since, to a very good approximation, each crystallite in a powder sample diffracts incident X-rays independently of any other particle, the observed diffraction pattern obtained from a mixture of crystalline compounds is a linear sum of the individual patterns from the compounds present in the mixture, weighted according to the amounts of each phase present. It follows that, provided the separate patterns are known for the components, it is, in principle, possible to unravel the combined diffraction patterns and identify the separate components and estimate their approximate relative concentrations.

It is important to remember that the 'raw' data from a powder study comprise a set of numbers relating to *d*-spacings and relative intensities with no indication of the complexity of the crystalline mixture involved. It is wise, therefore, to try to obtain as much additional information as possible at this stage, and optical microscopic examination can be invaluable in suggesting possible phases, either from the crystalline habits or the optical properties, birefringences, extinctions, anisotropies, etc., of the materials present. An idea of the chemical nature of the material is also useful, although it need be no more than qualitative and confined to a list of the elements represented. X-ray diffraction methods do suffer slightly from the fact that they are somewhat 'blind' in nature, despite their inherent power, and it is well to remember this and not to rely solely on X-ray evidence in carrying out an analysis. It must never be forgotten, of course, that X-ray methods do not detect glassy phases at all.

The observed pattern from an unknown sample may, therefore, be produced by

(a) a single crystalline phase whose diffraction pattern is known;
(b) a single crystalline phase whose diffraction pattern is not known;
(c) a mixture of phases whose diffraction patterns are all known;
(d) a mixture of phases some of whose diffraction patterns are known, the others being unknown;
(e) a mixture of phases whose diffraction patterns are all unknown.

Clearly, cases (a) and (c) can be relatively easily handled provided that there exists a scheme for filing and retrieving known diffraction patterns, and case (d) may be partly solved by the same means. Case (b) obviously cannot be handled as an identification procedure, but techniques now exist which make it possible to use such data to characterise the phase, and these will be discussed in section 7.7.2. Case (e) represents a currently insoluble problem in the absence of other data to distinguish it from (b), for example evidence of a multi-phase system from optical microscopy, or the presence of more than one unit cell, from electron diffraction (Chapter 8), and additional information would be required before the data could be used at all.

The J.C.P.D.S. File

In 1945 the American Society for Testing Materials began publishing all known powder diffraction patterns in a file, and this continued until 1969 when responsibility for the task passed to the Joint Committee on Powder Diffraction Standards, a body independent of the A.S.T.M. and based in Swarthmore, Pennsylvania. The change of name has not involved any change in the File, apart from the normal improvements that have been occurring over the years. The system is currently available in three forms, either as a set of printed cards, the J.C.P.D.S. Powder File, together with the Indexes (formerly the A.S.T.M. File and Indexes), or as the Matthews Key-Sort Index, or in computer compatible form. Since the printed card form is the oldest and still the commonest form in which the File is encountered, this version will be discussed in some detail.

The J.C.P.D.S. File and Indexes comprise three sections:

(a) The Hannawalt or the Fink indexes
(b) The alphabetical and the formulae sections
(c) The card file.

The Hannawalt is the older index and lists the three most intense lines in the pattern for each compound, while the Fink index, recently developed to help electron diffractionists who measure the same d-spacings as workers

3.39 – 3.32

										File No.	Fiche No.
	3.34x	4.26_3	2.13_8	7.40_6	2.57_6	2.03_6	3.49_4	2.24_4	$Ca_3Mn(SO_4)_2(OH)_6.3H_2O$	20- 226	1-119-C 1
	3.34x	4.26_7	2.13_6	7.40_5	3.49_5	2.58_5	2.24_5	2.21_5	$Ca_3Ge(SO_4)_2(OH)_6.4H_2O$	19- 225	1-104-F 11
*	3.34x	4.26_4	1.82_2	1.54_2	2.46_1	2.28_1	1.38_1	2.13_1	SO_2	5- 490	1- 18-C 2
	3.32x	4.26_8	8.47x	3.21_8	2.90_3	7.12_7	3.10_5	2.84_5	$(NH_4)_2V_{12}O_{29}$	23- 30	1-160-C 6
	3.39_9	4.25_7	2.81x	3.97_7	3.12_7	2.59_7	1.72_7	6.51_5	$Na_{21}MgCl_3(SO_4)_{10}$	12- 196	1- 45-E 9
	3.30_3	4.25x	3.15_5	2.04_9	5.40_2	2.69_2	2.13_2	1.89_2	$Sb_2(SO_4)_3$	1- 392	1- 3-D 3
	3.41_6	4.24x	3.86_8	3.28_8	3.52_5	3.08_3	4.17_3	2.93_3	$P_3N_3Br_{10}$	19- 420	1-106-E 7
*	3.37x	4.24_3	2.32_2	2.12_2	2.45_1	1.84_1	1.83_1	1.54_1	$GaPO_4$	8- 497	1- 30-C 7
	3.31_6	4.24x	2.62_3	2.17_8	2.85_5	2.97_4	1.87_3	1.75_3	$Fe_2H_3(TeO_3)_4Cl$	20- 536	1-121-F 12
o	3.36x	4.23_5	1.64_5	2.72_4	2.44_3	2.22_3	1.93_3	3.14_2	$Mo_3O_8.nH_2O$	21- 574	1-136-B 10
	3.31_8	4.23_6	6.95x	3.02_6	2.88_5	2.15_5	1.96_5	1.89_5	$Be_2AsO_4(OH).4H_2O$	15- 378	1- 66-F 7
o	3.41_8	4.20_7	2.88x	5.45_6	3.31_6	3.03_6	2.29_6	5.79_5	$C_2H_3LiO_2.2H_2O$	14- 840	1- 63-D 9
i	3.41_8	4.20_7	2.88x	5.45_8	3.31_6	3.03_6	2.29_6	5.79_5	$C_2H_3LiO_2.2H_2O$	23-1171	1-170-F 6
	3.37x	4.20_9	3.46x	2.48_9	7.87_8	6.93_5	4.37_4	3.28_4	$LiAlSi_2O_6.H_2O$	14- 168	1- 57-E 4
i	3.39x	4.19_5	3.67_4	2.81_2	2.53_2	2.17_1	1.89_1	1.94_1	$Li_4H_2Si_2O_7$	23-1185	1-171-B 3
*	3.41_3	4.18x	2.41_4	2.09_2	1.58_2	1.87_1	2.95_1	1.39_1	$Zn(CN)_2$	6- 175	1- 20-B 6
i	3.41_6	4.17x	5.98_8	4.21_8	5.85_4	2.98_3	4.81_2	2.95_2	$(NH_4)_3TiF_7$	23- 8	1-160-B 6
	3.39x	4.17x	4.04x	3.29_8	6.10_6	2.61_6	1.63_6	5.79_4	$BeSeO_4.4H_2O$	14- 55	1- 56-E 4
	3.33_8	4.16x	2.00x	1.14_8	8.36_4	3.05_4	2.96_4	2.44_4	InSe	12- 118	1- 45-B 5
	3.36x	4.16x	3.42x	2.86_8	2.39_8	2.36_8	2.24_8	1.96_8	$(CrO_3)16Q$	9- 47	1- 30-E 4
*	3.31x	4.16_8	4.04_8	3.84_6	3.49_5	3.44_5	2.35_5	2.13_5	Hg_2SbBr_2	19- 800	1-110-B 11
o	3.36x	4.15_9	2.70_5	2.02_5	1.73_5	3.90_4	2.98_4	2.84_4	$AgIO_2F_2$	19-1139	1-113-C 2
·	3.36_6	4.15x	2.07_8	1.77_3	7.42_3	3.39_4	3.42_3	1.51_3	$Na_3Zr_4F_{19}$	19-1194	1-113-E 11
	3.32x	4.15_8	3.60x	4.13_8	4.04_8	3.78_8	3.76_8	3.39_8	$(Ru_4Al_{13})102N$	18- 56	1- 88-F 6
	3.31x	4.15x	5.96_7	4.01_4	2.21_4	1.81_3	2.00_3	2.97_2	$RbUF_6$	19-1104	1-112-F 5
	3.38_9	4.13_6	3.44x	2.96_8	2.10_5	4.02_4	3.04_4	1.73_4	$Pb_{17}(Sb,As)_{22}S_{50}$	20- 564	1-122-C 6
	3.35_8	4.11x	3.13x	3.26_6	3.15_6	3.74_4	3.53_4	3.66_4	$Be_2P_2O_7$	19- 161	1-104-D 3
	3.40x	4.10x	3.75x	3.25x	1.45_8	5.50_6	4.70_6	2.42_6	$ZnSO_4.Ti(SO_4)_2$	18-1490	1-102-B 3
i	3.39_8	4.10_7	2.71x	2.84_7	2.82_7	2.06_7	9.36_5	3.90_5	$Ca_3SiO_4Cl_2$	23- 873	1-168-B 12
o	3.38x	4.10_8	2.08_8	1.78_8	1.96_7	1.87_7	2.92_6	1.58_6	KIO_2F_2	19- 952	1-111-D 8
	3.31x	3.02_8	4.32_7	2.87_7	6.60_6	4.70_6	2.67_6	2.51_6	$PbHPO_4$	6- 274	1- 20-E 3
*	3.31_4	3.02x	2.98_8	2.92_4	3.40_4	4.37_3	4.17_3	2.05_3	$Pb_5(VO_4)_3F$	23- 342	1-163-C 1
	3.41_8	3.01x	4.26_8	3.09_8	2.12_8	1.99_8	4.51_6	2.07_6	$Pb_5Br(VO_4)_3$	15- 226	1- 65-E 1
	3.41x	3.01_8	2.89_4	6.50_3	2.09_3	2.50_2	1.89_2	6.00_2	Pb_2SeO_5	11- 532	1- 42-D 8
	3.37_5	3.01_4	2.65x	3.03_4	2.25_4	2.24_4	3.93_3	3.86_3	$(SmSi_2)12P$	11- 49	1- 38-C 7
*	3.36x	3.01x	3.78_9	3.14_9	2.51_6	2.49_6	2.66_6	2.57_6	Ti_7O_{13}	18-1403	1-101-C 1
	3.35x	3.01x	2.81x	1.86_8	1.70_7	1.98_6	1.81_6	1.66_6	$(BaMg_2)12H$	6- 368	1- 21-C 4
*	3.33x	3.01x	4.69x	2.91x	2.06_9	1.83_9	1.84_8	1.82_8	$Y_4Al_2O_9$	22- 987	1-154-B 10
	3.33_4	3.01x	2.91x	4.71_2	2.07_2	1.83_2	1.82_2	2.62_2	$Al_2Y_4O_9$	14- 475	1- 60-C 8
	3.40_3	3.00x	3.71_4	2.89_3	2.28_2	2.45_2	2.15_2	1.86_2	$(NH_4)_2CO_3.H_2O$	1- 858	1- 5-C 2
	3.39x	3.00x	2.25x	4.24_6	3.18_6	5.16_4	2.52_4	2.15_4	$Ba_2(PO_3)_4$	12- 7	1- 44-B 8
*	3.38x	3.00x	3.97x	3.87_8	2.97_8	2.38_8	12.4_6	8.70_6	$Sb_{0.67}Nb_2O_6$	19- 80	1-103-E 8
	3.36x	3.00_7	2.83_7	8.61_6	3.30_6	3.28_5	2.85_4	2.04_3	$Pb_3(NO_3)(OH)_5$	22- 659	1-151-B 4
i	3.36_4	3.00x	1.94_6	2.14_4	1.84_4	2.60_4	2.12_3	2.89_2	CaB_2O_4	23- 407	1-163-F 3
*	3.35_8	3.00x	3.65_8	3.04_5	2.88_5	2.16_5	2.13_5	2.80_2	$BaSnS_2$	20- 150	1-118-D 6
*	3.33_9	3.00x	4.26_9	2.07_8	3.22_7	3.81_6	2.03_5	2.70_5	$PbSO_4$	5- 577	1- 18-E 2
	3.33_8	3.00x	3.98_8	2.55_6	1.70_5	3.60_5	2.80_5	2.71_5	$Ag_2Nb_8O_{21}$	21-1085	1-140-E 10
	3.33_9	3.00_9	2.07x	4.24_8	2.03_8	3.21_7	3.79_6	3.60_6	$Pb_3GeO_2(SO_4)_2(OH)_2$	12- 641	1- 49-E 2
*	3.31_9	3.00_7	3.24x	3.46_5	3.78_5	2.57_5	2.91_4	2.17_4	BaKNaCaAlSiO	19- 2	1-103-B 3
	3.31x	3.00_8	2.92_3	6.15_3	2.62_2	2.61_2	2.19_2	4.08_1	$Na_2Mn(SO_4)_2.H_2O$	20-1127	1-127-D 6
	3.31_3	3.00x	2.69_6	3.62_3	2.00_2	1.96_2	5.37_2	2.62_2	$(Sr_{0.90}Ca_{0.10})_3(PO_4)_2$	14- 205	1- 57-F 11
	3.30_3	3.00x	2.68_6	2.00_3	5.36_3	3.62_3	3.45_2	2.80_2	$(Sr_{0.95}Mg_{0.05})_3(PO_4)_2$	14- 491	1- 60-D 4
	3.39x	2.99_8	4.66_7	3.81_6	2.47_3	2.19_2	2.06_2	2.77_2	$Na_2S_3O_{10}$	21-1371	1-143-D 4
	3.39x	2.99x	2.93x	3.05_9	5.81_5	2.21_5	4.52_4	2.34_4	$BaBe_2Si_2O_7$	20- 119	1-118-B 12
*	3.39x	2.99_8	2.84_4	2.88_2	2.06_2	1.87_2	6.32_2	2.27_2	$Pb_2(OH)_2CrO_4$	8- 437	1- 29-E 12
	3.38x	2.99x	3.71x	3.52_9	2.56_8	2.53_8	2.48_8	2.42_8	$Zn_3O_2(SO_4)_3$	16- 305	1- 74-B 12
	3.38_6	2.99_6	3.40x	2.13_6	5.08_5	2.08_5	1.80_5	1.39_4	$RbSn_2F_5$	20-1011	1-126-D 4
*	3.38_6	2.99x	3.07_9	4.22_4	1.98_4	2.11_4	1.91_4	4.47_3	$Pb_5(VO_4)_3Cl$	13- 585	1- 56-B 6
	3.38x	2.99x	2.43x	3.69_6	2.31_6	2.01_6	2.15_4	1.91_4	PbTe	22- 397	1-148-D 11
	3.36_3	2.99_3	2.63x	3.91_3	3.84_2	3.02_2	2.25_2	2.23_2	$(GdSi_2)12P$	9- 91	1- 30-F 11
	3.33_9	2.99x	5.24_9	3.49_8	2.05_8	3.15_7	3.95_6	1.85_5	$Li_7U_6F_{31}$	10- 121	1- 35-B 3
	3.32x	2.99_5	3.45_4	2.68_2	2.17_2	2.12_2	6.41_1	2.02_1	$CuBCl_2N$	22- 228	1-147-B 4
*	3.31x	2.99x	4.24_8	3.78_8	2.05_8	2.02_7	3.19_5	2.74_5	$SmSO_4$	17- 775	1- 86-F 6
*	3.31_6	2.99_6	3.96x	2.54_6	3.69_6	2.84_6	2.80_6	2.70_6	$NaNb_6O_{15}OH$	19-1223	1-114-B 5
*	3.31_8	2.99_8	3.95x	2.54_8	3.68_6	2.83_6	2.79_6	2.69_6	$NaNb_6O_{15}F$	19-1222	1-114-B 5
	3.41_6	2.98x	2.90_6	3.23_6	4.03_5	2.79_5	2.32_5	2.38_4	$Cd_3(BO_3)_2$	9- 130	1- 31-C 4
	3.40x	2.98_4	3.86_3	4.01_2	3.48_2	2.46_2	2.38_2	5.18_1	Te_2O_5	21-1205	1-141-F 5
	3.40x	2.98_6	2.77_4	4.48_3	4.09_2	3.17_2	3.70_1	2.51_1	$NaHSO_4$	1- 624	1- 4-C 10
*	3.39_5	2.98x	4.80_9	3.40_5	2.53_5	2.54_4	2.40_3	2.15_3	TlCNS	8- 65	1- 27-C 12
	3.37x	2.98x	4.39x	2.23x	5.48_6	2.68_6	2.11_6	2.07_6	$BaB_2O_4.12/3H_2O$	21- 58	1-131-D 8

(a)

with X-rays, but who find quite different relative intensities due to the different dynamical model for the diffraction of electrons, lists the six longest and strongest lines with no direct reference to intensity. The sets of three or six lines are rotated to take account of possible preferred orientation effects and of different relative intensity values due to different

5-0490 MINOR CORRECTION

d	3.34	4.26	1.82	4.26	SiO_2
I/I₁	100	35	17	35	SILICON IV OXIDE ALPHA QUARTZ

Rad. CuKα₁ λ 1.5405 Filter Ni
Dia. Cut off Coll.
I/I₁ G.C. DIFFRACTOMETER d corr. abs.?
Ref. SWANSON AND FUYAT, NBS CIRCULAR, 539, VOL. III (1953)

Sys. HEXAGONAL S.G. D_3^4 - $P3_121$
a₀ 4.913 b₀ c₀ 5.405 A C 1.10
α β γ Z 3
Ref. IBID.

εα nωβ 1.544 εγ 1.553 Sign +
2V D_x 2.647 mp Color
Ref. IBID.

MINERAL FROM LAKE TOXAWAY, N.C. SPECT. ANAL.: <0.01% AL; <0.001% CA, CU, FE, MG.
X-RAY PATTERN AT 25°C.

REPLACES 1-0649, 2-0458, 2-0459, 2-0471, 3-0419, 3-0427, 3-0444

d Å	I/I₁	hkl	d Å	I/I₁	hkl
4.26	35	100	1.228	2	220
3.343	100	101	1.1997	5	213
2.458	12	110	1.1973	2	221
2.282	12	102	1.1838	4	114
2.237	6	111	1.1802	4	310
2.128	9	200	1.1530	2	311
1.980	6	201	1.1408	<1	204
1.817	17	112	1.1144	<1	303
1.801	<1	003	1.0815	4	312
1.672	7	202	1.0636	1	400
1.659	3	103	1.0477	2	105
1.608	<1	210	1.0437	2	401
1.541	15	211	1.0346	2	214
1.453	3	113	1.0149	2	223
1.418	<1	300	0.9896	2	402,115
1.382	7	212	.9872	2	313
1.375	11	203	.9781	<1	304
1.372	9	301	.9762	1	320
1.288	3	104	.9607	2	321
1.256	4	302	.9285	<1	410

1919

23-407 23-408

d	3.00	1.94	3.36	5.78	CaB_2O_4 $CaO \cdot B_2O_3$
I/I₁	100	55	40	10	Calcium Borate

Rad. CuKα λ 1.5418 Filter Ni Dia. 114.6mm
Cut off I/I₁ Visual I/I cor.
Ref. Fletcher et al., J. Am. Ceram. Soc., 53 95-97 (1970)

Sys. Orthorhombic S.G.
a₀ 6.219 b₀ 11.621 c₀ 4.287 A C
α β γ Z 4 Dx
Ref. Ibid.

εα nωβ εγ Sign
2V D mp Color
Ref.

d A	I/I₁	hkl	d A	I/I₁	hkl
5.78	10	020	1.381	5	113
3.36	40	111	1.370	5	440
3.10	15	200	1.339	5	033
3.00	100	121,210	1.309	10	133,262
2.886	20	040	1.229	5	233,422
2.860	10	031	1.211	10	191
2.735	10	220	1.192	5	290
2.604	35	131	1.169	10	-
2.235	15	141	1.119	5	-
2.137	40	002			
2.117	25	240			
2.043	10	051			
2.005	10	022			
1.935	55	151,060			
1.902	10	122,241			
1.838	40	311			
1.742	5	212			
1.683	10	222,331			
1.643	10	260			
1.503	20	242,171			

(b)

Figure 7.16 Typical pages and cards from the J.C.P.D.S. X-ray powder data file, to illustrate the interpretation of the powder data listed in table 7.2 (Reproduced by kind permission of the Joint Committee on Powder Diffraction Standards, Swarthmore, Pa.)

experimental conditions, or to allow for missing lines. Recent Hannawalt indexes have included the next five strongest lines, in order of diminishing intensity without rotating the values, to help in the initial search procedure. Figure 7.16a shows a typical page from such a recent index, from which it will be seen that the sorting scheme is such that the data are grouped between limits for the first line quoted, and listed in numerical length order for the strongest remaining line within the group. Each compound is listed three times, once for each of the three strongest lines. Also listed are the chemical composition, the chemical or mineral name, and a cross-reference to the full pattern listing in the card file.

In the Fink index the six lines are listed six times, initially in the order longest to shortest, i.e. in the order a,b,c,d,e,f; then in the order b,c,d,e,f,a; then c,d,e,f,a,b; and so on until each line has appeared as the initial line.

The alphabetical and formula indexes contain the same information as the Hannawalt and Fink indexes, but arranged in alphabetical order of the compound name or chemical formula. In all cases, separate indexes are now provided for organic and for inorganic, including mineral, compounds.

Typical cards from the card file are reproduced in figure 7.16b, and show that all the observed lines are listed together with their relative intensities, and these may be indexed if the unit cell is known. Further data relating to the unit cell, optical properties, density, etc., are included if these are known. It should be noted that the A.S.T.M. Index has grown over the years, and the J.C.P.D.S. Index continues to grow, and currently a new set of about 2000 cards are added each year, on average, as new data are generated from laboratories throughout the world. But, in addition to presenting new data for new compounds, new cards may also contain data which supersede older data, as in the card for quartz in figure 7.16b. Naturally, updated copies of all the indexes are provided with each new set of data cards.

The use of the J.C.P.D.S. File to analyse an unknown sample can be simply illustrated with an example. The X-ray powder data quoted in table 7.2 were obtained from an amorphous mixture of CaO, B_2O_3 and SiO_2 heated at 900°C for 7 days. From it we may expect calcium silicate, calcium borate, and silica phases to crystallise out. An initial search was made in the Hannawalt index using the lines at 3.35Å, 3.02Å and 2.13Å, the three strongest lines in the pattern, and these indicated the presence of calcium borate, CaB_2O_4, although the 3.35Å line was anomalously strong relative to the 3.04Å line. Comparison of the observed lines with those listed in the calcium borate card, figure 7.16b, showed that all the lines quoted were present in the unknown pattern, but that the quoted line at 3.40Å appeared to be at 3.35Å and was anomalously strong in the observed pattern, as were the lines at 2.24Å and 1.55Å. Calcium borate was thereby confirmed. Certain observed lines, however, could not be indexed in terms of calcium borate, and the problem of the anomalously strong lines remained. It was assumed that the 3.35Å line was the strongest line in a second phase and a second search was made using the remaining strong lines at 3.35Å, 4.25Å and 1.82Å. A match was found with quartz and comparison with the card, figure 7.16b, again showed that all the expected lines were present in the observed data, that all the remaining lines could be attributed to this phase, and that the anomalously high intensities were due to contributions from both phases. The observed relative intensities in the

TABLE 7.2

	Unknown Sample			Indexing	
	2θ(°)	d(Å)	I/I_0		
2	20.90	4.25	40		Q 35
1 2	26.60	3.35	100+	CB 42	Q 100
1	29.62	3.02	60	CB 100	
	30.94	2.89	20	CB 23	
	32.56	2.75	15	CB 7	
	34.41	2.61	40	CB 33	
	36.48	2.46	30		Q 12
	39.50	2.28	25		Q 12
	40.25	2.24	25	CB 7	Q 6
1	42.43	2.13	50	CB 50	Q 9
	47.09	1.93	25	CB 42	
	49.11	1.85	30	CB 33	
2	50.10	1.82	40		Q 17
	59.82	1.55	30	?CB 13	Q 15
	67.85	1.38	35		Q 11

Indexing of unknown powder pattern using Hannawalt index. Numbers on left indicate choice of lines for first and second searches. Columns on right show assignment of lines to possible cells, and relative intensities of lines in the test cells. CB = CaB_2O_4; Q = quartz, SiO_2.

pattern then matched well with the quoted data, indicating that the search was complete and that the crystalline phases present were calcium borate and quartz.

Of course, the above interpretation could be simplified by using a little forethought, coupled with some knowledge of X-ray patterns. Since the system contains SiO_2 it is probable that quartz or cristobalite could be present in the final product, and the line at 3.35 Å confirms this since it is the strongest line in the quartz pattern. A check in the alphabetical index will confirm quartz, and the quartz lines can be eliminated from the pattern using the card data as before. In the same way, a check in the calcium borate region of the alphabetical index will indicate calcium borate as a probability, and the card can again be consulted to confirm its presence. The extent to which the alphabetical index can be used does depend, of course, on the extent to which one can guess the likely phases present, but its use can help considerably in cutting down the amount of searching required to identify the components of the mixture.

Computer Aided Searches of the J.C.P.D.S. Index

The above example represents a simple, straightforward search involving only two components. As the number of components increases, the amount of work involved and the difficulty of sorting out the various possible phase assemblages increases rapidly. Vand and Johnson [85] have prepared a suite of programs (available free from the authors) which allow the user to search the whole of the powder diffraction file, stored on magnetic tape and available from the Joint Committee, and output probable compounds based on the observed d-spacings, intensities, and likely chemical formulae of the unknown phases. The program initially searches for all possible matches then estimates a 'likelihood factor' for each match depending on the number of coincidences encountered for the lines of that phase. It finally presents up to ten most likely phases which the user can finally check for correctness. The system is quite accurate and provides a welcome aid to harassed analysts at a not unreasonable cost, since one can either lease the tapes from the Joint Committee or have one's data run commercially by the University Computing Company.

Graphical Methods of Identifying Phases

When only a limited number of crystalline phases can occur in a series of analyses, for example in the analysis of the products of syntheses using the same set of elements or compounds but in varying ratios, a graphical method can be used to sort out the phases present. As shown in figure 7.17,

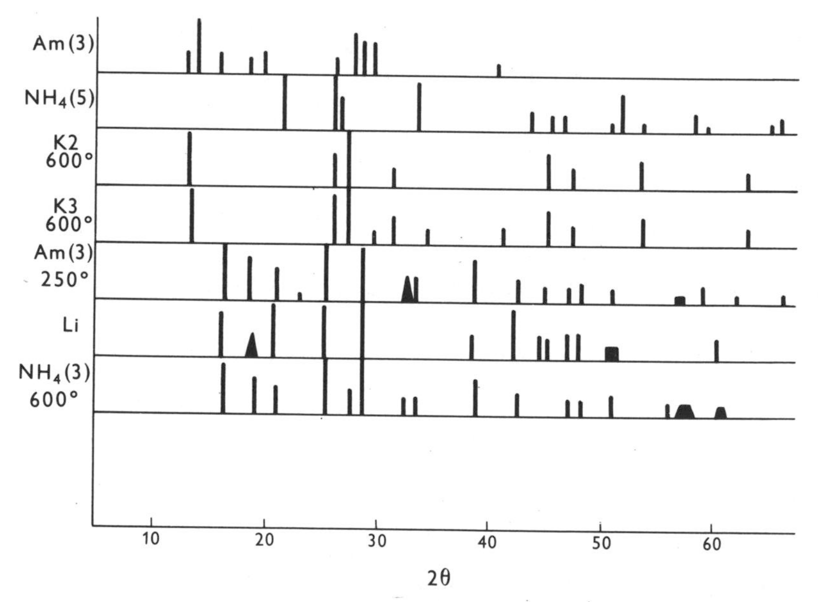

Figure 7.17 Line positions and intensities for a set of unknown phases, for comparison by the graphical method. The diagram shows that the mixtures K2(600°) and K3(600) contain the same component, possibly with a second phase in the latter mixture, and that mixtures Am(3)(250°), Li and NH_4(3)(600°) have very similar compositions. Am(3) and NH_4(5) are quite different from the others. The sharpness of the diffraction peak is indicated by the breadth and shape of the line on the diagram.

the observed lines for the unkown mixtures are plotted, in terms of their 2θ and relative intensity values, on graph paper and compared with similar representations of possible phases. The method is very simple, rapid, and surprisingly accurate, since quite small changes in d-value or relative intensity can show up remarkably clearly between the patterns. Plotting against the d-value in Å is not recommended, since variations in this parameter do not show up so well, but Johnson (private communication) has pointed out that excellent resolution can be obtained by plotting the relative intensity against the logarithm of the d-value. Plotting on this logarithmic scale has the added advantage that solid solutions, in which the d-values are shifted in a constant ratio from the corresponding values for the pure end member, can be simply related to the end member by a translation of the whole pattern parallel to the log d axis, since the patterns are related by the equation

$$\log d_{hkl} \text{ (solution)} = \log d_{hkl} \text{ (pure)} + \text{constant}$$

Johnson is presently preparing a new index for the J.C.P.D.S. File in which all the patterns are plotted on this logarithmic scale and arranged according to the value of the strongest line.

7.7.2 Computer Aided Interpretation of Powder Data

The computer can aid the X-ray diffraction analyst in two principal ways when handling powder data. On the one hand it can help him to index his data in terms of a known unit cell, with the corollary that one can identify the presence of a known crystalline phase from a powder pattern, and on the other it allows him to tackle the problem of deriving the unit cell of a phase directly from powder data.

Indexing and Search Procedures

The value of the d-spacing for the set of (hkl) planes in a crystal can be calculated using the formula

$$d_{hkl} = 1/\sqrt{Q_{hkl}} \tag{7.12}$$

where

$$Q_{hkl} = h^2a^{*2} + k^2b^{*2} + l^2c^{*2} + 2hka^*b^*\cos\gamma^* + 2klb^*c^*\cos\alpha^* + 2hla^*c^*\cos\beta^*, \tag{7.13}$$

and $a^*, b^*, c^*, \alpha^*, \beta^*, \gamma^*$ are the reciprocal cell parameters of the crystal, as discussed in section 7.5.3. Equation (7.13) is quoted for the triclinic case and it simplifies for the higher symmetry classes, especially orthogonal cells. Such a problem is well suited for computer applications, because of the large number of tedious calculations required, and many programs have been written to calculate either the d_{hkl} or the Q_{hkl} values for a given unit cell. Nicol [86, 87] has described a program to calculate the full list of permitted d-values from a given unit cell and space group, compare these with a set of observed d-spacings, and print out a list of possible indices for each line. The program also lists the lines which could not be indexed in terms of the given cell, and the process can be repeated using several unit cells so that the user is given an indication of which lines will index in terms of which cells. The program has proved to be of considerable use as an aid in identifying phases. Parthe [88] has written a similar program which combines equations (7.5) through (7.8) with equations (7.12) and (7.13) to generate computer simulated powder diffraction traces of the theoretical pattern from a given structure. The program is very useful for comparing an observed pattern with theoretical patterns from possible structures, and it has the advantage over Nicol's program of making the comparison in terms both of d-spacing and of intensity, and not just in terms of the former.

Procedures for deriving the unit cell directly from powder data have been attempted using computers, with mixed success. McMaster [89] has written a suite of programs based on Ito's method of analysis [90], which involves deriving initially a triclinic cell from the diffraction data and applying Delauney's reduction method [91] to derive a high symmetry cell therefrom. More recently, Visser [92] has tackled the same problem, also with a program to solve the most general case. These methods are quite successful provided that the data supplied are good. Unfortunately, they cannot operate effectively with poorly measured data or with incomplete data, and it will be obvious that patterns with contributions from more than one phase are quite unsuitable.

Refinement of Cell Parameters

The derivation of accurate unit cell parameters from powder data is easy only in the case of cubic crystals, which have but one variable to be refined. Provided that the observed lines in the pattern can be indexed, the value of a_0 can be calculated and either plotted as a function of the Nelson-Riley extrapolation function (section 7.6.4) or refined using a least squares method, in both cases using the simplified form of equation (7.13), i.e.

$$d_{hk} = \frac{a_0}{(h^2 + k^2 + l^2)^{1/2}}$$

The lower symmetry groups are best handled using a least squares method to refine the reciprocal cell parameters utilising the form of equation (7.13) relevant to the crystal class involved. Nicol [86, 87], Lindquist and Wenglen [93] and Elliott [94] have described computer programs for such routines.

Refined unit cell parameters are useful in analysing systems involving crystalline solutions of various types, since it is often possible to relate the size of the unit cell parameters to the chemical composition, particularly if the parameters of the solid solution series obey a Vegard's law relationship. Thus, the composition of an alloy may be determined in this way, and Katz [95] used the method to identify the composition of the spinel-gallia crystalline solution remaining in a single crystal after part of the originally dissolved gallia had been precipitated by annealing. Accurate determination of the cell parameters for a feldspar allows its nature to be determined very precisely so that one may determine, for example, where on the albite-anorthite join the specimen lies. Zussmann [11] has summarised many of the important uses to which X-ray diffraction studies may be put in mineral anlysis.

7.8 QUANTITATIVE POWDER STUDIES

Quantitative measurement of powder reflexion intensities provides information which may be used in two principal ways, either for the determination of the amount of any crystalline phase present in the mixture or for the determination of the average sizes of the crystallites in the sample.

7.8.1 Quantitative Phase Analysis of Mixtures

Brindley and Spiers [96] were among the first to discuss the parameters involved in the quantitative estimation of the amount of a crystalline compound present in a mixture, although the idea of using the intensity of reference peaks in the patterns from multi-component materials for this purpose had been proposed somewhat earlier [97, 98]. Alexander and Klug [99] and Brindley [100] further refined the basic relationships and discussed experimental procedures.

Since the composite pattern obtained from a mixture of crystalline phases is a linear sum of the separate patterns from the individual phases, with their intensities weighted according to the amounts of each present, equation (7.5) can be modified to take account of this further factor in calculating the observed intensity of a given reflexion from any phase, to read

$$I(hkl) = w \times p \times F^2(hkl) \times T \times A \times G \tag{7.5a}$$

where w is the weight fraction of the component under study together with a correction for the overall absorptivity of the sample. Hence, if all other factors are kept constant by standardising the conditions under which the line intensities are measured and due allowance is made for the absorptivity of the sample, differences in the relative intensities of chosen lines from the different phases will clearly be related to the different proportions of the phases in the mixture.

Although the basic concept of the method is very simple, practical details are more complex, with absorption of the diffracted radiation by the matrix from which it originated providing the major problem. The temperature and geometrical factors in equation (7.5a) have relatively little effect, and are amenable to simple correction, while the multiplicity factor and the structure factor are, of course, constant.

The question of sample absorption is vital in all quantitative studies, and Klug and Alexander [101] have emphasised the importance of understanding the absorption coefficient of the matrix in all such studies. In particular they point out that admixture of a weakly absorbing material to a strongly absorbing substance results in the intensities of reflexions from the weak absorber appearing anomalously low, and those from the strong absorber appearing anomalously high, when compared with the intensities from the pure materials when these are weighted in the ratio of the two admixed phases. They point out that the observed intensity of the SnO_2 pattern from a lung tissue contaminated with SnO_2 would indicate 20-25% SnO_2 by direct comparison with the pattern from pure SnO_2 when the actual concentration, determined after allowing for the very much lower absorptivity of the lung tissue, was only 1% of SnO_2. A simple comparison method can, however, be used when analysing mixtures in which all the components have the same mass absorption coefficient, for example in analysing mixtures of polymorphs or allotropes, and Klug and Alexander [101] have shown that an excellent straight line relationship exists for quartz-cristobalite mixtures.

In general, however, the absorption coefficients of the components of the mixture are not the same, and three major variants have been discussed to cover the three principal situations. Klug and Alexander [101] point out that any mixture can be considered in terms of two components, one being the phase to be determined and the other being the rest of the mixture, which they designate the matrix. Then, if the component to be determined is indicated by the subscript 1 and the matrix by M, the basic equation on which all quantitative work is based is

$$I_1 = \frac{K_1 x_1}{\rho_1 [x_1(\mu_1{}^* - \mu_M{}^*) + \mu_M{}^*]} \tag{7.14}$$

where K_1 is a constant depending on the nature of the component 1 and on the geometry of the X-ray diffraction apparatus used, $\mu_1{}^*$ and $\mu_M{}^*$ are the mass absorption coefficients for the component and the matrix respectively, ρ_1 is the density of the component, and $\mathbf{x}_1$ is the weight fraction of the component in the mixture.

The simplest case, that in which the matrix is the same as the substance to be measured has been discussed above. The next simplest comprises a mixture of two components, in which case $\mu_M{}^*$ is equal to $\mu_2{}^*$, the mass absorption coefficient for the non-analysed component. Klug and Alexander [101] show that the reflexion intensity for the analysed component in the pure state, under the same experimental conditions of analysis, is given by

$$(I_1)_0 = \frac{K_1}{\rho_1 \mu_1{}^*} \tag{7.15}$$

and combining this with equation (7.14) gives the theoretical relationship between intensity and weight fraction in the form

$$\frac{I_1}{(I_1)_0} = \frac{x_1\mu_1^*}{x_1(\mu_1^* - \mu_2^*) + \mu_2^*} \qquad (7.16)$$

Figure 7.17 shows the excellent fit which Klug and Alexander found between their experimental data and the theoretical curves for quartz-cristobalite (Case 1, $\mu_1^* = \mu_2^* = \mu_M^*$) and for quartz-beryllium oxide and quartz-potassium chloride (Case 2, $\mu_2^* = \mu_M^*$) mixtures.

The third and more general case, however, involves more than two components and in such a case the value of μ_M^* becomes the total mass absorption coefficient for the phases other than that being determined in the mixture [see section 3.2.2, equation (3.7)], and so is indeterminate. The difficulty can, however, be circumvented by adding a known amount of an internal standard material to specially prepared mixtures containing known amounts of the phase to be determined and drawing a calibration curve to relate the intensity of the chosen reflexion from the required component, I_1, to that of the standard, I_s. Alexander and Klug [99] have shown that

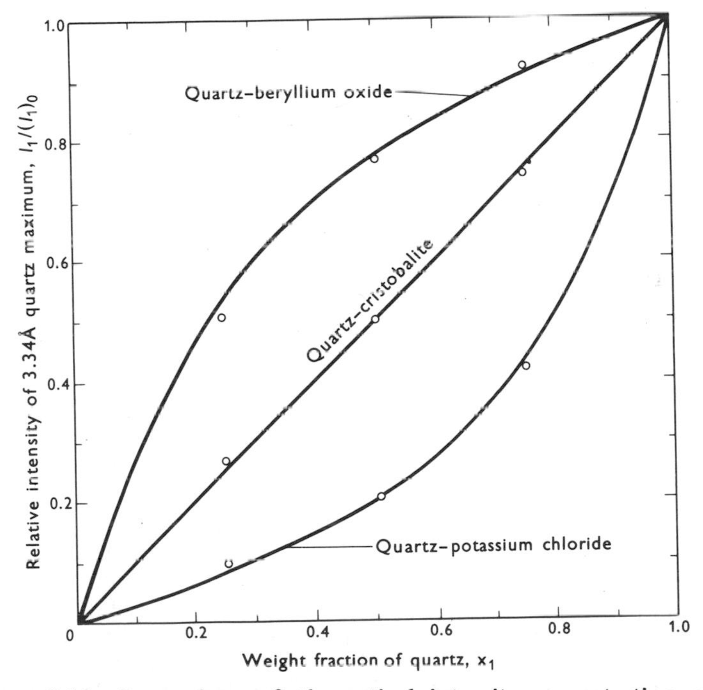

Figure 7.18 Comparison of theoretical intensity-concentration curves (solid lines) with experimental measurements (open circles) for several binary mixtures. (Reproduced from Alexander and King, *Anal. Chem.*, **20**, by kind permission.)

$$x_1 = k\frac{I_1}{I_2} \tag{7.17}$$

provided that the internal standard is added as a constant weight fraction of the total sample. The derivation does presuppose that $\mu_M{}^*$ does not change significantly for different ratios of the other components of the mixture, and the excellent straight lines found experimentally bear out this assumption. Zussmann [11] has pointed out that addition of a constant amount of a non-crystalline substance may help to maintain the constancy of the matrix absorption coefficient.

A modification of the last situation has been discussed by Barry, Stubican and Roy [102] in their study of the decomposition of $CaYb_2O_4$, into $CaO + Yb_2O_3$, which they followed by monitoring the diminution of a chosen $CaYb_2O_4$ peak and the simultaneous increase in a corresponding Yb_2O_3 peak intensity. Here no internal standard could be added, but it was possible to assume a constant matrix absorption, since no material was lost during the decomposition. They showed that the relative intensities were related to the ratio of the weights of the two components by

$$\frac{I_a}{I_b} = K\frac{W_a}{W_b} \tag{7.18}$$

Again excellent linearity was obtained, even when the measurements were made at temperatures above 1000°C, the temperature of the reaction.

Practical Considerations

Quantitative studies may be made using either film or diffractometric methods to collect the intensity data, although the latter is the more convenient in practice. In both cases it is important to use a sample of small particle size, passing 350 mesh, but with a crystallite size sufficiently large that line broadening does not occur (see section 7.8.2). Absolute standardisation concerning the conditions under which the data are collected is, of course, essential. If a film method is used, a large diameter camera is to be preferred because of its better resolving power and its ability to give a better estimate of the true background around the separate lines, whose intensity relative to this background must be measured using an optical microdensitometer. Naturally, lines chosen for quantitative work must be free from contributions from components other than the one to be determined. If a diffractometer is used, the associated X-ray generator must be stabilised, as must all the electronics in the recording circuits, to ensure that the incident intensity does not vary with time, i.e. with position in the recorded pattern. The detector should be a scintillation or a flow-proportional counter, to take advantage of the energy discrimination potential of these units and so reject extraneous noise as much as possible. The intensities themselves can be found either by measuring the area under the curve from the recorder trace or by integrating the counts in a predetermined angular range across the required peak and correcting for the average integrated background on both sides of the peak.

The sensitivity of quantitative X-ray powder diffraction analysis varies widely depending on the nature of the material being studied. A well crystallised phase is readily detected, especially if the pattern contains a strong reflexion, as in the case of quartz or calcite, and concentrations as low as 1% may be found, but poorly crystallised phases may remain undetected even up to quite high concentrations, 25% or more. The accuracy of the method is probably at best about ± 5-10% of the amount present, and at worst an order of magnitude.

Recently Visser [103] has described a 'universal scale' of intensity values for reference peaks in a number of pure materials and from it has developed an internal standard method based on the expression

$$x_1 = 100\frac{s}{w}\frac{I_s}{I_{sm}}\frac{I_N}{I_{Nm}} \tag{7.19}$$

where x_1 is the weight percentage of the component being measured, w is the weight of the unknown sample used, s is the weight of the standard added, I_{Nm} and I_{sm} are the measured intensities of the chosen reflexions from the component and the standard, and I_N and I_s are the 'universal scale' intensities of the reference lines in the component and the standard respectively. Tables of 'universal scale' intensities for some 300 substances are now being published as an appendix to the J.C.P.D.S. Indexes, based on their intensities relative to the (113) line for corundum, Linde A product, taken at $2\theta = 43.36°$ using Cu $K\alpha$ radiation ($d_{113} = 2.085$Å).

7.8.2 Crystallite Size Determination

The technique of crystallite size determination by X-ray powder diffraction methods is based on the relationship given earlier in equation (7.11) between the observed line width of the X-ray reflexions from a sample and the mean diameter of the crystallite in the sample, and depends on an ability to measure accurately the profiles of the powder lines produced. In general, X-ray methods are used to measure sizes in the sub-micron range, although Henry, Lipson and Wooster [104] point out that an estimate of grain size in the range greater than 10^{-3} mm can be made qualitatively by considering the degree of 'spottiness' of reflexions collected on film, especially if the sample is not rotated during the exposure. The larger the grain size, the more 'spotty' the powder lines appear, but the size estimate so made is no better than very approximate.

For a smooth powder line, its breadth is commonly defined in one of two ways, irrespective of whether the line profile has been measured directly using a diffractometer or indirectly from film with a microdensitometer. The more generally accepted definition is that the line breadth is the width of the line, expressed as an angle in °2θ, at a level midway between the local baseline and the point of maximum intensity, as shown in figure 7.18a. This is quite easily determined, but Laue [105] has pointed out that such a

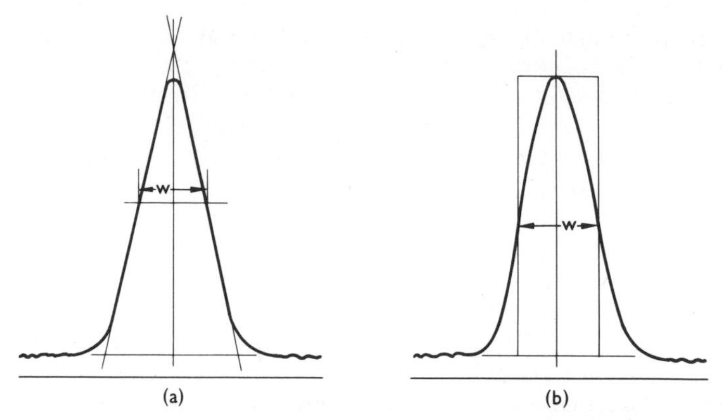

Figure 7.19 Line breadths in X-ray diffraction, showing (a) the line breadth derived as the width of the line at the half-height, and (b) the breadth as the width of the rectangle of area equal to that under the peak and height equal to that of the peak itself.

definition does not take into account the shape of the lower part of the curve and has proposed that the breadth be defined as the integrated area under the peak divided by the peak height, the construction being reproduced in figure 7.18b. Both methods have their advantages and disadvantages, mainly concerned with the ease of determining the limits of the peak, and both definitions are in use. In practice, the values obtained differ only by small amounts and it is probably more important to standardise on one or the other definition for all the work in a given laboratory, especially when comparative work is being done.

Scherrer's original equation [61] showed that the mean dimension, D, of the crystallites in a powder sample is related to the pure X-ray broadening of the diffracted lines, β, by the equation

$$D = \frac{K\lambda}{\beta \cos\theta} \tag{7.20}$$

where K is a constant with value approximately unity but related to the shape of the crystallites and to the way in which β and D are defined. Its value is often taken as 0.9, especially when little is known of the crystallite shape and size distribution, provided that D is defined as the crystallite thickness in the direction perpendicular to the reflecting planes and β as the corrected peak width at the half-maximum intensity position. Unfortunately, β is not necessarily the observed breadth of the diffraction line, since this experimental value also contains contributions from the various effects arising from instrumental factors (instrumental broadening) and possibly from strain and imperfections in the crystallites (strain broadening). Scherrer considered the effect of instrumental broadening and related it to the experimental and pure X-ray broadening values by

$$\beta = B - b \tag{7.21}$$

where B is the experimental broadening and b is the instrumental contribution. He did not consider the effect of strain.

The principal difficulties encountered in crystallite size determinations are associated with correcting observed line profiles for the above effects and hence deriving the pure size broadening component. Klug and Alexander [106] have discussed this problem in some detail, and more recent appraisals of advances in technique relating to different aspects of the subject can be found throughout the continuing series of 'Advances in X-ray Analysis' [107].

Klug and Alexander, following Stokes's [108] analysis, have shown that the correction given by Scherrer has no general validity and the correction for instrumental broadening is most generally, and most accurately, made by convolution analysis involving the application of Fourier transform theory to perform the subtraction of the pure diffraction profile, obtained on the same apparatus but without size or strain effects being present, from the experimentally observed, broadened profile to give the value of the size broadening parameter, β, conventionally in $°2\theta$. This can be substituted in Scherrer's equation (7.20) to give D. In practice the method entails measuring the diffracted intensity at small intervals across the relevant line both for the sample under study and for a sample with a crystallite size sufficiently large that the size broadening effect is negligible (usually greater than $1\mu m$, but sizes in excess of $3\mu m$ may be required with some parafocusing cameras). These observed intensities are than fed into the various Fourier series involved and from them are generated a new set of intensities which define the profile due solely to the size broadening.

The above method involves some rather time consuming computations, although the advent of programmable electronic computers has largely eliminated this problem, and various attempts have been made to simplify the method of correcting for instrumental broadening. Such simplified methods do generally require assumptions to be made about the shape of the line profile, usually that it is either Gaussian or Cauchy in form, and this can reduce their applicabilities. Thus Warren [109] has given a method assuming a Gaussian distribution across the line, and Jones [110] has devised a more general method based on an experimentally determined Cauchy distribution. These do not take account of the contributions due to strain in the crystallites.

The effect of strain has been discussed by Warren and Averbach [111] and by Warren [112] and computer methods have been developed to handle the mathematics involved. Rothman and Cohen [113] have discussed the applications of Fourier analysis to this problem and Gazzara and co-workers [114] have published a general computer program for the analysis of particle size and of strain.

More recently it has been shown that the X-rays scattered at very low angles, close to the incident beam can be used to determine particle, rather than crystal, sizes. Guinier [115] and Gerold [116] have discussed the basic theory of the method and Harkness, Gould and Hren [117] have given a critical appraisal of its usefulness. The technique has the advantage that it measures the size of the particle giving rise to the scattering, rather than the sizes of the individual crystallites inside the particle, and so is of considerable use in the study of precipitate particles, for example in alloy systems. The technique is somewhat difficult to apply, however, since it is necessary to make accurate measurements of scattered intensities close to the undeviated portion of the incident beam, and air scatter must be eliminated completely.

7.8.3 Automatic Data Collection

The tedious and time-consuming operations involved in data collection, both for identification procedures and, more especially, for quantitative studies, have now been simplified by the introduction of the automated powder diffractometer. Basically this is simply a standard diffractometer fitted with an accurate stepping motor, which allows the 2θ angle to be set accurately and rapidly under computer control in steps of $0.005^\circ 2\theta$, in place of the usual continous motor. With the addition of an automatic sample holder, typically holding up to 34 samples plus a standard, all aspects of data collection can be made totally automatic and also speeded up, since the maximum slewing rate of stepping motor is of the order of $70^\circ 2\theta$ per minute. Such an instrument has been described by Jenkins, Haas and Paolini [118] of North American Philips Co.

The diffractometer can be used in any of three principal modes. It can be programmed to perform a peak search for identification work, and in this mode it scans across the diffracted radiation until it encounters a reading differing from the local background by a factor greater than that pre-set by the operator. This it counts more carefully, on a second pass, and records as a peak, giving, in the print-out, the 2θ value, d-value in Å, and finally the intensity relative to the strongest line. Naturally, due account is taken of the peak profile in determining its 2θ value. The output can be displayed in a format compatible with the J.C.P.D.S. powder data file, or in any other format that the operator requires, and a graphical output is also produced.

In the continuous scan mode, the instrument can be set to accumulate data at pre-set intervals of 2θ and output these continuously. This differs from the peak search mode in which only 'relevant' data concerned with the actual peaks detected are output. Such data are of use in conjunction with the line profile mode, the mode used in quantitative work. In this last mode, the operator sets the 2θ range limits to cover only the peak to be counted, although more than one peak can be counted on each sample, and the instrument accumulates data at pre-set intervals again across the peak to provide both a line profile and an integrated intensity.

The instrument would appear to be of considerable use in laboratories processing large number of samples, despite its rather high cost. One major advantage of the instrument is the provision of a standard to which reference can be made automatically at any time for recalibration of the system during the course of a series of measurements.

7.9 SINGLE CRYSTAL DIFFRACTION

In single crystal studies a beam of X-rays impinges on a single crystal of a substance and the diffracted beams are detected and measured either on film or electronically. Since only one crystal is involved the patterns obtained are spot patterns, not line patterns as in powder work, and contain much detailed information about the size and shape of the unit cell, or repeat unit, of the crystal, from the relative positions of the diffracted beams in space, and about the arrangement of the atoms within this unit cell, from the relative intensities of the diffracted beams. Such patterns can be used for identification, but they are less used for this than are powder patterns, and their main use is in the characterisation of materials and in the description of new phases.

Single crystal patterns may conveniently be interpreted using the reciprocal lattice and the Ewald sphere constructions. As shown in section 7.5.3, a crystal has associated with it an imaginary 'reciprocal lattice' which bears a specific orientation relative to the crystallographic real cell axes and which moves as the crystal moves. Further analysis shows that this lattice comprises a set of points, corresponding to the set of (*hkl*) planes in real space, such that all the points with the same *h*-value – i.e. 0*kl*, 1*kl*, 2*kl*, etc. – lie in a plane which is perpendicular to the *a*-axis in real space, and similarly points with the same *k*- and *l*-values lie in planes perpendicular to the *b*- and *c*-axes respectively. The result is that the reciprocal lattice can be considered as a series of sheets of points oriented perpendicular to the three principal crystallographic axes, the separations between the sheets being

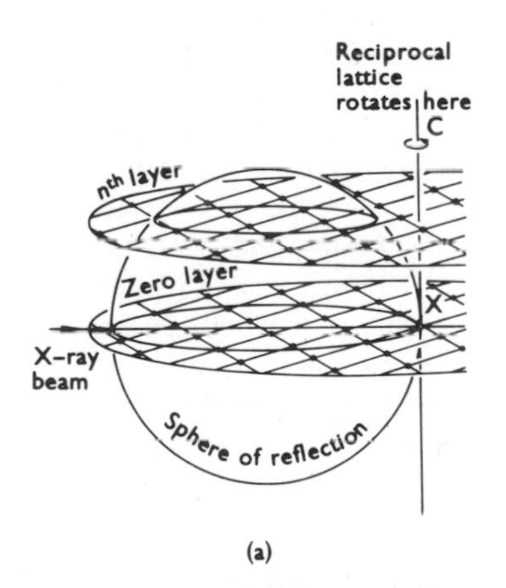

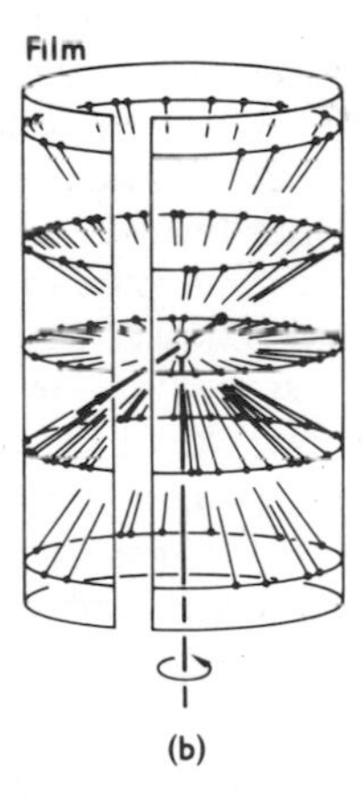

Figure 7.20 Schematic representations of (a) the interaction between the reciprocal lattice of a rotating crystal and the Ewald sphere of reflection, and (b) the resulting cones of diffracted radiation intersecting a cylindrical film mounted co-axially with the crystal. (Reproduced from ref. 119, by kind permission of Oxford University Press.)

inversely proportional to the repeat distances along the three axes. (Note that the separation does not give d_{100}, etc., but a_0, b_0 or c_0.)

The sheets will intersect the Ewald sphere in a series of parallel circles and the resulting sets of diffracted beams, each with the same *h*-, *k*- or *l*-value, will form cones of semi-angle equal to twice the Bragg angle for the axial spacing involved, as shown in figure 7.20. The various methods of detecting these beams provide means of obtaining information about different sections of the resulting set of three dimensional data and so build up a complete picture of the unit cell. Let us briefly consider the various methods available, the data which each provides, and the manner in which these may be combined to define the unit cell fully. Since Laue photographs have already been discussed in section 7.5.1 we shall only consider methods using monochromatic radiation and moving crystals.

7.9.1. Static Film–Rotating Crystal Methods

In this, the simplest type of single crystal diffraction study, the crystal is mounted at the centre of a cylindrical camera normally of 5.73cm or 6.00cm radius and rotated about an axis parallel to the axis of the camera. The diffracted beams are detected by a photographic film mounted on the inner surface of the camera. The crystal mount includes a pair of mutually perpendicular goniometer arcs which permit the crystal to be manipulated until a crystallographic axis lies parallel to the rotation axis. Three translational motions are also provided to move the crystal in directions parallel and perpendicular to the axis of the camera to centre the crystal in the collimated beam of X-rays. A motor and belt drive provides the rotational motion, and a series of cams with a follower attached to the central spindle permit the crystal to be oscillated through angles of 5°,10°, or 15° instead of being rotated through 360°.

Figure 7.21 shows a typical single crystal camera, and a film obtained from it. Note the way in which the spots are arranged into layers, whose separation is a measure of the repeat distance along the axis parallel to the rotation axis. Measurement of the *d*-spacings corresponding to the various spots is most easily made with the aid of the Bernal chart for the diameter of the camera used [21]. Since these charts are devised in terms of a construction using a Mercator-type projection of cartesian reciprocal space co-ordinates, the *d*-spacing is quickly obtained by noting the values of the vertical ζ and the horizontal ξ co-ordinates on the chart and applying the formula

$$d_{hkl} = \frac{\lambda}{\sqrt{(\xi^2 + \zeta^2)}} \tag{7.22}$$

Simple single crystal photographs collect three dimensional data on a two dimensional film surface, and so a rotational photograph can provide a set of *d*-spacings for the various (*hkl*) planes, but it cannot give any information about how these are oriented relative to one another in space,

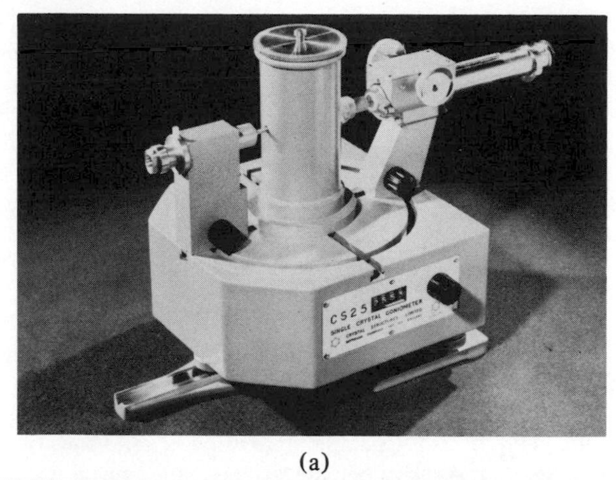

(a)

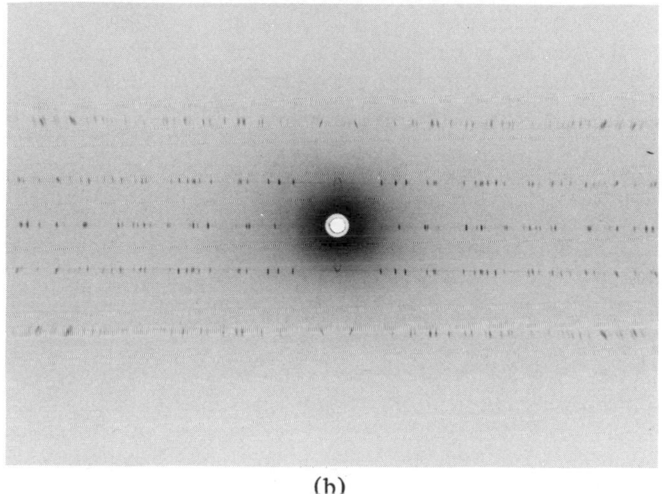

(b)

Figure 7.21 (a) The CS25, static film–moving crystal camera for single crystal studies. (Photograph by courtesy of Crystal Structures Ltd.) (b) A typical single crystal diffraction pattern obtained using the CS25 camera.

apart from indicating those with the same h-, k- or l-values. To obtain information about the relative angular positions of the spots within a layer we must add a constraint to the system and this can be done by restricting the range of angles through which the crystal rotates. By taking a photograph with the crystal oscillating through only 10°, only those reflexions will be found for which the reciprocal lattice point intersected the Ewald sphere in this region. A second photograph displaced by 10° from the first will provide a picture of the reciprocal lattice points in the next sector of the overall pattern, and if a series of such photographs is taken covering a total range of about 200°, sufficient data can be collected to define the unit cell, with a moderate degree of precision, by plotting the observed lattice points for each oscillation photograph within the angular range covered by the exposure. The resulting reciprocal lattice net constitutes a series of 10° arcs of circles from which one may recognise the principal axes and hence define the cell, and the method can be quite useful for orthogonal cells, or for monoclinic cells with β-angles not close to 90°. Better estimates of the cell shape and size can be made by using a 5° oscillation angle, which results in an uncertainty of only 5° in the position of the lattice point, or by taking successive pictures 1° apart and identifying where a given spot appears and disappears in the series of exposures. Techniques for the interpretation of oscillation photographs are discussed in detail by Bunn [119], both for parallel and tilted orientations of the crystal.

It is, of course, important to be able to orient the crystal with a principal axis parallel to the rotation axis of the camera, and this can be done by inspecting the oscillation photographs produced. Only when the crystal axis coincides with the rotation axis is a flat layer of spots produced, when they are at angle a curved layer is seen, the degree of curvature depending on the manner in which the crystal is tilted away from the rotation axis. Zussmann [11] has given a detailed description of the orienting of a crystal in a rotation camera.

7.9.2. Moving Film–Rotating Crystal Methods

Although there are some advantages to using oscillation photographs to define unit cells, in that the method uses inexpensive equipment and provides data about more than one layer in the reciprocal lattice simultaneously, the cell parameters so obtained are very approximate, and may be quite wrong if a poor orientation were chosen initially. Much better reciprocal cell data can be gathered using cameras in which the film is moved synchronously with the crystal, and only a limited part of the reciprocal cell data is gathered at any time. There are three principal varieties of camera in use.

The Weissenberg Goniometer

In the Weissenberg geometry, the crystal is mounted, as in the rotation camera, with a principal axis parallel to the rotation axis. The film is again held in a cylindrical cassette, but this time the cassette moves in a direction parallel to the rotation axis while the crystal rotates, and the movements of the crystal and the film are so coupled that the film moves through a distance of, normally, 1mm while the crystal rotates through an angle of 2°. In addition, a pair of screens are interposed between the crystal and the film to permit only one layer of the reciprocal lattice to reach the film at any time. By this means two-dimensional data can be collected on a two-dimensional film, and the spots in the same layer are angularly differentiated by their positions on the film. The resulting pattern of spots, which is shown in figure 7.22a, is a distorted representation of the plane of the reciprocal lattice perpendicular to the rotation axis.

The pattern can be interpreted in terms of a system of orthogonal cartesian co-ordinates either by plotting the spot positions using circular co-ordinates with the distance of the spot from the centre-line of the photograph, the line of the back-stop, as the radius and the distance along the film, parallel to the direction of translation, as the angular position from some arbitrary zero line, which may be the edge of the photograph. Alternatively, the special charts supplied by the Institute of Physics may be used to convert the spot positions into rectangular co-ordinates directly. Either method gives the reciprocal lattice directly and quite accurately.

The different layers can be photographed separately by tilting the rotation axis of the whole camera relative to the collimated X-ray beam by an amount which depends on the distance of the layer line in question from the zero layer, and moving the slits to compensate for this movement. Normally the camera is turned until the X-ray beam becomes a generator of the relevant cone of diffracted beams in figure 7.20, and the slits are moved to intercept the cone on the same side of the crystal. This is the 'anti-equinclination' setting, and it provides non-zero layer photographs to the same scale for each layer. The layers may also be photographed by leaving the rotation axis perpendicular to the X-ray beam, as in the zero-layer position, and moving the slits to the correct position, but the resulting photographs are on different scales for the different layers and so they cannot be compared by superposition as in the anti-equiinclination method. Henry, Lipson and Wooster [120] have described the method in some detail.

The Precession Camera

The geometry of a precession camera differs markedly from that of a rotation or a Weissenberg instrument. The crystal is mounted with a principal axis generally parallel to the X-ray beam and the film is held on a flat cassette mounted behind a screen with an annular slit cut in it. The crystal is then moved so that its axis precesses about the X-ray beam to describe a cone with a pre-determined semi-angle. The screen and film undergo a coupled, rather complex motion designed to permit only the desired diffracted beams to pass through the annular slit and simultaneously maintain the plane of the film perpendicular to the direction of the diffracted beams. The resulting photograph, which is shown in figure 7.22b, gives an undistorted representation of the reciprocal lattice, from which the cell dimensions can be calculated, knowing the distance from the crystal to the film and the various angles employed in the setting used.

Once again, the separate layers of the reciprocal lattice can be photographed separately, by varying the settings of the instrument, in particular the angle between the direction of the X-ray beam and the crystal to film vector, which is defined by the position of the annular slit. The

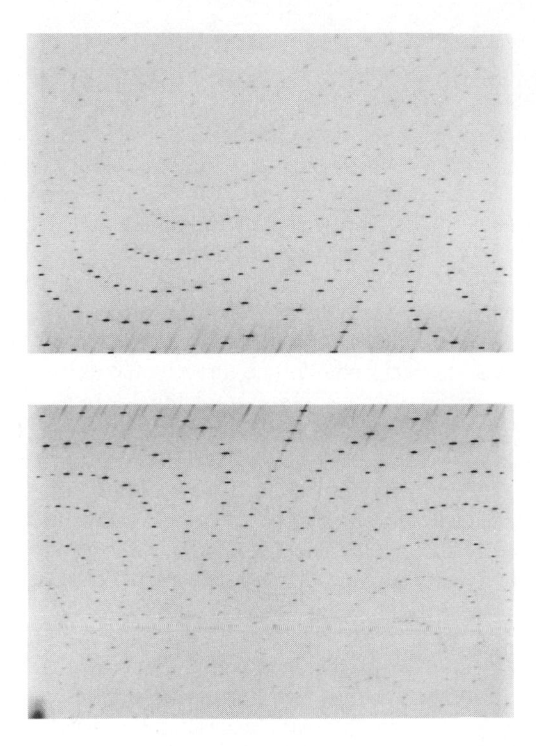

Figure 7.22 (a) A typical single crystal pattern obtained using the Weisenberg camera, and showing the $(h0l)$ layer of a monoclinic crystal. Note how the pattern is heavily distorted from its true, roughly orthogonal shape.

instrument has the further advantage over the Weissenberg goniometer that the magnification factor for the reciprocal lattice can be changed by varying the distance from the crystal to the film. Buerger [121] has discussed the theory and practice of the precession method in some detail.

The Rotation-Retigraph Method

This is a variation of the Weissenberg method, in that the crystal mounting is similar, but the reflexions are photographed on a flat film mounted generally perpendicular to the rotation axis of the camera. An annular screen again allows only the desired reflexions through to the film, and the various levels are chosen by varying the angle between the X-ray beam and the crystal rotation axis. In many ways this instrument was the predecessor of the precession camera, especially in that it was the first instrument to provide an undistorted picture of the reciprocal lattice.

After being largely superseded by the precession camera, however, the De Jong-Boumann method appears to be making a comeback in the Reciprocal Lattice Explorer, marketed by the Stoe Company [122]. In this instrument, the motions of the precession and the rotation-retigraph

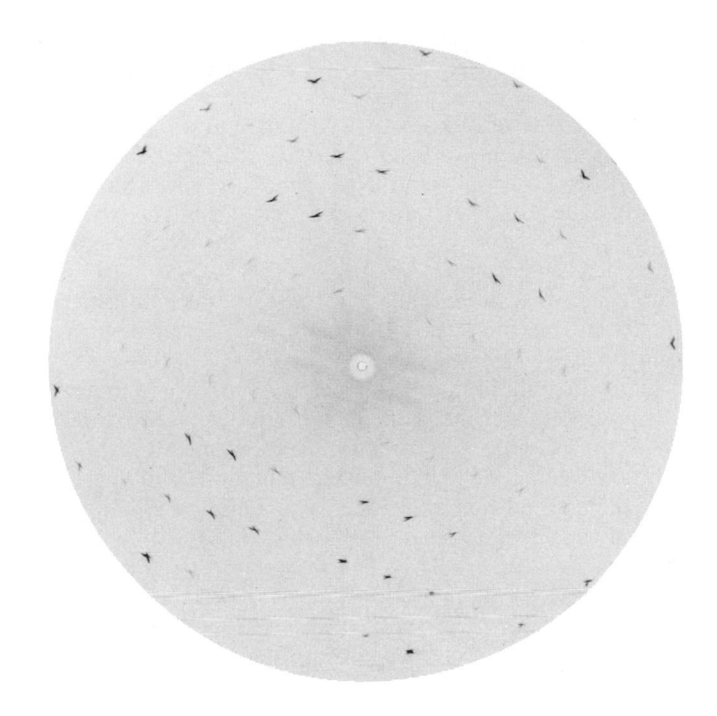

Figure 7.22 (b) A Buerger precession pattern from an orthorhombic crystal, showing the undistorted nature of the observed net.

methods have been combined to permit both types of photograph to be made without changing the setting of the crystal. The instrument permits reciprocal lattice sections to be prepared about all three axes in a very convenient manner.

7.9.3 High Temperature Modifications

Various designs of high temperature furnaces have been proposed for use with single crystal cameras. These initially took the form of small furnaces which fitted around the crystal in the simple rotation camera, and Lefkowitz and Megaw [123] and Rickson, Hall and McConnell [124] have described resistance-heating microfurnaces which permit photographs to be taken at high temperature in the Unicam S25 instrument. Such furnaces proved difficult to adapt to the Weissenberg goniometer, and heating systems were first based on blowing hot air over the crystal and monitoring the temperature with a thermocouple mounted close to the crystal. Such a system suffered from the dual disadvantages of overheating the crystal support system, especially the goniometer arcs, and of the limited temperature range that could be covered, typically up to about 300°C.

Recently, two designs have appeared which extend the temperature range considerably. In one, available from the Institute of Inorganic Chemistry, Slovak Academy of Sciences, Czechoslovakia, the crystal is mounted on a thermocouple which is used both as a heating and a temperature sensing device, and temperatures up to 1600°C are claimed. The other, marketed by LeMont Scientific, Lemont, Pennsylvania, mounts a hemispherical micro-furnace between the goniometer arcs and the crystal, which is mounted on a thermocouple at the centre of the furnace, and the whole is covered by a thin mica cone. Temperatures of up to 1000°C without temperature gradients are claimed. Water cooling is provided in both designs, to protect the goniometer arcs, and the whole of reciprocal space can be investigated in each case.

X-ray diffraction studies on single crystals at high temperatures can provide much useful information. Thus, more detailed studies of the anisotropy in, for example, the expansion coefficients of a material can be made than are possible with powder methods. But perhaps more importantly, the method makes it possible to study the reactions occurring in a material in considerable detail. On the one hand, the technique shows up phases which are stable only at high temperatures and sometimes these phases can be detected more easily than by powder methods, since the appearance of a new layer of spots, even weak spots, can often be seen more clearly than the presence of weak additional lines in a powder pattern. On the other hand, the so-called topotactic reactions, in which the product is oriented crystallographically relative to the original material, can be studied more easily and with more precision if all the necessary photographs can be taken without disturbing the mounting of the crystal.

7.9.4 Electronic Counter Diffractometry

Diffractometric methods can be applied to single crystal studies, but the use of diffractometers is limited to the determination of the intensities of reflexions for use in crystal structure analysis. Single crystal diffractometers are much more complex than their powder equivalents, although their principle is exactly the same. The difficulty in single crystal work is that the crystal must be correctly oriented in space relative to both the X-ray beam and the counter to permit the diffracted beam for only the required reflexion to enter the counter and this requires extra movements in the crystal mount to allow for this. Many instruments achieve the required motions by mounting the crystal head on a circle and arranging for the head to rotate on this circle and also rotate on its own axis at any orientation on the circle, while the circle itself can rotate about an axis coinciding with one of its diameters. Finally the counter head can rotate about an axis parallel to this last crystal-moving axis to record reflexions at different 2θ values.

Obviously, the angular motions required to bring all the parts of the instrument into the correct positions simultaneously and accurately are complex, and the instrument is normally under some form of computer control, either direct on-line control or via punched tape. Such instruments are expensive, and are not used in routine analytical laboratories. They are very specialised instruments and, as mentioned earlier, are used entirely for gathering data for full scale, high accuracy crystal structure determinations.

Although most instruments require the cell parameters of the material under study as part of the input data for the control systems, the most recent instruments can accept an unknown crystal and, by detecting a set of perhaps 20 random reflexions, can determine the unit cell of the material and proceed to collect the intensity data completely automatically. These techniques have arisen from advances in computer technology, especially in the development of the relevant programs to handle the situations. The prototype instrument produced by Philips is capable of this and the Enraf-Nonius Company have produced a diffractometer which incorporates this ability in a very compact design in which they have done away with the cumbersome 'Eulerian cradle' which normally carries the crystal head, as mentioned above, and replaced it by a series of four axial movements all controlled from the base of the mounting. This design has the great advantage of giving an uncluttered region about the crystal, and also of eliminating the inevitable dead areas incumbent on the presence of a large mass of metal around the actual crystal.

7.9.5 The Information Available from Single Crystals

As shown above, studies of single crystals can provide information relating to the shape and size of the unit cell of the crystal, information which is vital to the complete characterisation of any crystalline substance, and these data are readily collected and interpreted. However, a more detailed consideration of the data can provide yet more data. Thus, by identifying 'systematic absences' or types of reflexion which are consistently absent, the space group of the crystal can be identified and this can give some preliminary indication of the possible crystal structure. Thus, if it is found that all reflexions in the pattern from a monoclinic crystal for which $(h + k)$ is odd are absent, then we can say that the crystal is *C*-face centred, and if all reflexions belonging to the $(h0l)$ set for which l is odd are also absent, the crystal contains a *c*-glide, and the space group is monoclinic $C2/c$. This immediately tells us where groups of atoms must be located in the unit cell, with obvious advantages in structure determination work. Bunn [125] and Buerger [126] have discussed the determination of space groups in some detail, and Lonsdale [6] has collected all the space group absence criteria together for convenient reference.

If we further analyse the relative intensities, we can finally determine the complete crystal structure of the substance under study, by applying Fourier series summation methods to the 'structure factors' derived from the intensities. This procedure generates graphical representations of the electron density within the unit cell, as has been shown in section 7.5.5, and so permits us to identify both the positions of the atoms within the cell, from the regions of high electron density, and, in most cases, the chemical identity of the atom at each position, from the relative values of the density maxima. These studies are more difficult to carry out than are the simpler studies giving the unit cell and space group, since they require accurately determined relative intensities for all the reflexions. Various methods are used to obtain these and both film and counter techniques have their adherents. Photographic film data are usually collected on specially modified Weissenberg and precession cameras, in which the film is moved slightly relative to the diffracted beams between successive passes through a given set of reflexions. By this means, the intensity is 'integrated' over an area of the film and makes it possible to estimate the degree of blackening with a microdensitometer more accurately than with the sharp spots obtained in non-integrated photographs. The method is relatively cheap, the integrating attachments costing little more than the basic instrument in both cases, and the data obtained are reasonably accurate. Counter methods, as shown in section 7.9.4, can provide more accurate data and collect it automatically, but at a much higher cost. The subsequent mathematical processing is normally carried out on an electronic digital computer using one of the many program suites available. Mention may be made of the 'X-ray System, Version of 72' suite compiled by Stewart and co-workers [127] and of the suite available from Ahmed [128].

7.10 ELEMENTAL ANALYSIS USING LAUE PHOTOGRAPHS

The discussion of Laue diffraction given in section 7.5.1 shows, from equations (7.2), that if the angles of incidence and diffraction can be determined for a given spot in a Laue diffraction pattern, the wavelength of the X-rays producing that spot can be determined. Henriques and Milledge [129] have coupled this theory with the absorption characteristics of the elements (Chapter 3, section 3.2.2) to give a novel method of estimating the elemental composition of a substance.

The technique involves taking a crystal, the above workers prefer diamond because of the simplicity of its Laue pattern, and mounting it in an accurately known crystallographic orientation relative to the X-ray beam, so that the angle between any plane and the incident beam can be determined. They first obtain a Laue diffraction pattern from the crystal with a collimated beam of X-rays, and then a second pattern with the identical crystal orientation but with a thin sheet of the unknown sample inserted into the incident X-ray beam. Since different wavelengths in the beam will be absorbed to different extents depending on the chemical composition of the sample, the intensities of the various Laue spots obtained from the test crystal will be changed in the two photographs to different extents also. By measuring the intensities of a group of reflexions with and without the sample in the X-ray beam, it becomes possible to set up a series of simultaneous equations involving the absorptivities and concentrations of the elements present and the intensity ratios for the spots, at the wavelengths of the various reflexions, which can be solved to give the required concentrations. Henriques and Milledge claim that the method is fast, accurate and inexpensive, but point out that the crystal must be oriented accurately in order that the wavelengths relevant to the spots may be known with certainty. Figure 7.23 shows a typical multiple exposure set of Laue patterns from diamond using a beam filtered through (a) $(Mg,Fe)SiO_4$ powder, (b) Kimberlite garnet, and (c) with no absorber in the beam, and shows how the pattern of intensities changes for each absorber relative to the unabsorbed beam.

A microdensitometer is a necessity for measuring the spot intensities to the required degree of accuracy, and a qualitative knowledge of the composition, at least, is also needed. The development of the new automatic

ACCURATE WAVELENGTHS NECESSARY
STRAUMANIS MOUNTING
FOR ACCURATE ORIENTATION DETERMINATION

MG.FE.S1.04 POWDER 38.5 MINS
KIMBERLITE GARNET PLATE 0.2 MM 77 MINS
NO SPECIMEN 10 MINS

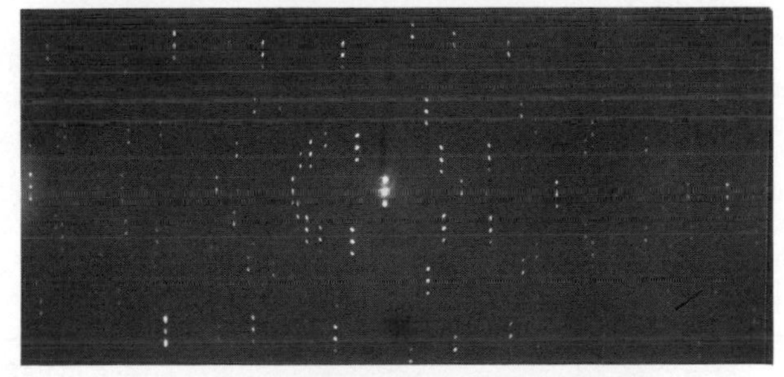

LAUE PHOTOGRAPHS OF DIAMOND
USED AS A RADIATION ANALYSER

Figure 7.23 Laue patterns from an oriented diamond crystal using beams filtered through $(Mg,Fe)SiO_4$ powder, Kimberlite garnet, and unfiltered radiation, for the determination of elemental compositions. Note how the relative intensities of different reflexions change with changes in the intensity spectrum of the incident beam, i.e. on the concentrations of the different elements in the absorbing medium. (Photograph by courtesy of Dr. J. Milledge and Mr. R. C. Hendriques.)

microdensitometer scanning facility at Harwell should ease this problem, however, and the method appears to be potentially very useful.

7.11 SUMMARY

X-ray diffraction is one of the oldest of the non-destructive methods of analysis but it is still very useful in all branches of solid state analysis. To the minerals analyst its most important ability is that of identifying crystalline phases rapidly and relatively simply. Its drawbacks lie in its inability to detect glassy phases and in the difficulty sometimes encountered with complex patterns, namely that of trying to assign the various reflexions to the different possible phases in the mixture. These difficulties, coupled with the fact that X-ray diffraction methods normally cannot give any direct indication of the elemental analysis of the sample, mean that the technique should be used in conjunction with other methods whenever possible. One should especially feed in chemical analytical data, to give the stoichiometry of the system, and optical microscopic findings, to give some indication of the presence of glassy phases or of the probable number of separate crystalline phases in the mixture.

REFERENCES

1. W. FRIEDRICH, P. KNIPPING and M. von LAUE, *Sitz. math.-phys. Klasse bayer Akad. Wiss. München,* 303 (1912).
2. M. von LAUE, *Ann. Physik.,* 971 (1913).
3. W. L. BRAGG, *Proc. Cambridge Phil. Soc.,* **17**, 43 (1913).
4. W. L. BRAGG, *Proc. Roy. Soc. London (A),* **88**, 43 (1913).
5. M. SIEGBAHN, 'Spektroskopie der Röntgenstrahlen', Julies Springer, Berlin, 2nd ed., 1931.
6. J. A. BEARDEN, *Phys. Rev.,* **48**, 385 (1935).
7. DEPARTMENT OF EMPLOYMENT AND PRODUCTIVITY, 'Code of Practice for the protection of persons exposed to ionising radiations in research and teaching', H.M.S.O., London, 1968.
8. Report of International Union of Crystallography, *Acta Cryst.,* **16**, 324 (1963).
9. H. P. KLUG and L. E. ALEXANDER, 'X-ray Diffraction Procedures', John Wiley and Sons, Inc., New York and London, 1954, Chapter 2.
10. K. NORRISH and B. W. CHAPPELL, 'Physical Methods in Determinative Mineralogy', J. Zussmann (ed), Academic Press, London and New York, 1967, Chapter 4.
11. J. ZUSSMAN, 'Physical Methods in Determinative Mineralogy', Academic Press, London and New York, 1967, Chapter 6.
12. G. W. BRINDLEY, 'X-ray Diffraction by Polycrystalline Materials', H. S. Peiser *et al.* (eds), Chapman and Hall, London, 1960, Chapter 4.
13. K. LONSDALE (ed), 'International Tables for X-ray Crystallography' Vols. I–III, International Union of Crystallography, Kynoch Press, Birmingham, 1962.

14. R. W. GOULD, S. R. BATES and C. J. SPARKS, *Appl. Spectr.*, **22**, 549 1968).
15. M. von LAUE, *Enzyklopadie der math. Wiss.*, **24**, 359 (1915).
16. N. F. M. HENRY, H. LIPSON and W. A. WOOSTER, 'The Interpretation of X-ray Diffraction Photographs', Macmillan, London, 1960, Chapter 6.
17. P. EWALD, *Zeits. Krist.*, **56**, 148 (1921).
18 M. J. BUERGER, 'X-ray Crystallography', John Wiley and Sons, Inc., New York and London, 1942, Chapters 6 and 7.
19. J. D. BERNAL, *Proc. Roy. Soc. London (A)*, **113**, 117 (1926).
20. R. C. EVANS and H. S. PEISER, *Proc. Phys. Soc.*, **54**, 457 (1942).
21. Copies, printed on transparent plastic sheet, are available from the Institute of Physics, 47 Belgrave Square, London.
22. N. F. M. HENRY, H. LIPSON and W. A. WOOSTER, 'The Interpretation of X-ray Diffraction Photographs', Macmillan, London, 1960, Chapter 3.
23. C. W. BUNN, 'Chemical Crystallography' Oxford, Clarendon Press, Oxford, 1961, p. 154.
24. A. H. COMPTON, *Phys. Rev.*, **21**, 715 (1923).
25. K. LONSDALE, (ed) 'International Tables for X-ray Crystallography' Vol. 1, International Union of Crystallography, Kynoch Press, Birmingham, 1962.
26. M. J. BUERGER, 'Crystal Structure Analysis', John Wiley and Sons, Inc., New York and London, 1960.
27. M. J. BUERGER, 'Vector Space and its Application in Crystal Structure Determination', John Wiley and Sons, Inc., New York and London, 1959.
28. G. RAMACHANDRAN (ed), 'Advanced Methods of Crystallography', Academic Press, London and New York, 1964.
29. H. LIPSON and A. J. C. WILSON, *J. Sci. Instrum.*, **18**, 144 (1941).
30. N. F. M. HENRY, H. LIPSON and W. A. WOOSTER, 'Interpretation of X-ray Diffraction Photographs', Macmillan, London and New York, 1960, 170ff.
31. A. E. van ARKEL, *Physica*, **6**, 64 (1926).
32. A. J. BRADLEY and A. H. JAY, *Proc. Phys. Soc.*, **44**, 563 (1932).
33. I. IEVINS and M. STRAUMANIS, *Z. Phys. Chem., (B)*, **33**, 165 (1936).
34. F. A. HILDEBRAND, *Amer. Min.*, **38**, 1050 (1953).
35. G. BROWN, G. C. DIBLEY and R. FARROW, *Clay Miner. Bull.*, **3**, 19 (1956).
36. H. S. PEISER, H. P. ROOKSBY and A. J. C. WILSON, 'X-ray Diffraction by Polycrystalline Materials', Chapman and Hall, London, 1960, 72ff.
37. J. E. SEARS and A. TURNER, *J. Sci. Instrum.*, **18**, 17 (1941).
38. ANON, *University Equipment*, September, 1972.
39. W. PARRISH and M. MACK, 'Data for X-ray Analysis', 2nd ed., Vols. 1–3, Philips Technical Library, Eindhoven, Netherlands, 1963.
40. A. TAYLOR, *J. Sci. Instrum.*, **26**, 201 (1951).
41. C. E. NOCKOLDS and R. H. KRETSINGER, *J. Phys. E.: Sci. Instrum.*, **3**, 842 (1970).
42. A. GUINIER, *Proc. Phys. Soc. (London)*, **57**, 310 (1945).
43. P. M. de WOLFF, *Acta Cryst.*, **1**, 206 (1948).

44. J. C. M. BRENTANO, *Proc. Phys. Soc. (London)*, **49**, 61 (1937).
45. R. LINDEMANN and A. TROST, *Z. Phys.*, **115**, 456 (1940).
46. H. FRIEDMAN, *Electronics*, **18**(4), 132 (1945).
47. D. P. le GALLEY, *Rev. Sci. Instrum.*, **6**, 279 (1935).
48. W. P. DAVEY, F. R. SMITH and S. W. HARDING, *Rev. Sci. Instrum.*, **15**, 37 (1944).
49. U. W. ARNDT, Ph.D. Dissertation, University of Cambridge, 1948.
50. W. SOLLER, *Phys. Rev.*, **27**, 158 (1924).
51. U. W. ARNDT, 'X-ray Diffraction by Polycrystalline Materials', H. S. Peiser *et al.* (eds), Chapman and Hall, London, 1960, Chapter 7.
52. H. W. KING, C. J. GILLHAM and F. G. HUGGINS, 'Advances in X-ray Analysis', **13**, 550 (1970).
53. H. P. KLUG and L. E. ALEXANDER, 'X-ray Diffraction Procedures', John Wiley and Sons, Inc., New York and London, 1954, p. 290ff.
54. G. L. McCREERY, *J. Amer. Ceram. Soc.*, **32**, 141 (1949).
55. H. P. KLUG and L. E. ALEXANDER, 'X-ray Diffraction Procedures', John Wiley and Sons, Inc., New York and London, 1954, p. 315ff.
56. F. CHAYES and W. S. MACKENZIE, *Amer. Miner*, **42**, 534 (1957).
57. A. J. C. WILSON, 'Mathematical Theory of X-ray Powder Diffractometry', Philips Technical Library, 1963.
58. W. PARRISH, J. M. TAYLOR and M. MACK, *Advances in X-ray Analysis*, 7, 66 (1964).
59. I. G. EDMUNDS, H. LIPSON and H. STEEPLE, 'X-ray Diffraction by Polycrystalline Materials', H. S. Peiser *et al.* (eds), Chapman and Hall, London, 1960, Chapter 15.
60. G. KETTMAN, *Z. Phys.*, **53**, 198 (1929).
61. P. SCHERRER, *Gottinger Nachrichten*, **2**, 98 (1918).
62. J. B. NELSON and D. P. RILEY, *Proc. Phys. Soc. (London)*, **57**, 160 (1945).
63. O. D. McMASTERS and W. L. LARSEN, U.S.A.E.C. Rept. IS-683, Office of Tech. Serv., U.S. Dept. of Commerce, Washington 25, D.C. (1963).
64. E. STRUM and W. LODDING, *Acta Cryst.*, **A24**, 650 (1968).
65. H. J. GOLDSCHMIDT, 'High-temperature X-ray Diffraction Techniques, Bibliography 1', I. U. Cr., Commission on Crystallographic Apparatus (1964).
66. J. W. EDWARDS, R. SPEISER and H. L. JOHNSTON, *Rev. Sci. Instrum.*, **20**, 343 (1949).
67. H. J. GOLDSCHMIDT and J. CUNNINGHAM, *J. Sci. Instrum.*, **27**, 177 (1950).
68. M. J. BUERGER, N. W. BUERGER and F. G. CHESLEY, *Amer. Miner.*, **28**, 285 (1943).
69. I. F. FERGUSON, U.K.A.E.A. Reactor Group Rept., TRG Rept. 2004(S) (1973).
70. H.-UDO LENNÉ, *Z. Krist.*, **116**, 316 (1961).
71. D. B. McWHAN, *Trans. Amer. Cryst. Assoc.*, **5**, 39 (1969).
72. J. C. JAMIESON and A. W. LAWSON, *J. Appl. Phys.*, **33**, 776 (1962).
73. G. J. PIERMARINI and C. E. WEIR, *J. Res. Nat. Bur. Stand.*, **66A**, 325 (1962).
74. J. C. JAMIESON and B. OLINGER, Presented at Conf. on Accurate Characterisation of High Pressure Environment, N.B.S., Gaithersburg, Md. (1968).

75. E. A. PERCY-ALBUERNE, K. FORSGREN and H. G. DRICKAMER, *Rev. Sci. Instrum.*, **35**, 29 (1964).
76. P. J. FREUD and C. B. SCLAR, *Rev. Sci. Instrum.*, **40**, 434 (1969).
77. C. E. WEIR, G. J. PIERMARINI and S. BLOCK, *Trans. Amer. Cryst. Assoc.*, **5**, 105 (1969).
78. J .D. BARNETT and H. T. HALL, *Rev. Sci. Instrum.*, **35**, 175, (1964).
79. J. D. BARNETT, J. PACK and H. T. HALL, *Trans. Amer. Cryst. Assoc.* **5**, 113 (1969).
80. J. S. KASPER, J. E. HILLIARD, J. W. CAHN and V. A. PHILLIPS, WADC Tech. Rept. No. 59–747, General Electric Coy., Schenectady, N.Y. (1960).
81. L. F. VERESCHAGIN, 'Physics of Solids at High Pressures', C. T. Tomizuku and R. M. Emrich (eds), Academic Press, London and New York, 1965.
82. H. O. A. MEYER, *Indust. Diamond Rev.*, **25**, 443 (1965).
83. B. L. DAVIA and L. H. ADAMS, *J. Phys. Chem. Solids,* **25**, 379 (1964).
84. P. J. FREUD and P. N. MORI, *Trans. Amer. Cryst. Assoc.*, **5**, 155 (1969).
85. V. VAND and G. G. JOHNSON, 'FORTRAN IV Programs (Version X) for the identification of multiphase powder diffraction patterns', ASTM, Philadelphia, 1969.
86. A. W. NICOL, *Nature,* **218**, 674 (1968).
87. A. W. NICOL, Res. and Dev. Rept., Publ. No. 68–18300/15–001, University of Birmingham (1970).
88. K. YVON, W. JEITSCHKO and E. PARTHE, Tech. Rept., Lab. for Res. on Structure of Matter, Univ. of Pennsylvania (1969).
89. O. D. McMASTERS and W. L. LARSEN, U.S.A.E.C. Rept. IS-839, Office of Tech. Serv., U.S. Dept. of Commerce, Washington 25, D.C. (1964).
90. T. ITO, *Nature,* **164**, 755 (1949).
91. B. DELAUNAY, *Z. Krist.*, **84**, 132 (1933).
92. J. W. VISSER, *J. Appl. Cryst.*, **2**, 85 (1969).
93. O. LINDQVIST and F. WENGELIN, *Ark. Kemi,* **28**, 179 (1967).
94. R. P. ELLIOTT, *Advances in X-ray Analysis,* **8**, 134 (1965).
95. G. KATZ, Ph.D. Thesis, Pennsylvania State University, 1965; 'Univ. Microfilms', Ann Arbor, Mich., Order No. 66 4821, 184pp.; Dissertation Abstr., **26**, 6806 (1966).
96. G. W. BRINDLEY and F. W. SPIERS, *Proc. Phys. Soc.*, **50**, 17 (1938).
97. A. W. HULL, *J. Amer. Chem. Soc.*, **41**, 1168 (1919).
98. G. L. CLARK and D. H. REYNOLDS, *Ind. Eng. Chem., Anal. Ed.*, **8**, 36 (1936).
99. L. ALEXANDER and H. P. KLUG, *Anal. Chem.*, **20**, 886 (1948).
100. G. W. BRINDLEY, *Phil. Mag.*, **35**, 638 (1944); **36**, 347 (1945).
101. H. P. KLUG and L. ALEXANDER, 'X-ray Diffraction Procedures', John Wiley and Sons, Inc., New York and London, 1954, Chapter 7.
102. T. BARRY, V. STUBICAN and R. ROY, *J. Amer. Ceram. Soc.*, **50**, 375 (1967).
103. J. J. VISSER, JCPDS Powder Defraction File, Search Manual Alphabetical Listing, Inorganic Compounds, SMA-73, p. 655, 1973.
104. N. F. M. HENRY, H. LIPSON and W. A. WOOSTER, 'The

Interpretation of X-ray Diffraction Photographs', Macmillan, London, 1960, Chapter 16.
105. M. von LAUE, *Z.Krist.*, **64**, 115 (1926).
106. H. P. KLUG and L. ALEXANDER, 'X-ray Diffraction Procedures', John Wiley and Sons, Inc., New York and London, 1954, Chapter 9.
107. Advances in X-ray Analysis, Procs. Annual Confs., Denver, 1957–72, Plenum Press, New York, Vols. 1–18.
108. A. R. STOKES, *Proc. Phys. Soc., (London)*, **61**, 382 (1948).
109. B. E. WARREN, *J. Appl. Phys.*, **12**, 375 (1941).
110. F. W. JONES, *Proc. Roy. Soc.*, **166A**, 16 (1938).
111. B. E. WARREN and B. L. AVERBACH, *J. Appl. Phys.*, **23**, 497 (1952).
112. B. E. WARREN, *Prog. in Metal Phys.*, **8**, 147 (1959).
113. R. L. ROTHMAN and J. B. COHEN, *Advances in X-ray Analysis*, **12**, 208 (1968).
114. C. P. GAZZARA, J. J. STIGLICH Jr., F. P. MEYER and A. M. HANSEN, *Advances in X-ray Analysis*, **12**, 257 (1968).
115. A. GUINIER and G. FOURNET, 'Small Angle Scattering of X-rays', John Wiley and Sons, Inc., New York and London, 1955.
116. V. GEROLD, 'Small Angle X-ray Scattering', H. Brumberger (ed), Gordon and Breach, New York, 1955, p. 277.
117. S. D. HARKNESS, R. W. GOULD and J. J. HREN, *Phil. Mag.*, **19**, 115 (1969).
118. R. JENKINS, D. J. HAAS and F. R. PAOLINI, *Norelco Reporter*, **18**, 12 (1972).
119. C. W. BUNN, 'Chemical Crystallography', Oxford, Clarendon Press, Oxford, 1961, Chapter 6.
120. N. F. M. HENRY, H. LIPSON and W. A. WOOSTER, 'The Interpretation of X-ray Diffraction Photographs', Macmillan, London, 1960, Chapter 7.
121. M. J. BUERGER, 'The Precession Method in X-ray Crystallography', John Wiley and Sons, Inc., New York and London, 1964.
122. E. R. WOLFEL, *J. Appl. Cryst.* **4**, 297 (1971).
123. I. LEFKOWITZ and H. MEGAW, *Acta Cryst.*, **16**, 453 (1963).
124. K. O. RICKSON, C. B. HALL and J. D. C. McCONNEL, *J. Sci. Instr.*, **40**, 420 (1963).
125. C. W. BUNN, 'Chemical Crystallography', Oxford, Clarendon Press, Oxford, 1961, Chapter 7.
126. M. J. BUERGER, 'X-ray Crystallography', John Wiley and Sons, Inc., New York and London, 1942, Chapter 22.
127. J. M. STEWART, G. J. KRUGER, H. L. AMMON, C. DICKINSON and F. R. HALL, Tech. Rept. TR192, Univ. Maryland Computer Sci. Center, College Park, Md., June 1972.
128. F. R. AHMED, 'Crystallographic Computing', Munksgaard, Copenhagen, 1970, p. 309.
129. R. HENDRIQUES, Ph.D. Thesis, London, 1972.

CHAPTER 8

Electron Microscopy

M. H. Loretto

Department of Physical Metallurgy and Science of Materials
University of Birmingham
Birmingham B15 2TT
England

8.1 INTRODUCTION

Considered in its simplest terms, the electron microscope is an extension of the transmitted light microscope in which the superior resolving power of a beam of electrons is used to investigate the shapes and sizes of crystals at a much finer level than is possible with the light microscope. However, the electron microscope is capable of more sophisticated investigation of crystalline materials than simply habit studies, important though these are, and it is the intention of this chapter to discuss some of these techniques and show how they may be applied to the investigation of minerals and related materials.

One of the most useful, and most used, capabilities of the electron microscope is the ability to form electron diffraction patterns from very small regions of a sample, typically about 1μm in diameter. Combination of electron diffraction evidence from a substance with morphological evidence from electron micrographs of the same region can be invaluable in relating, for example, growth habits to crystal structure features or in studying epitaxial or topotactical relationships between crystalline phases. The application of electron microscopy using these and related techniques has recently been very fully discussed in a monograph edited by Gard [1], and so the aim of this chapter will be rather to discuss some further techniques used widely in the metallurgical field, which might find wider applicability in the study of minerals also.

We may begin by considering the physical features common to most electron microscopes and the ray diagram for a typical instrument, so that the methods of obtaining an electron micrograph and a diffraction pattern may be appreciated before considering specific applications.

8.2 OPERATING PRINCIPLES OF AN ELECTRON MICROSCOPE

The ray diagram for a typical electron microscope with three electromagnetic lenses is shown diagrammatically in figure 8.1. The electron microscope comprises a column, maintained at a pressure of about 10^{-4} torr, which can be conveniently divided into four principal sections. At the top there is the electron gun, which includes a heated filament and grid at a potential of typically 100kV negative relative to earth, and a system of two electromagnetic lenses forming a double condenser system to provide a focused beam of electrons at the second section, which is the sample holder. The sample is held on a suitable support in the electron beam, and it is usual to provide translational movements in the plane of the sample, with added rotational movements about two mutually perpendicular axes lying in the plane of the sample being provided in many of the more modern instruments. Such movements permit stereoscopic-pair electron micrographs to be taken, and also allow the operator a great deal of control over the exact orientation of the sample in electron diffraction work. After passing through the sample, the electrons enter the enlargement section, which usually comprises at least three electromagnetic lenses to focus and enlarge the images generated by transmission of the electron beam through the sample. Finally, the beam enters the detection area, which normally comprises a fluorescent screen for direct viewing and a camera for taking a permanent record. The associated electronics provide very smooth ripple-free currents to the electron gun and the various lenses, to ensure stable operating conditions throughout the instrument.

Reference to figure 8.1 shows all the above sections diagrammatically. The electrons produced by the heated filament are accelerated by the negative voltage on the surrounding grid, and the double condenser system produces a beam of electrons at the sample. These electrons are both scattered by and transmitted directly through the sample, and the objective lens both brings them to a focus in its back focal plane, and also forms an enlarged inverted image at I_1. If the next lens, the intermediate lens, is focused on the plane of this first image at I_1, a second intermediate image is formed at I_2 and this image can be further enlarged by the final lens, the projector lens, to give an image that can be photographed or viewed on the fluorescent screen. This image is the electron micrograph, and shows a small

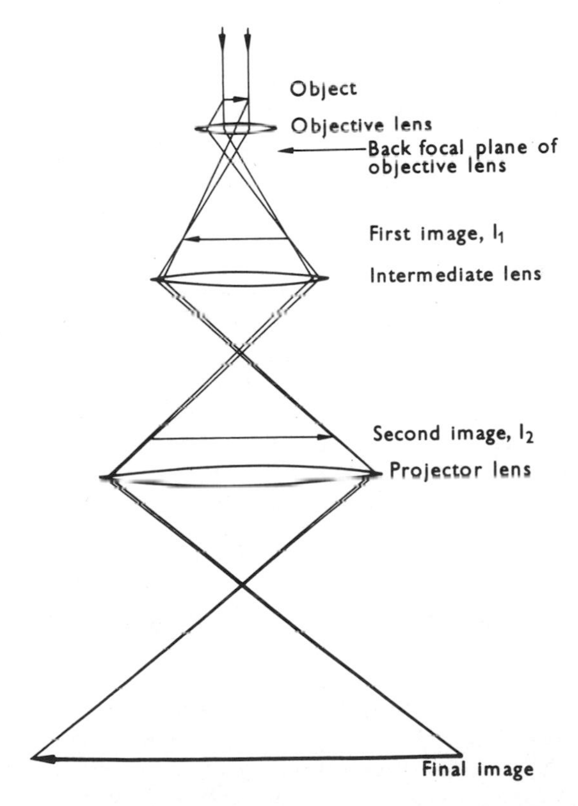

Figure 8.1 Ray diagram for an electron microscope. For further details see text.

part of the sample enlarged typically 2×10^4. If the image is photographed, the final magnification can be further increased by simple optical magnification of the negative, although no further improvement in the resolution will be obtained at this stage. Most modern electron microscopes will give resolutions of 10Å quite readily, and 2.5 to 3Å can be obtained with careful sample preparation and with accurate alignment of the electron optics.

If the intermediate lens is focused onto the back focal plane of the objective lens, however, by reducing the current to the intermediate lens and so increasing its focal length, this lens will then be focused on the diffraction pattern formed in this plane. This diffraction pattern is enlarged and projected onto the viewing screen, where it may again be photographed for comparison with the micrograph from the same region. Electron diffraction patterns will be discussed in more detail in section 8.4.

8.3 SAMPLE PREPARATION AND USE

The sample used in electron microscopy consists either of a thin section of the material under study, or a fine powder supported on a thin plastic or carbon film. The sample should be typically about 0.5μm thick, for conventional electron microscopes operating at voltages up to 200kV, but the exact thickness through which the electron beam can penetrate depends on the atomic weights of the elements comprising the sample, and will be thicker for low atomic weight materials. In both cases, the thin sample is supported, typically on a 2.3 or 3.05mm-diameter suitable grid, which is mounted in the sample holder of the microscope itself.

Studies of the surface morphologies of massive samples may be made by replication techniques involving the deposition and subsequent removal of thin carbon or plastic films onto the surface, so that they take up the shape of the surface, and investigation of these films. Electron diffraction studies cannot be made on these replicas, of course, and scanning electron microscopy (cf. Chapter 11) may provide better surface data, provided that the somewhat lower resolution and magnification levels are acceptable. Finally, in all cases where the sample is a non-conductor, the surface must be coated with a thin layer of a conducting medium. Carbon is usually used, a layer about 50Å thick being deposited by evaporation and condensation under vacuum, but other materials may be used in special circumstances.

It is necessary to operate the electron microscope under a vacuum of better than about 10^{-4} torr, to reduce scattering and absorption of the electrons by gas molecules, and these vacuum conditions can present problems, particularly in studying hydrated minerals. Most hydrates, including the majority of hydrated minerals, are quite stable under these vacuum conditions at temperatures close to room temperature but unfortunately the electron beam can have an appreciable heating effect on the samples and temperatures in excess of 500° C may be reached. Under these conditions many minerals will lose water and care must be exercised in studying them. In particular, low beam currents should be used to minimise heating and good thermal conduction to an adequate heat sink should be provided. The A.E.I. Million volt microscope has been modified so that it can accept an environmental stage and it is therefore possible to examine specimens which are in a wet atmosphere. This may well be a major field of application for very high voltage electron microscopy in the study of minerals.

The need to coat non-conducting specimens can be turned to good use, in the technique of shadowcasting, to emphasise surface features on the specimen itself or on a replica. A heavy metal, often platinum, is used instead of carbon and is evaporated onto the surface at a known low angle of incidence. Low lying areas of the surface, which lie in the shadow of higher regions with respect to the incoming atoms, are not coated and so appear dark on the negative of the resulting photomicrograph, since they transmit more electrons than the regions covered by the heavier platinum atoms. The lengths of the shadows can be used to measure the heights of the associated projections, for example cleavage step heights may be measured in this way.

Early electron microscopes held the specimen with its plane normal to the electron beam, but modern instruments permit the sample holder to be tilted through angles of up to 60° about an axis or axes in the plane of the sample (and also, less usefully, to rotate the sample through 360° in the plane of the sample) to permit studies to be made with the sample at different angles to the incident electron beam. Stereomicrographs may be made using this facility, and electron diffraction patterns may be obtained from various reciprocal lattice zones to help in the deduction of unit cell parameters and in the study of order-disorder phenomena.

Apart from the unwanted heating effects arising from the interaction of the electron beam with the sample, studies may be made at various temperatures using special stages. Temperatures up to 1200°C and down to 200°C are quite easily achieved and Cartz and Tooper [3] have studied the dehydration of phlogopite at temperatures up to 625° using such a stage. In addition, other specialised stages are available, for example to carry out strain studies in situ.

8.4 SELECTED AREA ELECTRON DIFFRACTION

De Broglie [4] first postulated the wave nature of the electron when he showed that an electron of mass m moving with a constant velocity v should be associated with a waveform of wavelength λ, given by

$$\lambda = hp^{-1} = h(mv)^{-1} \tag{8.1}$$

where h is Planck's constant. For an electron accelerated through a potential of the order of 100Kv, the wavelength in angstroms is

$$\lambda = 12.27\, V^{½}(1 + 0.978 \times 10^{-6}\, V)^{-½} \tag{8.2}$$

where V is the accelerating potential in volts. Davisson and Germer [5] and Thomson and Reid [6] have confirmed this wave nature, and Lande [7, 8] has discussed the interpretation of diffraction patterns.

Since electrons possess a wave nature they will be diffracted by crystalline materials in a manner similar to X-rays, although by a somewhat different mechanism. The very much shorter wavelength associated with electrons, about 0.037Å for 100kV electrons, coupled with the ability to focus the beam and obtain data from areas less than 1μm in diameter, however, make electron diffraction a much finer tool for studying small crystals than are X-rays. In particular, the very low Bragg angles at which diffraction occurs, the 2θ angle being less than 5° for d-spacings greater than 0.5Å, make it possible to obtain good undistorted representations of the reciprocal lattice directly on a flat plate camera without the need for the complicated mechanisms, such as in the precession camera, required with X-rays.

The mechanism whereby electrons are diffracted by a crystalline solid differs from that of X-rays, however, due mainly to the higher energy of the electrons. Whereas X-rays are principally diffracted through interaction with the electron clouds of the atoms present, the electrons tend to interact with the nuclei of the atoms as well as with the electron clouds. Electrons are generally more strongly scattered than are X-rays, and this can give rise to added complications due to the rediffraction of already diffracted electron beams, as will be shown in section 8.4.3. For this reason, the intensities of electron diffraction reflexions are not usually used for structure determination, because of the inevitable uncertainty arising from this 'double diffraction' in all but the thinnest crystals, although Sturkey [9] has shown that the intensities obtained from polycrystalline specimens can be used when X-ray intensities are unavailable, for example in work on clay minerals.

The actual size of the image of the diffraction pattern obtained in the back focal plane of the objective lens is very small and if this were to be projected *in toto* onto the viewing screen, the resulting pattern would contain contributions from a large area of the specimen and so would be difficult to interpret. It is usual to restrict the area of the specimen which can contribute to the pattern by inserting a 'selected area aperture', usually an earthed plate with an aperture of about 50μm, in the plane of the first image, at I_1 in figure 8.1. The area from which the diffraction pattern is obtained is thus effectively limited to a radius of about 1μm diameter around the primary axis. Hence the name 'Selected Area Electron Diffraction', often referred to by its initials SAED or SAD. If larger apertures are used, composite patterns from many crystals are obtained, which form powder rings if the material is a polycrystalline sample with truly random orientation. Such 'General Area Electron Diffraction' studies are of particular use in showing up texture and preferred orientations, for example in platy minerals such as clays. Vainshtein [10] and Zvyagin [11] have discussed such applications in some detail.

Rotation of the Image during Focusing

Electrons being brought to a focus by an electromagnetic lens follow a spiral path through the magnetic field set up between the pole pieces of the lens, the number of turns undergone by an electron being determined by the strength of the lens. Thus, when the strength of the intermediate lens is altered to focus on the diffraction pattern at the back focal plane of the objective, the resulting image produced by the intermediate lens will be rotated through some angle in the plane of the final image relative to the corresponding micrograph. Moreover, since the micrograph image has undergone one inversion before being focused by the intermediate lens, whereas the diffraction pattern has not been so inverted, the projected pattern on the screen may have undergone one fewer inversion than the micrograph image. Hence, when comparing a micrograph and its diffraction pattern, it is important to make allowance for this angular rotation, and it may also be necessary to allow for the 180° rotation due to the unequal inversions.

In practice, the angular rotation, which varies with the degree of magnification used, may be determined by using a crystal with a well developed, known morphology, for example molybdenum trioxide which has long straight edges perpendicular to the [100] direction, and doubly exposing a plate to superimpose the micrograph and its diffraction pattern. The angular rotation is simply given by the angle between the normal to the crystal edge and the [100] direction in the diffraction pattern.

As far as the inversion is concerned, some microscopes have an additional lens between the objective and the intermediate lens, and in some cases one lens may be turned off when obtaining the diffraction pattern, so that the diffraction pattern and the micrograph undergo the same number of inversions. The maker's handbook should always be consulted for the various modes of operation possible with any given instrument.

This facility of 'selected area electron diffraction' is fundamental to the success of transmission electron microscopy, since it makes it possible to obtain both a visual image and a diffraction pattern from the same small volume of the specimen, typically 1μm in diameter. In order to extract the maximum information from the technique, however, it is necessary

(a) to be aware of the instrumental errors inherent in selecting the area

(b) to be able to solve the electron diffraction patterns obtained

(c) to be aware of the complications arising from the large scattering of the electrons

(d) to appreciate the influence of twins and of precipitates in producing extra diffraction maxima

(e) to make use of Kikuchi lines present in diffraction patterns, and hence make use of Kikuchi maps

(f) to be able, in the study of crystal defects, to appreciate the importance of the many physical parameters which appear in the two-beam dynamical equations, and to be able to obtain the experimentally determinable parameters, such as crystal thickness and habit planes of precipitates or stacking faults.

The following sections will briefly cover each of the above topics, but no special emphasis is placed on minerals simply because the principles are applicable to crystals in general.

8.4.1 Errors in Selected Area Electron Diffraction

As shown in the previous section, in order to obtain a 'selected area' diffraction pattern from a specimen an aperture is inserted in the first image plane to define the 'selected area'. The power of the diffraction lens is then reduced so that it is focused onto the back focal plane of the objective lens, i.e. on the diffraction pattern, and the image formed by the diffraction lens is finally enlarged onto the viewing screen by the projector lens. Since the selector aperture is always inserted into the same plane in the microscope column then, to ensure that only the area defined by this aperture contributes to the diffraction pattern, it is necessary to ensure that the first image, I_1, is actually formed in the plane of the aperture.

Experimentally this is achieved by focusing the image of the selector aperture onto the viewing screen with the diffraction lens only, an operation most easily carried out with the objective aperture removed from its position, and then, with the objective aperture replaced, focusing a specimen in the stage also onto the screen using the objective lens only. This procedure ensures that, within focusing error, the first image and the selector aperture are coplanar. It should be noted that this procedure, which is essential in a microscope with only two lenses after the objective lens, means that a 'selected area' diffraction pattern can be obtained only at one setting of the diffraction lens.

Even after observing the above precautions there are still two factors which will cause diffracted rays, originating from outside the selected area, to contribute to the observed diffraction pattern. These are

(a) the unavoidable uncertainty in the focusing of the specimen and the selector aperture

(b) the unavoidable spherical aberration associated with the objective lens.

Let us consider the importance of each of these in turn.

Focusing Errors

In figure 8.2 the beam marked *hkl*, travelling at an angle α to the direct beam, will actually come from the area BB' of the specimen, the sense of the error from the selected area being in the sense of the diffracting vector *hkl*. This shift, x, is simply given by

$$x = d\alpha \tag{8.3}$$

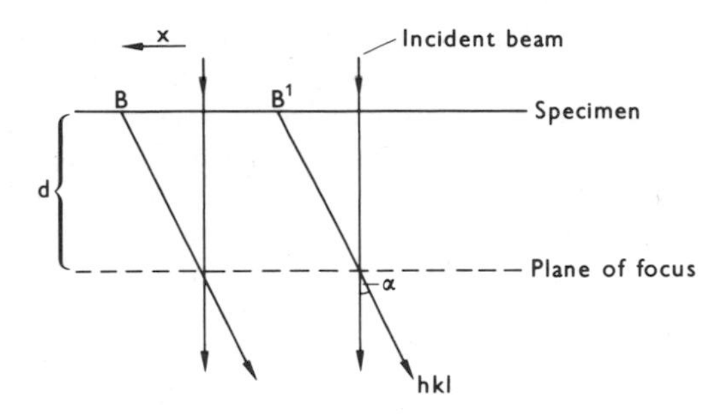

Figure 8.2 Ray diagram illustrating the origin of errors in selected area diffraction due to overfocusing of the objective.

where d is the focusing error. Taking α to be equal to 0.02 radians, a typical value for a low order hkl, reflexion, and d to be of the order of 10μm, the level found experimentally for thick specimens, x has a value of about 0.2μm.

Spherical Aberration Errors

The spherical aberration exhibited by electromagnetic lenses always leads to off-axis rays being brought to a focus closer to the lens than are near-axis rays, as shown diagrammatically in figure 8.3. Therefore, even if the first image were focused accurately at the plane of the selector aperture, and hence the image of the directly transmitted beam from area AA' in the specimen corresponded exactly to the area of the selector aperture, the beam hkl, which is at an angle α to the axial direction, would be displaced horizontally by an amount $Mc\alpha^3$ where M is the magnification and c is the spherical aberration coefficient. This corresponds to a displacement of $c\alpha^3$, between the areas in the specimen from which the directly transmitted and the diffracted beams originate. Typically, c = 3mm and so with α = 0.02, for

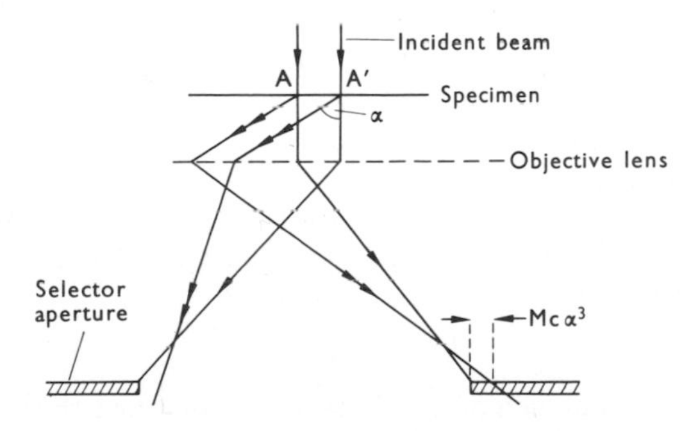

Figure 8.3 Ray diagram illustrating the origin of errors in selected area diffraction due to spherical aberration in the objective lens.

a small angle diffracting vector, the error is about 0.025μm. The error varies as α^3, however, and soon approaches 1-2μm for higher index reflexions.

The above errors can be additive and if correlations are made between electron micrographs and their corresponding SAD patterns it should be borne in mind that displacements of up to 2–5μm from the selected area are very likely. This displacement is quite easy to demonstrate experimentally by using a small selector aperture and just excluding the specimen from the selected area before reducing the power of the intermediate lens to obtain a diffraction pattern. In the absence of the above two effects only the direct beam will be seen but, in general, small changes in the objective current, corresponding to under- and over-focusing, will cause a diffraction pattern to appear and disappear.

8.4.2 Interpretation of Electron Diffraction Patterns

As indicated previously, the electron diffraction pattern obtained from a given specimen may be produced by a single crystal in the material, the most common case in SAD work, or it may be produced by a random polycrystalline aggregate, especially if the selected area is large, or it may be produced by a polycrystalline aggregate showing preferred orientation. The pattern from a single crystal comprises a series of regularly spaced spots corresponding to a section of the reciprocal lattice of the crystal, the pattern from a random polycrystalline aggregate comprises a series of circular rings similar to an X-ray powder pattern, and the pattern from a polycrystalline aggregate with preferred orientation comprises a series of rings with superimposed intensity maxima corresponding to the 'texture' of the specimen.

Powder Diffraction Patterns

Since the geometry of the diffraction of electrons by a crystalline lattice is the same as that of X-rays, despite the differences in mechanism, the positions of the maxima for the cones of diffracted radiation from a random powder are again given by the Bragg equation [equation (7.3)]

$$\lambda = 2d\sin\theta$$

The very small values of λ encountered with electrons, however, restrict 2θ to a maximum value of $\sim 5°$ and so it is permissible to make the approximation that $\sin\theta = \theta$ in radians. Now, if the diameter of the powder ring measured on the photographic plate is $2R$, and L is the effective camera length or the effective distance from specimen to camera after allowing for the magnification M, simple geometry shows that

$$R = L\tan 2\theta = L.2\theta$$

for 2θ small. Combination of the above approximations shows that

$$d = \frac{\lambda L}{R} \tag{8.4}$$

The factor λL is known as the 'camera constant' and is determined experimentally by using a standard material such as gold or aluminium or thallous chloride to calibrate the instrument. Provided that this constant is measured accurately on a number of lines of the standard pattern, *d*-spacings correct to about 1% of their absolute values may be obtained. (*Note:* Gard [12] recommends the use of an internal standard, such as gold evaporated onto the specimen, for high accuracy work.)

The set of *d*-spacings and intensities can again be used in conjunction with the JCPDS File and Indexes to identify crystalline phases either separately or as mixtures, as discussed in Chapter 7, section 7.7.1, but it is necessary to use the Fink index rather than the Hannawalt index in the initial stages. This arises from the different mode of diffraction of electrons already mentioned in section 8.4, which results in different relative intensities from those obtained with X-rays, although the *d*-spacing values are not affected. The three strongest lines in an X-ray diffraction pattern are not usually the three strongest lines in the electron diffraction set. Moreover, the high magnifications used in the electron microscope tend to favour identification of the longest spacings in comparison with X-ray methods.

Simple powder electron diffraction data can be used to provide a certain amount of structural information in the case of simple structures, mainly belonging to the cubic class, but the amount of information is severely restricted in the more complicated structures associated with most minerals. Textured patterns can provide further data both about the relative preferred orientations in the sample and about the atomic structure in some cases [10, 11].

Single Crystal Patterns

The very short wavelengths associated with the electrons used in electron diffraction make the Ewald sphere and the reciprocal lattice construction of particular use in interpreting single crystal electron diffraction patterns. A brief introduction to the theory of reciprocal space has been given in Chapter 7, section 7.5.3, and a formal description of reciprocal space and of some reciprocal lattice relationships is given in an appendix to this chapter. Of particular importance in the application of Ewald's construction is the fact that the very small wavelengths involved make the Ewald sphere radius, which is here taken to be equal to $1/\lambda$, very large, so large in fact that the section of the Ewald sphere involved in generating the electron diffraction pattern is effectively planar and so its intersection with the reciprocal lattice occurs virtually in a single plane without the marked curvature encountered with X-rays. The magnified image obtained on the photographic plate is thus a virtually undistorted projection of the reciprocal lattice as it appears to the electron beam. Figure 8.4 shows schematically the Ewald sphere construction used in generating the diffracted beams and figure 8.5a shows a typical electron diffraction pattern from the hydrated calcium silicate mineral kilchoanite [13].

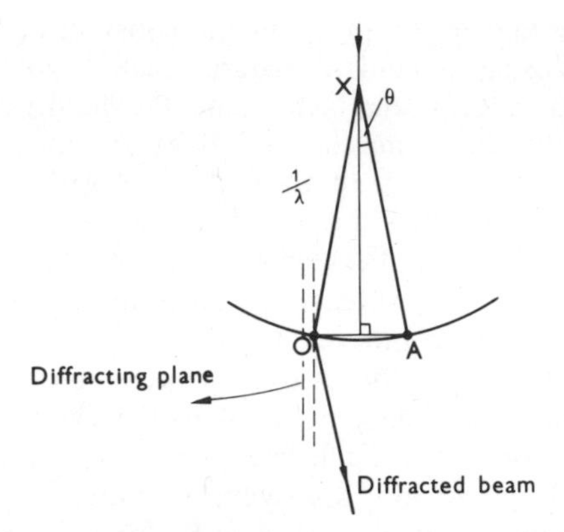

Figure 8.4 Detail of Ewald sphere construction for determining the direction of diffracted rays in selected area electron diffraction.

The appearance of the diffraction pattern from a single crystal depends on the orientation of the crystal relative to the electron beam. In figure 8.5a the beam lies almost parallel to a principal reciprocal axis and so the pattern shows clearly the orthorhombic nature of the reciprocal lattice of kilchoanite, from the undistorted equatorial lattice plane spots. Note that the reciprocal cell parameters, and hence the real cell axial parameters, can be estimated by measuring the distances between pairs of rows on either side of the central rows in the two principal directions and substituting these for R in equation (8.2). λL must again be measured by calibration, either with an external standard or, better, by plating the specimen with an internal standard.

In general, however, a crystal will not lie with an axis parallel to the electron beam, although platy crystals or crystals with marked cleavages or well developed basal planes do tend to lie with their normal axes closely parallel to the beam. Figure 8.6 shows two more general cases, from which it can be seen that the reciprocal lattice section is usually less clearly defined. Such a pattern could be used either to determine the orientation of the crystal relatively to the electron beam, if the cell parameters of the crystal are known, or to estimate the reciprocal lattice spacing along an axis lying nearly parallel to the beam.

Figure 8.6a is an electron diffraction pattern from a body centred cubic single crystal and the reciprocal lattice section to which this corresponds may be obtained in the following way. The two spots A and B (which have been chosen because they are the two smallest spacings in the pattern) are at distances R_1 and R_2 from the origin O of the reciprocal net. These distances can easily be converted into d-spacings using equation (8.2) and the indices of the spots identified. They correspond to a 110 and a 211 vector in the cubic lattice, respectively. Now the angle between the directions OA and OB corresponds to the angle between the respective (110) and (211) planes

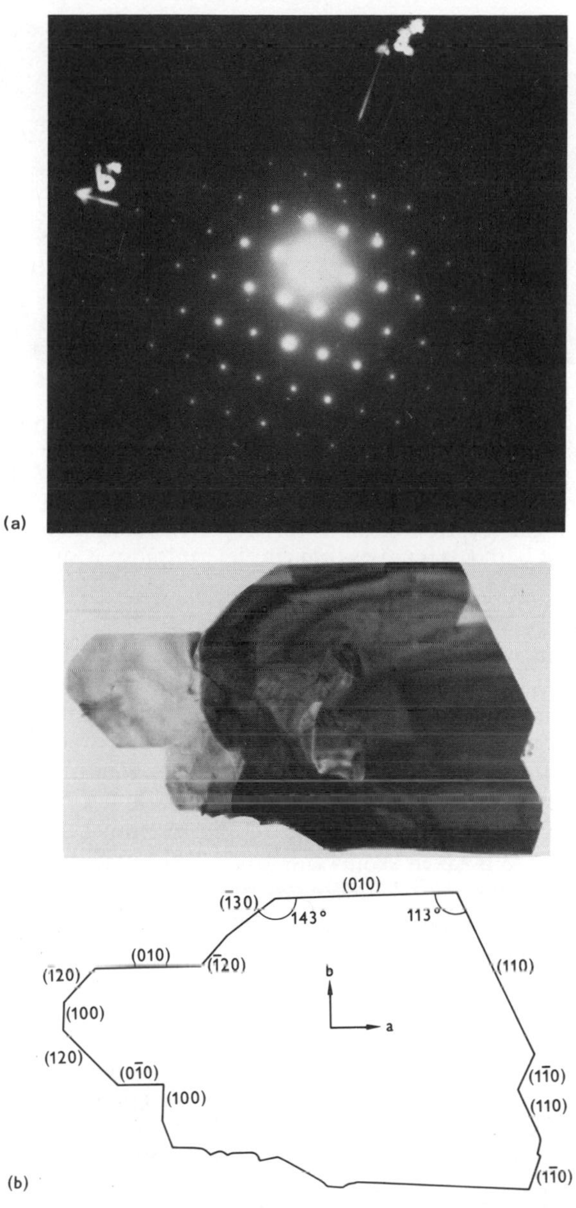

Figure 8.5(a) Selected area diffraction pattern obtained from the orthorhombic mineral kilchoanite, taken with the electron beam parallel to the *c*-axis. The space group is *Ibmm* and requires that reflexions with $(h + k + l)$ odd be systematically absent. (b) Electron micrograph of the crystal giving rise to the electron diffraction pattern shown in figure 8.5a with a tracing showing the indices of the various edges of the crystal. (Courtesy of Dr. A. W. Nicol.)

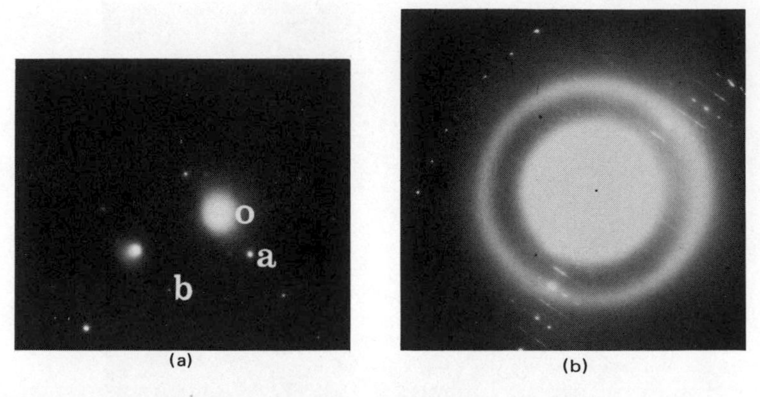

Figure 8.6(a) Portion of electron diffraction pattern from body centred cubic crystal showing two reflexions, a and b, from which the orientation of the crystal relative to the electron beam can be derived. (b) Portion of pattern from a synthetic wollastonite taken with a principal axis (the b-axis) at an angle to the electron beam. Note the curved zones and the manner in which the zone with $k = 0$ passes through the central spot.

which the spots represent. The measured angle is 73° and interplanar angles for the cubic class show that the actual planes can be (110) and ($\bar{1}$21). Thus the reciprocal section in this pattern is (1$\bar{1}$3) and, because in the cubic class the reciprocal and the real axes are colinear, the electron beam direction, referred to the crystal axes, is parallel to (1$\bar{1}$3). Where the sense of the electron beam is important, as in Kikuchi line work, it is necessary to use a convention to describe the 'direction' of the electron beam. In the above example the sense is taken as upwards from the crystal, so that the upward drawn normal to the crystal surface is acute to the electron beam direction. This convention will be used throughout this chapter.

Figure 8.6b shows the pattern obtained from a crystal whose unit cell is unknown. As can be seen, it comprises zones which represent sections of layers of the reciprocal lattice with, for example, different k values. Gard [12] has shown that the zone passing through the undeviated beam position is always the zero-order zone and that the other zones in the suite, the so-called Laue zones, are circular with their common centre at the point where the b-axis intersects the plane of the photographic plate. The radii of the zones will depend on the angle ϕ between the b-axis and the electron beam and also on the camera constant, according to the equation

$$b^* = 2L(\lambda L)l/(r_0 + r_l)(r_0 + r_l) \tag{8.5}$$

where r_0 and r_l are the radii of the zero- and the l-order zones respectively. He has further shown that a good estimate of b_0 can be obtained by taking a series of diffraction patterns using the tilting stage to vary the angle between the *b*-axis and the electron beam from nearly zero, the case illustrated in figure 8.5a, to quite high angles.

It is this ability to generate a reasonably accurate unit cell for an unknown material using SAED methods that is one of the major strengths of the technique. Whereas X-ray methods require crystals of the order of 1mm size, SAED can use even the smallest crystal to give its estimate. Accurate cell parameters can then be derived using the rather more accurate *d*-spacing parameters available from an X-ray powder diffraction photograph, as described in Chapter 7, section 7.7.2. Several unit cells have been derived by this means, among them those of foshagite [14] and synthetic $FeAlO_3$ [15], and the method may show up unexpected modifications of known minerals, as in Gard's studies on xonotlite [16] and on kaolinite [12].

Single crystal diffraction patterns may provide data other than simple unit cell parameters, however. Thus, figure 8.5b shows the micrograph of the crystal from which the diffraction pattern in figure 8.5a was obtained and gives the indices of the edges of the crystal based on the diffraction evidence. In principle, the data in the pattern could be used to determine the space group of the unknown crystal, from the systematic absences in the pattern, but care must be taken when attempting to extract this information from the observed spots.

8.4.3 Double Diffraction

Electrons are so much more strongly scattered than are X-rays that it is important to take note of an effect in electron diffraction that has no real parallel in X-ray diffraction. This is the phenomenon of 'double diffraction' in which a beam of electrons which has already undergone diffraction from a crystallographic plane undergoes rediffraction at another plane in the same crystal. Figure 8.7 shows diagrammatically how this may happen. The reciprocal lattice point with vector $\mathbf{g}_1$ (see Appendix), corresponding to the planes $h_1k_1l_1$, lies on the Ewald sphere and an intense diffracted beam of electrons is produced along the vector B_1. The corresponding reciprocal lattice point for the planes $h_2k_2l_2$, with vector $\mathbf{g}_2$, may also intersect the Ewald sphere and produce a beam along the vector B_2. If now a third lattice point simultaneously intersects the Ewald sphere with a vector $\mathbf{g}_2 - \mathbf{g}_1$, the diffracted beam along B_3, which corresponds to rediffraction of B_1 by the planes $h_3k_3l_3$ where

$$h_2 = h_3 - h_1$$
$$h_2 = k_3 - k_1$$
$$l_2 = l_3 - l_1$$

will be indistinguishable from the B_2 beam in direction and so will contribute to the intensity of the reflexion apparently arising only from the $h_2k_2l_2$ planes, with an accompanying reduction in the observed intensity of the $h_1k_1l_1$ spot. Three principal cases may be considered.

(a) Diffraction may occur from a permitted beam into another permitted

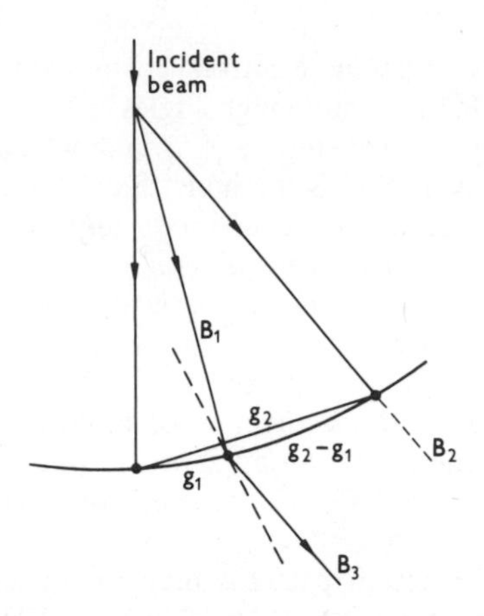

Figure 8.7 Ewald sphere construction illustrating the phenomenon of double diffraction. The beam B_3 has been generated by double diffraction of the already diffracted beam B_1 and is parallel to, and contributes to the intensity of, the beam B_2.

beam. This occurs for all crystals more than a few atoms thick and, since the only effect is to transfer intensity between the various beams, is almost undetectable in practice. Its existence makes it impossible to rely on the observed relative intensities of the spots to provide information on the relative structure factor values for the various crystalline planes, in the way that one can with X-ray intensities, and lies at the basis of all the problems associated with the derivation of crystal structures from electron diffraction patterns.

(b) Diffraction may occur from a permitted beam into a beam whose intensity is zero from space group requirements. In this case a spurious reflexion, not permitted by the space group, will appear in the pattern and may cast doubt on the validity of the space group assignment. The effect occurs only when the relevant pairs of planes, $h_1k_1l_1$ and $h_3k_3l_3$, are in the correct relative orientation and it is usually possible to check the occurrence of 'double diffraction' by rotating the crystal about an axis which includes a row of spots in which 'double diffraction' is suspected. If along this row some (weaker) spots disappear and perhaps reappear during the rotation, this constitutes proof of 'double diffraction' since the spots would only appear when a suitable primary beam was being excited.

A simple example of this is afforded by considering rotation about the c^* axis of a hexagonal close packed crystal, space group $P6_2$ to show the (100) and $(1\bar{1}0)$ sections, thus

$0\bar{1}4$	004	014		004	114
$0\bar{1}3$	003	013			
	x				
$0\bar{1}2$	002	012		002	112
$0\bar{1}1$	001	011			
	x				
$0\bar{1}0$	000	010		000	110

(100) section (1$\bar{1}$0) section

The reflexions denoted by **x**, the 001 and 003 reflexions, are not permitted by the space group but arise in the (100) section because there is a suitably placed primary beam, e.g. the 010, which can interact with a correctly positioned plane, the 011 or the 013 in this example, in the (100) section. No such pairing occurs in the (1$\bar{1}$0) section and so the non-permitted reflexions are absent.

Note that the possibility of producing these spurious reflexions could be predicted from the reciprocal lattice plot for the hexagonal close packed crystal, with due allowance being made for expected space group absences, by making a tracing of the pattern and translating the copy to a position in which another permitted reflexion coincides with the origin of the master plot. If any of the permitted reflexions in the copy coincide with space group absences in the master, a spurious reflexion can be formed experimentally. The only permitted movements in this case are translations parallel to the reciprocal cell axes; rotations are not permitted.

Finally, it should be noted that, since forbidden reflexions can occur only if there is more than one scattering centre in the most primitive unit cell, metallic face centred cubic and body centred cubic crystals cannot have extra spots introduced in this way in perfect single crystals. This arises from the form of the structure factor equation for the amplitude, and hence the intensity, of the reflexion (hkl)

$$F(hkl) = \Sigma_j f_j(\theta) \exp[-2\pi i(hu_j + kv_j + lw_j)] \qquad (8.6)$$

as given in equations (7.5) to (7.8), where u_j, v_j, w_j are the co-ordinates of the jth atom in the unit cell expressed as fractions of the unit cell edges, and $f_j(\theta)$ is the effective atomic scattering factor for the jth atom for the (hkl) reflexion. If there is only one atom in the unit cell, the origin can be chosen to coincide with it and so $u_j = v_j = w_j = 0$ and $F(hkl)$ becomes independent of (hkl), which means that all reflexions are permitted. This relationship does not necessarily hold for other, more complex face centred structures containing more than one scattering centre, such as spinel.

(c) Diffraction may occur from a permitted beam by a plane in a twin or precipitate which is suitably oriented relative to the parent lattice. In such a case diffraction will not only occur separately from the lattices of the parent and its twin, or the coexisting precipitate phase, to give a pattern with contributions from the two separate phases, but diffracted beams from one phase can be subsequently rediffracted by the second phase to produce quite unexpected maxima. Such patterns tend to be more complex and more difficult to solve, at first glance, but the positions of the anomalous reflexions can again be derived analytically by allowing the diffracted beams from, e.g., the parent lattice to act as direct beams for rediffraction in the second lattice. Again this can be demonstrated by superimposing the diffraction pattern of the second lattice onto various diffraction maxima in the parent lattice and looking for maxima in the second pattern that do not coincide with maxima in the parent pattern. Of course, the superposition must be made with the reciprocal lattices in the correct relative orientations.

If spurious reflexions that clearly arise from the relative orientations of two separate phases can be identified they may, in appropriate circumstances, be of help in deriving the relative orientation of the phases. Moreover, diffracted beams produced in the twin or precipitate phase can equally well be diffracted by the parent lattice to give a centrosymmetric character to the double diffraction spots. This symmetry may also help to indicate the relative orientations. Very complex patterns are obviously possible, and a schematic example of the pattern from a twinned crystal is shown in figure 8.8. A more detailed discussion of twinning follows in the next section.

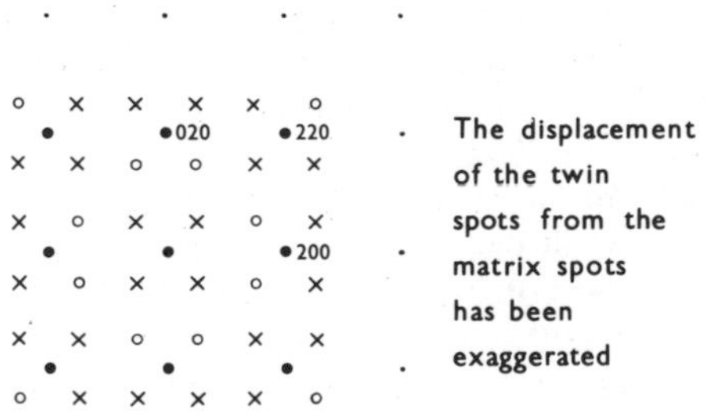

020
220
200
The displacement of the twin spots from the matrix spots has been exaggerated

Figure 8.8 Schematic diffraction pattern of a twinned structure showing matrix spots (●), twin spots (○), and twin spots due to double diffraction (x). The matrix is face centred cubic and the twinning is on $\{111\}$ The displacement of the twin spots from the matrix spots has been exaggerated.

8.4.4 Influence of Twinning on Diffraction Patterns

Since twinning involves a change in orientation of part of a crystal relative to the parent crystal, the diffraction pattern obtained from a twinned crystal is of necessity a composite pattern. The simplest case to consider is that in which the twin axis is perpendicular to the direction of the electron beam, since the diffraction pattern of such a twinned crystal can be obtained from the pattern of the untwinned parent by a rotation through 180° about the twin axis, which is here contained in the reciprocal lattice section. An example of twin spots associated with twinning about [111] on a (110) reciprocal lattice section of a face centred cubic crystal is shown in

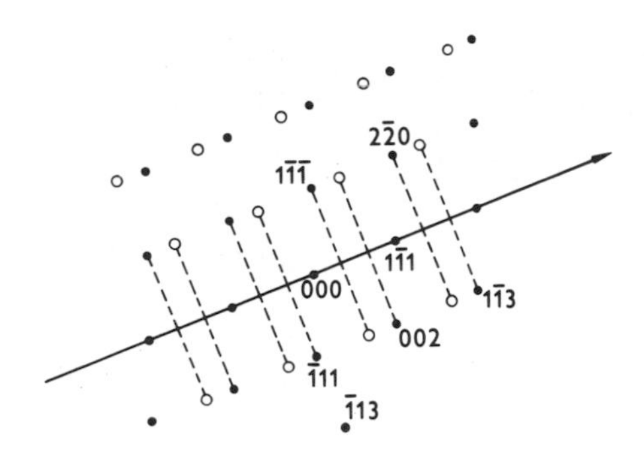

Figure 8.9 Schematic representation of electron diffraction pattern showing matrix spots (●) and twin spots (○) due to twinning about $[1\bar{1}1]$ in a (110) section of a face centred cubic matrix.

figure 8.9. Note how the spots pair up across the twin axis, which is denoted by the solid line, with the dashed lines joining equivalent reflexions in the parent and the twin crystal.

The construction of diffraction patterns when the twinning axis is not contained in the particular reciprocal lattice section under study is clearly more complex. In such cases it is helpful to calculate the indices of the twin spots with reference to the parent indices, i.e. to use the reciprocal cell axes of the parent crystal to index spots from both parts of the crystal. Such calculations are particularly straightforward in cubic systems but no new principles are introduced in their extension to other crystal classes, simply more computation. In the case of a cubic crystal the indices of the diffraction spots referred to the parent reciprocal lattice can be deduced as follows.

Consider the twinning plane to have Miller indices (*uvw*), then *uvw* is the reciprocal lattice vector normal to the plane (*uvw*). Twinning on the plane (*uvw*) will rotate the plane represented by the reciprocal vector $h_1k_1l_1$ about the vector *uvw* to the vector direction $h_2k_2l_2$, and we wish to know $h_2k_2l_2$ in terms of the parent reciprocal lattice indices.

Since only the direction, and not the magnitude, of the lattice vector has been changed by the twinning operation,

$$|h_1k_1l_1| = |h_2k_2l_2| \tag{8.7}$$

whence

$$|h_1k_1l_1| + |h_2k_2l_2| = n|uvw| \tag{8.8}$$

where n is an integer. It follows that

$$h_1 + h_2 = nu \tag{8.9a}$$

$$k_1 + k_2 = nv \tag{8.9b}$$

$$l_1 + l_2 = nw \tag{8.9c}$$

From the derivation of the reciprocal lattice, given in the Appendix, it is clear that, since d_{hkl} is the same in both parent and twin,

$$\frac{h_1^2 + k_1^2 + l_1^2}{a_0^2} = \frac{h_2^2 + k_2^2 + l_2^2}{a_0^2} \tag{8.10}$$

where a_0 is the real cell parameter. Substitution for h_2, k_2, and l_2 from equation (8.8) gives

$$\frac{h_1^2 + k_1^2 + l_1^2}{a_0^2} = \left\{\frac{nu - h_1}{a_0}\right\}^2 + \left\{\frac{nv - k_1}{a_0}\right\}^2 + \left\{\frac{nw - l_1}{a_0}\right\}^2 \tag{8.11}$$

We may solve for n from equation (8.11) and substitute the result back into equations (8.9) to give the relationships

$$h_2 = \frac{u(uh_1 + 2vk_1 + 2wl_1) - h_1(v^2 + w^2)}{u^2 + v^2 + w^2} \tag{8.12a}$$

$$k_2 = \frac{v(2uh_1 + vk_1 + 2wl_1) - k_1(u^2 + w^2)}{u^2 + v^2 + w^2} \tag{8.12b}$$

$$l_2 = \frac{w(2uh_1 + 2vk_1 + wl_1) - l_1(u^2 + v^2)}{u^2 + v^2 + w^2} \tag{8.12c}$$

Thus, in the example quoted in figure 8.9, for twinning on the (110) section with $1\bar{1}1$ as the twinning axis, if $h_1k_1l_1 = 1\bar{1}3$, substitution in equations (8.12) gives

$$h_2 = [1(1\text{x}1 + 2\text{x}\bar{1}\text{x}\bar{1} + 2\text{x}1\text{x}3) - 1(1 + 1)]/(1 + 1 + 1)$$
$$= 7/3$$

$$k_2 = [\bar{1}(2\text{x}1\text{x}1 + \bar{1}\text{x}\bar{1} + 2\text{x}1\text{x}3) - \bar{1}(1 + 1)]/(1 + 1 + 1)$$
$$= \bar{7}/3$$

$$l_2 = [1(2\text{x}1\text{x}1 + 2\text{x}\bar{1}\text{x}\bar{1} + 1\text{x}3) - 3(1 + 1)]/(1 + 1 + 1)$$
$$= 1/3$$

and the reflexion corresponding to the twinned $(1\bar{1}3)$ appears at the position $(\frac{7}{3}\,\frac{\bar{7}}{3}\,\frac{1}{3})$ relative to the parent axes, being clearly in an irrational position in the associated reciprocal lattice. In principle, the nature and orientation of the twin plane can be deduced by a reverse calculation using the observed positions of equivalent reflexions from a number of twin pairs to calculate the indices of the twin plane. A direct deduction of a twinning relationship by this method is possible only in simple cases, but a twin relationship can be deduced from the spot positions in an electron diffraction pattern in many cases, especially when there are additional clues to the probable twinning type. Thus Bhatty, Gard and Glasser [17] have deduced the presence of albite plus Carlsbad twinning in an anorthite from Niigata, Japan.

Once the relative orientations of the parent and the twin lattices have been confirmed and the combined spot pattern calculated using equations (8.12) it is possible to use this combined pattern to predict the occurrence of spurious reflexions arising from double diffraction by the superposition method, as in figure 8.8, and the technique has been discussed by Hirsch *et al* [18]. Dark field techniques will be discussed in section 8.6, suffice it here to say that such techniques are very useful in interpreting double diffraction phenomena in twinned crystals, as is also the method discussed in section 8.4.3 whereby the crystal is rotated about a reciprocal lattice row. Once again changes in the spots along that row imply that double diffraction is responsible for the variable spots.

Shape Effects

The shape of a diffracting body can play an important part in determining the nature of the beam diffracted from planes within it, and James [19] has discussed this in conjunction with X-ray diffraction. Moreover, the broadening of X-ray diffraction maxima by very small crystallites has been discussed in Chapter 7, section 7.8.2. Twins and precipitated phases are often present in the form of thin needles, layers, discs, etc., and their different shapes and sizes affect the shapes of the diffraction maxima in different ways. In general, the intensity distribution is extended along the direction parallel to the smallest dimension of the crystal. Thus for a plate-like precipitate inclusion, the reflexions would appear as spots with rod-shaped streaks running through them parallel to the normal to the plane of the disc. Clearly, careful observations of the 'shape-effects' in diffraction patterns can provide much additional information concerning the habits of small plates or twins and is an area in which the facility of selected area electron diffraction is of great use, with its ability to correlate the shape indicated from the diffraction pattern with that actually observed on the corresponding micrograph.

The interpretation of streaks or other 'shape effects' is greatly aided if the parent crystal is rotated in a controlled manner about some prominent direction in its reciprocal lattice so that changes in the intensity, position, and projected directions of the streaks can be observed. For example, the intersection of the Ewald sphere by an 'inclined streak', i.e. a streak which is not parallel to the direction of the electron beam, can lead to incorrect interpretation of the effect and to the measurement of incorrect *d*-spacings. Such streaks can be identified, however, if a series of diffraction patterns in different directions through the crystal is obtained. Systematic analysis of these shape effects is helped by obtaining systematic rotations of the parent crystal, and this is best achieved by using Kikuchi maps, which will be discussed in section 8.5.

Polytypism and Related Phenomena
Streaks in the diffraction pattern can also arise from polytypism, interstratification of two phases, or other order-disorder in the crystal. Such phenomena are often missed in X-ray studies of crystals, because of the relative 'coarseness' of the X-rays used, but Gard [16, 20] has demonstrated the power of SAED methods in elucidating fine effects. In particular he has shown how the different polytypes in the fibrous calcium silicates, particularly xonotlite, can be identified by considering streaked reflexions, both with and without included intensity maxima. He has also shown [17] that a second phase present as a thin interstratification within a major phase can also be identified using SAED where X-rays gave no evidence of the phase, simply because line broadening due to the thinness of the lamellae caused the lines to be lost in the general background. The multiple spot patterns which appear on some diffraction patterns, often in conjunction with streaking, can also be very useful in elucidating these order-disorder effects. Such spots can usually be differentiated from the spurious spots arising from double diffraction because they cannot be formed by the superposition of the simple lattices, as discussed in section 8.4.3.

8.5 FORMATION OF KIKUCHI LINES

As well as the diffraction maxima due to the elastically scattered electrons, contributions to the patterns arise from inelastically scattered electrons also. The most important way in which these inelastically scattered electrons appear is as Kikuchi lines and because these lines are so useful, and because Kikuchi maps, which are made up of families of Kikuchi lines, are so powerful an aid to analysis, the origin of Kikuchi lines and some uses of Kikuchi maps will be discussed.

Figure 8.10 illustrates the method of formation of Kikuchi lines [21] which, although inadequate to explain the observed intensities, predicts the positions of the Kikuchi lines accurately, and it is the positions of the lines

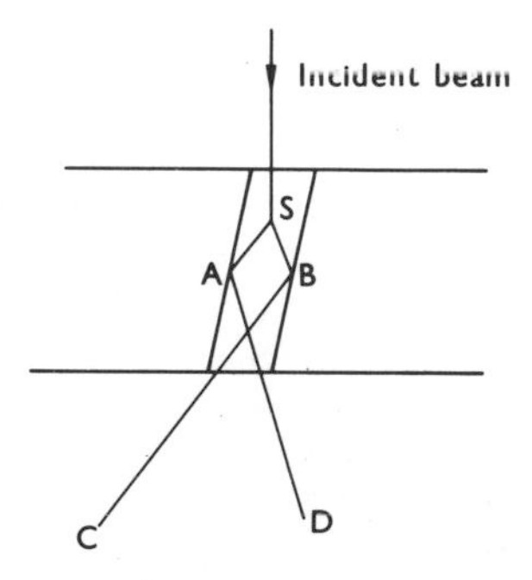

Figure 8.10 Illustration of a simple model for the production of Kikuchi lines. For explanation see text.

which is of most use in the analysis of crystal defects (section 8.6). In figure 8.10 it is assumed that electrons are inelastically scattered, with a negligible loss of energy and so no noticeable change in their wavelengths, at the point S. The intensity of these electrons decreases rapidly as the angle through which they are scattered increases. Rays travelling in certain directions from S will be at the Bragg angle for reflection by a set of crystal planes, as shown in figure 8.10. Those electrons originally travelling along SB are diffracted along BC whereas those originally travelling along SA are diffracted along AD, where SB and SA are oriented at the Bragg angle with respect to the planes shown. However, since the electrons inelastically scattered along SB have been scattered through a smaller angle than those along SA, the intensity of the diffracted beam along BC will be greater than that along AD. If all possible directions are considered, for those electrons which are initially inelastically scattered and subsequently Bragg diffracted from a given set of planes, it can be seen that these two types of direction, BC and AD, describe two sets of cones and the intersection of these cones with a screen or photographic plate produces two portions of conic sections which are good approximations to straight lines. These straight lines are known as pairs of Kikuchi lines. They are separated by the angle 2θ, where θ is again the Bragg angle for the diffraction planes, and are situated on either side of the intersection between the trace of the planes giving rise to the lines and the screen or photographic plate, as suggested by figure 8.10.

Pairs of lines occur for all reflecting planes and, as discussed in the next section, these lines can be used to solve electron diffraction patterns with a much higher degree of accuracy than solutions obtained from spot patterns. This arises basically because Kikuchi lines behave as if they were rigidly attached to a crystal when the crystal is tilted, for example in measuring the out-of-plane reciprocal axis length, which is a consequence of the fact that the lines are always situated symmetrically on either side of the planes generating them.

8.5.1 Kikuchi Maps

One of the main applications of Kikuchi lines is as 'navigational aids' when reorienting a crystal in a controlled manner. In crystals of low symmetry where diffraction patterns cannot be recognised on sight as easily as in crystals with cubic symmetry, it is virtually essential to use Kikuchi maps as an aid to orient the crystal, and even in cubic crystals it is highly desirable to eliminate possible ambiguities by using these maps. An example of an indexed, schematic Kikuchi map is shown in figure 8.11. Only the low index lines are shown for clarity. Many examples of Kikuchi maps obtained on actual crystals have been published and the reader is referred to publications in various scientific journals [22–24].

If a crystal is aligned with the electron beam precisely along [001] then the Kikuchi line pattern would appear as shown schematically in figure 8.12a. If, however, the electron beam direction were a few degrees off the [001] direction towards the [$\bar{1}$03] direction, the pattern would appear as in

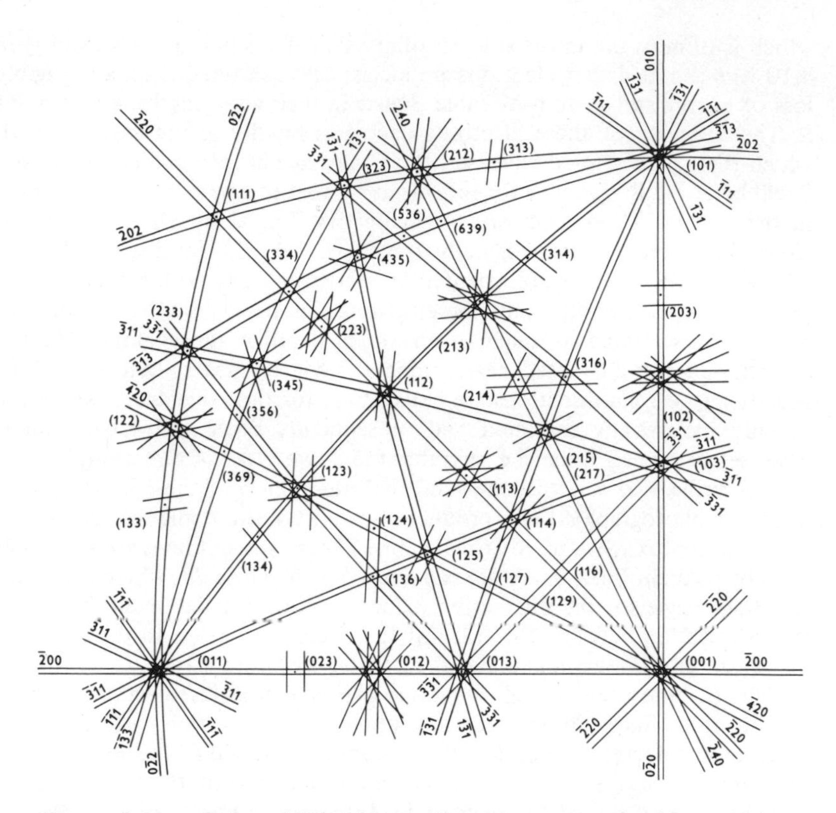

Figure 8.11 Indexed Kikuchi map for a face centred cubic crystal.

figure 8.12b. Thus, if the [001] pattern is displaced towards the top of the screen by rotating the crystal clockwise about [020], the electron beam direction changes in the sense towards [$\bar{1}$03]. This can be checked easily with a crystal model and leads to the indexing of the Kikuchi map being different from, but consistent with, that of a stereogram, i.e. [$\bar{1}$03] and [103], etc., are interchanged. Failure to appreciate this simple fact can lead to erroneous conclusions when analysing crystal defects or precipitate habit planes and so on.

Note that the precise deviation of the electron beam direction from [001] for a diffraction pattern such as that illustrated in figure 8.12b can easily be measured by finding where the centres of the two non-parallel pairs of Kikuchi lines meet. Since the Bragg angles for the Kikuchi line pairs are known, the distance from their meeting point and the direct beam spot can be expressed as an angular deviation of the beam from [001]. Thus the spot pattern of figure 8.13, solved in the manner described in section 8.4.2, indicates that the electron beam is parallel to [001] but the more accurate method based on the Kikuchi lines visible therein shows that the crystal is

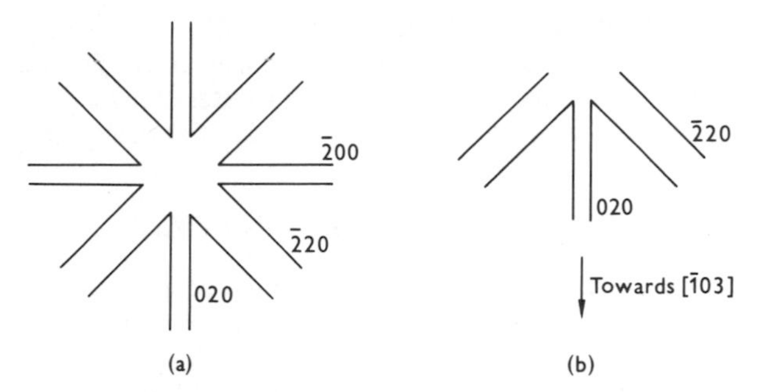

Figure 8.12(a) Diagrammatic representation of Kikuchi line pattern for crystal with electron beam aligned precisely along [001]. (b) Kikuchi line pattern for crystal with electron beam a few degrees from [001] toward [$\bar{1}$03].

Figure 8.13 Electron diffraction pattern for chromium showing Kikuchi lines, which indicate that the electron beam is aligned at an angle of 6° from [001] in a sense toward [111]. The simple diffraction spot pattern suggests that [001] is aligned parallel to the electron beam.

actually rotated 6° away from [001] in a sense towards [111] relative to the beam.

Ambiguities in indexing a set of diffraction patterns can arise since, for example, the diffraction patterns for the electron beam parallel to [113] and parallel to [$\bar{1}\bar{1}$3] can be superimposed but they contain different 211 reflexions, and the information contained in the Kikuchi maps provides the most reliable way of removing the attendant uncertainty. The way in which the diffraction pattern changes as the crystal rotates along a pair of Kikuchi lines enables one to distinguish between otherwise identical patterns.

8.6 ANALYSIS OF DEFECTS IN CRYSTALS

The standard method of using an electron microscope in the transmission mode involves the insertion of a small aperture into the objective lens so that only the central direct beam or a selected, diffracted beam contributes to the final image, to provide bright field and dark field illumination respectively. On the final screen, or photographic plate, we are therefore looking at a highly magnified picture of the distribution of these selected electrons as they leave the bottom surface of the crystal. There are basically two theories of diffraction which are used to calculate this distribution of electrons. The first of these, the kinematical theory, is really a special case of the second, the dynamical theory.

The kinematical theory is based on the premise that the direct beam intensity is very little affected by the removal of those electrons which are diffracted away, and is therefore only strictly applicable to very thin crystals or to crystals so oriented that no reciprocal lattice vector lies close to the Ewald sphere. Moreover, the refractive index of the crystal for electrons is ignored. The more general dynamical theory takes into account dynamical interactions between the various beams as they propagate through the crystal. Because the kinematical theory is of such limited applicability it will not be further discussed here, although it should be emphasised that the theory is useful because by its very simplicity it is possible to keep the physical basis for contrast clearly in mind. On the other hand, the two methods of formulating the dynamical theory (wave optical and wave mechanical) completely involve equations with so many variables that it is difficult to maintain an awareness of the physical situation to which the equations refer.

The approach adopted here is simply to present the basis for the wave optical approach so that the importance of the many experimental variables can be appreciated.

8 6.1 Wave Optical Dynamical Theory

The solution to electron diffraction patterns presented in section 8.4.2 and subsequent sections makes the explicit assumption that the radius of the Ewald sphere is infinite and therefore all reflexions lying on a particular reciprocal lattice plane are simultaneously excited. This is a very useful concept for solving diffraction patterns but clearly if, say, the 111 reflexion lies precisely on the Ewald sphere, then the 222 and other *hhh* reflexions on that 'systematic row' cannot also lie precisely on the sphere. For contrast calculations (at least for 100kV electrons), therefore, we make a two-beam approximation, i.e. we consider only the direct beam and one intense diffracted beam in making the contrast calculations. Reciprocal lattice points not lying on this 'systematic row', which contains the strong diffracted beam, need not be considered since they can be eliminated experimentally by rotating the crystal about the row in question.

Just as in the simpler kinematical theory, the basic method for formulating the mathematics of the dynamical theory lies in the division of the crystal into Fresnel zones, which are discussed by Jenkins and White [25], and in order to include the effect of the refractive index, 'forward scattering' is introduced, again using Fresnel zones to calculate its importance.

The formulation of the equations takes into account the fact that at any point, r, in the crystal both the amplitude of the direct wave, ϕ_0, and of the diffracted wave, ϕ_g, vary with depth due to scattering from one beam to the other. Thus on propagating through a slab of crystal of thickness d_z, ϕ_0 changes by an amount $d\phi_0$ and ϕ_g changes by $d\phi_g$ in such a way that

$$d\phi_0 = (\text{Forward scattering from } \phi_0) + (\text{Rediffraction from } \phi_g)$$

$$d\phi_g = (\text{Forward scattering from } \phi_g) + (\text{Rediffraction from } \phi_0).$$

Mathematical formulation of these changes in ϕ_0 and ϕ_g results in the derivation of a pair of coupled differential equations. Since we are interested in the application of the dynamical theory to crystals containing imperfections, the form of the two-beam equations which will be quoted, after Hirsch *et al* [18], includes a term which takes into account the displacement of atoms from their positions in a perfect crystal associated with the imperfection. In addition, terms are included in the two-beam equations which take account of the experimental observation that some electrons are inelastically scattered (i.e. their wavelengths are changed) during passage through the crystal. These terms are generally referred to as absorption terms, although absorption is not really a true description of what is happening to the inelastically scattered electrons – they simply do not contribute to the final image, with the exception of those electrons which have been inelastically scattered and yet pass through the objective aperture. These do contribute and this explains why the absorption parameter is related to the size of the objective aperture.

One form of the coupled differential equations is

$$\frac{dT}{dz} = \frac{-\pi}{\xi_0'}T + \pi\left\{\frac{i}{\xi_g'} - \frac{1}{\xi_g'}\right\}S \tag{8.13a}$$

$$\frac{dS}{dz} = \pi\left(\frac{i}{\xi_g} - \frac{1}{\xi_g'}\right)T + (-1)\left\{\frac{\pi}{\xi_0'} + 2\pi i(s + \beta')\right\}S \tag{8.13b}$$

where ξ_0 and ξ_g are constants, termed extinction distances, which make the equations dimensionally correct;

ξ_0' and ξ_g' are factors which take into account anomalous absorption;
s is the deviation from the Bragg reflecting condition;

β' is equal to $\mathbf{g}\cdot(d\mathbf{R}/dz)$, where $\mathbf{g}$ is the diffracting vector and $d\mathbf{R}/dz$ is the rate of change of $\mathbf{R}$, the displacement of atoms from their positions in a perfect crystal;

T is the direct wave amplitude; and

S is the diffracted wave amplitude.

This coupled equation is integrated down columns of the crystal, with the assumption made that the amplitude of the electron wave at the bottom surface of the crystal for a column of radius about 10Å is unaffected by adjacent columns. The justification for this 'column approximation' lies in the agreement between results computed in this way with results from more complex computations not using this simplifying approximation.

8 6.2 Contrast from Dislocations

The only term involving the displacements associated with crystal defects is the term $\beta' = \mathbf{g}\cdot(d\mathbf{R}/dz)$ and this immediately leads to one of the most important and powerful applications in the analysis of defects. Thus when the atomic displacements associated with a defect are perpendicular to the diffracting vector, i.e. when $\mathbf{g}\cdot(d\mathbf{R}/dz) = 0$, the crystal will transmit electrons as if no defect were present. This is the basis of determining the Burgers vector, **b**, which measures the direction and the extent of the displacement introduced into the lattice by the imperfection. For example, in an elastically isotropic crystal the only displacements associated with a screw dislocation are those displacements which lie parallel to the direction of the dislocation so that if any reflecting plane which contains **b** is used to image the screw dislocation, $\mathbf{g}\cdot\mathbf{b}$ will be zero and the dislocation will be invisible.

For dislocations with more complex displacement fields, for example mixed dislocations or dislocations in elastically anisotropic media, the dislocations will not in general go out of contrast when $\mathbf{g}\cdot\mathbf{b} = 0$. Determination of the Burgers vector in such cases has to a great extent been done by assuming that weak images correspond to the condition $\mathbf{g}\cdot\mathbf{b} = 0$. This can, in some cases, be misleading and for rigorous analysis it is desirable to use an image-matching technique in which computed images are compared with experimental images taken under a variety of diffracting conditions. This technique is inevitably very tedious, especially when compared with the comparatively simple method of determining which diffracting vectors make the dislocation invisible, because in order to compute the displacement field associated with a dislocation (and hence the contrast to be expected in the microphotograph) it is necessary to determine, among other things, the crystallographic direction of the dislocation. This is most easily done by stereographic techniques. Clearly the techniques involved in the determination of the crystallographic directions of dislocations are precisely those that would be used in the determination of the habits of precipitates or stacking faults, and the general technique for these will be discussed in section 8.7 under the heading of trace analysis.

The determination of the other factors which influence the images of

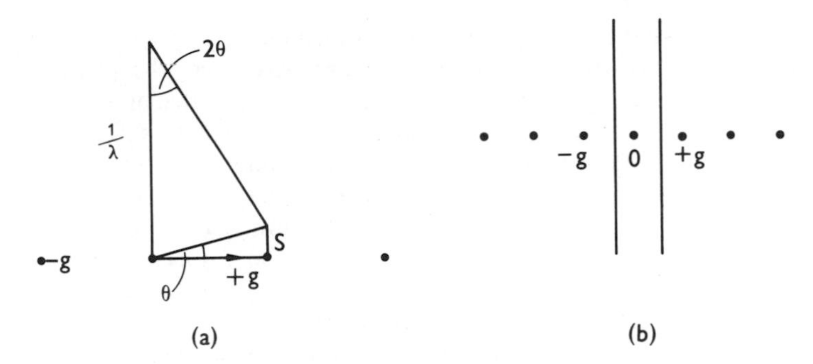

Figure 8.14 Diagram illustrating method of determining s, the deviation parameter. (a) Ewald sphere and reciprocal lattice construction showing s emphasised. (b) Corresponding electron diffraction pattern showing relation of Kikuchi lines and diffraction spots.

dislocations are straightforward and may be dealt with briefly. The crystallographic direction in which the defects are projected on a micrograph is, of course, obtained from the solution of the corresponding diffraction pattern. The deviation parameter, s, is also obtained directly from the diffraction pattern, as illustrated schematically in figure 8.14 where it can be seen that when the Kikuchi lines are oriented symmetrically with respect to the spot pattern, the value of s is given by

$$s = \frac{g^2\lambda}{2} = \frac{\lambda(h^2 + k^2 + l^2)}{2a^2} \qquad (8.14)$$

where hkl are the Miller indices of the spot and a is the cell parameter. Hence the total distance between g and the origin corresponds to twice this value of s, since the angle involved corresponds to 2θ and not to θ. Thus the value of s is deduced directly from any diffraction pattern by measuring the distance between g and its Kikuchi line and scaling the value of s given from equation (8.14) appropriately.

The values of the thickness of the sample and of the orientation of the normal to the specimen, which are also needed, can be obtained by methods to be discussed in the following sections.

8.6.3 Contrast from Stacking Faults and Precipitates

With the aid of a computer it should be possible to identify any dislocation uniquely as to type. The situation is not so satisfactory in the case of distinguishing between faults and thin precipitates within a crystal. Solution of the dynamical equations shows that for an inclined fault the contrast seen on a micrograph consists of a series of fringes parallel to the direction in which the fault intersects the surface of the specimen. This contrast arises because of the phase change which the electrons suffer when they reach the plane of the fault and the change can be introduced into the dynamical equations as a term $\mathbf{g}\cdot\mathbf{R}_F$, where the displacement across the fault is $\mathbf{R}_F$.

Stacking faults can be envisaged as being formed either by translation of one part of the crystal with respect to the other part along the plane of the fault, or by displacing one part of the crystal perpendicular to the fault plane. (Clearly the dislocations surrounding the fault will be different in the two cases.) The similarity between a planar precipitate and a stacking fault can be better seen in terms of the displacement perpendicular to the fault, or precipitate, plane. If the atoms making up the precipitate are smaller than the matrix atoms, the top of the crystal will be displaced towards the bottom half by an amount $\mathbf{R}_F$ perpendicular to the precipitate plate. Again the contrast will depend on $\mathbf{g}\cdot\mathbf{R}_F$. The form of the two beam equation most useful in discussing the contrast associated with stacking faults is [18]

$$\frac{d\phi_0}{dz} = \pi\left(\frac{i}{\xi_0} - \frac{1}{\xi_g'}\right)\phi_0 + \left(\frac{i}{\xi_g} - \frac{1}{\xi_g'}\right)\phi_g \exp[2\pi i(sz + \mathrm{g}\cdot\mathbf{R}_F)] \quad (8.15\mathrm{a})$$

$$\frac{d\phi_g}{dz} = \pi\left(\frac{i}{\xi_0} - \frac{1}{\xi_0'}\right)\phi_g + \left(\frac{i}{\xi_g} - \frac{1}{\xi_g'}\right)\phi_0 \exp[2\pi i(sz + \mathrm{g}\cdot\mathbf{R}_F)] \quad (8.15\mathrm{b})$$

From these it can be seen that, if $\mathbf{g}\cdot\mathbf{R}_F = 0$, the faults (or precipitates) will be invisible since the equations then reduce to those for a perfect crystal. However, as can also be seen from the equations, the fault (or precipitate) will be invisible if $\mathbf{g}\cdot\mathbf{R}_F$ = any integer, and this affords a possible way of distinguishing between a stacking fault and a precipitate. This follows from the fact that if $\mathbf{g}\cdot\mathbf{R}_F$ is an integer, it is of course necessary for $\mathbf{R}_F$ to be a simple fraction of a lattice vector. This is always truc for a stacking fault but is unlikely to be true for a precipitate. Hence, if a diffracting vector can be found for which the fringes disappear, in addition to any for which $\mathbf{g}\cdot\mathbf{R}_F = 0$, then the magnitude of $\mathbf{R}_F$ can be determined [from equations (8.15)] and will presumably correspond to a stacking fault. On the other hand, if no reflexion can be found for which $\mathbf{g}\cdot\mathbf{R}_F$ = an integer then the fringes would be presumed to arise from a precipitate. It is possible to determine the approximate value of $\mathbf{R}_F$ for a precipitate by using successive reflexions along a systematic row to image the precipitate. The reflexion for which $\mathbf{g}\cdot\mathbf{R}_F$ is closest to an integer will then correspond to the weakest set of fringes. Alternatively, images can be computed for these diffraction conditions for assumed values of $\mathbf{R}_F$ in the model used.

Clearly, if the value of $\mathbf{R}_F$ for a precipitate is close to that for a fault, as would be the case for a close packed hexagonal precipitate in a face centred cubic crystal, then $\mathbf{g}\cdot\mathbf{R}_F$ can have a value close to an integer and the distinction between a precipitate and a stacking fault cannot be made using transmission electron microscopy.

8.6.4 Other Defects

The two previous sections have dealt with line defects and planar defects. This section will deal briefly with three dimensional defects. Three dimensional voids can be observed in crystals and, if the voids are small, the contrast shown is a very sensitive function of the diffraction conditions. If the voids do not have a large strain field, caused, for example, by the presence of trapped gas, then their contrast is essentially due to the effective change in the thickness of the crystal in section across the void. Depending, therefore, on their depths in the crystal and on the deviations from the Bragg angle, the voids can appear darker than, brighter than, or equal to the background intensity. It should be noted that a similar contrast effect can be obtained from small precipitate particles.

Small defects which show black-white lobes along the sense of **g** are small strain centres due either to clusters of vacancies, or self interstitials, or to small impurity centres. Again differentiation between impurity clusters and defect clusters will not always be possible, although the sense of the strain around the precipitate or defect can be obtained using stereoanalysis [18].

8.7 TRACE ANALYSIS

Provided that the angular rotation between a micrograph and its electron diffraction pattern is known (cf. section 8.4) it is quite a straightforward task to transfer crystallographic information to the micrograph. For example, we can expect to index the habit faces of the crystal, as shown in figure 8.5. However, because all the directions in the micrograph are directions in the crystal projected onto a plane perpendicular to the direction of the electron beam, some further information is required to determine fully the crystallographic orientation of any feature. This information is best obtained by taking micrographs using widely different electron beam directions relative to the sample axes, since the true direction along which, for example, a line feature lies must be common to the great circles defined by the pairs of electron beam directions and the observed directions of the feature for the various beam directions. The selection and identification of these electron beam directions is again best done using Kikuchi maps.

While carrying out the necessary tilting operations in order to determine the true direction of a feature, it is extremely useful to take stereo pair micrographs which can subsequently be viewed in a stereoviewer to give a three-dimensional picture of the feature. It must, however, be remembered that the contrast from a defect is sensitive to the diffracting conditions and so it is necessary to use the same diffracting vector in obtaining the stereo pairs, which must, of course, have completely equivalent contrasts in the two exposures. Again the necessary movements are best made with the help of Kikuchi lines as navigational aids, and a rotation of 15°–20° between the exposures is about the optimum for good depth resolution, depending on the exact conditions applying for a given observation.

Once the top and the bottom surfaces of a crystal have been identified by stereo pairs it becomes quite simple to identify on which surface a dislocation emerges. Lines can then be drawn joining the ends of dislocations in one surface and the foil normal is then defined by the cross product of the true directions of any pair of these lines. Note that the foil normal and the electron beam can easily be 45° apart with modern tilt stages and this is an essential fact to bear in mind when carrying out trace analysis of, for example, faults which intersect the foil surface. The identification of the types of crystal plane on which the faults lie can be made more complex when the foil is not perpendicular to the beam.

The foil thickness is also of interest during contrast analysis, and this can be obtained from a series of micrographs in which the true direction $[h_1k_1l_1]$ of a line defect of projected length P is viewed in an electron beam direction $[h_2k_2l_2]$ if the foil normal $[h_3k_3l_3]$ has also been determined. Thus the foil thickness, t, is given by

$$t = \frac{P(h_1h_3 + k_1k_3 + l_1l_3)}{(h_3^2 + k_3^2 + l_3^2)^{1/2}} \left\{ (h_1^2 + k_1^2 + l_1^2) - \frac{(h_1h_2 + k_1k_2 + l_1l_2)^2}{(h_2^2 + k_2^2 + l_2^2)} \right\}^{-1/2}$$

The ability to determine the thickness of the specimen can be of considerable use in studies on platy crystals, such as clays and clay minerals.

8.8 FINAL COMMENTS

These notes have given a brief qualitative account of some of the aspects of transmission electron microscopy of interest to research workers studying crystals in general and minerals in particular. A thorough treatment of much of the underlying mathematics will be found in the 'Electron Microscopy of Thin Crystals' by Hirsch *et al.* [18] and readers are referred to this work for further reading in this area. (Gard's monograph [1], as mentioned earlier, is probably the best book for practical details on the operation and use of electron microscopy in the minerals field and is strongly recommended.)

Electron microscopic studies of minerals are very useful indeed, despite the thinness of the specimen that one is constrained to use. The recent development of microscopes capable of operating at 1000kV, however, appears to hold out hope of improvements in this respect since they are able to accept much thicker specimens due to the extra penetrating power of the more energetic electrons. Particular mention was made of the study of minerals at the recent conference on High Voltage Electron Microscopy held in Oxford [26]. It is possible that specimens as thick as 10μm may be acceptable in some cases, and specimens 5μm thick should be quite acceptable in all cases. As already mentioned, the possibility of maintaining crystals in a wet atmosphere during examination at high voltage will increase the range of materials that can be examined very significantly. Finally, the added accuracy of the selected area diffraction patterns obtained with the resulting very short wavelength electrons, of the order of 10^{-3} Å, should prove most useful in the field of analysing complex phases, such as phases with several stacking modifications or showing order-disorder effects.

The most profitable application in the field of metallurgy has perhaps been in the field of radiation damage, when it has proved possible to study defects introduced by the high energy electrons and also how these defects interact with one another whilst under study in the microscope. It is probable that this technique could be applied to similar problems in mineral species. At the moment, the field is open and the applications are there to be discovered.

8.9 APPENDIX

8.9.1 The Reciprocal Lattice

Definition

Let the primitive translations of a primitive space lattice be **a**, **b**, **c**, then the vector **k**(uvw) to any lattice point is given by

$$\mathrm{k} = u\mathrm{a} + v\mathrm{b} + w\mathrm{c} \tag{8.16}$$

The definition of the reciprocal lattice is such that the translations **a***, **b***, **c*** which define the reciprocal lattice fulfil the following relations

$$(1) \quad \mathbf{a}^* \cdot \mathbf{a} = \mathbf{b}^* \cdot \mathbf{b} = \mathbf{c}^* \cdot \mathbf{c} = 1 \tag{8.17a}$$

$$(2) \quad \mathbf{a}^* \cdot \mathbf{b} = \mathbf{b}^* \cdot \mathbf{c} = \mathbf{c}^* \cdot \mathbf{b} = \text{etc} = 0 \tag{8.17b}$$

and from relations (8.16) it follows that

$$\mathbf{a}^* = \frac{(\mathbf{b} \wedge \mathbf{c})}{\mathbf{a} \cdot (\mathbf{b} \wedge \mathbf{c})}; \quad \mathbf{b}^* = \frac{(\mathbf{c} \wedge \mathbf{a})}{\mathbf{b} \cdot (\mathbf{c} \wedge \mathbf{a})}; \quad \mathbf{c}^* = \frac{(\mathbf{a} \wedge \mathbf{b})}{\mathbf{c} \cdot (\mathbf{a} \wedge \mathbf{b})};$$

It is possible to deduce some extremely important properties of a reciprocal lattice from these relations;

(a) The vector **g**(hkl), defined by $\mathbf{g} = h\mathbf{a}^* + k\mathbf{b}^* + l\mathbf{c}^*$ where h and k and l are integers, to the point hkl in the reciprocal lattice is normal to the plane in the real space primary lattice having Miller indices (hkl). Thus, if ABC in figure 8.15 is the (hkl) plane that passes nearest to the origin, then OA, OB, OC equal a/h, b/k, c/l respectively. Therefore, AB = $(b/k - a/h)$ and the dot product of this with **g**(hkl) equals zero, since

$$(h\mathbf{a}^* + k\mathbf{b}^* + l\mathbf{c}^*) \cdot \left(\frac{b}{k} - \frac{a}{h}\right) = \mathbf{b} \cdot \mathbf{b}^* - \mathbf{a} \cdot \mathbf{a}^* = 1 - 1 = 0$$

It can similarly be shown that g is perpendicular to BC and CA and so the vector **g**(hkl) is perpendicular to the plane (hkl) in the primary real space lattice.

(b) The spacing $d(hkl)$ between planes with these Miller indices (hkl) is

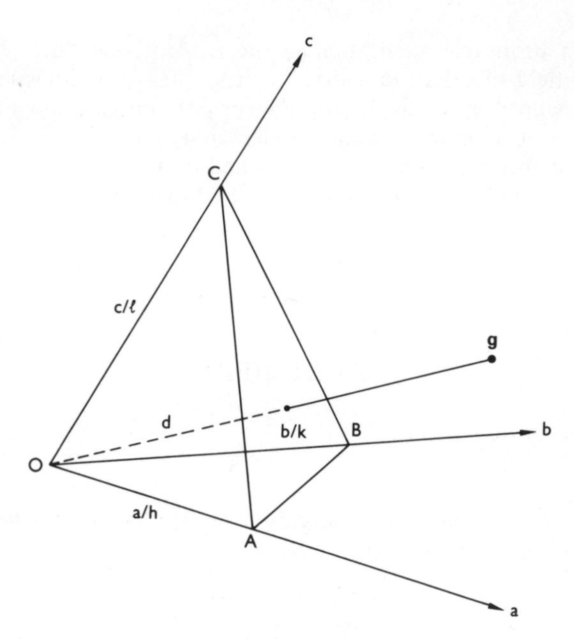

Figure 8.15 Diagram to show the construction of the reciprocal lattice vector g for a general crystallographic plane. The plane ABC has Miller indices (hkl) and is the (hkl) plane that cuts the a, b, and c axes (at a/h, b/k, and c/l respectively) nearest to the origin, O. The vector g is proportional in length to $1/d$, where d is the perpendicular distance from O to the plane ABC. (Compare figure 7.3.)

given by the length of ON in figure 8.15, which has been drawn perpendicular to the plane ABC, through the origin. Then, if $\mathbf{n}$ is a unit vector along ON, the length of ON is the scalar product of OA, which equals a/h, with $\mathbf{n}$. Since it has also been shown above that $\mathbf{g}(hkl)$ lies along ON,

$$\mathbf{n} = \frac{\mathbf{g}}{|\mathbf{g}|} \tag{8.18}$$

It follows that

$$d(hkl) = \mathbf{n} \cdot \frac{a}{h} = \frac{g}{|g|}\frac{a}{h} = \frac{1}{|g|}$$

or

$$d(hkl) = \frac{1}{|g|}$$

REFERENCES

1. J. A. GARD (ed), 'The Electron-Optical Investigation of Clays, Monograph No. 3', Mineralogical Society, London, 1971.
2. J. J. COMER, *Ibid.*, Chapter 3.
3. L. CARTZ and B. TOOPER, Proc. 3rd European Reg. Conf. Electron Microsc., Prague, Vol. A, 335 (1964).
4. L. de BROGLIE, *Phil. Mag., ser.* 6, **47**, 446 (1924).
5. C. DAVISSON and L. H. GERMER, *Phys. Rev.*, **30**, 705 (1927).
6. G. P. THOMSON and A. REID, *Nature*, **119**, 890 (1927).
7. A. LANDE, *Brit. J. Phil. Soc.*, **15**, 307 (1956).
8. A. LANDE, 'New Foundations of Quantum Mechanics', Cambridge University Press, Cambridge, 1965.
9. L. STURKEY, Symp. Techniques in Electron Metallography, *Spec. tech. Publns. Amer. Soc. Test. Mater.*, **339**, 31 (1962).
10. B. K. VAINSHTEIN, 'Structure Analysis by Electron Diffraction' transl. E. Feigl and J. A. Spink, Pergamon Press, Oxford, 1964.
11. B. B. ZVYAGIN, 'Electron-Diffraction Analysis of Clay Mineral Structures' transl. S. Lyse, Plenum Press, New York and London, 1967.
12. J. A. GARD (ed), 'The Electron-Optical Investigation of Clays, Monograph No. 3', Mineralogical Society, London, 1971, Chapter 2.
13. A. W. NICOL, Ph.D. Thesis, Aberdeen University, 1962.
14. J. A. GARD and H. F. W. TAYLOR, *Amer. Min.*, **43**, 1 (1958).
15. R. R. DAYAL, J. A. GARD and F. P. GLASSER, *Acta Cryst.*, **18**, 574 (1965).
16. J. A. GARD, Proc. 2nd European Reg. Conf. Electron Microsc., Delft (1960), Vol. A, 333 (1961).
17. M. S. Y. BHATTY, J. A. GARD and F. P. GLASSER, *Miner Mag.*, **37**, 780 (1970).
18. P. B. HIRSCH, A. HOWIE, R. B. NICHOLSON, D. W. PASHLEY and M. J. WHELAN, 'Electron Microscopy of Thin Crystals', Butterworths, London, 1965.
19. R. W. JAMES, 'Optical Principles of X-ray Diffraction', Bell, London, 1958.
20. J. A. GARD, *Nature*, **211**, 1078 (1966).
21. S. KIKUCHI, *Jap. J. Phys.*, **5**, 83 (1928).
22. G. THOMAS, *Trans. A.I.M.E.*, **233**, 1608 (1965).
23. E. LEVINE, W. L. BELL and G. THOMAS, *J. App. Phys.* **37**, 2141 (1966).
24. M. von HEIMENDAHL, W. L. BELL and G. THOMAS, *J. App. Phys.*, **35**, 3614 (1964).
25. F. A. JENKINS and H. E. WHITE, 'Fundamentals of Optics', McGraw-Hill, New York and London, 1957.
26. P. R. SWANN, C. J. HUMPHREY and M. J. GOVINGE (eds), 'Proceedings of the Third International Conference on h.v.e.m., Oxford, 1974', Academic Press, London and New York, 1974.

CHAPTER 9

Infra-red Spectroscopy in Mineral Chemistry

V. C. Farmer

Department of Spectrochemistry
Macaulay Institute for Soil Research
Craigiebuckler, Aberdeen AB9 2QJ
Scotland

9.1 INTRODUCTION

Since the introduction of commercial infra-red spectrometers, some twenty years ago, this technique has become the hand-maiden of the organic chemist, but we are only now approaching the situation where the mineral chemist will, as a routine procedure, examine the infra-red spectra of his specimens. In some ways this is a little surprising, for even the earliest surveys were sufficient to show that more than a few mineral specimens had been wrongly identified, and that infra-red spectra could readily distinguish, for example, structural hydroxyl groups from water of crystallisation.

The slow progress of mineral spectroscopy, compared with organic spectroscopy, derives from the physical nature of the material under study since minerals have to be studied as solid samples which contain rather complex atomic arrangements whose stoichiometry cannot always be guaranteed. By comparison, organic spectroscopists normally work with compounds of well defined composition in which the atoms are arranged in relatively simple molecular groupings, and these they can study either in solution or in the gas phase, as independent molecules. Even in solids, vibrational interactions between organic molecules linked through Van der Waals forces are weak, whereas in most minerals the weakest bonding is usually ionic, and the interaction between structural units correspondingly stronger. Table 9.1 summarises these differences. In particular, it has proved very difficult to assign the features seen in the infra-red spectrum of a mineral to atomic groupings within the crystal structure, except for special cases such as water or discrete anions such as carbonate or sulphate, although general assignments can be made in some situations.

In spite of these difficulties, a body of empirical information has been accumulated about mineral spectra, and a depth of understanding achieved, which makes the routine study of minerals well worthwhile. Above all, it should be remembered that to obtain an infra-red spectrum requires less time and skill than the determination of refractive indices, and the results will usually be more informative.

Texts which deal specifically with the spectra of inorganic compounds mostly [1-3] reflect the current interest of academic scientists in co-ordination compounds and the simpler inorganic crystals, their principal concern being with the theoretical interpretation of such spectra and with the information to be derived from them on stereo-chemistry and bonding in co-ordination compounds. Lazarev's monograph [4] on the vibrational spectra of silicates emphasises the theoretical relationships between

TABLE 9.1

Organic	Mineral
Weak interactions between 'isolated' molecules	Strong solid state interactions
Isolated vibrations assignable to separate bonds (C—H, O—H, C=O etc.)	Coupled vibrations within condensed structures (e.g. —Si—O—Si—O—Si—)
Model compounds easily available	Models rare, synthesis and purification difficult
Well defined compositions	Range of compositions in many cases
Many sharp bands in spectrum	Few broad bands

structure and spectra. This chapter will concentrate on the use of infra-red spectra in characterising minerals and mineral products and in following their reactions. Applications of infra-red spectroscopy to the characterisation of mineral surfaces and surface reactions have been the subject of recent books [5, 6] and reviews [7-9] and will not be discussed here. Lyon [10] and Tuddenham and Stephens [11] have discussed the use of infra-red spectroscopy in determinative mineralogy.

Before turning to applications, however, it will be necessary to consider some consequences of vibrational coupling in crystals and the effects on the spectra to be expected from the isomorphous substitutions that are common in natural minerals. Only a broad sketch of the theoretical background of crystal vibrations can be given here, and texts by Hadni [12] and Mitra [13] should be consulted for a more detailed treatment.

9.2 THE VIBRATIONS OF CRYSTALS

The basic principles underlying infra-red spectroscopy, namely that the vibrational states in a molecule due to the different ways in which the constituent atoms move relative to one another are quantised and hence that transitions between the different possible states correspond to well defined energies, i.e. to the absorption or emission of radiation of well defined frequencies, have been discussed in Chapter 1, section 1.1.2. In practical terms this means that an almost infinite number of vibrations may occur in a crystal, but spectroscopic measurements in the infra-red detect only those vibrations that result in a change in the dipole moment in the structure over a distance comparable with the infra-red wavelength, which is in the range 2-200μm. Thus in a crystalline compound we are concerned only with those motions in which the unit cells all vibrate in phase, that is with, at most, $3N - 3$ separate vibrations, where N is the total number of atoms in the unit cell. Of these $3N - 3$ vibrations, moreover, some may produce no change in the dipole moment and so these will be inactive in the infra-red and will not contribute to the observed spectrum. Let us now consider how these criteria must be developed to allow for the special factors applying to the solid state.

9.2.1 Two Model Crystals

As an introduction to the vibrations of crystals we may consider the vibrations of a linear crystal containing two different atoms, A and B, in the unit cell. Two extreme examples will be compared, as shown in figure 9.1, of which one comprises an array of diatomic A-B molecules with a strong primary bond between each pair of A and B atoms but with weak Van der Waals forces between the separate molecules, and the other comprises a chain of A and B atoms in which the -A-B-A- bonds are all equally strong, being either covalent or ionic. We shall consider the crystals to possess, at

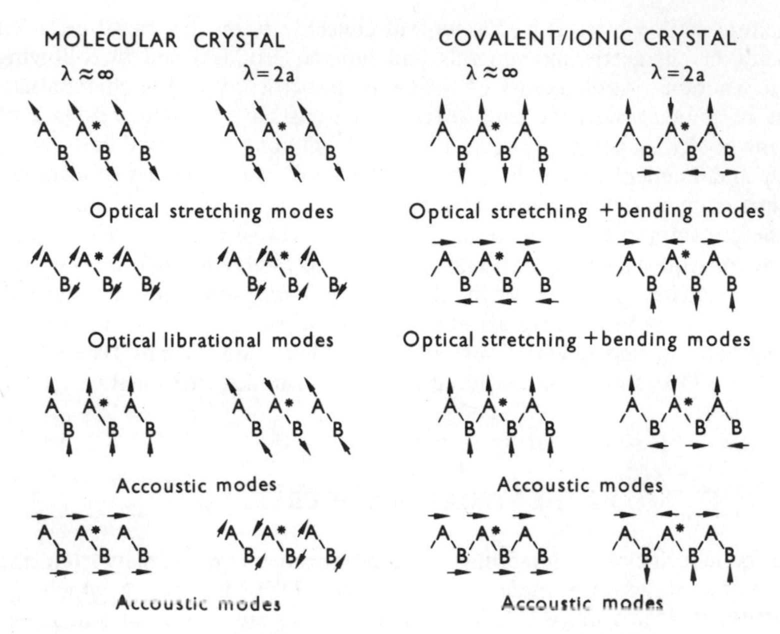

Figure 9.1 The vibrations of molecular and chain-like linear crystals.

most, two dimensions and confine our attentions to vibrations in the plane of the crystal only.

As stated above, infra-red radiation has a wavelength long compared with the unit cell dimensions of the crystals and so it will excite vibrations of long wavelengths and involving oscillating dipole moments, in which adjacent cells vibrate in phase. For each of the model crystals here considered there are two modes of vibration, the optical modes with $\lambda \approx \infty$ in figure 9.1, which can interact with infra-red radiation, plus further modes, the acoustic modes, which are infra-red inactive. Inspection of figure 9.1 will show that the optical modes involve relative motion of the A and B atoms within each molecule or chain in directions which result in changes in the size or orientation of the various interatom vectors, whereas the acoustic modes involve motion of the A-B molecules or chains as units without relative motion of the separate atoms. The former motions result in changes in the dipole moments of the crystals relative, for example, to their long axes, while the latter motions do not.

The Molecular Crystal

In the molecular crystal, one of these optical modes is an A-B stretching vibration, little changed in form or frequency from that given by the A-B molecule in the gas phase, since the perturbing effect of the very weak Van der Waals forces acting on the molecule is slight. The other is a rotary vibration, known as a libration, whose frequency, controlled as it is by the weak force constant of the Van der Waals interaction, is much lower than that of the A-B stretching mode. Associated with these long wavelength optically-active vibrations are two families of similar vibrations with wavelengths ranging down to *2a*, where *a* is the unit cell dimension. The two modes involving this shortest wavelength are also sketched in figure 9.1, and from the diagrams it will be obvious that they cannot be optically active, since the dipole moments in adjacent cells are opposed and so cancel out when a number of cells are considered. Nevertheless, the whole set of crystal vibrations which involve motion of atoms within a unit cell relative to one another are called *optical modes*, although only those of long wavelengths can be infra-red or Raman active.

Two families of acoustic mode vibrations can also be identified in the crystal, and these involve translation of each molecule as a whole (figure 9.1). The name acoustic mode derives from the ability of the long wavelength members of these families to carry sound waves. Again, the shortest wavelength possible is *2a* and, at this limit, the frequency of these modes could approach or exceed that of the librations, since they again involve Van der Waals forces, but they are infra-red inactive. As the wavelength increases, the frequency of the associated vibration decreases and reaches zero in the limit with $\lambda \approx \infty$, which corresponds to translation of the crystal as a whole.

The resulting infra-red spectrum obtained from this crystal would, therefore, contain two quite narrow absorption bands corresponding to the quantised energies for the stretching and the librational modes of vibration. Of these two, the higher energy band will correspond to the stretching mode.

If, now, a proportion of foreign molecules, A*B, were introduced into this molecular crystal (for example, A could be hydrogen and A* deuterium, but A and A* need not be isotopic) then the infra-red spectrum of the resulting mixed crystal would show an absorption corresponding to the A*B stretching mode in addition to the band for the AB stretching mode. If the coupling forces between adjacent molecules are weak, with respect to stretching vibrations, the behaviour of the surrounding AB molecules will be little affected and the character of their stretching absorption will not change. Similarly, a new absorption band due to A*B libration could appear but, if its frequency should happen to lie close to that of any of the libration or the acoustic modes of the pure crystal, the A*B molecule could couple with these vibrational modes to give a rather broad absorption band. Moreover, the librations of adjacent AB molecules in such a close-packed array would be noticeably perturbed by the presence of the A*B molecules, and this would also result in broadening of the libration absorption.

This model of an isolated vibrator in a solid crystal can be well approximated by the hydroxyl ions in the talc structure. Each hydroxyl group is co-ordinated to three divalent cations and, in a mixed Mg–Ni talc, four slightly different hydroxyl groups can be identified, co-ordinated to Mg_3,

Mg_2Ni, $MgNi_2$, and Ni_3 respectively. Such a talc does show four discrete absorption bands, as shown in figure 9.2, whose values vary little in frequency or band width as the ratio of magnesium to nickel changes [14]. Note, however, how the relative intensities of the bands change, as the relative proportions of the four co-ordination groups change.

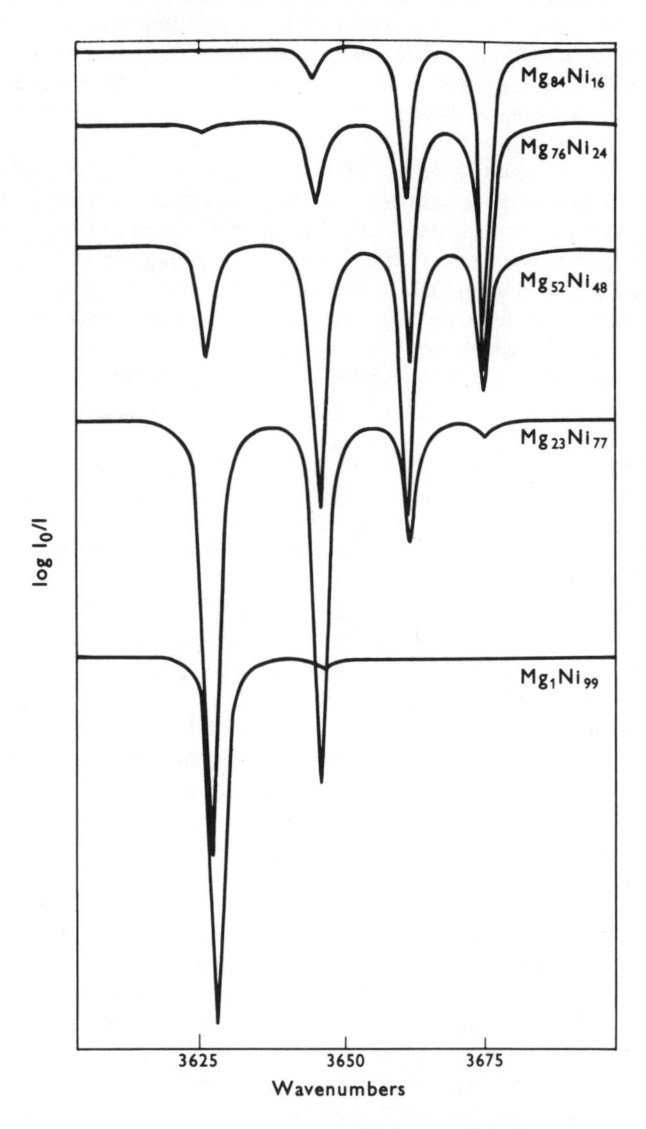

Figure 9.2 O—H stretching vibrations in mixed Ni–Mg talc. (From Wilkins and Ito [14], by permission of the *American Mineralogist*.)

The Covalent/Ionic Crystal
If we now turn to the condensed chain structure (figure 9.1), it can be seen that the low-frequency libration of the molecular crystal has here become a vibration that involves an A-B stretching mode. In fact, coupling between the two possible stretching modes gives two infra-red active vibrations, one with a dipole oscillation direction (or transition moment) oriented perpendicular to the chain, i.e. parallel to the family of 2-fold rotation axes which pass through the A and the B atoms, and the other with its transition moment perpendicular to these 2-fold axes, i.e. parallel to the long axis of the chain. The two modes are represented by the upper and the lower 'optical stretching + bending modes' respectively in figure 9.1, and it may be noted that this situation provides an example of the way in which the symmetry can determine both the form of the crystal vibrations and the directions of their transition moments. The infra-red inactive optical modes at the short wavelength limit ($\lambda = 2a$) could have even higher frequencies, since they are purely A-B stretching vibrations.

The acoustic modes again range in frequency from zero at $\lambda = \infty$ to a limit, at $\lambda = 2a$, determined by the masses of the atoms involved and by the force constants which characterise ABA and BAB angular deformations and non-bonded A-A and B-B interactions. These frequencies will again be lower than the optical modes, although they will be much higher than those of the corresponding modes of the molecular crystal.

The presence of a foreign atom (A*) in such a crystal results inevitably in interactions between the vibrations of the foreign atoms and other lattice vibrations, since the presence of such an atom no longer noticeably affects only the one B atom to which it is linked, in the molecular crystal, but it now affects at least the two B atoms adjacent to it and the two nearest A atoms, and it may also affect the next nearest pairs of B and A atoms along the chain. For an isomorphous replacement series ranging from AB to A*B, the frequency of a maximum absorption may vary almost linearly with composition from the frequency characteristic of one end member to that of the other, provided that these frequencies are not too different in value. Figure 9.3c shows diagrammatically such a regular series, exemplified by the solid solution series $Na_xK_{1-x}Cl$. If the frequencies of the pure end members are further separated, two absorption maxima may be present at intermediate compositions to give the spectra shown diagrammatically in figure 9.3a and 9.3b, which represent the behaviour of the absorption maxima in the spectra of CdS_xSe_{1-x} solid solutions.

A further possibility may be noted, that the crystal may become ordered at the composition AA^*B_2, with A and A* occupying alternate sites in a superlattice arrangement. Under these circumstances the four modes with wavelength $2a$, both optic and acoustic in figure 9.1, will all become infra-red active, since adjacent transition moments no longer cancel out. They must now be reclassified as optic modes with $\lambda \approx \infty$, based on unit cells of composition AA^*B_2 which are undergoing identical vibrations. Thus for this more complex regular structure we should obtain six discrete

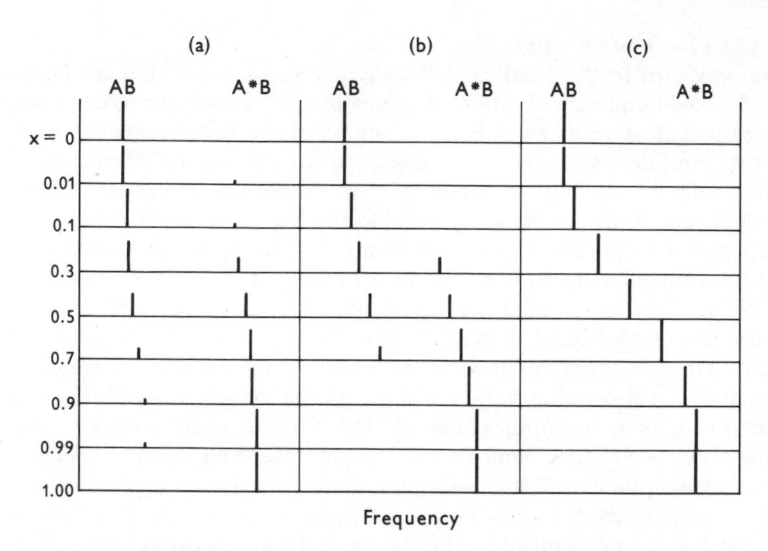

Figure 9.3 Diagrammatic representation of the changes in absorption pattern in three series of solid solutions of the type $(A_{1-x}A^*_x)B$, as x varies from 0 to 1. The hypothesis of separable vibrations is acceptable in (a), a rough approximation in (b), and invalid in (c). (After Tarte [15].)

infra-red active in-plane vibrations, some involving A-B and A*-B stretching motions, and some BA*BA bending motions. It will be realised, therefore, that many optical and acoustic modes can become infra-red active in disordered structures, leading to broad absorption bands in the general regions of the optical and acoustic vibrations of the pure AB and A*B crystals.

A series which illustrates some of the features associated with ordered and disordered substitutions in condensed structures is shown in figure 9.4. The layer silicate mineral, pyrophyllite, contains regular planar anions of unit cell composition Si_2O_5 and gives sharp absorption bands in the infra-red. Random replacement of one in four of the silicons by aluminium produces muscovite, which gives much more diffuse absorption bands and an additional band near $750 cm^{-1}$ (13.3μm) which involves Al-O vibrations. In the related aluminium-rich mica, margarite, half of the silicons have been replaced by aluminium to give the anion the ideal composition $SiAlO_5$, with the silicon and the aluminium occupying alternate sites. The absorption bands from such an ideal, synthetic margarite are significantly sharper than those from muscovite. Natural margarites, however, depart to a greater or lesser extent from the ideal composition, and the resulting departures from the perfectly ordered arrangement of silicons and aluminiums causes the spectra of such materials to be again more diffuse. Finally, particularly diffuse spectra result [16] when part of the aluminium content is randomly replaced by beryllium, but no new absorption maxima appear in this case

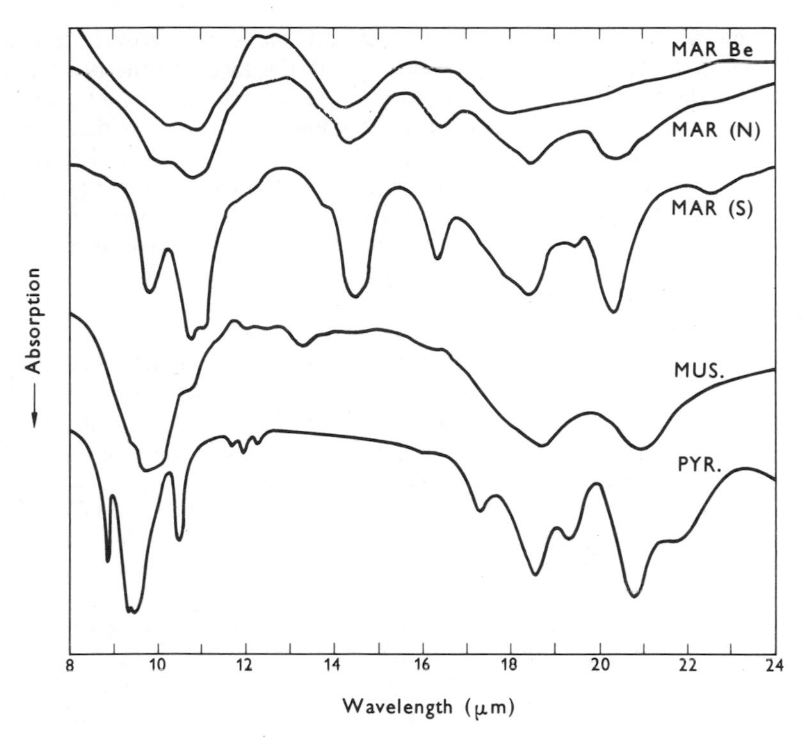

Figure 9.4 Order-disorder effects in layer silicates illustrated by the spectra of pyrophyllite (PYR), muscovite (MUS), synthetic margarite of ideal composition [MAR(S)], natural margarite [MAR(N)], and beryllian margarite [MAR(Be)].

since the Be-O stretching frequencies are close to those of Si-O and so the contributions overlap.

9.2.2 More Complex Crystals

As the number of atoms, N, per unit cell increases so the number of possible optical modes of vibration increase rapidly. We can quickly derive the number of optical modes possessed by these atoms, as mentioned in section 9.2, by remembering that each atom can be considered to have three vibrational degrees of freedom, giving a total of $3N$ for all the atoms in the unit cell. From this number the three vibrational modes which result in translations of the unit cell as a whole, and which contribute to the acoustic modes, must be subtracted to give a grand total of $3N - 3$ optical modes. As may be expected, some of these modes may produce no change in the dipole moment, and so these will be infra-red inactive. It may be noted that the complementary technique of Raman spectroscopy can detect vibrations that produce a change in the polarisability of the crystal and so will detect certain others of the optical modes to provide further information about the crystal. Moreover, if the local symmetry of each atom in the unit cell is known in addition to the crystal symmetry of the unit cell as a whole, one can predict the number of vibrations that can be active in the infra-red and the Raman regions, and also the directional properties of the infra-red absorption and the Raman scattering [17, 18]. Such knowledge permits more detailed studies of the crystal structure to be made; for example, a single crystal of the material can be investigated with polarised infra-red radiation to detect changes in absorption with changes in orientation of the absorbing groups [19].

None of the crystals that concern us here are simple molecular crystals of the type discussed in the previous section. The closest approach to this type, in fact, are crystals of the $M_a^{n+}(AB_x^{m-})_b$ type, in which the bonds within the AB_x^{m-} complex ion are considerably stronger than the electrostatic forces between the separate ions. In such a case the internal vibrations of the AB_x^{m-} complex anions can often be considered as more or less independent of the vibrations of the cations relative to them, the so-called external vibrations. These external vibrations lie at much lower frequencies, provided that M^{n+} is not very light, and show the characteristics of the ionic crystal discussed in the previous section. The forces exerted on such a complex ion by the surrounding ions are not negligible, however, and these forces affect the detailed movements of the various atoms so that, although the frequency range in which a given ion absorbs may not vary greatly from one crystal to another, as shown in table 9.2, the pattern of absorption bands within this range may vary considerably from one crystal type to another. Within an

TABLE 9.2

Vibrational Frequencies (cm^{-1}) of Atomic Groups in Minerals

Group	Stretch	Bend
XOH	3750-2000	1300-400
H_2O	3660-2800	1650-1590
NH_4	3330-2800	1500-1390
CO_3; HCO_3	1650-1300	890-700
$(BO_3)_n$	1460-1200	800-600
$(BO_4)_n$	1100-850	800-600
$(SiO_4)_n$	1250-900	<500
$(SiO_6)_n$	950-600	–
SO_4	1200-1100	700-600
$(PO_4)_n$	1200-900	600-500
VO_4	915-730	<500
CrO_4	870-700	<500
WO_4; MoO_4	850-740	<500
$(MoO_6)_n$	1000-750	<500
$(WO_6)_n$	900-700	<500
AsO_4	850-730	<500

isomorphous series in which the M^{n+} cation varies, however, the pattern of absorption of the AB_x^{m-} complex will be fairly stable. This is true, for example, for the talc structure in which Mg^{2+} is replaced by Ni^{2+}. Little change is found in the vibrations of the Si_2O_5 layer anions at frequencies greater than 450cm^{-1} (22μm), although there are profound differences below this frequency where vibrations involving the associated cation occur. This is well illustrated in figure 9.5 where the patterns of absorption in the frequency range greater than 450cm^{-1} are closely similar in the two pairs of spectra, whereas the spectra obtained from the nickel-containing talc shows a considerably more complex structure at very low frequencies than do those from the magnesian talc. Replacement of the three Mg^{2+} cations in the talc unit cell by two Al^{3+} cations, giving pyrophyllite, on the other hand, causes substantial changes in the symmetry and the strengths of the forces exerted on the Si_2O_5 anions, which now show a very different pattern of absorption in the range 450–1200cm^{-1}, as can be seen by comparing the spectra in figure 9.5 with the lowest spectrum in figure 9.4.

Other spectral features found in infra-red spectra may arise because crystals frequently contain a set of two or more complex ions in the unit cell, each set occupying crystallographically equivalent sites in the unit cell. Thus, the two hydroxyl ions shown in figure 9.6, related by a plane of mirror symmetry perpendicular to the plane of the hydroxyl ions and an axis of two-fold symmetry in the plane of the ions, exemplify such a set. This combination could give rise to two distinct OH stretching frequencies, corresponding to the symmetric (figure 9.6a) and anti-symmetric (figure 9.6b) vibrational modes, which have transition moments respectively parallel and perpendicular to the two-fold axis. The separation between the two frequencies would depend on the coupling forces acting between the two hydroxyls. Such coupling is usually mediated through the oscillating electric fields generated by the dipolar vibrations, and so is largest when the absorption bands are potentially intense. The coupling can be recognised by examining the spectrum of a crystal of the substance which contains a low concentration of an isomorphous replacement. In the example quoted, a proportion of the OH groups might be replaced by OD whereupon only a single OD frequency would be found, and not a doublet as for the OH vibrations, since the OD groups could not couple with their OH neighbours due to the marked difference in the frequencies for the vibrations of OH and OD groups.

A further point to note is that, in an isomorphous series in which only the cation changes, the vibrations of coupled anions usually change little, but in a series in which the anion changes, i.e. where GeO_4 replaces SiO_4, there can be marked differences in spectra between those of the pure salt and the dilute crystalline solution, since coupling is absent in the latter case [15].

In most oxides, halides and sulphides, no isolated ionic groupings can be distinguished. Nevertheless, some general regions of absorption can be associated with X-O stretching vibrations in oxide structures, particularly in

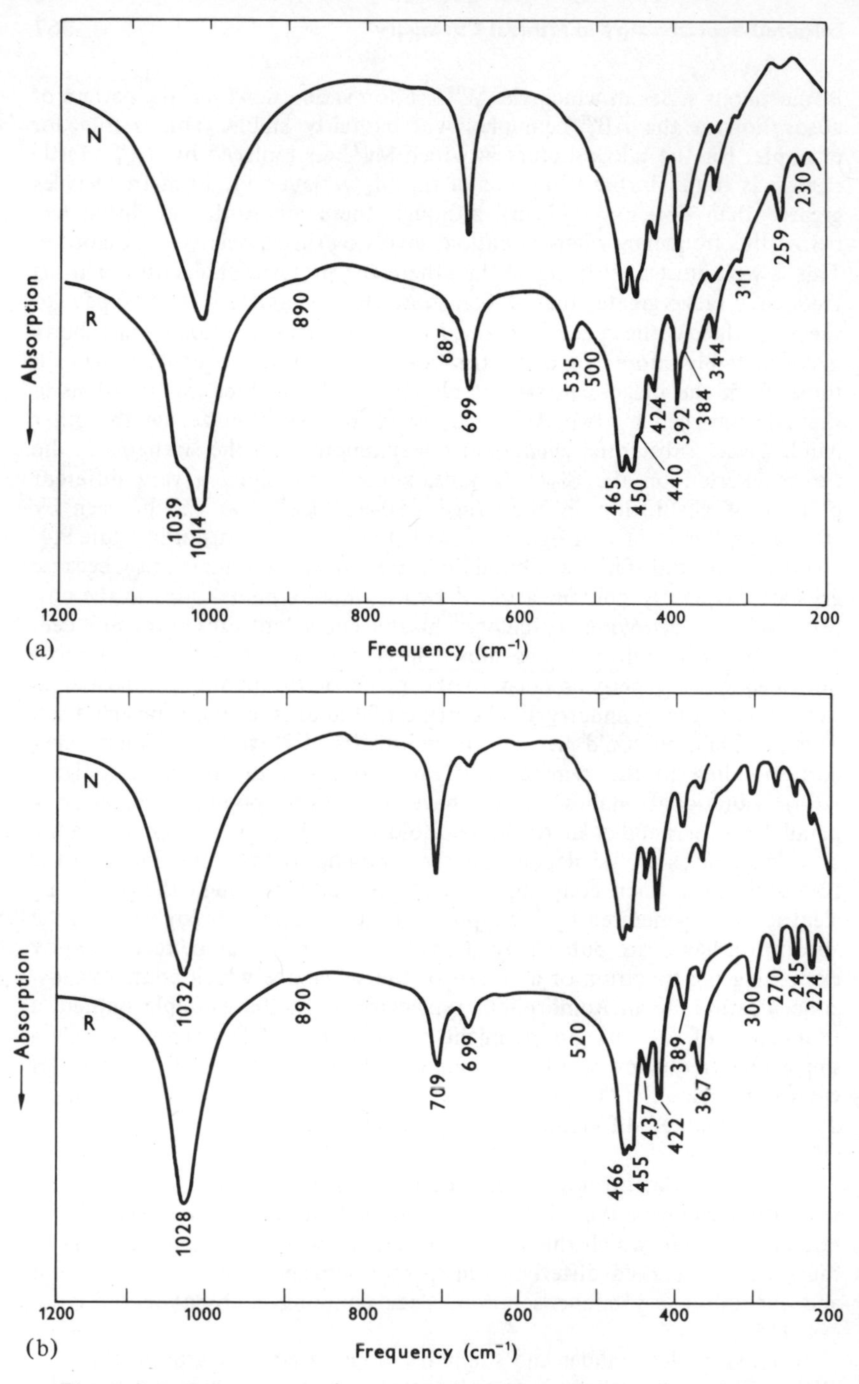

Figure 9.5 Infra-red spectra of (a) Mg–talc and (b) Ni-talc. Comparison spectra taken using pressed discs (R) and oriented films (N). (From Russell *et al.* [19].)

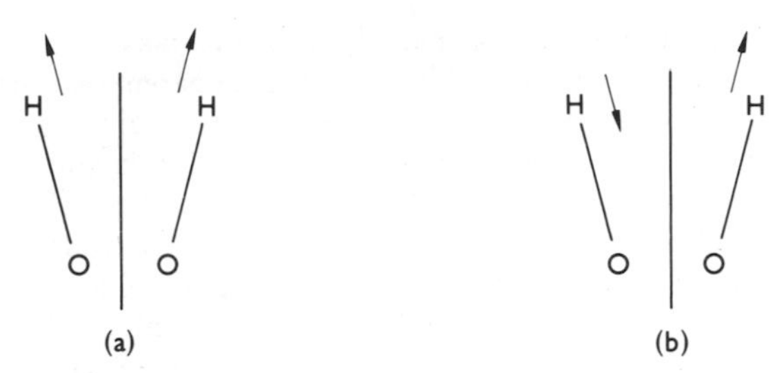

Figure 9.6 Schematic representations of the symmetric (a) and anti-symmetric (b) vibrations of two hydroxyls related by a plane of symmetry.

structures where the cation-oxygen link predominates over any other bond to the oxygen. Table 9.3 lists the absorption ranges for a number of metal-oxygen polyhedral arrangements in isolated and condensed forms, and from these data the general factors which control vibrational frequencies can be deduced to be:

(a) Cations of lower mass absorb at higher frequencies (compare Be^{2+} with Zn^{2+}, and Al^{3+} with Fe^{3+}), but with increasing cation mass, the vibrational movement becomes localized on the oxygens and so becomes insensitive to the cation mass;

(b) Cations of higher charge absorb at higher frequencies (compare Fe^{3+} with Fe^{2+});

(c) Cations in tetrahedral co-ordination absorb at higher frequencies than when in octahedral co-ordination;

(d) If an 'isolated' XO_n polyhedron can be distinguished in the crystal, its absorption will usually lie at lower frequencies than when the co-ordinated oxygens are shared with cations of the same type to give condensed polyhedra.

TABLE 9.3

Regions of Absorption (cm^{-1}) Corresponding to Metal-Oxygen Stretching for Tetrahedral and Octahedral Co-ordinations. Isolated and condensed polyhedra are distinguished.

Metal	XO_4		XO_6	
ion (X)	isolated	condensed	isolated	condensed
Ti^{4+}	650-800			600-650
Al	650-800	700-900	400-450	500-650
Fe^{3+}	550-650	550-750	300-450	400-550
Fe^{2+}				~400
Zn	400-500	400-650		~400
Mg	500-550			350-480
Be		700-950		
Li	400-500	400-600		<300

The effect of these factors, namely valency, mass, and co-ordination number, are also obvious in table 9.1. The apparent anomaly whereby tungsten and molybdenum in 4-fold co-ordination apparently absorb at lower frequencies than when in 6-fold co-ordination is due to the marked distortion of the oxygen octahedron in the latter case, resulting in some X-O distances being shorter than those in the tetrahedral co-ordination.

Similar tables can be drawn up for halides and sulphides [2]. The greater mass of the sulphide ion, of course, causes a displacement of the vibrational bands to lower frequencies for each cation, compared with the values for the corresponding oxide.

9.2.3 Effects of Particle Shape

In inorganic materials [20, 21], the oscillating dipoles often generate considerable electric fields, which account for much of the coupling between adjacent oscillators. In crystals that are smaller than the wavelength of the infra-red radiation used, the synchronous vibration of these oscillators corresponds to an electrical polarisation of the crystal, which is in turn equivalent to the generation of positive and negative charges on opposite faces of the crystal, the signs and magnitudes of the charges changing in phase with the atomic vibrations. There is, therefore, an electric field in a crystal which can be regarded as being generated by these surface charges. This field is negligible for vibrations in the plane of a thin plate, or along the axis of a thin needle, as shown in figure 9.7b, but it can be

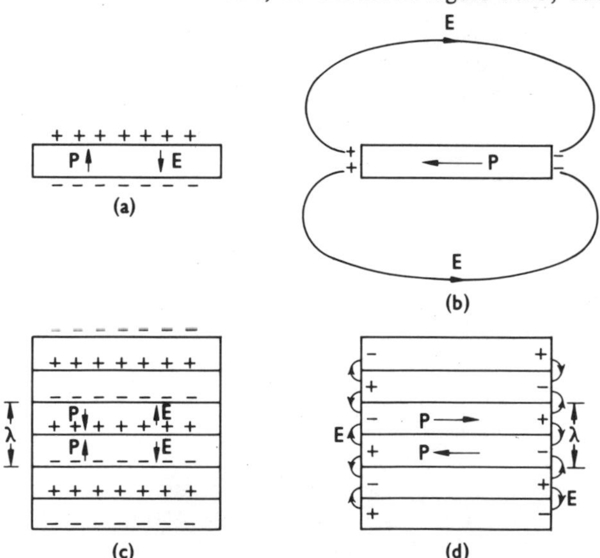

Figure 9.7 The dielectric polarisation and associated electric fields produced by vibrations in thin plates or needles (a, b), in the longitudinal mode (c) and the transverse mode (d) of large isotropic crystals. (From Farmer and Russell [21], by permission.)

considerable for oscillations perpendicular to these directions, as in figure 9.7a. This electric field opposes the dipole vibrations, since it is generated by the vibrations, and so it increases the effective force constant for the vibrations and hence their frequencies [21].

The effect of these fields can clearly be seen in spectra obtained from very fine needles of silicon carbide and shown in figure 9.8. Here the same pairs of atomic vibrators absorb at 792 and 928cm^{-1} depending on whether the vibration vector is oriented parallel to or perpendicular to the needle axis, respectively [22]. The needles are, moreover, coated on their surface by SiO_2 and here, too, there is a marked difference between the two sets of vibrators depending on whether their vibrational directions are oriented parallel to the needle axis (giving absorptions at 1080 and 455cm^{-1}) or perpendicular to this axis (1165 and 490cm^{-1}). It is found that non-fibrous particles of SiC give broad absorptions at frequencies intermediate between the doublets quoted above [22].

In isotropic crystals which are large compared with the infra-red wavelengths there exists a range of crystal vibrations associated with a single unit cell mode extending from those with zero electric field (the transverse optical modes shown in figure 9.7d) to those with an associated field $4\pi P$, where P is the polarisability of the material (the longitudinal modes shown in figure 9.7c). Such crystals are opaque in the infra-red between these frequencies, and this phenomenon accounts for much of the breadth of the

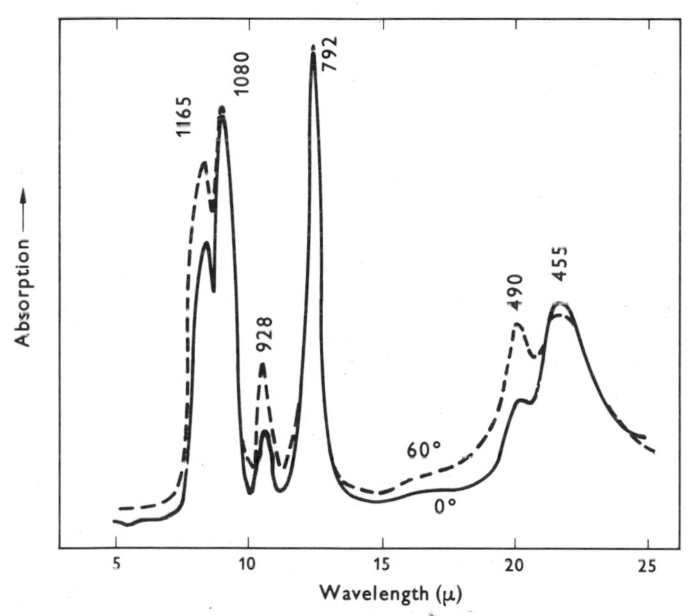

Figure 9.8 Infra-red spectra of a film of fibrous SiO_2–SiC taken at 0° (solid line) and 60° (dotted line) incidence to the infra-red beam. (From Russell *et al.* [22].)

absorption of simple oxides, such as MgO, and halides, NaCl, which possess only a single infra-red active mode of vibration.

The consequence of the foregoing observations is that care must be taken not to ascribe differences in spectra which arise by virtue of the crystal shape to differences in composition or structure.

9.3 APPLICATIONS TO MINERALS

The above discussion will have shown that infra-red spectroscopy is capable of identifying atomic groupings in a substance, and hence that it can be used both for the identification of unknown minerals, or for the characterisation of the constituent groupings in a material, or as an aid in following the course of a solid state reaction.

9.3.1 Identification of Minerals

Infra-red spectroscopy is largely complementary to X-ray diffraction (Chapter 7) as a tool for the identification of mineral species. The method has the great advantage over X-ray diffraction, however, of being able to detect and characterise non-crystalline compounds, since these absorb as strongly as crystalline materials. Moreover, the infra-red spectrum points more directly to the general nature of an unknown substance, such information being obtainable from the presence or absence of characteristic absorption bands, as listed in table 9.2. Because the positions of the absorption bands in a spectrum are sensitive to the mass, charge, and bonding characteristics of the constituent ions, infra-red spectroscopy can often place a mineral species within the range of compositions over which it exists, indicating, for example, the iron content of montmorillonites [19], as in figure 9.9. In mixtures it may be found that some components are more readily recognised in the X-ray diffraction pattern, others may show up more clearly in the infra-red spectrum.

An example of the power of infra-red spectroscopy to distinguish between two mineral species is afforded by the case of reyerite and truscottite. The X-ray diffraction patterns of these two minerals are sufficiently similar to have led to a proposal that they be regarded as a single mineral species. The infra-red spectra, however, show that the silicate anions in the two minerals must differ either in structure or in environment, as the patterns in figure 9.10a-c indicate [23]. Moreover, the reversible change in absorption pattern shown by reyerite on dehydration (figure 9.10d) is clear evidence that this mineral contains water in zeolitic channels, whereas truscottite does not [23].

Of course, the empirical identification of a mineral from its infra-red spectrum requires access to collections of infra-red spectra prepared from well characterised specimens, and it is not easy to locate even a fraction of those that have been published. Partly this arises from the fact that lists of the frequencies of absorption maxima are seldom adequate for identifi-

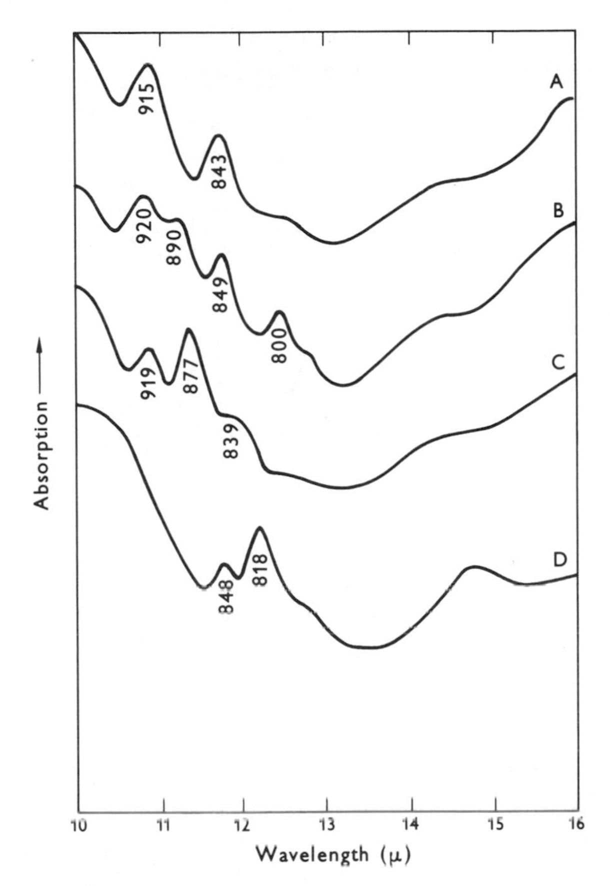

Figure 9.9 Infra-red absorption of smectite minerals in the region of OH bending vibrations: montmorillonites of low (A), medium (B), and high (C) iron content, and nontronite (D).

cation purposes, since band contours and relative intensities are often important distinguishing features. Hence there is no system for infra-red spectra which is equivalent to the J.C.P.D.S. File Index for X-ray diffraction patterns (Chapter 7, section 7.7.1).

Bibliographies compiled by Lawson [24] and Lyon [25] cover the older literature, but this is mostly useful only in the range 2000-700cm^{-1}. Valuable collections of mineral spectra extending to 400cm^{-1} have been published by Lyon [10] and Moenke [26]; Miller and co-workers [27, 28] have recorded spectra of a variety of inorganic salts down to 300cm^{-1}; Sadtler's Spectra [29] include 1500 inorganic compounds, the data ranging down to 200cm^{-1}; while Nyquist and Kagel [29a] give 900 spectra covering 3800-45cm^{-1}. Some collections of interest to specific industries have been

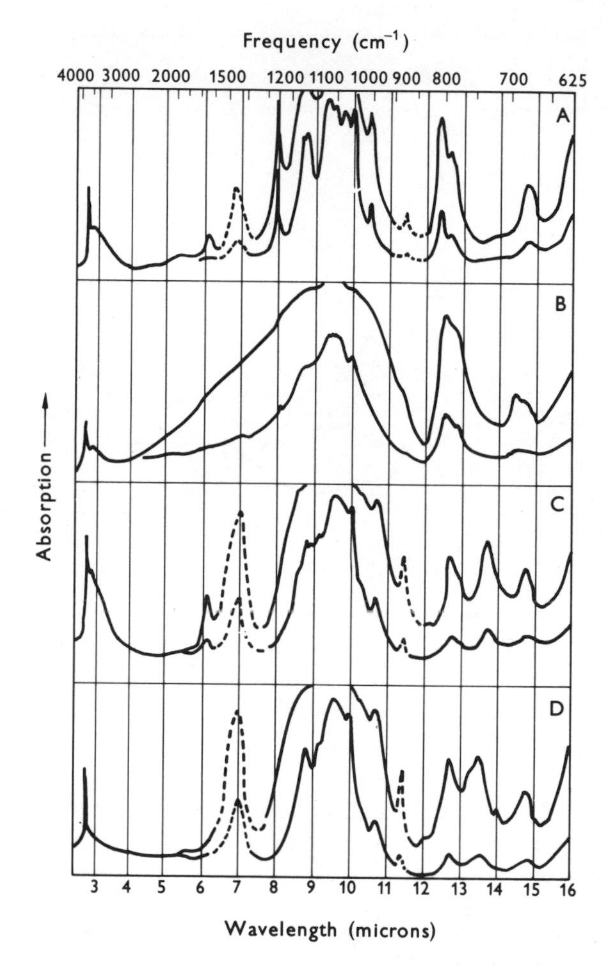

Figure 9.10 Infra-red spectra of synthetic truscottite (A), natural truscottite (B), natural reyerite, hydrated form (C), and natural reyerite, dehydrated at 250°C (D). Carbonate absorption bands are shown dotted. Spectra are shown for two sample concentrations of 2mg (upper) and 0.33mg (lower) in 12.7mm KBr discs in each case. (From Chalmers *et al.* [23], by permission of the *Mineralogical Magazine*.)

published, including spectra of pigments [30], phosphates [31], and mineral dust components [32]. Some selected references to mineral spectra and data are given in table 9.4.

Generally, however, the most valuable collection of spectra is that built up in a laboratory over the years, since this is usually more detailed and more specifically concerned with subjects of direct interest to the laboratory than any published collection can hope to be. Moreover, the

TABLE 9.4

References to Mineral Spectra

Oxides and Hydroxides

Simple Oxides: 50, 51
Spinels: 52
Rare Earth Garnets: 53
Hydroxides: 54
Alumina and Aluminates: 55, 56, 57

Miscellaneous Salts

Borates: 26, 27, 28, 58, 59, 60
Carbonates: 26, 61, 62, 63, 64, 65
Germanates: 4
Halides: 2
Molybdates: 66
Niobates and Tantalates: 67, 68
Nitrates: 1, 2
Oxyhalogen Salts: 2, 27, 28
Phosphates, Arsenates and Vanadates: 26, 31, 36, 69, 70, 71
Polyphosphates: 4, 31, 71, 72
Sulphates and Selenates: 26, 73, 74, 75, 76
Sulphides: 2
Tungstates: 26, 66
Uranyl: 77

Silicates

Orthosilicates: 26, 78, 79
Pyrosilicates: 4, 26, 78
Cyclosilicates: 4, 26, 80, 81, 82
Chain Silicates: 10, 26, 83, 84, 85, 86
Layer Silicates: 39
Silica: 26, 32, 87
Felspars: 10, 26, 88
Zeolites: 26, 89
Other 3-dimensional Silicates: 26
Cements: 90, 91
Glasses: 92, 93

spectra will have been recorded under known conditions from samples whose purity and general physical state are at least reasonably well known, and it is worth noting that large collections of spectra, such as those of Moenke or of Miller and co-workers, have often been obtained from labelled specimens from museums or from a bottle on a shelf. The labelling under these conditions is not always accurate, and certainly no guarantee of purity could possibly be given!

9.3.2 Characterisation of Constituent Groupings

The most distinctive of all infra-red absorption bands are those arising from water of crystallisation and from structural hydroxyl groups. The range of stretching frequencies of water in crystals overlaps that of structural hydroxyl (table 9.2), but the bending vibration of the water molecule near 1630cm^{-1} is a very distinctive feature which allows the ready recognition of water molecules in a mineral in the presence of structural hydroxyl groups, as illustrated in the previous section for reyerite. The loss of the zeolitic water in the dehydrated mineral is easily seen by the disappearance of the peak near to the 6μm line, in figure 9.10c, from the spectrum shown in figure 9.10d. Yet another good example of how structural information can be derived on this basis is provided by a study [33] of α- dicalcium silicate hydrate, which has the empirical formula $Ca_2SiO_5H_2$ and could be formulated either as $Ca_2SiO_4 \cdot H_2O$, with separate silicon-oxygen tetrahedra and water molecules, or as $Ca_2(HOSiO_3)(OH)$, in which the water is present as two types of hydroxyl group, one being associated with the cations as a free hydroxyl and the other existing as part of the silicon-oxygen tetrahedra, as hydroxyl bound to silicon. The absence of a band near 1630cm^{-1} in the observed spectrum shows that discrete water molecules are not present. On the other hand, a sharp band at 3532cm^{-1}, in the region typical of OH stretching absorption, can be assigned to a weakly hydrogen-bonded free hydroxyl group and two broad bands at 2847cm^{-1} and 2450cm^{-1} are typical of strongly hydrogen-bonded acidic hydroxyl groups, in this instance the SiOH groups. Thus the spectrum is consistent only with the formulation $Ca_2(HOSiO_3)(OH)$.

It should be remembered that water-containing structures fall into a special category in relation to isotopic replacement of hydrogen by deuterium. It has been shown (Chapter 1, section 1.2.3) that the vibrational frequency associated with a particular group is affected by the mass numbers of the isotopes present in the group and replacement of hydrogen, mass number 1, by deuterium, mass number 2, effectively doubles the reduced mass of the system O-H without noticeably affecting the force constant between the atoms. Two consequences of this are, firstly, that the displacement of the OD vibrations from those of OH is greater than for any other isotopic pair, and, secondly, that the OH and OD groups do not exhibit vibrational coupling. Thus, many structural studies that concern water or hydroxyl in minerals are studied in greater detail by examining synthetic preparations containing small proportions, typically 10-20%, of deuterium substituting for hydrogen, to avoid the complications arising from vibrational coupling. Thus, the presence of two different OH stretching vibrations in the spectrum of $Ba(NO_3)_2 \cdot 2H_2O$ does not in itself disprove the conclusion drawn from X-ray structural studies that all the hydrogens occupy equivalent positions. On the other hand, the fact that isolated HDO molecules in the structure still show two OH stretching and two OD stretching vibrations is conclusive proof that there are two different environments for hydrogen in the crystal [34] and the two OH vibrations from the original material do not arise from coupling alone. In this case, the infra-red spectrum has provided information which X-ray diffraction could not give.

This isotopic substitution procedure cannot be so readily applied to other molecular or ionic groupings in crystals, since the changes in the reduced mass will be less marked. Nevertheless, careful examination of the absorption pattern can permit useful conclusions to be drawn concerning the types of anion present and their environments in minerals.

Thus Moenke [35] was the first to recognise, from its spectrum, that the hydrothermal silicate mineral thaumasite contained silicon in 6-fold co-ordination. Again, the pattern of carbonate absorption given by some apatites [36] is clear evidence that the carbonate ion is present in solid solution in these minerals, and not as a separate calcite phase. An impressive record of combining an empirical approach with sound theoretical understanding in the characterisation of new synthetic silicates and germanates is to be found in the work of Lazarev [4].

9.3.3 Characterisation of Solid-State Reactions

Infra-red spectroscopy is particularly appropriate to the study of reactions involving hydroxyl groups, and can give unique information which would be difficult to obtain by any other method. Many examples could be cited from the literature, but one of particular note is the study made by Rouxhet [37] on the exchange of hydrogen and deuterium in muscovite mica. Treatment of mica at high D_2O pressure in the temperature range 580-670°C results in replacement of OH by OD in the sheets, and several studies have been made of this and of the related dehydroxylation reaction [38]. Rouxhet has obtained more detailed information by studying the behaviour of mica single crystals through monitoring the changes in the degree of absorption of a polarised beam of infra-red radiation, directed normally to the sheets, in the OH and OD stretching band regions. His results indicate that both reactions can proceed by a diffusion path perpendicular to the mica layers with an activation energy of 47kcal.mole^{-1}.

The technique is not, however, limited to the study of hydroxyl reactions, and Russell has shown that ferric ions in the centre of the mica-like layers of montmorillonite can be reduced to the ferrous state by exposing the mineral to hydrazine vapour [39], since a vibration at 877-890cm^{-1}, associated with ($Fe^{3+}Al^{3+}OH$) groups in the original mineral, is lost in the process, as shown in figure 9.9. Similarly, in expanded biotites, or vermiculites, the increase in positive charge associated with the oxidation of ferrous to ferric ions has been shown to be compensated first by the reversible loss of protons from structural hydroxyl groups, and then, when the available protons are exhausted, by an irreversible expulsion of ferric ions from the structure [40].

Infra-red studies can also be very advantageously coupled with differential thermal analytical and thermogravimetric studies, allowing weight losses to be correlated with loss of water molecules, structural hydroxyl, or carbon dioxide from carbonate groups. Figure 9.11 illustrates some of the conclusions that can be drawn from studies of the thermal transformations of a mineral, in this case the hydrous calcium silicate mineral, tobermorite, whose thermal dehydration has been studied by Farmer and co-workers [41]. The low temperature phases (figure 9.11a and 9.11b) contain separate water molecules, as evidenced by the band at 1615-1630cm^{-1}, but this is lost around 250°C, leaving a band at 3490cm^{-1}, which corresponds to the absorption of SiOH groups left in the structure (figure 9.11c). These are largely lost by 700°C (figure 9.11d). The presence of carbonate is shown by the absorption near 1440cm^{-1} (figure 9.11a-c) but this decreases on heating above 500°C as carbon dioxide is lost (figure 9.11d). The thermally stable absorption at 1340-1400cm^{-1} indicates the presence of boron. Changes in the silicate absorption pattern in the region below 1200cm^{-1} indicates that the structure of the silicate anion is changing as the water and the structural hydroxyl are lost, evidence which is in accord with the X-ray diffraction data, and the absorption patterns clearly define the separate phases formed at the different stages of the reaction (figure 9.11a-d). Finally, by 800°C, a wollastonite-like phase has formed, (figure 9.11e) which transforms to wollastonite itself at 940°C (figure 9.11f). The appearance of wollastonite is associated with the separation of silica, which can be identified in figure 9.11f by the broad absorption between 1100cm^{-1} and 1250cm^{-1}.

The use of infra-red spectroscopy to complement X-ray diffraction methods in following the crystallisation of glasses, and the subsequent ordering processes that occur have been well illustrated by the work of Langer and Schreyer [42] on phases lying in the cordierite composition field. In the transformation of high cordierite to low cordierite, the spectra of the two forms indicated that short-range ordering of aluminium and silicon ions in the structure was achieved before long-range order was fully established. Moreover, in studies of the products of hydrothermal synthesis, infra-red spectra have proved to be sensitive indicators of structural change associated with variations in composition, and Rutstein and White [43] have applied the technique successfully to pyroxenes and pyroxenoids.

9.4 TECHNIQUES AND INSTRUMENTATION

Although the procedures and equipment necessary to record and study mineral spectra are not fundamentally different from those that are well documented in the numerous texts available on infra-red spectroscopy, it must be emphasised that spectrometers and sample-preparation procedures that are adequate for routine use in an organic chemistry laboratory are not necessarily adequate for obtaining mineral spectra. For example, a spectrometer operating in the range 4000-650cm^{-1} will cover the vast majority of strong absorption bands in organic materials, whereas it is often essential to examine the region 4000-200cm^{-1} in order to characterise

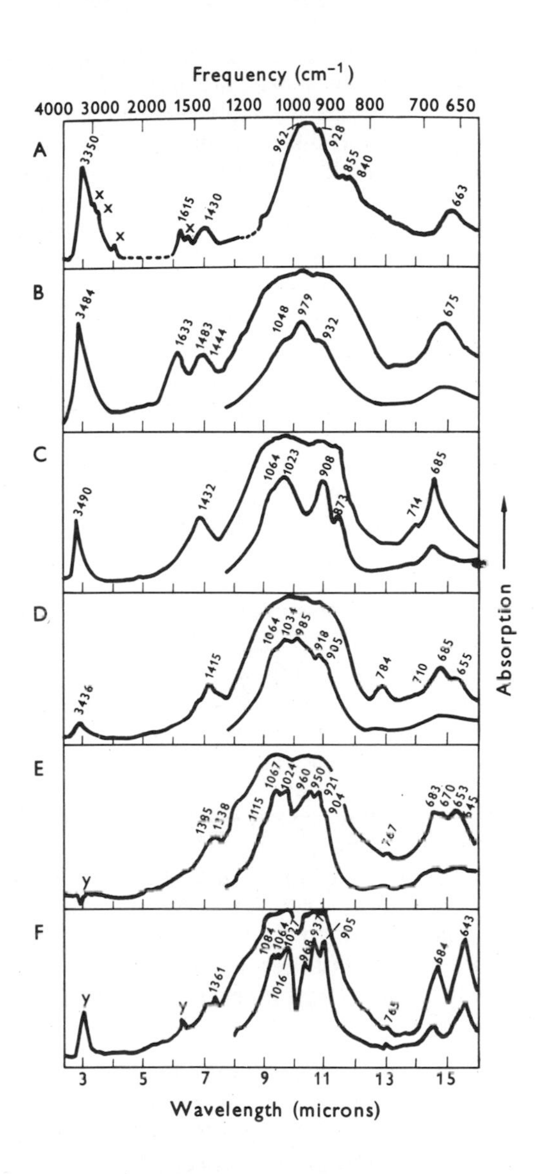

Figure 9.11 The thermal transformations of tobermorite, showing (A) the unheated mineral (14Å phase); (B) heated at 90°C (11.3Å phase); (C) heated at 520°C (9.3Å phase); (D) heated at 710°C (9.7Å phase); (E) heated at 800°C (intermediate phase); and (F) heated at 940°C (wollastonite). Where two curves are shown these are for different sample concentrations (2mg and 0.33mg in 12.7mm KBr discs). Bands labelled X arise from a mulling agent, those labelled Y from adsorbed water. (From Farmer *et al.* [41], by permission of the Highway Research Board.)

inorganic compounds adequately. Again, in most practical applications, minerals, like solid organic substances, must be examined in powder form; but the higher refractive indices of most minerals require that they be ground more finely than organic compounds to avoid excessive light scattering with the attendant loss of energy from the infra-red beam, and their hardness makes the necessary fine particle size more difficult to achieve. It may be noted that excessive particle size results in spectra of low relief with broad, distorted absorption bands, and these features are, unfortunately, too commonly seen in published spectra.

It is possible here to mention only briefly some of the factors which must be considered in selecting a spectrometer and ancillary equipment for mineral investigations. The reader is referred to books by Ferraro [2], Cross and Jones [44], and Stewart [45] for further information on spectrometers; to texts by White [46] and Potts [47] for general guidance on points of technique; and to reviews by Lyon [10], Tuddenham and Stephens [11], and Farmer [39] for discussions of some practical aspects that are particularly relevant to infra-red spectroscopic studies of minerals.

9.4.1 Choice of Spectrometer

Infra-red spectra are obtained using a double-beam instrument similar to the varieties used for optical spectroscopy, as described in Chapter 2, section 2.4.5. Earlier instruments used a prism, usually an alkali-halide prism, to disperse the spectrum but recently a number of filter-grating spectrometers in the median price range, that are particularly suitable for studies of minerals, have appeared on the market. These cover a wide spectral range, typically from $4000cm^{-1}$ to $250cm^{-1}$ or $200cm^{-1}$, with a higher resolution and better energy transmission than the older instruments. They, therefore, cover the strongest absorption bands of the oxygen-containing minerals, which have been the most productive field for infra-red studies to date, and their resolving powers ($1\text{-}2cm^{-1}$) are quite adequate since mineral absorption bands are very seldom narrower than this. In choosing an instrument, in practice, the factors to be considered tend to fall more into the area of operating facility, rather than into the field of instrumental optics, and may include

(a) the provision of adequate reserves of energy in the beam to cope with very small or highly scattering samples;
(b) the provision of sufficient space in the sample compartment to insert vacuum cells, low and high temperature cells, polarisers, and other ancillary equipment;
(c) the facility to purge the sample of water vapour in order to obtain trustworthy spectra in regions where water vapour absorbs strongly;
(d) and, finally, overall convenience of operation.

Those who would attempt a complete interpretation of mineral spectra must also have access to frequencies down to $50cm^{-1}$, but it is not yet clear how many practical advantages there are, if any, in having access to this 200-50cm^{-1} region. One possible application may be for the complete characterisation of anhydrous halides and the sulphide-group minerals. Oxide minerals that contain alkali metal or alkaline earth metal cations have strong absorption bands in this region, otherwise absorption is weak; for example quartz is quite transparent. The absorption is often featureless for minerals that are not perfectly ordered. Therefore the additional expense involved would not be justified in most routine applications.

9.4.2 Sample Preparation

Many techniques are available for presenting a mineral sample to the spectrometer, and these have been reviewed by Lyon [10]. As may be expected, the generally used methods of sample preparation are relatively simple and easy to apply, since, as in all such cases, the spectrometer will be little used if the sample preparation is tedious or ineffective.

In the first place it must be remembered that only a very small amount of sample, of the order of 0.1-1.0mg, is actually involved in absorbing the infra-red beam, and so very precise sampling techniques are required from the very beginning. Fortunately, the small amounts of material involved makes hand-picking a feasible, if not essential, technique for ensuring purity when investigating a single mineral species, and this does not often present major problems. More problems may arise in preparing a mixed sample and ensuring complete homogeneity of the sample at all stages, especially during the crushing and grinding procedures. It must be noted that some substances, notably quartz, carbonates, etc., absorb very strongly and as little as 1-2% of these as impurities can present difficulties in interpreting the resulting spectra.

Grinding the Sample

It is necessary to grind the sample to a size less than the wavelength of the incident radiation in order to ensure the sharpest possible absorption bands and to reduce scattering. In general this involves grinding to a particle size below 2μm. Grinding, however, presents its own problems, particularly because local temperatures of up to 500°C can be generated during the grinding process in air, resulting in the dehydration of many hydrated minerals or the reforming of some anhydrous species. Fortunately, however, the problem can almost always be avoided by grinding the mineral under an inert liquid, such as absolute alcohol, but it is imperative that all grinding procedures be carried out as carefully and as gently as possible. Agate and mullite mortars are preferred.

Sample Preparation

Minerals are normally studied in transmission and the spectra recorded as absorption bands, although reflection spectra have advantages for theoretical studies. Minerals are usually presented to the infra-red spectrometer for absorption measurements either pressed into a transparent disc with a large excess of an alkali halide or mixed into a mull with a suitable organic oil and held between alkali halide discs. In both cases, the surrounding material plays a double role. On the one hand it binds and supports the loose powder of the sample, and on the other it reduces scattering of the infra-red beam by virtue of the close match between its refractive index and that of the powder.

The alkali halide disc method is the most popular technique currently in use, and is the most satisfactory for the majority of minerals. KBr, which is transparent down to $250cm^{-1}$, is the most commonly used disc material, but CsI is necessary to reach $150cm^{-1}$ and has advantages in reducing scattering from samples with high refractive indices [30]. KC1 and KI may be used, and Pytlewski and Marchesain have described the use of AgC1 [48]. Selected analytical grade KBr is perfectly suitable after grinding and heating to 500°C, and can be stored at room temperature if the relative humidity is kept below 60%.

Typically, about 1mg of the sample, which has previously been crushed by hand in an agate mortar slurried with absolute alcohol, is added to about 350mg of KBr and mixed in an efficient, low volume ($5cm^3$ or less) vibratory grinder, of the Wig-L-Bug type, fitted with a steel or agate vial and beads. A mixing time of about 10 minutes normally suffices to give a homogeneous product with a sample particle size of $<2\mu m$. The mixture is then charged into a vacuum die and subjected to a pressure of about 130,000psi to give a clear disc which can be stored permanently for subsequent re-use, provided that surface and adsorbed water is removed by warming overnight in a vacuum oven.

With quite simple equipment, therefore, including a torsion balance to weigh 0.5-5mg samples, a standard balance for 250-500mg KBr, the grinder and the press, a technical assistant should be able to produce quantitatively reproducible spectra every 20 minutes or so. The method is very fast, and, as indicated, readily gives reproducible results. It is also flexible, and it is often advantageous to run more than one spectrum using different concentrations of the sample in order to emphasise different features in the spectrum, as shown in figure 9.10. Such discs can also be used to follow the course of dehydration and thermal transformation in a single small mineral sample, by heating a disc containing it step-wise up to 700°C. Of course, this last method would not be applicable to systems in which reaction could occur with the KBr itself.

The pressed disc technique must not, of course, be used blindly. Any grinding operation can destroy the crystallinity of the sample, and so care must be taken in using the vibratory mixer with, for example, easily dehydrated materials. Water-soluble salts cannot be incorporated into alkali halide discs without danger of some interaction leading to new chemical species, and ion exchange reactions are also possible with minerals containing zeolitic channels. Highly hydrated minerals can also dehydrate in the vacuum used while pressing the discs. Whenever such possibilities arise, alkali halide disc spectra should be confirmed by using paraffin oil mulls, or discs pressed from polyethylene powder.

The other popular method involves immersing the ground sample in an oil of comparable refractive index and supporting the resulting mull between two clear alkali halide plates. Nujol, a medicinal paraffin oil with a minimal number of absorption bands in the region of interest, is commonly used, with plates made from NaCl or KBr. Fluoro-carbon oil is used to examine regions, near $3000 cm^{-1}$ and $1400 cm^{-1}$, that are obscured by paraffin absorption. The method is again fast, but uses more sample, and it suffers with respect to the KBr method in that interferences can occur between sample and oil absorption bands, and in that the specimen control is poorer and quantitatively reproducible spectra are less easily produced.

Discs pressed from polyethylene are transparent at all frequencies below $700 cm^{-1}$ but interferences can occur at higher frequencies and their low refractive index results in severe scattering losses unless meticulous care is taken in reducing the particle size of the sample.

In special circumstances a number of other sample preparation techniques may be used. These include the deposition of a film of dry material from alcohol suspension onto a transparent rock-salt window [49], which is useful for rapid studies but suffers from scattering problems and from preferred orientation of the particles; the formation of self-supporting films, particularly useful for producing a 'paper' of oriented clay particles to permit studies to be made specifically of the O-H vibrations [49] in the clay; discs pressed from pure mineral powders without the addition of a binder; and cleaved or ground sections. These last are especially applicable to reflection studies, which are of great importance for quantitative studies of absorptivities and refractive indices, but which are of limited use in routine work.

9.4.3 Measurement of Spectra

The use of double-beam spectrometers in infra-red spectrometry provides much flexibility in the way in which absorption spectra may be recorded. In particular, the instrument may be used in a differential mode by placing the sample in the 'recording' beam and a blank or a modified sample in the 'reference' beam. The resulting spectrum will represent the differences in absorptivities of the substances in the two beams, over the wavelength range of the scan.

In its simplest form the technique can be used to 'remove' absorption due to the support material used. Thus, when using a KBr disc or a Nujol mull, it is usual to place either a disc of pure KBr or a sample of Nujol, in both cases adjusted to match the amount present in the sample, in the reference beam in order to provide a flat baseline for the spectrum. In the case of Nujol, this also markedly reduces the interference from the absorption bands of the oil and makes the spectrum much clearer and easier to read. A similar comment applies to discs made from polyethylene.

From this it is but a small step to techniques which involve 'removing' the absorption bands due to a component from the overall spectrum. Sufficient of the substance involved is incorporated into the blank disc to balance exactly the absorptions due to the component in the two beams, and hence give a zero reading in the final signal to the recorder. The technique is very useful both for removing the contribution of a major compound in order to emphasise the spectrum of a minor component, and also to remove a 'contaminant' or unwanted phase. Thus quartz, which absorbs very strongly, can be removed from soil and mineral samples, or the loss of water and the changes in the pattern of silicon-oxygen and cation-oxygen absorption bands can be followed by using discs containing the mineral dehydrated to varying degrees as blanks against a disc of the original mineral, or vice versa.

It must be remembered, however, that the intensity of absorption is directly proportional to the amount of the absorber present. Therefore, two factors must be borne in mind when using the differential method. In the first place, if too much energy is absorbed from the two beams, the differential signal will be very weak and the recorder reponse will be very sluggish. In the second, if there is more of the absorber in the blank than in the sample, a negative absorption peak will be seen in the spectrum, indicating that the concentration in the blank should be reduced.

9.4.4 Quantitative Studies

The ease with which reproducible discs can be produced would suggest that it is, in principle, a straightforward matter to apply the technique of infra-red spectroscopy quantitatively. The separate minerals in a mixture will absorb independently of one another, according to their separate patterns, and the intensity of a given absorption band relative to the intensity from the same band for the pure mineral under the same conditions of measurement is proportional to the amount of the mineral in the mixture, in the usual Bouguer-Beer law relationship. Hence it should be possible to apply the same methods as are used in optical spectroscopy to determine the relative amounts of the components in a mixture. The technique does depend, however, on the ability to distinguish an absorption band which is characteristic of the mineral in question and to measure its intensity independently of contributions from any of the other minerals in the assemblage, and this can prove difficult because of the breadth of the absorption bands and the extent of overlap or partial overlap in many spectra. Moreover, considerable care is needed in preparing the sample, to give a very small particle size, and differential recording with KBr discs should be used.

The method is useful with minerals having easily distinguished bands, or in the study of changes in the concentration of a group with a strongly characteristic band or bands, as listed in table 9.2. Lyon [10] has pointed out that determination of the components in a mixture may be made quite accurately by feeding the intensities of these absorptions into the relevant set of simultaneous equations and solving. It is probably most useful, however, in monitoring the loss of a volatile component on heating. As shown in section 9.3.3 and figure 9.11, the course of dehydration or the rate of carbonate loss can be followed quite closely from the changes in intensity of the water bands at around 3700-3600 and 1640cm^{-1} or of the carbonate bands at 1450-1410 and 880-800cm^{-1}, and, especially in conjunction with thermal analytical studies, infra-red spectroscopy can give a wealth of information about the reactions occurring, which would be quite inaccessible by other means.

Generally, however, infra-red spectroscopy is applied as a semi-quantitative method at most, and usually it is used solely as a means of qualitative identification.

9.5 SUMMARY

Infra-red spectroscopy provides a rapid, simple and convenient non-destructive means of characterising and identifying minerals. Moreover, since infra-red absorptions arise from the interactions between atoms within the crystal structure of the mineral, it can readily indicate the presence of specific atomic groupings within the crystal structure in a way that other methods, such as X-ray diffraction, cannot. In particular, it is probably the best technique for detecting the presence of water in a mineral, and for indicating the form in which the water is present. It is also a powerful means of detecting carbonate and of providing clues to the nature of the silicate anion in the mineral structure. The technique has the great advantage over X-ray diffraction of being able to detect non-crystalline phases, and it has been shown that some minerals are more easily identified from their infra-red spectra than from their X ray diffraction patterns.

As the amount of good data available on minerals grows and is correlated and compiled, so the method will grow in importance as a means of identification. At present, however, it appears that infra-red spectroscopy will not replace the older methods of mineral identification, but it can prove to be a very potent tool when used in conjunction with X-ray, optical, and thermal analysis. It is probably the best single technique to use in a preliminary examination with a view to defining the general nature and purity of a mineral specimen.

REFERENCES

1. K. NAKAMOTO, 'Infra-red Spectra of Inorganic and Co-ordination Compounds', 2nd. ed., John Wiley and Sons, New York and London, 1970.
2. J. R. FERRARO, 'Low-frequency Vibrations of Inorganic and Co-ordination Compounds', Plenum Press, New York and London, 1971.
3. D. M. ADAMS, 'Metal-ligand and Related Vibrations', Arnold, London, 1968.

4. A. N. LAZAREV, 'Vibrational Spectra and Structure of Silicates', Plenum Press, New York and London, 1971.
5. L. H. LITTLE, 'Infra-red Spectra of Adsorbed Species', Academic Press, London and New York, 1966.
6. M. L. HAIR, 'Infra-red Spectroscopy in Surface Chemistry', Arnold, London, 1967.
7. D. J. C. YATES, *Catal. Rev.*, **2**, 113 (1968).
8. M. R. BASILA, *Appl. Spectros. Rev.*, **1**, 289 (1968).
9. V. C. FARMER, *Soil Science*, **112**, 62 (1971).
10. R. J. P. LYON, 'Physical Methods in Determinative Mineralogy', J. Zussman (ed), Academic Press, London and New York, 1967, Chapter 8.
11. W. M. TUDDENHAM and J. D. STEPHENS, 'Modern Methods of Geochemical Analysis, R. E. Wainerdi and E. A. Uken (eds), Plenum Press, New York and London, 1971, Chapter 6.
12. A. HADNI, 'Essentials of Modern Physics Applied to the Study of the Infra-red', Pergamon, Oxford and New York, 1967.
13. S. S. MITRA, 'Optical Properties of Solids', S. Nudelman and S. S. Mitra (eds), Plenum Press, New York and London, 1969, Chapter 14.
14. R. W. T. WILKINS and J. ITO, *Amer. Min.*, **52**, 1649 (1967).
15. P. TARTE, *Bull. Soc. fr. Ceram.*, **81**, 63 (1968).
16. V. C. FARMER and B. VELDE, *Miner. Mag.*, **39**, 282 (1973).
17. D. M. ADAMS and D. C. NEWTON, *J. Chem. Soc. A*, 2822 (1970).
18. W. G. FATELEY, N. T. McDEVITT and F. F. BENTLEY, *Appl. Spectros.*, **25**, 155 (1971).
19. J. D. RUSSELL, V. C. FARMER and B. VELDE, *Miner. Mag.*, **37**, 869 (1970).
20. R. RUPPIN and R. ENGLMAN, *Rep. Progr. Phys.*, **33**, 149 (1970).
21. V. C. FARMER and J. D. RUSSELL, *Spectrochim. Acta*, **22**, 389 (1966).
22. J. D. RUSSELL, W. J. McHARDY and A. R. FRASER, *Clay Miner.*, **8**, 87 (1969).
23. R. A. CHALMERS, V. C. FARMER, R. I. HARKER, S. KELLY and H. F. W. TAYLOR, *Miner. Mag.*, **33**, 821 (1964).
24. K. E. LAWSON, 'Infra-red Absorption of Inorganic Substances', Reinhold, New York, 1961.
25. R. J. P. LYON, 'Minerals in the Infra-red', Stanford Research Institute, California, 1962.
26. H. MOENKE, 'Mineralspektren', I and II, Akademie-Verlag, Berlin, 1962 and 1966.
27. F. A. MILLER and C. H. WILKINS, *Anal Chem.*, **24**, 1253 (1952).
28. F. A. MILLER, G. L. CARLSON, F. F. BENTLEY and W. H. JONES, *Spectrochim. Acta*, **16**, 135 (1960).
29. Sadtler Inorganic Spectra 200-4000cm^{-1} (1517 Vine Street, Philadelphia).
29a. R. A. NYQUIST and R. O. KAGEL, 'Infra-red Spectra of Inorganic Compounds', Academic Press, New York, 1971.
30. L. C. AFREMOW and J. T. VANDEBERG, *J. Paint Technol.*, **38**, 169 (1966).
31. J. R. LEHR, E. H. BROWN, A. W. FRAZIER, J. P. SMITH and R. D.

THRASHER, 'Crystallographic Properties of Fertilizer Compounds', Tenn. Val. Auth., Chem. Eng. Bull. No. 6, (1967).
32. D. G. TAYLOR, C. M. NENADIC and J. V. GRABLE, *Amer. Ind. Hyg. Assoc. J.*, **31**, 100 (1970).
33. YA. I. RYSKIN, *Inorg. Mater.*, **7**, 375 (1971).
34. G. BRINK and M. FALK, *Spectrochim. Acta,* **27A**, 1811 (1971).
35. H. MOENKE, *Naturwiss.*, **51**, 239 (1964).
36. R. Z. LE GEROS, J. P. LE GEROS, O. R. TRAUTZ and E. KLEIN, 'Developments in Applied Spectroscopy', Plenum Press, New York and London, 1970, 7B, Chapter 1.
37. P. G. ROUXHET, *Amer. Min.*, **55**, 841 (1970).
38. G. L. GAINES and W. VEDDER, *Nature,* **201**, 495 (1964).
39. V. C. FARMER, *Clay Miner.*, **7**, 373 (1968).
40. V. C. FARMER, J. D. RUSSELL, W. J. McHARDY, A. C. D. NEWMAN, J. L. AHLRICHS and J. Y. H. RIMSAITE, *Miner. Mag.*, **38**, 121 (1971).
41. V. C. FARMER, J. JEEVERATNAM, K. SPEAKMAN and H. F. W. TAYLOR, Symp. Structure Portland Cement Paste and Concrete (Spec. Rept. 90, Highways Res. Board, Washington, D.C., 1966). p. 291.
42. K. LANGER and W. SCHREYER, *Amer. Min.*, **54**, 1440 (1969).
43. M. S. RUTSTEIN and W. B. WHITE, *Amer Min.*, **56**, 877 (1971).
44. A. D. CROSS and R. A. JONES, 'Introduction to Practical Infra-red Spectroscopy', Plenum Press, New York and London, 1969.
45. J. E. STEWART, 'Infra-red Spectroscopy: Experimental Methods and Techniques', Dekker, New York, 1964.
46. R. G. WHITE, 'Handbook of Industrial Infra-red Analysis', Plenum Press, New York, 1964.
47. W. J. POTTS, 'Chemical Infra-red Spectroscopy', John Wiley and Sons, New York and London, 1963.
48. L. L. PYTLEWSKI and V. MARCHESAIN, *Anal. Chem.*, **37**, 619 (1965).
49. J. M. SERRATOSA and W. F. BRADLEY, *J. Phys. Chem.*, **62**, 1164 (1958).
50. O. KAMMORI, N. YAMAGUCHI and K. SATO, *Bunseki Kagaku,* **16**, 1050 (1967).
51. N. T. McDEVITT and W. L. BAUN, *Spectrochim. Acta,* **20**, 799 (1964).
52. J. PREUDHOMME and P. TARTE, *Spectrochim. Acta.* **27A**, 845, 961, 1817 (1971).
53. N. T. McDEVITT, *J. Opt. Soc. Amer.*, **59**, 1240 (1969).
54. E. SCHWARZMANN, *Z. Naturforsch.*, **24b**, 1104 (1969).
55. G. A. DORSEY, *Anal. Chem.*, **40**, 971 (1968).
56. O. HENNING, *Z. Hochsch. Architekt, Bauw., Weimar,* **14**, 639 (1967).
57. P. TARTE, *Spectrochim. Acta,* **23A**, 2127 (1967).
58. S. D. ROSS, *J. Mol. Spectros.*, **29**, 131 (1969).
59. M. C. VALYASHKO and E. V. VLASSOVA, *Jena Review,* **No. 1**, 3 (1969).
60. C. E. WEIR, *J. Res. Natl. Bur. Std.*, **A,70**, 153 (1966).
61. H. H. ADLER and P. F. KERR, *Amer. Min.*, **48**, 124, 839 (1963).
62. G. BARON, S. CAILLERE, R. LAGRANGE and T. POBEGUIN, *Bull. Soc. Miner. Cristallog.*, **82**, 150 (1959).

63. R. CHESTER and H. ELDERFIELD, *Sedimentology,* **9**, 5 (1967).
64. C. K. HUANG and P. F. KERR, *Amer. Min.,* **45**, 311 (1960).
65. G. E. KALBUS, E. A. BERG and L. H. KALBUS, *Appl. Spectros.,* **22**, 497 (1968).
66. R. G. BROWN, J. DENNING, A. HALLETT and S. D. ROSS, *Spectrochim. Acta,* **26A**, 963 (1970).
67. C. ROCCHICCIOLI, T. DUPUIS, R. FRANK, M. HARMELIN and C. WADIER, *C.R. Acad. Sci., Paris,* **270**, Ser. B., 541 (1970).
68. A. I. BOLDYREV and A. S. POVARENNYKH, *Zap. Vses. Mineral. Obshchest.,* **97**, 3 (1968).
69. E. Z. ARLIDGE, V. C. FARMER, B. D. MITCHELL and W. A. MITCHELL, *J. Appl. Chem.,* **13**, 17 (1963).
70. B. O. FOWLER, E. C. MORENO and W. E. BROWN, *Arch. Oral Biol.,* **11**, 477 (1966).
71. D. E. C. CORBRIDGE, *Top. Phosphorus Chem.,* **6**, 235 (1969).
72. E. STEGER and B. KAESSNER, *Spectrochim. Acta,* **24A**, 447 (1968).
73. H. H. ADLER and P. F. KERR, *Amer. Min.,* **50**, 132 (1965).
74. R. G. BROWN and S. D. ROSS, *Spectrochim. Acta,* **26A**, 945, 955, 1149 (1970).
75. K. D. SCHUBERT, *Int. Kalisymp. Vortr.,* 3rd. 1965, **4**, 433 (1967).
76. E. WIEGEL and H. H. KIRCHNER, *Ber. Deut. Keram. Ges.,* **43**, 718 (1966).
77. J. I. BULLOCK, *J. Chem. Soc.* A, **5**, 781 (1969).
78. P. TARTE, *Mem. Acad. Roy. Belg. Claise des Sciences,* **35**, 4a,b (1965).
79. R. K. MOORE and W. B. WHITE, *Amer. Min.,* **56**, 54 (1971).
80. I. I. PLYUSNINA and B. G. GRANADCHIKOVA, *Kristallografiya,* **14**, 450 (1969).
81. I. I. PLYUSNINA and E. A. SURZHANSKAYA, *Zh. Prikl. Spektrosk.,* **7**, 917 (1967).
82. D. L. WOOD and K. NASSAU, *Amer. Min.,* **53**, 777 (1968).
83. R. G. BURNS and R. G. T. STRENS, *Science,* **153**, 890 (1966).
84. P. J. LAUNER, *Amer. Min.,* **37**, 764 (1952).
85. J. H. PATTERSON and D. J. O'CONNOR, *Aust. J. Chem.,* **19**, 1155 (1966).
86. W. R. RYALL and I. M. THREADGOLD, *Amer. Min.,* **51**, 754 (1966).
87. E. R. LIPPINCOTT, A. V. VALKENBURG, C. E. WEIR and E. N. BUNTING, *J. Res. Natl. Bur. Std.,* **61**, 61 (1958).
88. C. S. THOMPSON and M. E. WADSWORTH, *Amer. Min.,* **42**, 334 (1957).
89. E. M. FLANIGEN, H. KHATAMI and H. A. SZYMANSKI, *Adv. Chem. Ser.,* **101**, 201 (1971).
90. V. C. FARMER, 'The Chemistry of Cements, Vol. 2', H. F. W. Taylor (ed), Academic Press, London and New York, 1964, Chapter 23.
91. A. BARON, *Bull. Soc. fr. Ceram.,* **81**, 73 (1968).
92. N. NEUROTH, *Glastech. Ber.,* **41**, 243 (1968).
93. J. WONG and C. ANGELL, *Appl. Spectros. Rev.,* **4**, 155 (1971).

CHAPTER 10

Thermal Analysis

R. C. Mackenzie

Macaulay Institute for Soil Research
Craigiebuckler, Aberdeen AB9 2QJ
Scotland

10.1 INTRODUCTION

Thermal analysis has been defined [1] as a general term covering 'a group of related techniques whereby the dependence of the parameters of any physical property of a substance on temperature is measured', and techniques coming under this heading are now extensively used in mineralogical studies, particularly in clay mineralogy. This chapter provides a brief résumé of the most commonly used techniques, their application, and the information which each provides.

The most common thermoanalytical techniques depend on detecting and measuring changes, on heating a sample, in its weight, in its energy, in its dimensions, and in the nature and amounts of evolved volatiles, and the various techniques dependent on these variables can be related to one another by a network like the 'family tree' shown in figure 10.1. Differential thermal analysis (DTA), thermogravimetry (TG), and derivative thermogravimetry (DTG) are undoubtedly the methods most often used for the analysis and characterisation of minerals, but evolved gas analysis (EGA) and dilatometry, along with other less well-established variants, are assuming increasing importance. Clearly, it is impossible to do justice to all techniques within the confines of this chapter, but information on a wide range will be found in the books by Wendlandt [2], Garn [3], Harmelin [4], and Berg [5], and these should be consulted for an overall assessment of the scope of thermoanalytical methods. Books treating individual methods in more detail should also be used to supplement the account given below, and special mention may be made of dissertations on DTA by Smothers and Chiang [6], Schultze [7], and Mackenzie [8], on TG by Duval [9] and Keattch [10], and on EGA by Lodding [11]. Much information can also be garnered from the Proceedings of the International Conferences held triennially and published under the editorships of Redfern [12], Schwenker and Garn [13], and Wiedemann [14].

10.2 BASIC THEORY*

In Chapter 1 it has been shown that all matter consists of atoms which are bonded together into simple and complex molecules and three-dimensional structures. Within these the atoms, although formally held in well-defined positions relative to one another, in reality vibrate about these mean positions, the degree of vibration depending on the temperature of the body.

The fact that the atoms in a body undergo increasingly large vibrations as the temperature rises means that each atom occupies a larger volume of space as its effective atomic or ionic radius increases and, in consequence, the average distance between any two atoms becomes greater. This increase in bond length shows up on the microscopic level as the thermal expansion of the body and can be measured by dilatometry; on the microscopic level it is evidenced by changes in the infra-red absorption bands, which point also to a general weakening of the bonding forces between the atoms as the temperature rises. This weakening is to be expected since the force constant, which acts between the atoms to bind them together, depends on the distance apart of the atoms. Hence, as the temperature of the body is raised, the tendency for the material to undergo reaction increases, and this change in reactivity can be discussed from any of several points of view.

A material may, for example, be considered from a purely 'bonding' point of view. Within any material there will be a range of bond strengths between the various atoms and atomic groupings present which will depend not only on the nature of the atoms involved but also, to a lesser extent, on whether the atoms are in the bulk of the material or near a grain boundary

* Section added by Editor, with author's approval.

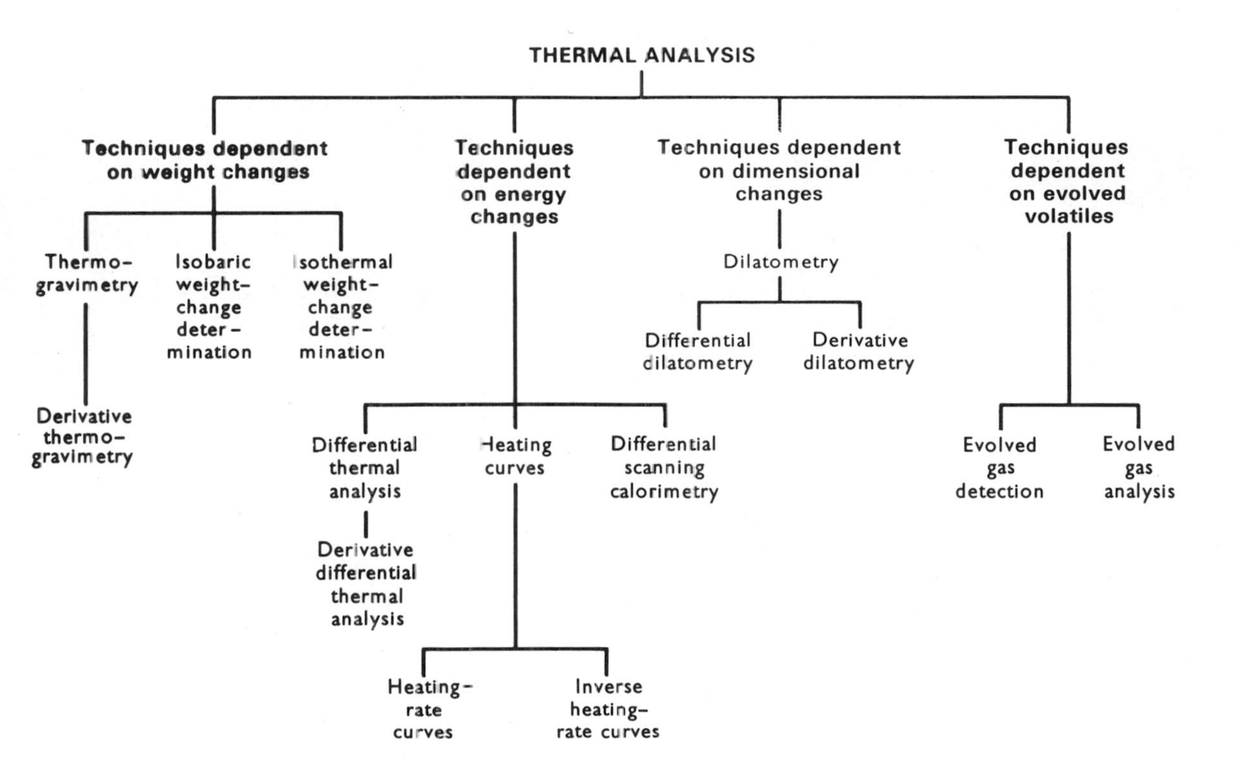

Figure 10.1 Common thermoanalytical techniques and their interrelationships.

or a surface. Thus, in a zeolite, the aluminosilicate framework and the separate water molecules are each held together by strong bonds, but the bond between the water molecules and the framework is relatively weak. Consequently, on heating to temperatures of about 100°C, this bond is the first to break and discrete water molecules are lost from the substance. The framework structure is little affected by this loss. In like manner, interlayer water molecules in montmorillonite particles are lost at a lower temperature than is the water represented by hydroxyl groups in the octahedral layer of the mineral.

A substance can also be considered as a mass of ions in which the cations are held in the interstices of the anion structure – so that a silicon ion is held in the tetrahedral hole between four oxygen ions – and the vibrational behaviour of such a system can be examined. At temperatures around room temperature, although the ions are all vibrating, the average separation between the three oxygen ions forming one tetrahedral face is not large enough to permit the passage of the silicon ion. However, as the degree of vibration increases, and the effective radius of the central space between the oxygens becomes larger, a point is eventually reached where the silicon can pass through the aperture, migrate through the structure, and so cause reaction to occur. Similar considerations apply to other cations, as shown by Taylor [15] in a discussion of the dehydroxylation of various silicate and non-silicate minerals. Dent-Glasser and Glasser [16] have also shown how such a mechanism can explain the transformation of rhodonite into wollastonite without the loss of any volatile component.

Finally, from a thermodynamic viewpoint, the phase assemblage that is thermodynamically stable at any temperature is that which gives a system its lowest total free energy, so that substances undergo reaction to form products if the net free energy change of the reaction is negative. Thus, a hydrated zeolite is stable relative to an assemblage containing a less highly hydrated zeolite plus free water at low temperatures, but the latter assemblage becomes relatively the more stable above some limiting temperature which depends not only on the nature of the zeolite involved but also on the pressure of water vapour in the surrounding atmosphere, as predicted by the law of mass action. A similar condition applies to xonotlite and foshagite, which are stable relative to wollastonite plus free water at low temperatures, but here another factor is involved. Both hydrated minerals lose water at about 700°C when heated in air but undergo dehydration at about 400°C when heated in the presence of water at high pressure, the water appearing to act as a catalyst. These minerals therefore, seem to be thermodynamically metastable in air at temperatures between 400°C and 700°C, no reaction occurring because insufficient energy exists in the material to overcome the activation energy of the reaction. Similar thermodynamic considerations control the relative stabilities of polymorphic forms of the same compound; thus, α-quartz is stable relative to β-quartz below 573°C, but its free energy rises above that of the β-form at higher temperatures.

It follows that the course of reactions occurring in a substance on heating can be monitored by measuring changes in weight or energy. Clearly, weight changes will occur only in certain reactions, such as those involving decomposition, oxidation, etc., but energy changes occur in all reactions, since finite amounts of energy are required to drive off a volatile, to institute a polymorphic change, to trigger off a solid-state reaction, etc. Both TG and DTG are based on weight changes whereas energy changes are measured by DTA: since the latter topic is of more widespread interest, it will be considered first.

10.3 DIFFERENTIAL THERMAL ANALYSIS

According to the Nomenclature Committee of the International Confederation for Thermal Analysis (ICTA) [1], differential thermal analysis is 'a technique of recording the difference in temperature between a substance and a reference material against either time or temperature as the two specimens are subjected to identical temperature regimes in an environment heated or cooled at a controlled rate'. Mackenzie [17] has also given a definitive summary of the use of terms relating to various aspects of thermal analysis.

In essence, differential thermal analysis involves placing small amounts, now seldom larger than 50 mg, of the sample and of the reference material (usually a substance thermally inert over the temperature range of interest) into two matched specimen holders, heating these in a furnace specially designed to provide identical temperature environments around the two holders, and monitoring the heating rate and the temperature difference between sample and reference by some sensitive temperature-measuring device in contact with the two powders. The resulting curve of temperature difference (ΔT) against time (t) or temperature (T), illustrated diagrammatically in figure 10.2, shows a more or less straight line, called the base line, with peaks representing endothermic (conventionally downwards) or exothermic (upwards) effects superimposed thereon. The base line represents conditions when the sample is simply heating up without any reaction occurring and may, for simplicity, be considered to be straight and coincident with the zero line ($\Delta T = 0$) although, in practice, deviation, referred to as base-line drift, usually occurs because of differences in the thermal characteristics of the sample and the reference material and of changes in these characteristics with increasing temperature. Base-line drift can be greatly reduced by use of very small samples and by diluting the sample with reference material (section 10.3.3).

The physical significance of the points of departure from and return to the base line, at the beginning and end of a peak, and of the peak-tip temperature depend on the configuration of the sample with respect to the temperature-measuring device used, which is usually a thermocouple assembly. For example, for a phase transition in a solid, if the thermocouple

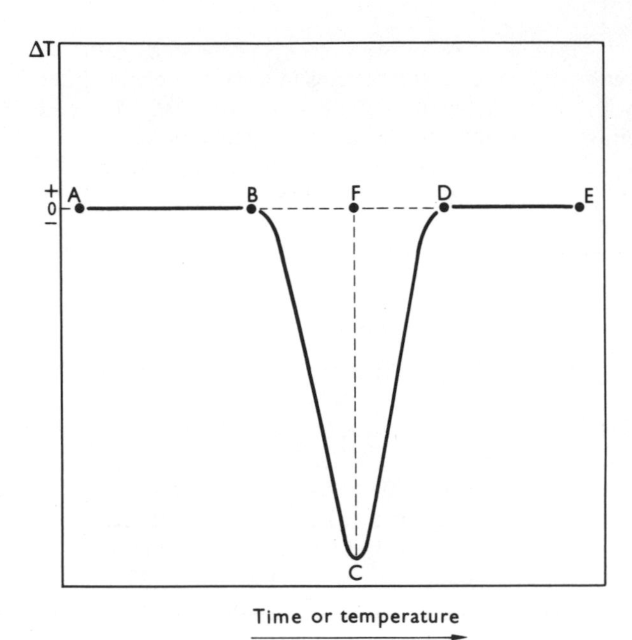

Figure 10.2 Formalised DTA curve showing: *AB* and *DE* – base line; *BCD* – peak; *CF* – peak height; *BCDB* – peak area. The temperature at point *C* is the peak temperature.

is embedded in the centre of the sample, as in figure 10.3a, the true temperature of the phase transition is close to the temperature of the peak tip (the 'peak temperature'), whereas if the sample is situated inside the thermocouple, as in figure 10.3b, the true temperature is close to that of the point of departure from the base line [17]. This distinction arises from the fact that reaction commences in the outer layers of the sample and proceeds inwards to the centre along the inevitable temperature gradient. In both instances the points indicated are those where the powder adjacent to the thermocouple reacts. Unfortunately, the point of departure from the base line is usually extremely difficult to detect with accuracy, because not only is it dependent on the sensitivity of the thermocouple and recording system used but the change can occur gradually, resulting in a slow build-up of ΔT. Consequently, the temperature most commonly quoted is the peak temperature, despite the fact that it rarely has a strict physical significance (the phase transition mentioned above is an exception), being merely the point at which, for an endothermic reaction, the rate of heat absorption by the sample is equal to the rate of heat supply from the furnace. The various critical points on the curve have recently been defined [18] and the significance of these points for different types of reaction with different sample-thermocouple configurations has also been considered in some detail [17].

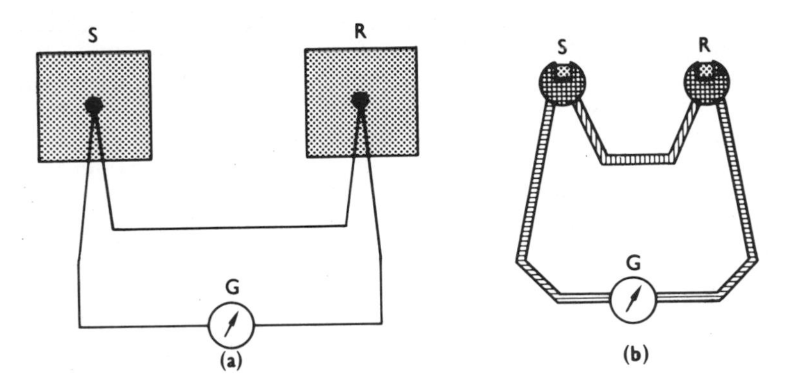

Figure 10.3 Diagrammatic sketch of sample-thermocouple configurations: (a) classical; (b) that of Mazières [75]. S – sample; R – reference material; G – galvanometer.

10.3.1 Theoretical Considerations

A simple theoretical consideration of differential thermal analysis by Speil [19] leads to the relationship

$$m\Delta H = g\lambda \int_{t_1}^{t_2} \Delta T \cdot dt = kS \tag{10.1}$$

where m is the mass of the reacting species, ΔH is the enthalpy change associated with the reaction, g is a geometrical constant determined by the equipment used, λ is the thermal conductivity of the sample, $\int_{t_1}^{t_2} \Delta T \cdot dt$ (= S) is the area of the peak on the curve recording ΔT against time (t), and k is a constant for one material packed to the same density in one sample holder. From equation (10.1) it is clear that the peak area on a curve for ΔT against t is independent of the heating rate and that either the mass of reactant present or the heat of reaction is proportional to the peak area when all other factors are constant. Proks [20] has shown that, when curves are recorded on a ΔT against T basis, the peak area becomes proportional to the heating rate, all other factors being equal.

The proportionality indicated by equation (10.1), however, depends on the thermal conductivity of the sample being constant throughout the reaction – an assumption that is unlikely to be realised in practice. Moreover, heat flow through a powder is a very complicated process so that equation (10.1) can be regarded only as an approximation based on many assumptions the validity of which depend on the design of the apparatus, the experimental technique used, and the sample characteristics. The relationship between m or ΔH and the peak area is, therefore, affected by changes in the thermal conductivity of the sample (λ), in its specific heat (c) and consequently in its thermal diffusivity (a) – since

$$a = \lambda/\rho c \tag{10.2}$$

where ρ is the bulk density of the sample – during the period of heating which, of course, includes the period of reaction. Heat transfer along the thermocouple wires and several other factors bear on this problem, which was first given a comprehensive theoretical treatment by Sewell [21], subsequently somewhat amplified by Cunningham and Wilburn [22]. Although the theoretical treatment cannot be reduced to the form of a simple equation, it can be employed as the basis for designing equipment in which the relationship between, for example, peak area and the mass of reactant is very nearly quantitative – as Wilburn [23] has recently shown. The simple line of reasoning also shows that the best equipment for obtaining curves with minimal distortion of the peaks from their theoretical shapes is not the best for quantitative studies [23], where peak area is critical.

10.3.2 Equipment for Differential Thermal Analysis

Until a little over a decade ago most DTA equipment was home-constructed and frequently suffered from the fact that the constructor was so unfamiliar with the technique that the design was faulty and the results obtained were unacceptable. With the increasing interest now being taken by chemists, however, many instrument manufacturers have become aware of the need for standardised equipment and many excellent instruments are currently on the market.

Since DTA is almost an empirical technique, no apparatus is universal and the equipment available can be divided into three categories:

(a) instruments for general use;
(b) instruments for certain specific problems or substances;
(c) instruments for highly specialised applications.

Category (a) includes the least refined instruments that can, nevertheless, give useful information on a wide range of substances or problems without necessarily being sufficiently accurate for quantitative studies on any. Such equipment is particularly useful for teaching purposes and is frequently adequate for general studies on mineral species where problems related to volatilisation or the provision of special atmospheres are unlikely to be severe. They also have the added attraction of low cost although, as always, care must be exercised in applying the cost criterion in choosing equipment. The instruments in category (b) are also available as commercial units and are very reliable when applied for the purpose for which they were designed, but they may be quite unsuitable for other studies. Thus, in a simple but extreme case, apparatus designed for low-temperature studies over the range −196°C to about 400°C would have little relevance in studies on refractory materials. On the other hand, many instruments in this category are constructed on a modular basis and can be applied to a variety of materials or problems by interchanging such components as the furnace and the specimen holder. Equipment in category (c) is obtained either by building the apparatus 'at home' or by modifying a commercial instrument, and the resulting unit is of very little use outside the field for which it was designed.

Mackenzie and Mitchell [24] have listed the commercial instruments available in 1970 and given accounts of the characteristics of representative units, but care must be taken when using such information, since some manufacturers have introduced completely new models or extended their ranges and other manufacturers have entered the field in the interim period. The annual publication by the American Association for the Advancement of Science of a complete list of American (and other) manufacturers [25] of laboratory equipment serves as a useful source of up-to-date information. Indeed, continuous development in equipment design is one of the keynotes of thermal analysis at the present time.

Apparatus suitable for use above ambient temperatures conforms to the general pattern of figure 10.4 in having a specimen holder located in a heat source, the temperature of which can be controlled by an energy regulator in such a way that a constant heating rate is maintained at the centre of the reference material. Specimen holders vary considerably in size; older instruments require 0.5–1.0g samples but most modern instruments employ the flat-pan type of holder designed for use with only a few milligrams. Small sample sizes have many advantages both theoretically and practically [8], but the lower limit is dictated by such factors as the ability to select a truly representative sample and physical limitations. Thus, irregularities can develop on curves for single crystals of microgram size [26] due to heat transfer and similar problems. The specimen holder is also provided with a means for detecting the actual temperature (T) of the sample or the reference material and the temperature difference (ΔT) between the sample and the reference material. As indicated above, this usually takes the form of a thermocouple arrangement.

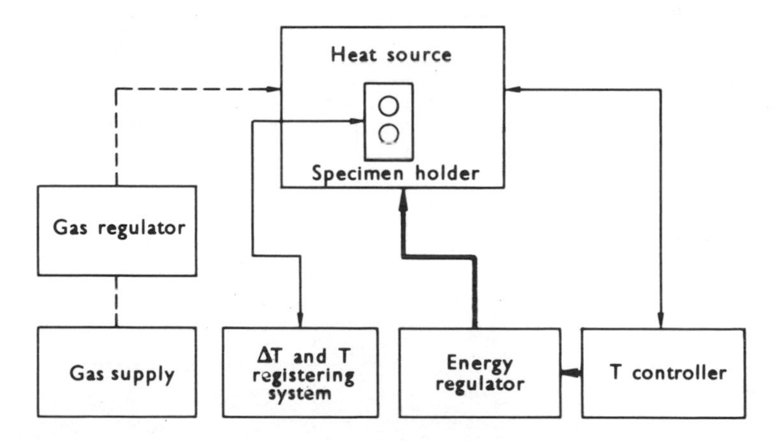

Figure 10.4 Basic components of a DTA apparatus: heavy lines denote the power supply, thin lines the thermocouple circuits, and broken lines the gas supply. (From Mackenzie and Mitchell [24].)

For recording the DTA curve, the signals from the T and ΔT thermocouple assemblages are, if necessary, amplified and then passed to the recording equipment, which varies, for different models, both in sensitivity and in type. Some instruments use a two-pen, strip-chart recorder to yield two traces, one for temperature against time (T/t) and the other for temperature difference against time ($\Delta T/t$), whereas others use an X-Y recorder to give a direct record of ΔT against T. Since only on a $\Delta T/t$ trace is the peak area directly related to the enthalpy change for the reaction (section 10.3.1), the former recording system is generally to be preferred and the latter facility should always be provided with some mechanism for simultaneously monitoring the heating rate.

Finally, all modern instruments would be expected to make some provision for controlling the gaseous environment of the specimens.

Low-temperature instruments that can be used down to the temperature of liquid nitrogen, or even liquid helium, are now commercially available [24] and are finding increasing use in organic chemistry [27] but have been little considered for use in connection with minerals, apart from some specialised applications, such as those reported by Kato [28] on clays. The basic system for a low-temperature unit is shown diagrammatically in figure 10.5.

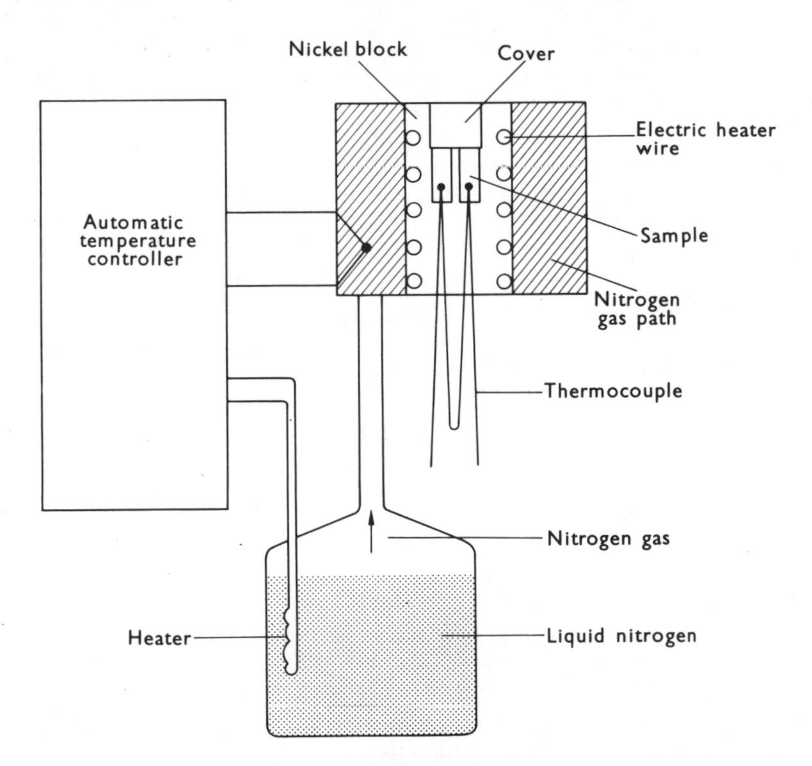

Figure 10.5 Basic system of the Rigaku-Denki low-temperature DTA apparatus. (From Mackenzie and Mitchell [24].)

Instrumental Characteristics
Various desirable features that instruments should possess, and criteria to be used in their assessment, have been discussed in detail elsewhere [24] and only a few points need be elaborated. An electric furnace provides the most convenient and most easily controlled heat source, and it is essential that its heat output should be greater than that required to raise the temperature of the furnace and its contents at the maximum desired rate over the whole temperature range; moreover, the sample and reference material should be so located that each receives identical amounts of heat. The furnace should preferably be wound non-inductively and earthed close to the specimen, or an earthed shield should be provided around the specimen holder and the thermocouple leads, to prevent spurious induced signals in the recording circuit. The temperature-regulating system should be capable of providing a constant and reproducible heating rate over the whole temperature range, and a proportional control system using a saturable reactor or thyristor, preferably with a three-term controller, is recommended. The control thermocouple may be located in the specimen holder, especially with the three-term controller circuit, although in other designs it is best located close to the furnace windings.

Various metallic and non-metallic specimen holders, both integral units drilled from a block of material and separate holders, are in common use and have their separate advantages and disadvantages [24]. In some designs, one arm of the ΔT thermocouple system provides the T values, whereas in others a separate well is provided for the T thermocouple. It should be reiterated, however, that, according to Wilburn [23], care must be taken in choosing an appropriate specimen holder, since a holder suitable for qualitative studies, i.e. one that gives minimum peak distortion, is not the best for quantitative studies, where the peak area must be related to the energy change.

One important criterion to be borne in mind when assessing the materials used in the construction of the specimen holder or the thermocouple system is that not only must they not react with the sample or with the products of reaction, but they must also not catalyse or otherwise interfere with the course of the reaction. Spurious effects due to such interferences are well known [29] and can lead to difficulties in the interpretation of curves for unknown materials. Silicate minerals are generally somewhat unreactive and can be examined in most instruments capable of attaining temperatures of 1000°C or above, but care must be taken with atmosphere control when samples that contain variable-valence cations, such as Fe^{2+}, or readily oxidisable compounds are involved. As against this, minerals such as chalcogenides raise particular difficulties associated with their corrosive nature, their high reactivity towards oxygen and their ready dissociation on heating; these have been reviewed by Bollin [30] and require such highly specialised specimen holders as those illustrated in figure 10.6. Salts, such as sulphates, are somewhat intermediate between these extremes, in that they or their decomposition products can react with certain metals [31] so that specimen holders and thermocouples for these must be carefully selected.

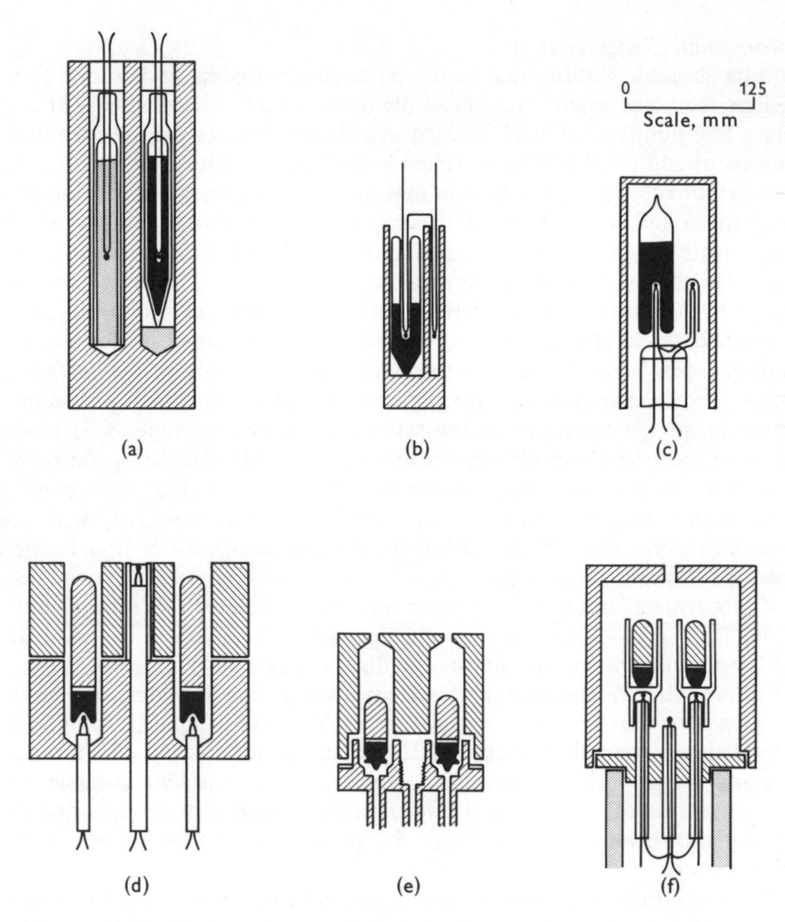

Figure 10.6 Various types of sealed silica vial specimen holders that have been used for chalcogenides. (From Bollin [30].)

The temperature recorded should be that of the sample and corrections should be applied if the temperature of the reference material or the specimen holder is monitored. With regard to thermocouple systems, chromel/alumel, platinum/platinum-rhodium, and tungsten/tungsten-rhenium pairs are commonly used, their maximum temperatures being about 1100°C, 1600°C, and 2500°C respectively. Chromel/alumel has the added advantage of a linear response to temperature change for much of its normal range. Recording is usually by potentiometric pen-and-ink recorder but photographic recording of galvanometer deflections has also been used [24]. Background noise level should be low and sensitivity as high as possible and a 'scale-expanding' facility, whereby the pen is returned to the other side of the chart and recording continued whenever the upper or lower limit position is reached, is a decided advantage in accurately recording large and small peaks at one sensitivity.

Errors in temperature measurement due to faulty calibration of thermocouples can now be readily avoided, since the U.S. National Bureau of Standards [32] are currently supplying materials, selected and tested by ICTA, for temperature standardisation of DTA equipment over the range 125-925°C. Despite this, differences will always be found in curves for nominally the same mineral examined in different instruments, because of variations in

(a) mineral composition,
(b) construction and characteristics of the equipment,
(c) sample environment in the apparatus.

The common use of only a limited range of commercial instruments, together with calibration using these temperature standards, should, however, help to eliminate the relatively enormous variations observed in the past and chronicled by Mackenzie [33].

Differential Thermal Analysis at High Pressures

The first to perform DTA under a pressure other than atmospheric appears to have been Orcel [34] who examined the changes induced in DTA curves for chlorites by examining them *in vacuo*, but the most intensive examination of the effect of pressure on the DTA curve is due to Stone [35] who designed and marketed an apparatus enabling investigations with a known gas flowing through the specimens at any pressure from vacuum to several atmospheres. This can be valuable in identification studies, in separating overlapping peaks originating from different processes, in examining reaction mechanisms, etc. An excellent brief review was given by Stone in 1960 [36].

DTA studies on minerals under high pressures in the kilobar region are valuable in defining the behaviour of rock-forming materials at depth within the earth. Harker [37] and Weber and Roy [38] have developed DTA specimen holders that fit into Tuttle cold-seal pressure vessels so that reactions can be followed under hydrostatic water-vapour pressures up to 15,000 psi at temperatures up to 900°C. The latter authors have shown that although the dehydroxylation temperature of kaolinite rises with increasing water pressure up to 400 psi – as would be expected from the Clausius-Clapeyron relationship – at higher pressures it falls due, they surmise, to the catalytic effect of water on the reaction.

10.3.3 Experimental Techniques

Experimental technique is just as important as apparatus design in determining the quality of the results obtained from the study of a material. Since the main factors operative have, again, recently been reviewed in some detail [39], only major aspects need be mentioned.

With all types of specimen holder, pretreatment and particle size of the sample and reproducibility and constancy of the heating rate are important.

Constancy of heating rate is essential for any quantitative or semi-quantitative work and is also advisable for qualitative studies since a variable heating rate can result in notable changes in the appearance of the curve. Thus, a material giving, with uniform heating, two peaks of almost identical size some hundreds of degrees Centigrade apart will, when heated at a non-uniform rate which decreases with increasing temperature, give a $\Delta T/t$ curve on which the high temperature peak is shallower and broader than the low-temperature peak. The heating rate is, of course, itself a significant factor in determining the appearance of the curve, and the rate chosen will depend on the nature of the sample and on the degree of accuracy required. Rates from less than 0.1°C to over 200°C per minute have been used and most commercial instruments provide rates in the range 2–20°C/min; the common range for normal work on minerals is 8–12°C/min, 10°C/min being the most usual value. At this rate the peaks are of a satisfactory size and reasonably well separated and the analysis time (1½–2 hours) is not excessively long. The slower the heating rate the better is the resolution of the peaks on the $\Delta T/t$ curve and the closer to equilibrium are conditions in the sample, although more sensitive recording equipment is necessary because of the broad shallow peaks obtained and the time needed for analysis is longer. Fast heating rates may be useful in, for example, rapid surveys.

As regards pretreatment, fine-grained materials such as clays have an affinity for moisture varying with both the nature of the mineral and the saturating cation. Consequently, all such samples should, for many purposes, be saturated with a single exchangeable cation and equilibrated as regards sorbed moisture under standard conditions, e.g. over a saturated solution of $Mg(NO_3)_2.6H_2O$ (55% relative humidity at 18°C), before examination. Particle size is particularly important for certain types of reaction, especially for decompositon reactions that are diffusion-controlled, but less so for others, such as phase transitions. The particle-size distribution will, as it affects density, have some effect on the heat-transfer characteristics, see equation (10.2), and it is also essential to have uniform particle-size fractions for comparative work, although particle size is not generally critical below 2μm. Should comminution be necessary, dry grinding should be avoided, since its use can completely alter the appearance of a DTA curve [40]. Ultrasonic dispersion is useful for aggregates; otherwise, careful crushing or filing can suffice. Should grinding be unavoidable, it should be carried out in the most gentle manner possible, with the sample immersed in an inert liquid to prevent local overheating (see Chapter 9, section 9.4.2).

Marked differences exist between the classical type of specimen holder, where the sample is compressed into a container or well, and the currently favoured flat-pan holders, in which a much smaller sample lies in contact with the thermocouple junction. Certain factors that do not affect results from flat-pan units can have a considerable influence on curves obtained using classical holders. These include the position of the thermocouple relative to the specimens, the size of the specimens, the degree of packing, the extent of dilution of the sample with reference material and the type of atmosphere control. It is impossible to give definitive rules here for the effect of each of these factors, except to stress that reproducibility is absolutely essential at all stages of pretreatment and sample preparation. In this connection, hard packing of the sample into its holder is normally recommended as the most reproducible, since it ensures good thermal contact with the thermocouple and with the walls of the container, but some reproducible form of loose packing has to be employed if atmosphere control is to be achieved by gas flow through the sample. Dilution of the sample with reference material, preferably to a sample concentration of 30% or less [41, 42], is essential for any quantitative studies, since this ensures that the thermal characteristics of the sample are close to those of the reference material during the whole heating regime and consequently that base-line drift during the initial heating period and change in position of the base-line after reaction are minimised. Since such considerations do not apply to the flat-pan type of holder, this type not only has the advantage of requiring very small amounts of material (a few mg) but its use also eliminates much of the tedium associated with sample preparation for classical holders.

10.3.4 Applications of Differential Thermal Analysis

For greater detail than can be given here, a recent comprehensive account by Hodgson and Robertson [43] of the use of DTA in the mineral industries ought to be consulted, as well as volumes dealing with particular groups of minerals [44–47].

The great merit of a DTA curve is that it shows up all the energy changes occurring in the sample during heating – provided the sensitivity of the equipment is adequate – irrespective of whether or not there is an associated weight change. Moreover, since the curve reflects the energy involved in reactions and the temperatures at which these occur, the DTA curve for a material should be characteristic of that material and therefore a criterion for identification. While this is true with an unlimited temperature range, over the limited temperature range possible in practice there are, for many materials, various difficulties.

For example, it is clear from the typical curves for various minerals in figure 10.7 that, in a mixture of kaolinite (curve A) and illite (curve C), the illite dehydroxylation peak at about 550°C would be occluded by the kaolinite peak and illite could remain undetected if one were unaware of the possibility of its presence. Moreover, not only can mixtures of certain minerals yield curves that could be mistaken for a single mineral species but the occurrence of solid-state reactions between the components of a mixture can give rise to completely new peaks, not belonging to any of the individual components and rendering the overall curve atypical for the species present [48]. Another difficulty arises from the fact that the nature of the saturating cation can alter the appearance of the curves for some minerals (as evidenced by montmorillonite in figure 10.8) and still another

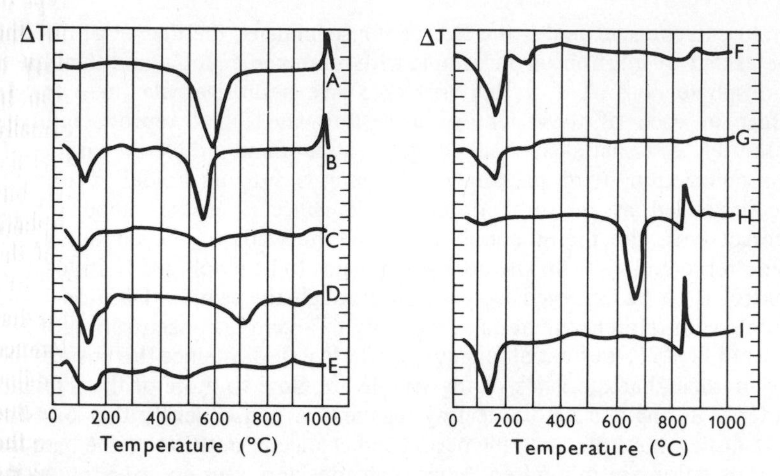

Figure 10.7 DTA curves for: *A* – kaolinite; *B* – partially dehydrated halloysite (with a peak at 320°C due to gibbsite); *C* – illite; *D* – montmorillonite: *E* – palygorskite; *F* – vermiculite; *G* – hydrobiotite; *H* – pennine; *I* – sepiolite. (From Mackenzie [61].)

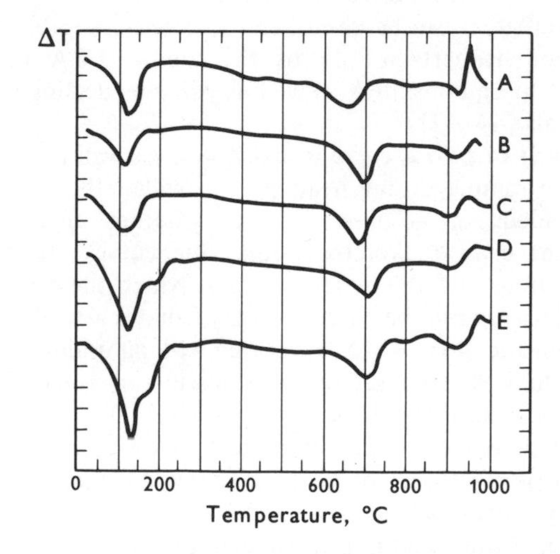

Figure 10.8 DTA curves for montmorillonite saturated with: *A* – NH_4^+; *B* – K^+; *C* – Tl^+: *D* – Ca^{2+}; *E* – Sr^{2+}. (From Mackenzie [50].)

from the observation that fine-grained minerals do not always yield a curve similar to that obtained from massive crystals of apparently the same composition. Thus, vermiculite in clays does not always give a curve similar to that for pure vermiculite (curve A in figure 10.9) and some clay chlorites, which are responsible for the small peak at about 600°C on curves C and D

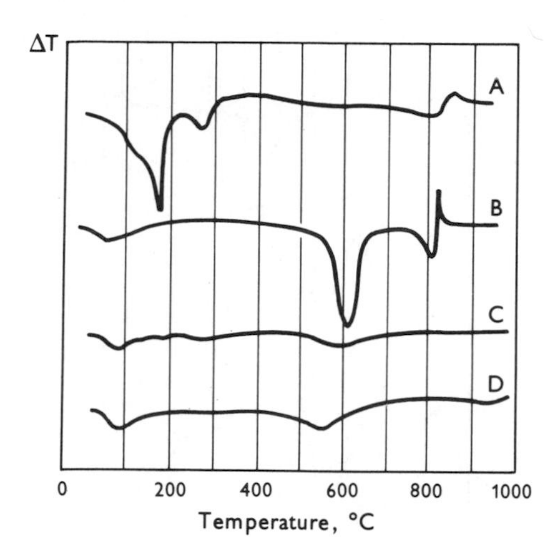

Figure 10.9 DTA curves for: *A* – vermiculite; *B* – chlorite (pennine); *C* and *D* – soil clays containing 40% and 20% chlorite respectively. (From Mackenzie and Mitchell [51].)

in figure 10.9, do not react like a conventional chlorite (curve B in figure 10.9), presumably due to the oxidation of ferrous iron in the lattice [49]. Complications can also arise when distinctions usually associated with degree of crystallinity can be confused with differences caused by particle size and shape, as is well exemplified in the kaolinite group [50] where a well crystallised and highly crystalline triclinic kaolinite with thick particles about 10μm across gives a curve almost identical with that for dickite; conversely, dickite of particle size less than 2μm can give a curve with a peak temperature approximately the same as that for kaolinite [50].

Despite such difficulties, however, it is often possible at least to attribute clay and other minerals to the correct group on the basis of the general appearance of their DTA curves [50] and the method is often valuable in carrying out quantitative, or semi-quantitative, assessments of the relative abundances of components once the peaks associated with the various minerals present have been identified [51]. It may also be possible, at times, to devise pretreatments that render DTA more diagnostic. Thus, mixtures containing gibbsite and goethite with overlapping peaks can be analysed quantitatively [52] by determining curves for

(a) the original material;
(b) the material after treatment with $Na_2S_2O_4$ solution to remove goethite;
(c) the material after treatment with NaOH solution to remove gibbsite.

In like manner, identification of minerals can sometimes be assisted by piperidine treatment, which is particularly applicable to clays [53, 54]

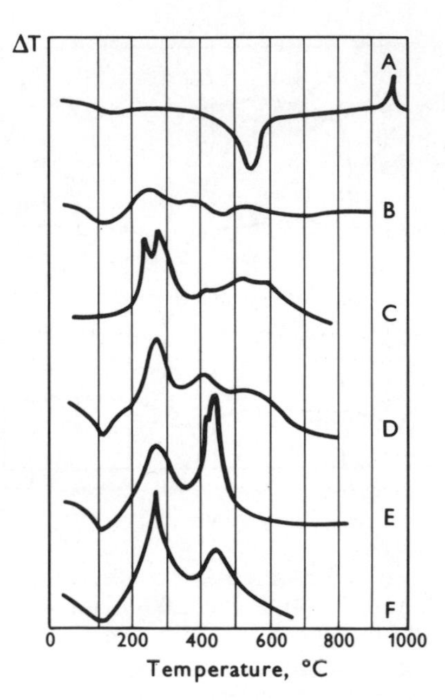

Figure 10.10 DTA curves for piperidine-saturated samples of: *A* – kaolinite; *B* – illite; *C* – 'normal' montmorillonite; *D* – 'abnormal' montmorillonite; *E* – nontronite; *F* – allophane. (From Mackenzie [54].)

(figure 10.10), by altering the composition and pressure of the furnace atmosphere [36], or possibly by extending the usual temperature range of room temperature to 1000° C either upwards or downwards, since no two materials can exhibit identical effects over a very wide temperature range.

Relationship to X-ray Diffraction

In general, fine-grained crystalline minerals are undoubtedly best identified by X-ray diffraction methods, but DTA can provide useful information when non-crystalline components are present [55]. Furthermore, DTA studies can be used to detect differences that cannot be observed by X-ray diffraction. For example, in so-called 'abnormal' species of such dioctahedral clay minerals as pyrophyllite, illite, and montmorillonite [50], dehydroxylation occurs at temperatures different from those associated with 'normal' species. Consequently, distinction is easy from DTA curves once the mineral has been identified [51]. Since X-ray diffraction patterns of the 'abnormal' and the 'normal' species are identical, the difference must be due to very minor structural details, such as the mode of binding of the hydroxyl groups in the lattice, but no definite theory as to the reasons has yet been accepted [56].

Interstratification in clay minerals can be observed at levels below those detectable by X-ray diffraction [50]. Thus, the presence of random brucite layers in some samples of reputedly pure vermiculite and saponite – which is virtually indeterminable by X-ray methods and which gives essentially interstratifications of chlorite with these minerals – can be readily seen from the presence of an endothermic peak at about 600°C and an associated exothermic at about 800°–850°C, as shown in figure 10.11.

DTA information can also supplement X-ray diffraction data on accessory minerals, since, being a particularly sensitive test for such minerals as hydrated sesquioxides and carbonates, it can indicate the amounts present in clays when these are too small for detection by normal X-ray methods [51].

The continuing accumulation of DTA information on minerals and other materials has led to the compilation of indexes of curves based on the temperatures of the principal and secondary peaks in the pattern, and whether these are endothermic or exothermic. Mackenzie [33] has compiled a punched card index for minerals and inorganic and organic compounds on this basis and the Sadtler Research Laboratories [57] and

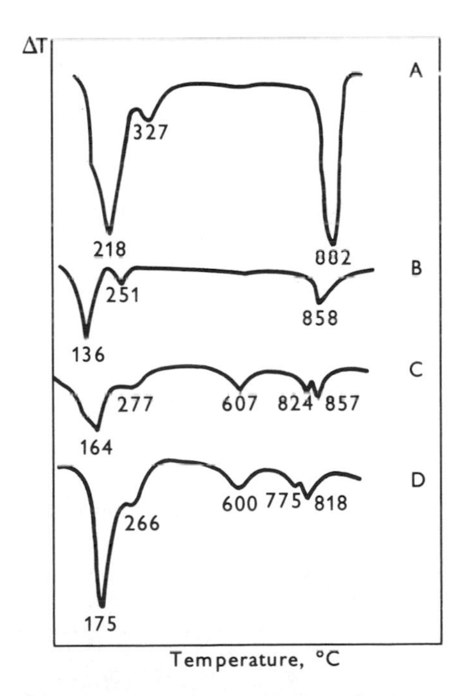

Figure 10.11 DTA curves for saponite samples from: *A* – South Africa; *B* – Scotland; *C* – England; *D* – Germany. Curves *C* and *D* show interstratification with chlorite. (From Mackenzie [50].)

Liptay [58] have produced libraries of reference curves for various materials. Such compilations are useful for making an initial selection of a limited number of possible substances having DTA characteristics similar to those of the unknown before subsidiary tests are used, if necessary, to complete the identification. Generally, the indexes act as indicators rather than complete identifiers (in the sense of the J.C.P.D.S. X-ray diffraction index – see Chapter 7, section 7.7.1), since the problems outlined above render DTA on its own a method of decidedly limited use in identification. On the other hand, the information that DTA provides can be extremely valuable when used in conjunction with the evidence from other techniques, such as chemical analysis, X-ray diffraction, infra-red absorption spectroscopy, electron-optical investigations, etc.

A topic closely allied to identification is the question of 'finger-printing' a substance – namely, checking samples for identity of constitution and provenance. DTA can be, and is being widely, used to check similarities and differences between apparently identical samples not only in mineral sciences [43] but also in organic chemistry [27, 59, 60], since very small differences in structure or composition, or the presence of small amounts of impurities typical of material from a given source, can readily be detected on DTA curves when it would be difficult if not impossible to observe them otherwise.

Quantitative determinations, as indicated above, are possible by DTA provided the peaks involved can be unequivocably identified and no overlap or occlusion occurs. Moreover, it has already been mentioned (sections 10.3.2 and 10.3.3) that the equipment and experimental technique used must be carefully chosen to ensure complete reproducibility at all stages in order that equation (10.1) may be applied in its simple form. Accuracies of better than 2–3% are probably rarely achieved under ordinary conditions,

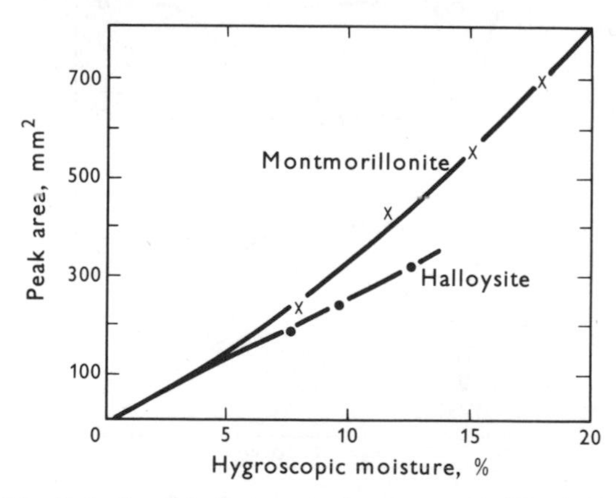

Figure 10.12 Relationship between the amount of sorbed moisture and the low-temperature peak area for montmorillonite and halloysite. (From Mackenzie [61].)

even with the most stringent precautions, and a value of about ±5% is more likely, so that results should generally be regarded as semi-quantitative rather than quantitative. Recently, however, Wilburn [23] has shown that very reliable results can be obtained if great care is taken to employ equipment and techniques based on theoretical considerations.

But, even with the utmost care in experimentation, other factors may intrude. Thus,. during an examination of the relationship between sorbed moisture and peak area for montmorillonite and halloysite, Mackenzie [61] has noted a departure from linearity for montmorillonite, as shown in figure 10.12. This departure is not due to inaccuracy of theory or of experimental technique, but to the water associated with the exchangeable cations being more tightly bound and requiring more energy for removal at low water contents than at high, a factor that does not influence the results for halloysite. One must, therefore, be careful not to reject non-linear relationships in quantitative DTA studies without giving some thought as to the mechanism of the particular reaction involved.

10.4 THERMOGRAVIMETRY AND DERIVATIVE THERMOGRAVIMETRY

Thermogravimetry (TG) is 'a technique whereby the weight of a substance, in an environment heated or cooled at a controlled rate, is recorded as a function of time or of temperature,' and the term derivative thermogravimetry (DTG) is used for 'any technique yielding the first derivative of the TG curve with respect to either time or temperature' [1]. It follows that these methods are applicable only to materials that undergo a weight change on heating – either a loss caused by decomposition or a gain caused by oxidation, although competing reactions, such as simultaneous dehydroxylation and oxidation in a ferrous amphibole, can complicate the result.

10.4.1 Equipment for Thermogravimetry

The definition of thermogravimetry implies the existence of a dynamic temperature regime with continuous monitoring of the weight of the sample. Two closely related techniques that also measure weight change with temperature, but under different conditions, are isobaric weight-change determination and isothermal weight-change determination [1].

The former of these is 'a technique of obtaining a record of the equilibrium weight of a substance as a function of temperature at a constant partial pressure of the product or products' [1] and involves maintaining the substance at each chosen temperature until it reaches equilibrium. The value of the technique in the study of minerals has been discussed by Nutting [62] and more recently Buckle [63] has reviewed its use in determining the dehydration behaviour of hydrated calcium silicate phases. The latter method, which involves bringing the sample rapidly to the required temperature and measuring the weight-change as a function of time, is useful in checking reaction mechanism. Although equipment for both techniques can be relatively simple – e.g. a suitable furnace and a balance – more complex systems involving quartz spring balances [63, 64] and modified thermobalances [65] can also be employed and may at times yield more meaningful results.

The value of simple weight-change determination has been shown by Nicol [66] in the dehydration of nekoite, and it can generally be concluded that a good case can be made for studying the dehydration of a mineral by several methods, especially where there is any possibility of unusual behaviour.

Since, as in DTA, a dynamic temperature regime is used in thermogravimetry, various factors have to be taken into account in assessing the quality of a thermobalance. According to Lukaszewski and Redfern [67] the principal requirements are that

(a) both weight and temperature be registered separately and continuously;
(b) the furnace be capable of reaching the maximum desired temperature easily;
(c) the heating rate be constant over the whole temperature range;
(d) the uniform-temperature zone around the sample be as long as possible;
(e) provision be made for
    (i) different heating rates,
    (ii) different recording chart speeds,
    (iii) atmosphere control,
    (iv) use in isothermal studies;
(f) there be regular heat flow to the sample;
(g) the reaction chamber be inert to attack by the sample or the products of reaction;
(h) the balance mechanism be adequately protected;
(i) it be possible to measure accurately the temperature of the sample.

Two basic types of thermobalance are in current use, the deflection type which can be sub-divided into several variants [10], and the null-point type. The Chevenard thermobalance – a simple deflection type illustrated in figure 10.13 – displays the movement of the beam by the deflection of a light beam reflected from a mirror mounted on the end of the balance lever on to a photographic drum driven synchronously with the temperature controller. The Stanton range of thermobalances [72] are also built on a deflection principle, but the movement of the beam varies the separation between the two plates of a parallel-plate condenser, the resulting capacitance change being displayed as a change in weight. In the Cahn electrobalance, which is of the null-point type and is readily converted into a thermobalance, the sample is maintained in a constant position by applying an e.m.f. to the central coil, as shown in figure 10.14, in order to counteract the torque produced as the sample weight changes. The value of this e.m.f. is again displayed as a weight change.

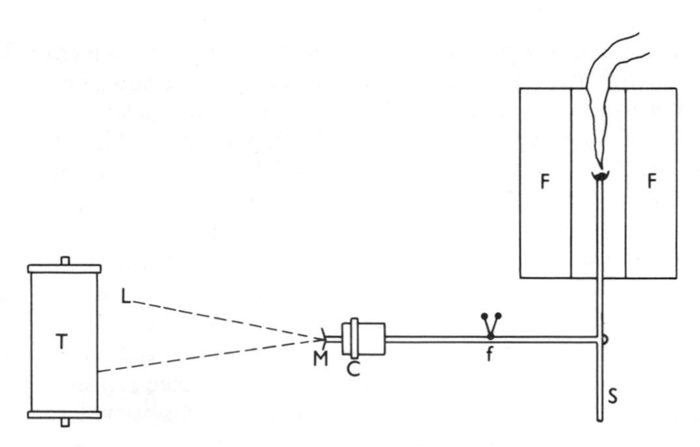

Figure 10.13 Diagrammatic sketch of the Chevenard thermobalance: *C* is the counterweight; *F* the furnace; *f* the fulcrum; *L* the light source; *M* the moving mirror; *S* the crucible support; *T* the recording drum. (From Mackenzie [61].)

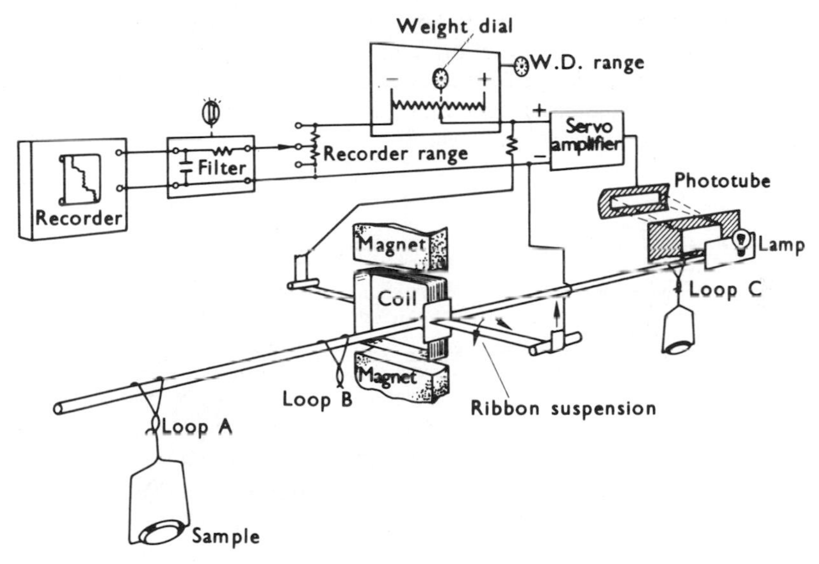

Figure 10.14 Diagrammatic sketch of the Cahn electrobalance. (From Redfern [72].)

Keattch [10] has briefly indicated the characteristics of thermobalances, and of those recording balances that can readily be converted, that were available in 1969. Selection of any one particular type must be determined by the type of problem being investigated, the sensitivity required, the necessary degree of refinement, the cost, etc., and therefore no firm recommendations can be made.

As with DTA, the present tendency in TG is to use samples of very small size – around a few milligrams; this has several advantages, since not only does it minimize temperature gradients but in certain systems, such as those examined by Brindley and Nakahira [68], it permits volatiles to be released with minimal overpressure effects. Very small samples require very accurate equipment and render it difficult to determine sample temperature, since it may not be possible to insert, say, a thermocouple into the sample. A possible solution to the latter difficulty involves calibrating the instrument by heating, as samples, materials that show Curie points at different temperatures. If the hot zone is in a magnetic field, the TG curve will show an abrupt deflection at the Curie point, owing to the sudden change in the magnetic properties of the sample. This principle is used in the Perkin-Elmer thermobalance, which is based on the Cahn electrobalance, and can be readily applied to several others.

DTG curves may be plotted manually from the TG record, by constructing tangents to the curve at closely spaced intervals, but it is more convenient, and more accurate, to differentiate the TG curve electronically. Many instruments now record both TG and DTG curves automatically; such a system is readily devised with a null-point electrobalance and one is in daily use at the Macaulay Institute. An alternative system, which can be applied to a deflection-type balance, is that shown in figure 10.15 for the Derivatograph [69], where the DTG curve is provided by the rate of movement of a coil in a magnetic field.

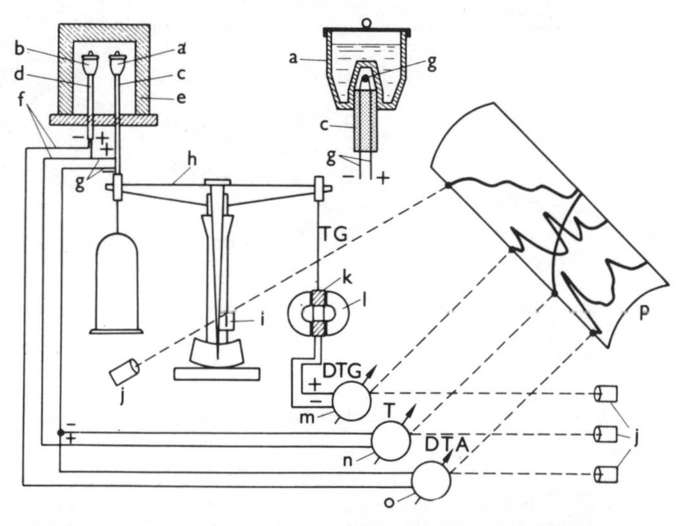

Figure 10.15 Diagrammatic sketch of the Derivatograph: *a* is the sample holder; *b* the reference holder; *c,d* the porcelain tubes; *e* the furnace; *f,g* the thermocouples; *h* the balance; *i* the optical slit; *j* the lamps; *k* the moving coil; *l* the permanent magnets; *m,n,o* the galvanometers; *p* the recording drum. (From Paulik *et al* [69].)

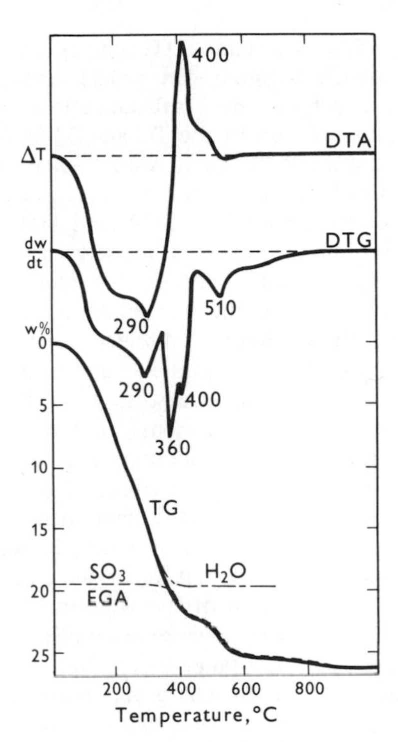

Figure 10.16 Thermal analysis curves for ferric oxide gel. (From Paulik *et al* [69].)

10.4.2 Applications

Since TG and DTG curves reveal only reactions in which weight changes occur, they consequently can give less information about the behaviour of a substance on heating than can a DTA curve. The information that they do give is, however, quantitative. DTG curves have some advantages over TG curves in that minor changes in slope, i.e. in reaction rate, which might be missed on the TG curve are immediately visible as peaks and, as figure 10.16 illustrates, they also show accurately the procedural initial and final temperatures in addition to the temperature of the maximum rate of reaction. Thus, the critical points on a DTG curve do have a physical significance and so can differ in temperature from those on the DTA curve for the same material heated under the same conditions. Weight change, on the other hand, is more readily measured from a TG curve, although the peak areas on a DTG curve are, by definition, also quantitative representations of weight changes.

Obviously, the DTA, TG and DTG curves for a sample form a complementary set which together can provide much more information about the reactions undergone by a substance than could any one trace alone. In particular, comparison of the TG and DTA curves shows which of the peaks on the latter corresponds to weight changes in the sample and which to processes involving no overall compositional changes, e.g. phase transition, exsolution, etc. Similarly, DTA and DTG curves can provide information about the relative amounts of energy and weight involved in a reaction and may shed some light on the mode of bonding of the component gained or lost and on the mechanism of the change.

Thermogravimetric studies have been widely used, at the research level, to determine the compositions of minerals and other species, both from weight-loss and from weight-gain measurements. The total weight loss occurring on heating obviously represents all the components originally present that are volatile at the temperature attained and can be used to derive, for example, the total water content of a hydrated species, provided that a complete analysis of the non-volatile components is also available. But a detailed assessment of the temperature at which the weight-loss occurs can also provide some evidence of the nature of the component being lost at that stage. This is especially true of hydrated minerals, from which, for example, adsorbed or zeolitic and semi-zeolitic water molecules are generally lost in the temperature range below 200°C, water from hydroxyl groups bound to silicon (in the Si–OH group, particularly in the hydrated calcium silicate minerals) is lost in the range 300-450°C, water present as hydroxyl groups bound to aluminium (in clays) is lost above 450°C, and water present as free hydroxyl groups co-ordinated to calcium is not lost until about 700°C [70]. The temperature of evolution of carbon dioxide from carbonates depends mainly on the mineral present but the amount of mineral also has an effect unless a carbon dioxide atmosphere is used. Interpretation of the curve can sometimes be helped by a knowledge of the overall composition from chemical analysis.

Since both techniques are quantitative, TG and DTG curves can be used to determine the amount of any particular mineral that is known to be present and that decomposes or oxidises in the temperature range studied. This type of analysis is particularly useful for carbonates. The results obtained from TG and DTG curves can also be used to derive information on the kinetics of reactions much more rapidly than by conventional isothermal techniques [71], although the values so obtained are not of the same level of accuracy.

As mentioned above, it is generally true, as it is with so many of the physical methods of analysis discussed in this book, that the information from TG and DTG results must be integrated with that from other techniques, both thermoanalytical and non-thermoanalytical, for a complete interpretation of the behaviour of the material under study. When such an integration has been made, however, thermoanalytical data often provide very important backing to support evidence obtained otherwise.

10.5 EVOLVED GAS ANALYSIS

Evolved gas analysis, or EGA, has been defined as 'a technique for determining the nature and/or amount of the volatile product or products formed during thermal analysis' [1]. It is, in a way, a subsidiary technique to TG and DTA in that the volatiles evolved during heating are swept from the specimen-holder assembly by a stream of inert carrier gas and analysed by gas-liquid chromatography, mass spectrometry or another relevant technique [11, 72]. Sometimes two techniques can be used in tandem, chromatography, for example, being employed to concentrate gases present in only minute amounts before they are passed to a mass spectrometer.

EGA has been discussed in some detail by Lodding [11] and can be extremely useful in mineral studies, since it immediately reveals, for example, whether any specific DTA or DTG peak is associated with the evolution of water, carbon dioxide, etc. [51] without the arbitrariness

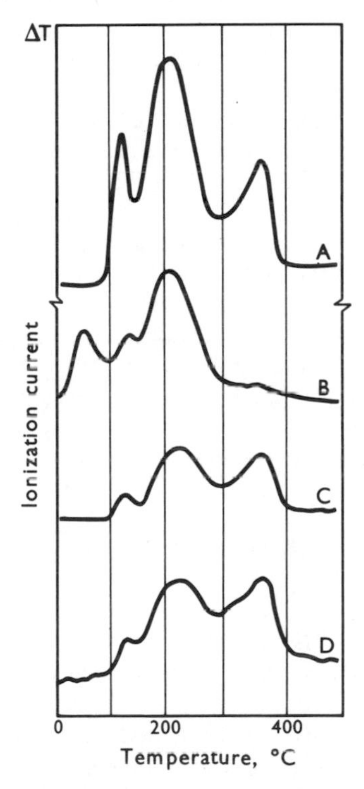

Figure 10.17 Curves for spores of *Lycopodium clavatum* in flowing oxygen: *A* – DTA; *B* – amount of water evolved; *C* – amount of carbon monoxide evolved; *D* – amount of carbon dioxide evolved. The water peak at 60°C is due to sorbed water swept into the mass spectrometer by the carrier gas. (From Mackenzie and Mitchell [51].)

associated with the temperature of reaction. The curves shown in figure 10.17, although not for a mineral, make it abundantly clear that examination of EGA curves along with the DTA curve enable much more information to be derived than would have been possible from DTA alone.

The application of the technique to mineralogical studies is still in its infancy, but there is no question that it can greatly increase the utility of TG, DTG and DTA curves.

10.6 SIMULTANEOUS TECHNIQUES

Since the application of two or more thermoanalytical techniques, especially DTA and TG or DTG, to the study of the same sample can yield more information than either technique on its own, there has developed a tendency over the past decade to apply several techniques simultaneously to the same sample. The thesis underlying this development is that the environment can play an important role in determining the details of a reaction and that more valid information can be drawn from curves that have been obtained under identical conditions; moreover, it is extremely difficult, if not impossible, to ensure that environmental conditions are exactly replicated on two separate instruments. Unfortunately, the optimum conditions for TG are not those for DTA, but such difficulties can be overcome, to a large extent at least, by employing small samples and by taking special care in the design of the equipment.

An early example of an apparatus for this purpose was described by Keler [73] in 1955, and several simultaneous DTA, TG and DTG instruments are now available commercially [69, 74]. While it is probably still true that the most reliable results can be obtained by using separate instruments specifically designed for each technique, where time is at a premium there is an undoubted advantage in employing equipment which permits simultaneous measurements to be carried out.

10.7 CONCLUSIONS

The thermoanalytical techniques most commonly applied in mineralogy are undoubtedly DTA and TG or DTG. Of these three, DTA has definite advantages, since it enables phase changes, solid-state reactions, etc., to be detected as well as decomposition or oxidation reactions. However, comparison of DTA and DTG curves immediately shows which reactions are associated with weight change and which are not. None of the techniques is particularly diagnostic on its own, but the value of all can be notably improved when they are combined with each other or with EGA, which gives additional information on the nature of the volatile products of reaction.

It should, in closing, also be mentioned that other thermal techniques as yet relatively little used for diagnostic purposes, such as dilatometry [4] and electrical conductivity [76], can all add their quota of information to provide a more complete picture of the material under study. Moreover, thermoanalytical results should always be considered in conjunction with information from X-ray diffraction, infra-red absorption spectroscopy, etc., since only by integrating the results from all available techniques can even a reasonably complete picture be built up.

REFERENCES

1. R. C. MACKENZIE, *Talanta,* **16**, 1227 (1969).
2. W. W. WENDLANDT, 'Thermal Methods of Analysis', Interscience, New York, 1964.
3. P. D. GARN, 'Thermoanalytical Methods of Investigation', Academic Press, New York and London, 1965.
4. M. HARMELIN, 'La Thermo-analyse', Presses Universitaires de France, Paris, 1968.
5. L. G. BERG, 'Vvedenie v Termografiyu' [Introduction to Thermal Analysis], Izd. Nauka, Moscow, 1969, 2nd edn.
6. W. J. SMOTHERS and Y. CHIANG, 'Handbook of Differential Thermal Analysis', Chemical Publishing Co., New York, 1966.
7. D. SCHULTZE, 'Differentialthermoanalyse', Verlag Chemie, Weinheim, 1969.
8. R. C. MACKENZIE (ed), 'Differential Thermal Analysis', 2 Vols., Academic Press, London and New York, 1970, 1972.
9. C. DUVAL, 'Inorganic Thermogravimetric Analysis', Elsevier, Amsterdam, 1963, 2nd edn.
10. C. J. KEATTCH, 'An Introduction to Thermogravimetry', Heyden and Son, London, 1969.
11. W. LODDING (ed), 'Gas Effluent Analysis', Marcel Dekker, New York, 1967.
12. J. P. REDFERN (ed), 'Thermal Analysis 1965', Macmillan, London, 1965.
13. R. F. SCHWENKER and P. D. GARN (eds), 'Thermal Analysis', 2 Vols., Academic Press, New York and London, 1969.
14. H. G. WEIDEMANN (ed), 'Thermal Analysis: Proceedings of Third ICTA, Davos, 1971', 3 Vols., Birkhäuser Verlag, Basel and Stuttgart, 1972.
15. H. F. W. TAYLOR, *Clay Miner. Bull.,* 5, 45 (1962).
16. L. S. DENT-GLASSER and F. P. GLASSER, *Acta Cryst.,* **14**, 818 (1961).
17. R. C. MACKENZIE, in 'Differential Thermal Analysis' (R. C. Mackenzie, ed), Vol. 1, Academic Press, London and New York, 1970, Chapter 1.
18. ZIE, C. J. KEATTCH, D. DOLLIMORE, J. A. FORRESTER, A. A. HODGSON and J. P. REDFERN, *Talanta,* **19**, 1079 (1972).
19. S. SPEIL, *Tech. Paper Bur. Mines, Washington, D.C.,* No. 664, 1 (1945).

20. I. PROKS, *Silikáty*, **5**, 114 (1961).
21. E. C. SEWELL, Research Notes, Building Res. Station, Watford, Herts., U.K. (1952-1956).
22. A. D. CUNNINGHAM and F. W. WILBURN, in 'Differential Thermal Analysis' (R. C. Mackenzie, ed), Vol. 1, Academic Press, London and New York, 1970, Chapter 2.
23. F. W. WILBURN, Ph.D. Thesis, University of Salford, 1972.
24. R. C. MACKENZIE and B. D. MITCHELL, in 'Differential Thermal Analysis' (R. C. Mackenzie, ed), Vol. 1, Academic Press, London and New York, 1970, Chapter 3.
25. ANON, 'Guide to Scientific Instruments, 1972–73', *Science,* **178A**, No. 4063A, 58 (1972).
26. C. MAZIÈRES, *Comptes Rend., Séanc. Acad. Sci., Paris,* **249**, 540 (1959).
27. J. P. REDFERN, in 'Differential Thermal Analysis' (R. C. Mackenzie, ed), Vol. 2, Academic Press, London and New York, 1972, Chapter 30.
28. C. KATO, *J. Ceram. Soc. Japan,* **67**, 243 (1959).
29. J. L. MARTIN VIVALDI, F. GIRELA VILCHEZ and P. FENOLL HACH-ALI, *Clay Miner. Bull.*, **5**, 401 (1964).
30. E. M. BOLLIN, in 'Differential Thermal Analysis' (R. C. Mackenzie, ed), Vol. 1, Academic Press, London and New York, 1970, Chapter 7.
31. G. LOMBARDI, *Periodico Miner.*, **36**, 399 (1967).
32. H. G. McADIE, in 'Thermal Analysis: Proceedings of Third ICTA, Davos, 1971' (H. G. Wiedemann, ed), Vol. 1, Birkhäuser Verlag, Basel and Stuttgart, 1972, p. 591.
33. R. C. MACKENZIE (Compiler), ' "Scifax" Differential Thermal Analysis Data Index', Macmillan, London 1962–1964, Index and First Supplement.
34. J. ORCEL, *Bull. Soc. fr. Minér. Cristallogr.*, **52**, 194 (1929).
35. R. L. STONE, *Bull. Ohio Engng Exp. Stn,* No. 146 (1951).
36. R. L. STONE, *Analyt. Chem.*, **32**, 1582 (1960).
37. R. I. HARKER, *Amer. Min.*, **49**, 1741 (1964).
38. J. N. WEBER and R. ROY, *J. Amer. Ceram. Soc.*, **48**, 309 (1965).
39. R. C. MACKENZIE and B. D. MITCHELL, in 'Differential Thermal Analysis' (R. C. Mackenzie, ed), Vol. 1, Academic Press, London and New York, 1970, Chapter 4.
40. R. C. MACKENZIE, R. MELDAU and V. C. FARMER, *Ber. dt. keram. Ges.*, **23**, 222 (1956).
41. R. W. GRIMSHAW and A. L. ROBERTS, *Trans. Brit. Ceram. Soc.*, **52**, 50 (1953).
42. G. SABATIER, *Bull. Soc. fr. Minér. Cristallogr.*, **77**, 953, 1077 (1954).
43. A. A. HODGSON and R. H. S. ROBERTSON, in 'Differential Thermal Analysis' (R. C. Mackenzie, ed), Vol. 2, Academic Press, London and New York, 1972, Chapter 35.
44. A. I. TSVETKOV and S. P. VALYASHIKHINA, *Trudy Inst. Geol. rudn. Mestorozh.*, No. 4 (1956).
45. R. C. MACKENZIE (ed), 'The Differential Thermal Investigation of Clays', Mineralogical Society, London, 1957.
46. A. I. TSVETKOV, S. P. VALYASHIKHINA and G. O. PILOYAN, 'Differentsialnyi Termicheskii Analiz Karbonatnykh Mineralov [Dif-

ferential Thermal Analysis of Carbonate Minerals], Izd. Nauka, Moscow, 1964.
47. V. S. RAMACHANDRAN, 'Application of Differential Thermal Analysis in Cement Chemistry', Chemical Publishing Co., New York, 1969.
48. R. T. MARTIN, *Amer. Miner.*, **43**, 649 (1958).
49. D. C. BAIN, *Nature, phys. Sci.*, **238**, 142 (1972).
50. R. C. MACKENZIE, in 'Differential Thermal Analysis' (R. C. Mackenzie, ed), Vol. 1, Academic Press, London and New York, 1970, Chapter 18.
51. R. C. MACKENZIE and B. D. MITCHELL, in 'Differential Thermal Analysis' (R. C. Mackenzie, ed), Vol. 2, Academic Press, London and New York, 1972, Chapter 36.
52. R. C. MACKENZIE and R. H. S. ROBERTSON, *Acta Univ. Carol., Geol., Suppl.* 1, 139 (1961).
53. W. H. ALLAWAY, *Proc. Soil Sci. Soc. Amer.*, **13**, 183 (1949).
54. R. C. MACKENZIE, *Pochvovedenie*, No. 4, 75 (1965).
55. J. M. BRACEWELL, A. S. CAMPBELL and B. D. MITCHELL, *Clay Minerals*, **8**, 325 (1970).
56. L. HELLER, V. C. FARMER, R. C. MACKENZIE, B. D. MITCHELL and H. F. W. TAYLOR, *Clay Miner. Bull.*, **5**, 56 (1962).
57. SADTLER RESEARCH LABORATORIES, INC., 'DTA Reference Thermograms', Sadtler Research Labs., Inc., Philadelphia, 1965.
58. G. LIPTAY (ed), 'Atlas of Thermoanalytical Curves', 2 Vols., Akadémiai Kiadó, Budapest, 1971, 1973.
59. D. A. SMITH, in 'Differential Thermal Analysis' (R. C. Mackenzie, ed), Vol. 2, Academic Press, London and New York, 1972, Chapter 40.
60. R. F. SCHWEKDER and P. K. CHATTERJEE, in 'Differential Thermal Analysis' (R. C. Mackenzie, ed), Vol. 2, Academic Press, London and New York, 1972, Chapter 41.
61. R. C. MACKENZIE, in 'Soil Clay Mineralogy' (C. I. Rich and G. W. Kunze, eds), Univ. of North Carolina Press, Chapel Hill, N.C., 1964, p. 200.
62. P. G. NUTTING, *Prof. Paper, U.S. Geol. Surv.*, No. 197E, 197 (1943).
63. E. R. BUCKLE, *J. Phys. Chem.*, **63**, 1231 (1959).
64. J. H. SHARP, G. W. BRINDLEY and B. N. N. ACHAR, *J. Amer. Ceram. Soc.*, **49**, 379 (1966).
65. J. TARNEY, A. W. NICOL and G. F. MARRINER, *Min. Mag.* **302**, 158 (1973).
66. A. W. NICOL, *Acta Cryst.*, **B27**, 469 (1971).
67. G. M. LUKASZEWSKI and J. P. REDFERN, *Lab. Pract.*, **10**, 469 (1961).
68. G. W. BRINDLEY and M. NAKAHIRA, Proc. 5th Nat. Conf. Clays and Clay Minerals (1956), 266 (1958).
69. F. PAULIK, J. PAULIK and L. ERDEY, *Talanta*, **13**, 1405 (1966).
70. H. F. W. TAYLOR, *Prog. in Ceram. Sci.*, **1**, 89 (1959).
71. A. W. COATS and J. P. REDFERN, *Analyst*, **88**, 906 (1963).
72. J. P. REDFERN, in 'Differential Thermal Analysis' R. C. Mackenzie, ed), Vol. 1, Academic Press, London and New York, 1970, Chapter 5.
73. E. K. KELER, in 'Trudy pervogo Soveshchanlya po Termografii,

Kazan, 1953' [Transactions of the First Conference on Thermal Analysis, Kazan, 1953], (L. G. Berg, ed), Izd. Akad. Nauk SSSR, Moscow-Leningrad, 1955, p. 239.
74. H. G. WIEDEMANN, *Chemie-Ingr-Tech.*, **36**, 1105 (1964).
75. C. MAZIÈRES, *Comptes Rend., Séanc. Acad. Sci., Paris,* **248**, 2990 (1959).
76. L. G. BERG, in 'Differential Thermal Analysis' (R. C. Mackenzie, ed), Vol. 1, Academic Press, London and New York, 1970, Chapter 11.

CHAPTER 11

Scanning Electron Microscopy and Microanalysis

H. N. Southworth

Department of Physical Metallurgy and Science of Materials
University of Birmingham
Birmingham B15 2TT
England

11.1 INTRODUCTION

This chapter deals with a number of techniques which have been specially tailored to provide a particular kind of information. The major emphasis has been laid on techniques for the quantitative chemical analysis of typically micron-sized volumes using principally electron beams as the exciting radiation, and also as the detected radiation in some cases. This concept of 'microanalysis' has, somewhat unconventionally, been extended to include techniques for investigating surfaces, which is a very new and important field. Associated with many of the techniques described are some novel ways of imaging the specimen using electrons, X-rays, or ions. This combination of elemental analysis with surface microscopy is especially important in enabling the composition of a phase to be related to the microstructure or texture of the material. Thus it may be important to know not only the percentage of a given metal present in an ore, but also how it is distributed and hence whether its extraction is likely to be commercially feasible. For example, knowledge of the way in which the sulphides of zinc and lead are intergrown in a mixed lead-zinc ore will determine the feasibility of separating the two phases by froth flotation.

Since a total of nine techniques will be discussed, the treatment given to each can only be superficial. In each case, after a brief outline of the type of apparatus involved and the principles upon which its operation depends, the type and quality of the information available from each method will be described and evaluated, and its relevance (in some cases still to be established) to various areas of minerals engineering indicated. Where possible, reference will be made to general review articles rather than to particular examples, and so the discussion of applications should not be taken as exhaustive.

11.2 PHYSICAL BACKGROUND

With one exception, the techniques to be described depend on the detection and analysis of some form of electromagnetic radiation which is emitted from the sample after excitation. The excitation can be caused in a number of ways, for example by electrons (sections 11.3 and 11.4), or by X-rays (section 11.6.3), or by ions (section 11.5). The two instruments currently of major importance, the scanning electron microscope (SEM) and the electron probe microanalyser (EPMA), both use a fine beam of electrons, or 'probe', to scan the specimen surface. Several different types of excitation may result [1], and it will be instructive to begin by considering the spectrum of electron emission itself. Figure 11.1 depicts the energy distribution of the electrons emitted from a metal surface using an incident primary beam of electrons of energy E_p = 2keV.

A portion of the incident electron beam will be scattered elastically, that is with no loss in energy, and such electrons give rise to the sharp elastic peak at the energy E_p. In fact, these are the electrons which have undergone Bragg diffraction, and hence on detection they may be used to provide structural information about the specimen (cf. LEED, section 11.6.2). Just below the elastic peak are several smaller peaks which correspond to diffracted primary electrons which have undergone specific energy losses due to plasmon interactions, but we shall not be concerned further with these.

Most of the back-scattered primary electrons are scattered inelastically and contribute to the general continuous spectrum of energies shown ranging downwards from E_p in figure 11.1. At the lower end of this energy range is a large secondary electron peak which represents electrons that were originally present in the solid and which have been emitted following the ionisation of the atoms in the solid due to the inelastic scattering of the primary beam. In practice, of course, it is impossible to distinguish between low energy back-scattered primary electrons and true secondary electrons, and so the latter are arbitrarily defined as those having an energy of less than 50eV. Image formation in the SEM makes use of both the back-scattered and of the secondary emitted electrons [2].

Electrons, however, form only a part of the total radiation emitted from the solid, and figure 11.2 shows diagrammatically the process of ionisation referred to above. If the incident electron beam possesses an intrinsic energy greater than a certain limit, which depends on the material under study, it

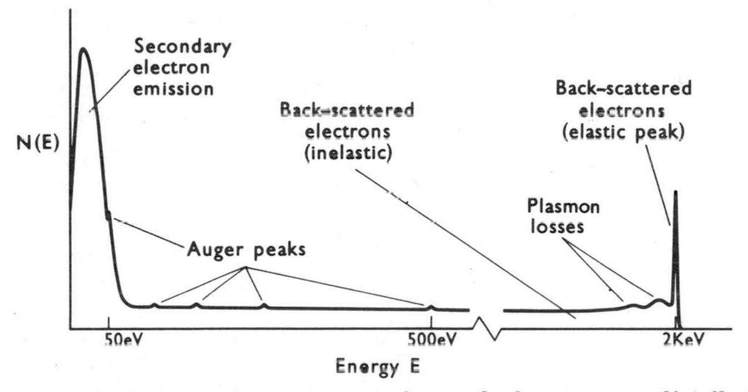

Figure 11.1 Schematic representation of the energy distribution of electrons emitted from a metal surface using an incident electron beam of primary energy E_p = 2keV.

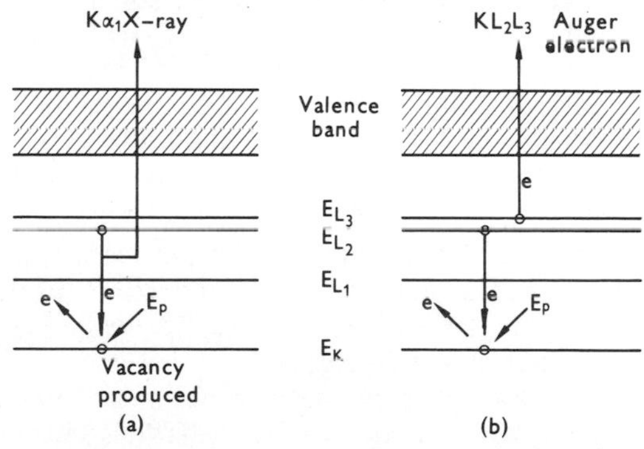

Figure 11.2 Energy diagram showing the transitions involved in the excitation of (a) characteristic X-ray emission, (b) Auger electron emission.

may displace an electron from one of the inner electron shells of an atom in the solid, e.g. from the K-shell. This situation is energetically unstable and an electron from a higher level, e.g. L_2, may drop down into the vacant position, thereby releasing an amount of energy, $\Delta E = E_K - E_{L_2}$, which may manifest itself in several ways.

Firstly, the transition may be accompanied by the emission of a photon of electromagnetic radiation of frequency ν, where $h\nu = \Delta E$, h being Planck's constant (figure 11.2a). Thus, for the transition $E_{L_2} \rightarrow E_K$, the characteristic $K\alpha_1$ X-radiation is emitted (note that the transition $L_1 \rightarrow K$ is forbidden) and such X-rays can be used for identification and analysis, as discussed in Chapter 3. Both the SEM and EPMA can take advantage of this emission for analysis, and may, in addition, use these X-rays directly to form an image. Alternatively, a photon of ultra-violet or of visible light may be emitted as well as, or instead of, the X-ray photon. This phenomenon of cathodoluminescence may also be used in image formation.

A further means by which the excess energy accompanying the $L_2 \rightarrow K$ electronic transition may be lost is illustrated in figure 11.2b. In this mode, the energy is transferred to yet another electron, in this case in the L_3 level, which is then ejected as an Auger electron. The nomenclature of Auger electrons differs slightly from that of X-rays, the electron in the above example being referred to as a KL_2L_3 Auger electron. Its energy, equal to $(E_K - E_{L_2} - E_{L_3})$, is again characteristic of the atom from which it is ejected, and so these electrons may be used for identification and analysis. The Auger electrons appear diagrammatically as small peaks in the electron energy distribution spectrum shown in figure 11.1, and their use in analysis is discussed in section 11.6.1.

11.3 THE SCANNING ELECTRON MICROSCOPE (SEM)

The technique of scanning electron microscopy was developed in the early nineteen-sixties, and has already found many uses in all fields involving the study of solids, and particularly of solid surfaces [2-4].

11.3.1 Design

Although superficially similar to the transmission electron microscope (TEM), the SEM differs radically in its method of image formation. In particular, the specimen for a TEM is in the form of a very thin section and is studied in transmission (cf. Chapter 8) whereas the specimen for an SEM is generally opaque and is studied in back reflection. Figure 11.3 shows schematically the general lay-out of a typical SEM, while Figure 11.4 shows a sophisticated commercial instrument.

In the TEM, as in the optical microscope, the area of the specimen to be studied is illuminated with a static beam of radiation, which takes the form of electrons in the TEM and light in the optical microscope. Subsequent to its interaction with the specimen, during its passage through it, the radiation

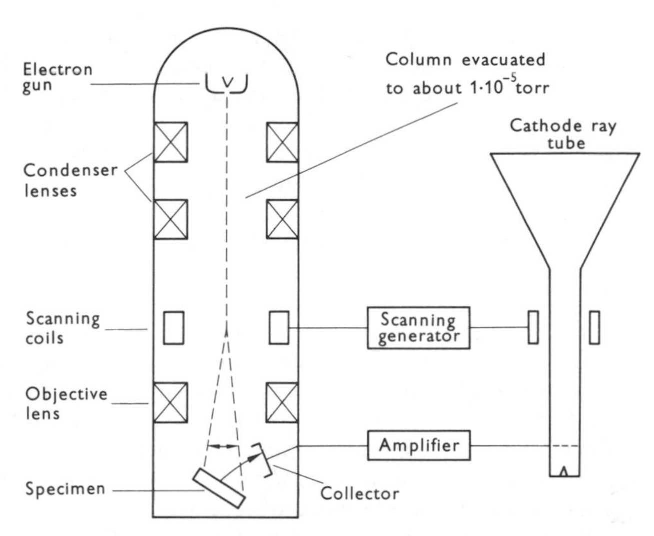

Figure 11.3 Schematic diagram of a typical scanning electron microscope.

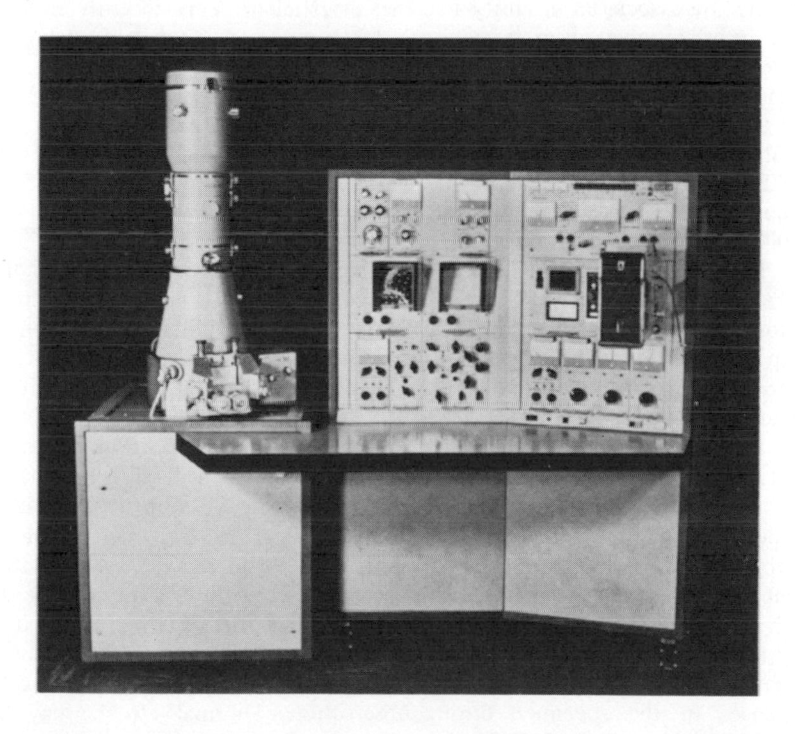

Figure 11.4 A modern commercial scanning electron microscope. (Reproduced by kind permission of Cambridge Scientific Instruments, Ltd.)

is focused by a lens system to form the magnified image. In the SEM, on the other hand, the lens system is designed to focus the incident beam of electrons to a very fine spot, about 100-300Å diameter, before it interacts with the specimen, and this fine spot is dynamically scanned across a square area of the specimen surface. Electrons scattered from the surface are collected by a Faraday cage and the resulting signal, after amplification, is used to modulate the brightness of a cathode ray tube display, which is being scanned synchronously with the incident beam, to generate a one-to-one image of the specimen surface. Thus no further focusing occurs after the electrons leave the specimen surface, and the magnification is simply the ratio of the cathode ray display tube area, which is constant, to the area of the specimen scanned, which is variable, and may be varied continuously between 20x to 100,000x without any need to refocus. In practice, two display tubes are used, one with a long persistence phosphor for visual observation, and the other with a short persistence coating, for photographic recording. A further discussion of the possible modes of operation of the instrument will be given in section 11.3.3.

The resolution obtained is dependent on the size of the spot used, and there are clearly restrictions on the extent to which its size can be reduced before the associated intensity becomes insufficient. Very recently an SEM has been designed using a field emitter type of electron gun, giving an intensity some 10^5 times greater than that given by a conventional heated filament gun, and this has provided resolution levels comparable with those of the TEM, using a spot 10Å in diameter [5]. In most existing commercial instruments, however, the optimum resolution is in the range 100-200Å, which is about one hundred times worse than the best TEM (range 2-5Å), about ten times worse than a routinely operated TEM (range 10-50Å), but still ten times better than the limit of the optical microscope.

The major problem attendant on the use of mineral specimens in the SEM is their low electrical and thermal conductivities. Their poor electrical properties result in a build-up of charge on the specimen surface, which repels the incident beam, causing wandering and loss of signal. Fortunately, the problem can be overcome quite simply, by coating the surface with a thin layer, typically about 500Å thick, of carbon, aluminium, gold, or a gold palladium alloy. If the coating is applied by evaporation onto the rotating specimen under vacuum, no appreciable loss in detail is detectable on the surface. The poor thermal conductivity can be more troublesome, especially if the specimen is hydrated. Local heating can result in dehydration of the specimen, and it is possible in some cases to watch bubbles of steam form under the conductive coating. This may be an advantage, if it is desired to study reactions in situ, but in general it is undesirable even in anhydrous materials to have appreciable localised heating, since thermal effects, such as expansion or diffusion, may lead to changes in the specimen during observation. Thermal effects may be minimised by using relatively thin specimens backed by heat sinks, or by decreasing the intensity of the incident beam.

11.3.2 Surface Topography

Perhaps the most visually dramatic application of the SEM is in the study of undulating surfaces. The depth of field of the optical microscope, at its limit of resolution, is about 1000Å, i.e. the specimen surface must be flat to within this rather fine tolerance for the whole of the image to be in focus. Operating under similar conditions of magnification and resolution, the depth of field in the SEM is about one thousand times greater, or about 0.1mm (10^6 Å). Even at the very high magnification of 10^4x, the depth of field of the SEM is about 1μm, while at low magnifications, around 20x, it is in excess of half a centimeter.

Although comparable depths of field may be obtained with the TEM, some form of replication technique must be resorted to in order to study surface topography, and such techniques are successful only with surfaces having relatively simple topographies.

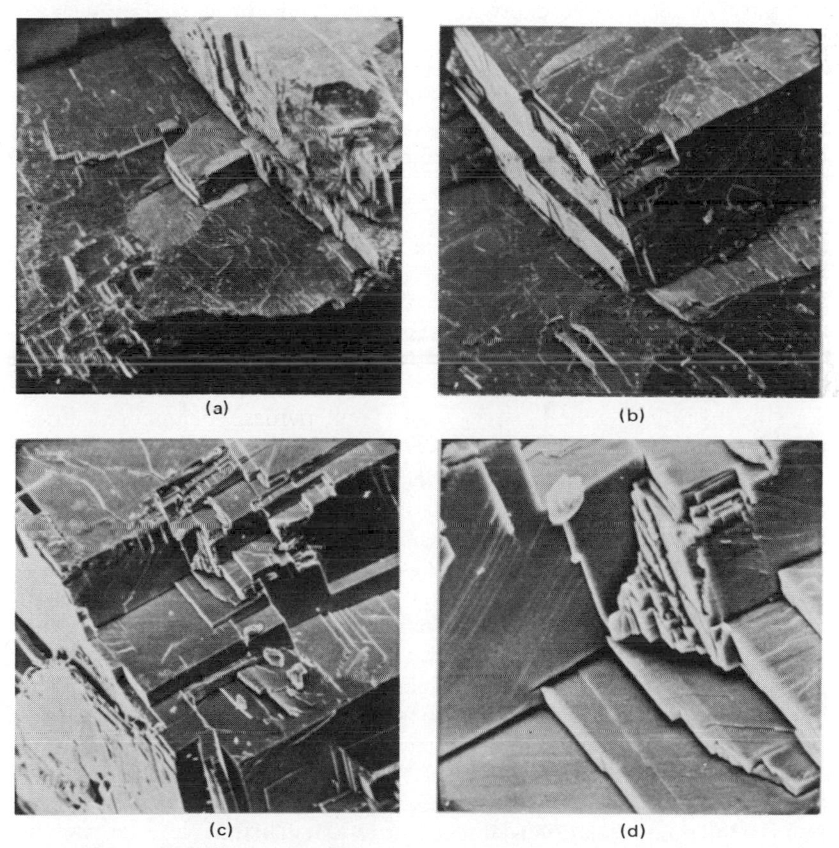

Figure 11.5 Scanning electron micrographs of the surface of a fractured galena specimen. Magnifications are (a) 16x, (b) 75x, (c) 380x, (d) 1500x.

Thus quite 'real' surfaces may be studied directly in the SEM, and figure 11.5 shows a sample of galena which had been fractured before insertion into the microscope. This series well illustrates the ability to 'zoom-in' onto a feature of interest, and successive photographs reveal increasing detail of the characteristic cleavage steps of which the fracture surface is composed.

These photographs have been taken using the secondary electrons emitted from the surface (cf. section 11.3.3), and their use helps to enhance the three dimensional appearance of the image since their low energies make it possible for a positively biased collector to capture them even if they originate outside the direct line of sight of the collector, for example inside holes and voids. They are also helpful in taking stereomicrographic pairs to aid in the interpretation and measurement of surface features. Vertical as well as horizontal surfaces can be detected in the two photographs, taken with the specimen at two slightly different angles of tilt relative to the beam, and these give a more natural appearance when viewed through the necessary viewer than do photographs in which only horizontal surfaces are clearly defined.

In many applications, therefore, the SEM can be operated at quite low magnifications, well within the capability of the optical microscope, but under these conditions it can be used to study variations in surface contour on a scale impossible with optical methods.

11.3.3 Modes of Operation

If the structure of the surface is itself not of great interest, the microstructure is best studied by taking a flat section through the specimen, but not necessarily polishing it before examination. The degree and appearance of the image contrast obtained depends on the type of radiation selected to form the image. The principal modes of operation are

(a) the emissive mode
(b) the reflective mode
(c) the absorptive mode
(d) the cathodoluminescent mode
(e) the X-ray mode.

Figure 11.6a-d shows a section through a pyrite mineral viewed successively in each of the first four modes.

In the emissive mode, figure 11.6a, the most commonly used mode of operation, the electron collector is positively biased in order to capture the low energy secondary electrons from the specimen, in addition to the primary back-scattered electrons. Since the secondary electrons originate from within 50Å of the specimen surface, the information they provide is most nearly characteristic of the actual surface. The image contrast depends quite strongly on variations in surface flatness, and hence shows up surface features quite clearly. It is also affected by the composition to a certain extent. Optimum resolution is obtained with this mode.

In the reflective mode, figure 11.6b, the collector is slightly negatively charged to discourage all but the high energy back-scattered primary

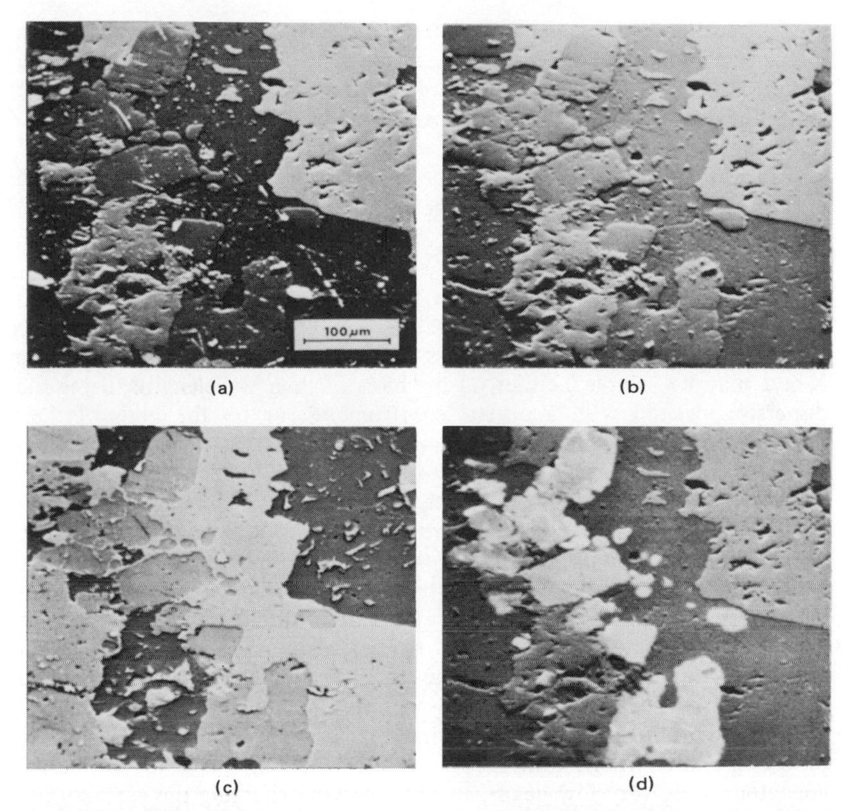

Figure 11.6 Comparison scanning electron micrographs of a mineral section taken in (a) the emissive mode, (b) the reflective mode, (c) the absorptive mode, (d) the luminescent mode. (Reproduced by kind permission of Cambridge Scientific Instruments, Ltd.)

electrons from being detected. These electrons originated from material within a few microns of the specimen surface. Since the back-scattered electron yield is strongly dependent on the atomic number of the elements comprising the surface, somewhat higher contrast can be obtained using this mode, although the resolution is generally poorer by a factor of two or three, due to the greater depth of penetration of the electrons. This 'atomic number contrast' may allow a qualitative discrimination to be made between different elements, and is analogous to the ability to distinguish between different phases in the optical microscope by their different colours, or in the TEM by differences in their absorption coefficients.

An electrical lead is attached to the specimen in order to use the absorptive mode, figure 11.6c, and the current flowing in it acts as the signal to the cathode ray display. In any given region of the specimen, the more secondary electrons there are generated, or the more primary electrons there are back-scattered, the smaller is the current generated in the specimen and so the signal in the attached lead is reduced and the display screen will be darker. Hence, the display in this mode is complementary to those obtained in the emissive or the reflective modes.

Certain solids will fluoresce under electron excitation, and when the SEM is used in this luminescent mode, the electron collector system is replaced by a photomultiplier tube and light guide. The output from this unit is again used to modulate the cathode ray display, to give the picture shown in figure 11.6d.

In the X-ray mode, the image is formed by the X-rays generated from near the surface. These are characteristic of the elements present and so show clearly their distributions across the area of the specimen scanned. The X-rays may be sampled either on the basis of their wavelengths, using the dispersive method with a crystal spectrometer, or on the basis of their photon energies, in the non-dispersive mode using an energy sensitive detection system. The latter method of analysis has tended to be favoured over the dispersive method, mainly because the low beam energies used in the SEM give very low X-ray intensities, leading to long counting times for crystal spectrometric detection. The superior performance of non-dispersive systems based on Si(Li) or Ge(Li) detectors permit these to gather sufficient counts in reasonable times, i.e. of the order of one minute. This high counting efficiency is important in studies of minerals, again because of their poor electrical properties. Since the majority of the X-rays generated come from the surface layers of the specimen, it is clear that a conducting coating can produce difficulties, by masking X-rays excited from the surface of the specimen itself. Minerals, therefore, generally have to be studied uncoated, using very low beam currents to avoid charging the surface, and this again tends to reduce the intensity of the X-rays produced. The output may again be used to modulate the cathode ray display using the output for each element separately, but X-ray units tend to be external to the main body of the SEM, and the unit made by the Ortec Company displays its output on a separate cathode ray tube after passage through a multi-channel analyser. X-ray studies may also be made on separate points in the specimen, by simply stopping the scan.

11.3.4 Applications

There are four main kinds of information that the SEM is able to supply concerning minerals. Firstly there is the study of surface topography, aided by the excellent depth of field obtainable, discussed above [6]. The morphology of the various phases exposed may be examined, either in the as-prepared condition or after controlled reaction with the environment. Alternatively the specimen may be fractured to expose internal interfaces, and to reveal faulting, pores, etc.

Secondly, the bulk microstructure may be investigated, in the form of chosen sections through it. Different modes of imaging may be used to emphasise different aspects of the structure and the information from all the modes may be combined to give a description that is very complete indeed. Minor surface features may be identified using the high resolution obtainable using the emissive mode and correlated with the improved 'atomic number' contrast obtained with the reflective or the absorptive modes, to indicate regions of different chemical composition. Subsequent X-ray studies may then be used to identify the elements present more definitively. Finally, cathodoluminescence may be used to reveal the presence of certain constituents which occur only as trace impurities.

Thirdly, quantitative studies may be made. A very approximate measure of the relative concentration of each element present can be obtained using the back-scattered electrons, but a much better estimate can be made using the characteristic X-rays. This method of analysis is essentially the same as that employed in the EPMA, and will be discussed in section 11.4. Accuracy in the SEM, however, is lower than in the EPMA, partly due to the spot size, as discussed above. The spot size can be increased until the X-ray signal is potentially as good as that in the EPMA, but the accuracy is still less because the analysing system is not designed to provide the same performance as that in the EPMA, and also because the electron beam is not incident normally onto the specimen.

Finally, it may be possible to obtain crystallographic information from single crystal specimens by varying the incident beam angle at a given spot. A pattern rather resembling a Kikuchi pattern (cf. Chapter 8) is produced, caused by preferential electron channelling in certain directions in the crystal structure.

11.4 THE ELECTRON PROBE MICROANALYSER (EPMA)

Superficially, the EPMA [7-10] and the SEM are fairly similar, both with regard to instrumental layout and to the kinds of information produced. Their differences stem from the fact that the EPMA is designed primarily to provide quantitative chemical analyses, image formation being of secondary concern, whereas in the SEM the emphasis is reversed. Operating in their respective optimum modes, each is markedly superior to the other and can give complementary information to improve our knowledge of the material under study.

11.4.1 Design

Like the SEM (figure 11.3), the EPMA consists of an evacuated column containing an electron gun, electron lenses, and deflection coils to scan the beam across the specimen. Electron and X-ray detectors are located close to the specimen surface, normally so that the 'take-off' angle of the X-rays detected is about 70° to minimise absorption effects due to surface irregularities, and the facility to view the specimen under low magnification optical microscopy is usually also provided. Since the main aim of the

EPMA is high sensitivity for X-ray detection, the primary electron beam is incident normally on the specimen surface and is over an order of magnitude larger than the beam in the SEM, being typically about 1μm in diameter. The image resolution is thus effectively restricted to this value, but the higher beam currents available mean that the intensities of the X-rays generated are also much higher. These X-rays are excited for a depth of about 1μm in the sample, and so the minimum volume which can be analysed is of the order of a cubic micron, which is still quite a tiny volume! It is possible in practice to use somewhat smaller beam sizes, but some loss in accuracy must then be accepted.

The principles of X-ray fluorescence analysis have been discussed in Chapter 3, and much of what has been said there is applicable to the EPMA. The characteristic X-rays, having been generated by an electron beam, are primary X-rays and so they are superimposed on a continuous background, from which they must be selected before measurement. Once again, the X-ray wavelength is characteristic of the element involved, and the observed intensity is proportional to its concentration, after correcting for absorption effects due to the matrix.

The X-ray wavelengths are initially sampled by means of a crystal spectrometer in order to select the required wavelength for transmission to the detector. In order to be able to examine all elements of atomic number between 4 and 92, a considerable range of wavelengths (100Å to 0.5Å) must be catered for, although the heavier elements are usually detected through their *L*-series X-radiations. This requires the use of a number of interchangeable crystals with different interplanar spacings in each spectrometer, and table 11.1 lists the normal range of crystals from which

TABLE 11.1

Range of Crystals Commonly Used in X-ray Spectrometers

Crystal	Interplanar Spacing, Å	Useful X-ray Range, Å
LiF_{200}	2.01	0.7 – 3.5
Calcite	3.04	1.0 – 5.0
Quartz	4.23	1.5 – 7.0
EDDT (Ethylene Diamine *d*-Tartrate)	4.35	1.5 – 7.0
PET (Pentaerythritol	4.39	1.5 – 7.0
ADP (Ammonium Dihydrogen Phosphate)	5.3	2.0 – 9.0
Mica	10.0	3.5 – 17
KAP (Potassium Acid Phthalate)	13.3	5.0 – 25
Clinochlore	14.1	5.0 – 27
Lead Stearate	49.0	15 – 100

the selection is made. Clinochlore [11] is a recent, and versatile, addition, which is particularly good for the detection of oxygen. Most instruments have at least two complete spectrometers, permitting the simultaneous imaging of more than one element, and the provision of three spectrometers, as in the Applied Research Laboratories EMX instrument, can amply repay the additional cost by virtue of the considerable increase in versatility attendant on the presence of the extra information channel. Thus Denny and Roy [12] have distinguished clays from zeolites in a mixture, and simultaneously monitored the distribution of calcium or potassium between the two phases, using this configuration.

The X-ray detector itself is usually a gas-proportional counter, and so is capable of differentiating between X-rays on the basis of their photon energies (cf. Chapter 1, section 1.3.3). This second stage of discrimination is vital to prevent interferences due to line overlap from preventing accurate estimates of line intensities being made. For example, in a specimen containing both fluorine ($\lambda_{K\alpha}$ = 18.31Å) and phosphorus ($\lambda_{K\alpha}$ = 6.155Å) very bad line overlap can occur using a lead stearate crystal, since the first order fluorine reflection and the third order phosphorus reflection differ by less than 0.002°. Thus the fluorine count (conventionally performed on the first order peak) could be masked by that of the phosphorus. However, since the energy of the fluorine radiation is only one third that of the phosphorus, they may subsequently be electronically separated, using a pulse height analysis circuit, to give the correct fluorine intensity.

In addition to the principal X-ray detectors, the instrument is usually also provided with a collector for the back-scattered electrons, and the signal from this may be used to give a rather poorly defined image on a cathode ray display tube, in addition to the displays available from the various X-ray channels. Some instruments also provide a detector for cathodoluminescence.

Summarising, therefore, the output channels on the EPMA include

(a) a light optical system, for directly viewing the specimen, and this may include a cathodoluminescence unit

(b) an electron optical system, for producing either a secondary or a back-scattered electron image, as in the SEM, on the cathode ray tube

(c) an X-ray identification system, again providing a direct image, with positive contrast at those sites containing the element for which the spectrometer is set (figure 11.7)

(d) an X-ray intensity measurement system, giving quantitative information for analysis via a pen recorder, tape punch, or other printout device.

11.4.2 Specimen Preparation

The preparation of a specimen for study in the EPMA requires more care than does preparation for the SEM. In particular, since the X-rays are generated from appreciable depths in the specimen, appreciable, that is, compared with electrons, it is imperative that the surface of the specimen be

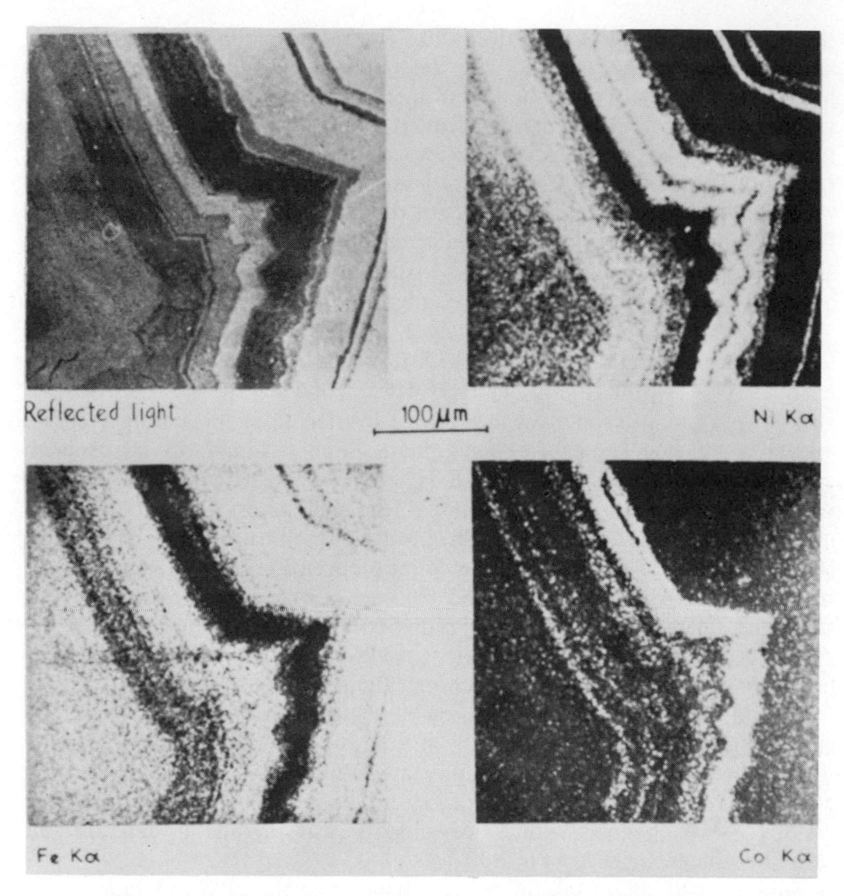

Figure 11.7 Optical and X-ray images from the electron probe microanalyser, showing the distribution of nickel, iron, and cobalt across a section of the naturally zoned mineral bravoite.

flat, to minimise errors due to differential absorption arising from different path lengths through the sample. The normal configuration of the instrument, with the spectrometers 'looking' nearly normally at the surface does help to obviate these difficulties, but confusion can still arise from the presence of 'hills' and 'valleys' in the surface. Bulk samples, therefore, can be mounted in epoxy resin and polished to an optical flatness, and powders can be mounted on a polished beryllium rod, by allowing a drop of dilute suspension to evaporate thereon [13].

The problem of surface charging is again present, and to a greater extent than in the SEM since the beam currents used are higher. A conductive coating is imperative, and graphite is the usually acceptable medium. Its low atomic weight makes it both a poor emitter and a poor absorber of X-rays, and the higher beam currents permit excitation of the characteristic X-rays from the sample quite easily. Once again, evaporation under vacuum is the preferred method of application.

11.4.3 Operation

Although analysis is the primary aim of the EPMA, it is important to be able to image the region being analysed in order to select significant features. This may be accomplished by optical, electron, or X-ray imaging. Operated at low magnification, the optical microscope may be used initially to select the general area of interest, and then at higher magnification, to locate small particles, specific phases, etc. It may be used either in reflection or in transmission, depending on the nature of the specimen, and often with the additional facility of polarised light.

Electron imaging does not permit markedly greater magnifications to be obtained, because the optimum resolution in the secondary emission mode is only about 1μm, but its greater depth of field makes it a valuable alternative should the specimen surface be rough, or where the X-ray intensity is very low and itself inadequate for image formation, since the efficiency for electron emission is markedly greater than for X-ray emission. Electron imaging also provides the opportunity to get the electron spot into its best focus, as judged by the sharpness of the resulting image.

For qualitative analysis, the X-ray spectrometer is set to the wavelength corresponding to the species of interest, and the resulting X-ray image generated on the cathode ray display reveals the distribution of that element over the area being scanned, as shown in figures 11.7a-d, which were obtained from a naturally zoned bravoite [14]. It is not too difficult to correlate the three elements identified with the zones shown in the optical image, but there is always some degree of uncertainty as to precisely how the elements are combined or segregated in the sample, in data presented in this manner. In particular, the occurrence of two or more elements in a given location can indicate that they are chemically combined, and so phase analysis becomes possible, but this requires quite accurate correlation of elemental positions to distinguish between a true compound and two interstratified phases. Jones, Gavrilovic and Beaver [15] have described a colour synthesis technique whereby the conventional micrographs, obtained from the cathode ray display as black and white negatives, are copied onto a single colour film using a different coloured filter for each negative, e.g. green for Fe, blue for Co, and red for Ni. In the resulting colour positive, a yellow area, for example, would represent an area rich in both Fe and Co but poor in Ni, and so on, and bands of green and blue would represent intergrowths of material rich in Fe and Ni respectively. In mineral and geological work very complex and intergrown assemblages can occur, and this technique would appear to be potentially of great value here.

If quantitative information is required, the electron spot can be held stationary over the point of interest while the spectrometer crystal is rotated through a small sweep about the required detection angle for the element of interest. By comparing the area recorded under the peak, using standardised counting conditions, with that from a pure sample of the same element, held in special positions in the specimen mount for easy reference, the absolute concentration of that element in the sample may be determined in the usual way. Figure 11.8 is a plot of the relative concentrations of the elements that occur in a section across the mineral shown in figure 11.7.

It is important to point out that the element in the standard should be in at least the same oxidation state as the element in the sample, and it should preferably also be in the same co-ordination state. The characteristic X-ray peak for an element is a very good approximation to a delta function and so a shift of the order of seconds of arc from the precise position of the peak maximum can reduce the apparent intensity due to the element concerned by an appreciable amount. The crystal field surrounding an atom can affect its X-ray spectrum in the same way as, although to a lesser extent than, it can affect the optical spectrum, due this time to shifts in the *K*- and *L*-levels in the atom. White and Gibbs [16] have shown that these small shifts can be detected by careful analysis, and that it is possible not only to identify the elements present in a sample but also to identify their oxidation states and their co-ordination numbers in many cases. Thus one can easily distinguish Al^{3+} in six-co-ordination with respect to oxygen from Al^{3+} in five-co-ordination or Al^{3+} in four-co-ordination. A corollary of this is that the peak position must be checked for every element in any given sample.

The accuracy of analysis depends on a number of fundamental factors [7, 8], although experimental limitations can often constitute the major sources of error. For maximum accuracy the beam must meet the specimen

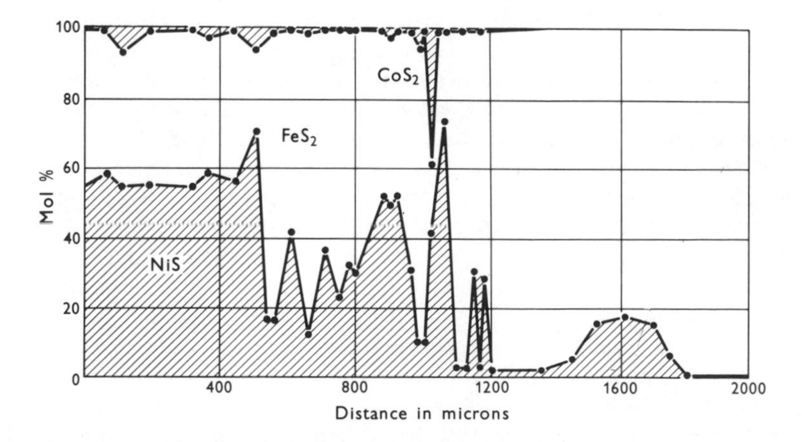

Figure 11.8 Electron probe microanalyser plot of the relative concentrations of the sulphides of cobalt, iron, and nickel, across a section similar to that shown in figure 11.7. [Figures 11.7 and 11.8 are from ref. 14, Academic Press, New York and London. (Courtesy of Dr. J. V. P. Long.)]

at normal incidence, which means that the specimen must be accurately positioned and, as mentioned above, the surface should be free from relief, to minimise absorption errors. The ability to polish to the accuracy required for optical microscopy may be difficult to achieve in multicomponent samples having phases of different hardness. The need for a conducting layer introduces problems, and care should be taken to ensure that the coating on both specimen and standard are the same thickness.

The assumption that the X-ray intensity is proportional to concentration is not strictly valid. Correction factors must be applied to the observed intensities to account for such factors as differences in the atomic numbers of the elements determined, for the matrix absorption effects mentioned above, and, although usually of lesser importance, for fluorescence. There is also the problem that the intensity of the incident beam may vary during the study of either or both specimen and standard, and this will have an appreciable effect on the X-ray intensities. Hence the degree of sophistication underlying these correction factors throws a considerable burden onto the reproducibility both of different areas of the reference standard and of the electron beam current, and considerable effort has been expended on attempts to calculate the intensity-concentration relationship directly [7, 17-19]. White [13, 20] has shown that elemental ratios, if not total analyses, can be carried out by simultaneously integrating the counts for two or more elements in the same volume of the specimen and comparing the ratios with those from standards of closely similar composition. Even without correction for absorption, elemental ratios within 2% of the expected value were obtained, for known samples, by averaging the ratios determined at about 30 points on each surface.

The analysis of light elements is particularly difficult for a number of reasons [7]. With decreasing atomic number, the wavelengths of the characteristic X-rays become longer, the effect accelerating below about Z = 10. Thus, for fluorine, oxygen, nitrogen, carbon, boron, and beryllium (Z = 9 through 4, respectively) their *Kα* wavelengths increase in the order 18, 24, 32, 44, 68, and 113Å. As has already been shown in table 11.1, these necessitate the use of special heavy metal stearate crystals in the spectrometer, whose preparation is still somewhat of an art. The introduction of clinochlore, however, has greatly improved the situation for oxygen at least. In addition, such soft X-rays are difficult to measure, and a special design of proportional counter is required. Even with this refinement the sensitivity remains poor, and when there is added to this the fact that the basic X-ray yield is low anyway for these light elements it is not surprising that their accuracy levels are considerably reduced compared with the heavier elements.

The light elements apart, the relative sensitivity of the technique is in the range 10-100ppm, while the absolute detection sensitivity is about 10^{-14}g. In the case of mineral specimens, an overall accuracy of 3-5% can be expected, which should be perfectly adequate for most applications.

To conclude the section on 'operation', we may look to the future, and to automation. The development of a control system which will automatically carry out the sequence of operations used to obtain a full multi-element analysis is already in progress [21]. This offers the opportunity for greater utilisation of the instrument, including over-night and weekend use, and may help to offset the very high initial cost of the facility.

11.4.4 Applications

In mineralogy, perhaps even more than in metallurgy, the range of problems ideally suited to EPMA study is considerable, and a number of good review articles exist [7-10, 22]. The virtue of the EPMA lies not in its sensitivity or accuracy, neither of which is remarkable, but rather in its ability to perform analysis on a very fine scale. If the microstructure is composed of very small particles or grains, then the EPMA is often the only available tool, since it may be impossible physically to separate the phases in order to carry out a 'standard' chemical analysis, and X-ray fluorescence, or even diffraction with its greater phase sensitivity, will be similarly powerless.

Thus mineral identification is a primary application and it is perhaps appropriate that one new mineral (castaingite) has already been named after the person first responsible for the development of the electron probe microanalyser [23]. In addition to the identification of unknown minerals, however, the technique has also been able to correct many erroneous identifications based on less sensitive techniques. For example, the mineral mackinawite, FeS, which usually occurs in ore deposits as very minute grains, was long thought to be identical with valleriite, $CuFeS_2$, on the basis of their similar strong optical anisotropy. Electron probe studies showed them to be chemically quite different [22].

In addition to identification, quite accurate compositional analyses can be made, leading to the investigation of solid solutions and of phase equilibria. Once again the EPMA has contradicted earlier results obtained by macroscopic chemical analysis, or explained anomalous findings, and in both cases it has shown that the discrepancies are attributable to minute inclusions of unsuspected minerals. Most minerals are compositionally inhomogeneous and the EPMA has been able to reveal the presence of compositional gradients not apparent under optical examination, as shown in figures 11.7 and 11.8 for the zoned bravoite discussed previously [14].

Cathodoluminescence can again, as in the SEM, provide quite vital information at times [17, 24], since its occurrence is so noticeable that the presence of certain trace elements, such as rare earths, can be detected at concentrations far below that of the normal detection capability of the EPMA. In favourable cases, a semi-quantitative measure of their concentrations may even be made.

So far, only the determination of compositional information allied to microstructural studies has been considered, but the EPMA can also give crystallographic information by using a Kossel camera [7], which is an X-ray diffraction technique on the micron scale. The sample in the EPMA is covered with a very thin foil of a metal which will emit X-rays of a known wavelength when irradiated by the primary beam. Some of these X-rays enter a very restricted region of the specimen and there are diffracted in the normal way (cf. Chapter 7) to form a selected area diffraction pattern. If this is photographed either as a transmitted or a reflected pattern, it will show the orientations of the crystallites in the specimen, and can also be used to give very accurate values of their lattice parameters.

11.4.5 Electron Microscope Microanalyser

The transmission electron microscope, by virtue of its ability to resolve at the 2-10Å level, its good contrast, and its facility for selected area electron diffraction, is the most powerful of all the electron-optical microscopes. The facility of microanalysis has recently been married to it, and the resulting hybrid is termed the 'electron microscope microanalyser', or EMMA [25].

To the microanalyst, EMMA offers the opportunity of detecting very small particles, less than one micron in size, together with those larger ones which give rise to poor contrast in either the electron back-scatter or the X-ray scanning modes in the EPMA. Added to this, of course, is the powerful facility of selected area electron diffraction, which permits identification of the crystalline phases present. Hence it is, in principle, possible to gather data concerning the habit, crystallography, and composition of sub-micron particles easily and accurately. However, although it is possible to locate and analyse particles down to $10^{-1}\mu m$ in size, for good quantitative work $1\mu m$ is still the effective lower limit. Hence the usefulness of EMMA is rather more to the existing electron microscopist, who now has the facility of obtaining some form of compositional information, which is at worst semi-quantitative, in addition to his major microstructural information.

11.5 ION PROBE MICROANALYSIS

The two techniques discussed above have used beams of electrons as the exciting radiation. The techniques discussed in this section use ion beams in place of the electrons, but the information which they can supply is of much the same type as before.

11.5.1 The Proton Probe

A major limitation of the EPMA is its comparatively low sensitivity, due mainly to the successive slowing down of the primary electrons as they penetrate the target, producing a continuous X-ray spectrum, the bremsstrahlung (cf. Chapter 3), upon which the characteristic lines are superimposed. The sensitivity, which depends on the ratio of the peak height to the background count, is, therefore, clearly limited. As shown in Chapter 3, no such bremsstrahlung occurs when primary X-rays are used as the exciting radiation, and high energy protons act in a very similar way.

Recent work at the Atomic Energy Research Establishment, Harwell, has led to the development of the proton probe [26], which uses a beam of protons, rather than electrons, to excite the characteristic radiations of the elements present. Using this technique, it is expected that, for certain elements, a sensitivity of 1ppm will be achieved.

11.5.2 The Ion Microprobe

A further limitation of the EPMA is its poor performance with elements of low atomic number. Another comparatively new instrument, the ion microprobe [27, 28], avoids this difficulty by relying on mass spectrometric analysis instead, which provides the additional facility of isotopic analysis.

Several versions of the technique exist, but basically they all use a primary ion beam directed at the specimen to produce secondary ion emission, or ion sputtering. These secondary ions are then analysed by means of a mass spectrometer. Thus the specimen surface is undergoing continual erosion at a rate which is variable between a few angstroms and some hundreds of angstroms per second, and so, since the analysis is being carried out on material actually removed from the sample, a fresh surface is being continually generated as the analysis proceeds. This continual removal of material is, of course, the major difference between this method and those based on electron beam excitation.

Different modes of analysis may be employed, depending on the information required;

(a) the mass spectrum from a selected area may be scanned in order to identify the various elements present, as discussed in Chapter 5;

(b) depth distributions for one or more elements may be obtained by measuring the variation with time of a selected emission over the whole surface, or of all the emission from a single point;

(c) a particular secondary ion may be selected to produce a near-instantaneous image of its distribution within the bombarded area.

The area of sample examined can vary in the range 1-250μm, with an image resolution of 1-5μm, and a depth resolution typically around 100Å. Relative sensitivities are highly variable, because they depend on the sputtering efficiency of the primary ion beam for the particular element considered, the bonding state of the element, and the surface topography of the specimen. The range is as wide as 1ppb through 1ppm, optimum values being obtained only when the maximum amount of sample is taken for analysis, i.e. when the area examined is a maximum. Absolute sensitivities are difficult to determine, since no adequate theory exists to relate signal strength to concentration. Accuracies within an order of magnitude are the best that can be expected at the present.

The advantages of the ion microprobe over, say, the EPMA lie in its ability to analyse low atomic number elements, including hydrogen; its considerably higher sensitivity; its better depth resolution (100Å compared with 1μm); its ability to produce depth profiles; and the measurement of isotopic ratios. Its disadvantages are poorer area resolution and direct image producing qualities, and its very much higher cost even than the EPMA. Thus, at present, the ion microprobe is less satisfactory for use on surfaces containing fine scale inhomogeneities.

11.6 SURFACE AND NEAR-SURFACE TECHNIQUES

A comparatively neglected area in materials research is the specimen surface itself. After all, this is the point at which the material meets and reacts with its environment. Two factors have hitherto restricted work in this area. One is the absence of techniques with sufficiently small depth resolutions, although at 100Å the ion microprobe is approaching this ability, while the other is the difficulty in obtaining a sufficiently good vacuum environment so that the surface can be maintained in a known condition, free from atmospheric contamination. The two factors are, of course, related. In recent years a number of instruments have been developed which combine to push the depth resolution down to include the surface layer of atoms only [29]. Thus, at least in one dimension, they are very much microanalytical techniques, and it has been felt worthwhile to point out their capabilities.

11.6.1 Auger Electron Spectroscopy

It was described, in section 11.2, how characteristic electron emission can occur as an alternative to X-ray emission, although the appearance of the Auger peaks in the $N(E)$ distribution curve shown in figure 11.1 does not seem, at first glance, to offer great analytical possibilities. However, if, as in figure 11.9, the derivative of the $N(E)$ versus energy curve is plotted, it can be seen that detection of the Auger peaks becomes much easier.

The apparatus used is shown in figure 11.10. It is very much simpler than that used in any of the foregoing methods, since no great effort is made to focus the beam to a very fine spot (about 1m diameter is usual), and no visual image of the specimen is produced. The specimen is situated within an ultra-high vacuum environment, facing a hemispherical system of grids. The middle two grids are slowly swept by a retarding potential, varying from zero volts to E_p, superimposed on which is a modulating signal which allows the $N(E)$ distribution to be obtained. This is then differentiated electronically to give $d(N(E))/dE$.

Since the escape depth of Auger electrons is in the region of 1-4 atomic layers, the technique is very much one of surface analysis, although concentration profiles into the bulk may be obtained by ion sputtering of the surface layers with argon. This removes material in a manner analogous to that occurring in the ion microprobe, although usually at a much slower rate. A relative sensitivity of from 0.3% of a monolayer, for carbon or sulphur, up to 5% of a monolayer for sodium and aluminium can be obtained, although absolute concentrations are more difficult to determine and the technique can be considered to be no more than semi-quantitative

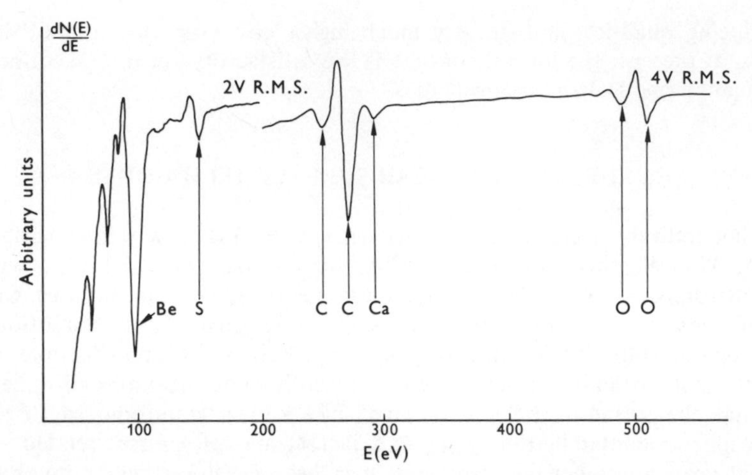

Figure 11.9 Auger electron spectra from beryllium, showing the presence of surface contamination. The unspecified low energy peaks are probably loss peaks corresponding to the main beryllium peak at 99eV.

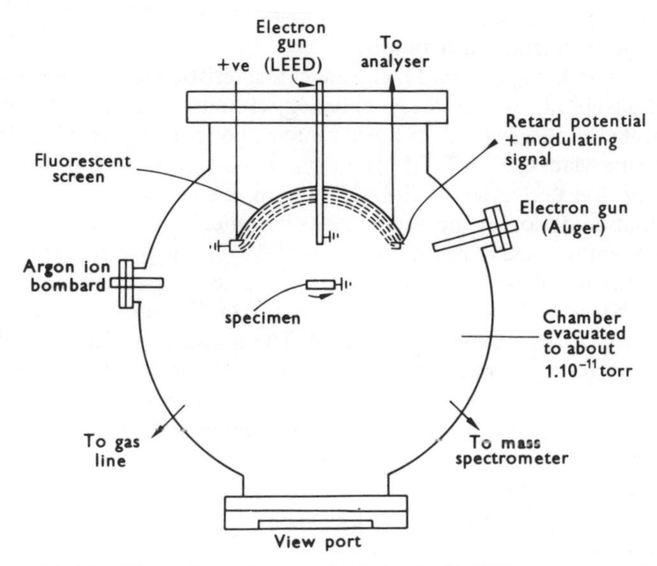

Figure 11.10 Plan view of a typical Auger/LEED system, using 4-grid optics.

at present. The electron energy resolution is, in some cases, sufficiently good to distinguish between different oxidation states of the same element.

Thus AES yields a chemical analysis whose depth resolution is restricted to only a few Å, although its area resolution is comparatively poor. Current applications, therefore, include the detection of surface segregants, the investigation of surface reactions, and the analysis of internal surfaces produced by *in situ* fracture or cleavage. All the elements, with the exception of hydrogen and helium, may be detected, and, in contrast to the EPMA, the yields from low atomic number elements are comparatively high. The technique is very new, but recent work of relevance in this sphere includes investigations of the spectra resulting from freshly cleaved silica and alumina [33], and mica [33, 34]. A potentially extremely important application of AES could lie in the field of investigating surface composition and its influence on froth-flotation techniques.

11.6.2 Low Energy Electron Diffraction

In the electron energy distribution shown in figure 11.1, only the elastically scattered electrons have not yet been considered. However, no form of radiation is ignored in chemical and structural analysis, and detection of these electrons forms the basis of the technique of low energy electron diffraction, or LEED. Using exactly the same apparatus, shown in figure 11.10, as for AES, which LEED actually predates by several years, and by retarding all but the elastically scattered electrons, a diffraction pattern may be formed on the fluorescent screen.

Thus LEED yields structural and crystallographic information about the atom positions in the first few layers of the specimen. A severe restriction is that the sample must be in the form of a single crystal, and the majority of LEED work has been on the study of adsorption onto selected crystallographic planes, and even in this simple type of system the results are difficult to interpret fully.

11.6.3 Electron Spectroscopy for Chemical Analysis

Auger spectroscopy is a relative newcomer to an already existing field of electron spectroscopy [35]. By irradiating a solid with some form of electromagnetic radiation other than electrons, for example ultra-violet light, normally called photoelectron spectroscopy, or X-rays, when the technique is called electron spectroscopy for chemical analysis or ESCA [36], ionisation of the atoms of the sample may again be induced, followed by emission of an electron, with an energy given by

$$E = h\nu - E_{binding}$$

where $h\nu$ is the photon energy of the incident radiation and $E_{binding}$ is the energy of the electron in the atom. The ejected electron may be detected and determined. Thus, an accurate knowledge of the energy of the incident radiation, and soft X-rays such as Al $K\alpha$ or Mg $K\alpha$ are usually used in ESCA, yields electron energies characteristic of the atoms present. It thus constitutes a variant on X-ray fluorescence, where X-ray excitation is also used, but it is the simultaneously emitted X-rays that are monitored in the latter method.

It will be clear that the technique can be used for analysis. More importantly, however, it can be used to distinguish between different binding or oxidation states, due to the narrow line-widths of the energies of the emitted electrons and the high resolution of the spectrometers employed. Insufficient data are as yet available to permit comparisons to be drawn with the identification of the above parameters from EPMA data, but the ESCA data should be at least as good as those from EPMA. ESCA has the decided advantage over photoelectron spectroscopy in that the higher energies involved make it possible to ionise electrons deep in the core of the atom, as well as the valence electrons, and so obtain more and better information on chemical shifts due to oxidation state and to environment.

Although Auger electrons may also be generated and detected using this technique, permitting at-surface analyses to be carried out, the greater emission energy of the photoejected electrons means that they originate from a depth of up to 20 atomic layers, about 50Å, below the specimen surface. Thus the analogous information is presented for a depth level intermediate between that of AES and that of the ion microprobe. The area resolution is similar to that of AES and again no specimen imaging facility exists. All elements, with the exception of hydrogen, can be detected.

Again this is a very recent technique, at least as far as commercial availability is concerned, and its application to the minerals field is unexplored. Its virtues lie in the origin of the information presented, which lies close enough to the specimen surface to be characteristic of surface effects rather than bulk and yet deep enough to be fairly insensitive to actual surface cleanliness, and in its ability to detect chemical shifts, and hence, for example, to be able to distinguish between sulphur present as a sulphide or as a sulphate, as in the EPMA [16, 37].

11.6.4 The Field-Ion Atom Probe

Just as this Chapter started with a technique, the SEM, whose function was primarily that of microscopy with an analytical facility added, so it ends with one. With the field-ion microscope (FIM) [38, 39] it is possible directly to resolve the individual atoms in the crystal structure of the surface of the specimen.

The form of the specimen is necessarily restricted to a fine needle whose tip has been shaped, for example by electropolishing, to a point of radius 100-1000Å. This is situated within a vacuum system, as shown in figure 11.11, facing a fluorescent screen. A small quantity of inert gas (He, Ne, or Ar) is leaked into the chamber and a high positive potential, in the range 0-30kV, is applied to the specimen. Sufficiently high fields are thereby generated at the sharply pointed tip for these gas atoms to become ionised over the surface of the specimen. The positive ions so produced are then repelled from the tip to excite an image on the screen, whose magnification is approximately the ratio of the screen radius to the tip radius, typically of the order of 5cm to 500Å, or 10^6 X.

So sensitive is the ionisation probability to the field strength that the

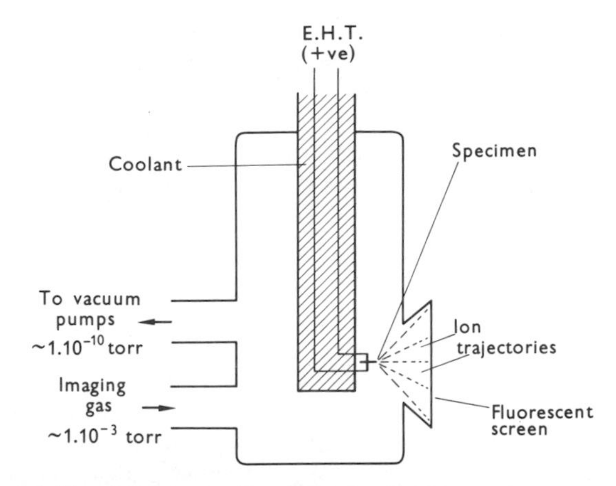

Figure 11.11 Schematic diagram of a field-ion microscope.

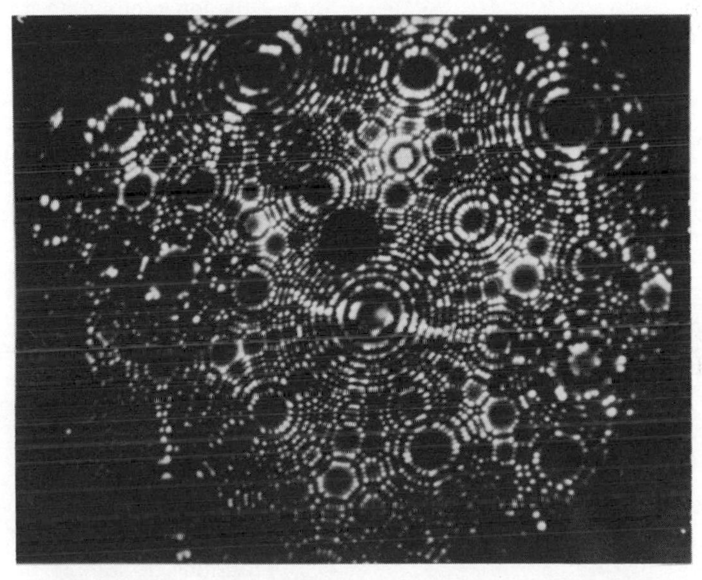

Figure 11.12 Field-ion micrograph of tungsten. The probe hole aperture is just below the centre of the image. (Courtesy of P. J. Turner.)

local field enhancement above each atom is sufficient to ensure that a separate beam of ions originates from each of these atom sites. Thus, in the resulting image, shown in figure 11.12, each white spot represents the image of a single surface atom.

If the applied potential is increased, the surface atoms can be made to field evaporate. This process is somewhat analogous to thermal evaporation or sublimation, but is controllable to the extent that the atoms can be removed two or three at a time. In this way the bulk microstructure may be revealed in the form of successive surface sections, at intervals either of a few atoms or of many atomic layers.

In addition to the unique structural information that is provided by this method, some distinction can be made between atoms of different atomic species on the basis of differences in image point intensity [40]. The newly developed 'field-ion atom probe', however, allows such distinctions to be made exactly [41]. In this instrument a time-of-flight mass spectrometer is coupled to the image screen of the FIM, into which a hole or aperture of a selected size has been introduced. To utilise the facility, the specimen is manipulated until the aperture covers, depending on its size, a selected atom or a group of atoms, which are then field evaporated using a high voltage pulse. After leaving the surface, the atoms follow the same trajectory as the imaging gas atoms, pass through the screen aperture, and are identified. In this way single atom isotopes can be positively identified and related to their original location in the microstructure – surely the ultimate in chemical analysis. The existence of a fairly large screen aperture is clearly visible near to the centre of figure 11.12.

Unfortunately, only pure metals give images of the perfection of figure 11.12. In figure 11.13, which is an image from LaB_6, the image regularity is very substantially reduced. Largely this is due to the fact that the field required for evaporation is less than that required for ionisation and imaging, resulting in a constantly changing surface.

However, even if the full atomic resolution of the technique were not to

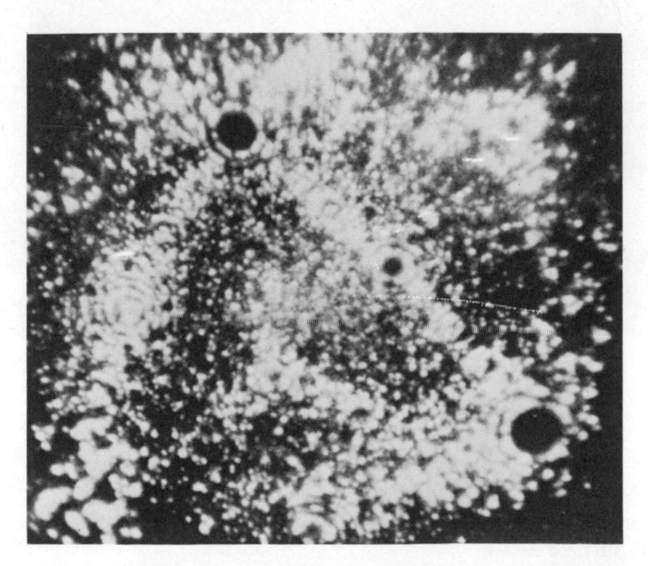

Figure 11.13 Field-ion micrograph of lanthanum hexaboride. Note the markedly reduced image quality compared with the micrograph of tungsten shown in figure 11.12. (Courtesy of D. A. Smith.)

be used with such materials, there is no reason why its analytical capability should not be used. Provided that the image quality is sufficiently good to permit areas of interest to be identified, then the mass spectrographic analysis could simply be carried out in a continuous manner as the surface underwent field evaporation spontaneously. A 'probe' hole whose diameter covered, for example, a 50Å area on the surface might be used in such cases, although in principle there is no reason why the entire image, which might be the surface of a hemisphere up to a few thousand Angstroms in diameter, could not be analysed at once.

Many similarities exist here with the ion microprobe, especially with regard to the 'information carrying species', ions in both cases, and to the method of detection, mass spectrometry. In this case, however, there is the advantage that both the area and the depth resolution are substantially better, of the order of a single atom instead of hundreds of layers. The major limitation of the FIM is probably the extremely restricted form that the specimen must adopt. In all of the other techniques described, it has been possible to make the specimen much more representative of the 'bulk' material, which is usually a solid object with dimensions in the centimetre range.

11.7 SUMMARY

To a very large extent the techniques described above have been complementary to one another, combining to provide a broad spectrum of information. This has taken one or more of three basic forms:

(a) elemental analysis,
(b) microstructural information, in the sense of a direct image revealing the distribution of the various phases, and
(c) crystallographic information, yielding information on the crystal structure, either directly, as in the FIM, or indirectly via a diffraction pattern,

and there are many instances where it is important to obtain information at more than one of these levels in order to build up a complete picture of the material under study. To this end, table 11.2 attempts to compare the parameters that are important in assessing the analytical capabilities of each of the techniques. Data for the field-ion atom probe are not included since this technique does not fit readily into such a scheme.

Thus the first need is to decide at what precise level information is required, and then to select the technique, or combination of techniques, that will provide the necessary data. For example, if maximum sensitivity is required then techniques such as optical spectroscopy or X-ray fluorescence rate more highly than EPMA. Where the latter gains is in its area resolution, i.e. in its ability to provide *in situ* analysis on the micro scale. In this sense, the SEM should be even better, but it usually fails to reach the accuracy levels of the EPMA.

TABLE 11.2

Comparison of the Various Analytical Techniques

	EPMA	SEM	Photon probe	Ion Microprobe	AES	ESCA
Primary Excitation	electrons	electrons	protons	ions	electrons	X-rays
Measured Emission	X-rays	X-rays	X-rays	ions	electrons	electrons
Analysis Depth	1μm	1μm	1μm	variable about 100Å*	10Å*	50Å
Area (diam) Resolution	1μm	200Å	5μm	1μm–250μm	500μm	500μm
Relative Sensitivity	100ppm	0.1–1%	1ppm	10ppm to 10% 1ppb to 1ppm	1% of monolayer	?

* Progressive, by successive removal of surface atoms.

Increasingly better depth resolution is obtained going from the EPMA to the ion microprobe, to ESCA and to AES. However, the area resolution gets progressively worse. In addition, neither ESCA nor AES provide specimen imaging facilities, making it difficult to select particular areas of the specimen for analysis. A start has been made to overcome this difficulty, however, by building an AES analyser directly into an ultra-high vacuum SEM [42].

In many ways the field-ion atom probe escapes these limitations, providing as it does image formation and a single atom analysis capability. However, the limitations already mentioned in regard to the form of the specimen, coupled with the extremely small volume that is analysed, means that the sort of microstructure suitable for examination is quite limited. In any case, the idea of analysis on an individual atom scale is probably still too novel to be regarded as particularly useful at the present time.

The SEM and the EPMA, therefore, are probably the two most versatile techniques, depending on whether qualitative or quantitative analyses are required, of all the techniques herein described, and they are certainly the most used techniques in the minerals field at the present moment. However, most of the other techniques are still very new, and they may turn out to have unexpected possibilities when they are applied to the study of minerals.

ACKNOWLEDGEMENTS

This Chapter was prepared during a short period spent at the Atomic Energy Research Establishment, Harwell, where most of the techniques described are in use, and valuable discussions were held with Mr. B. W. Mott and Drs. H. E. Bishop, J. P. Coad, J. C. Rivière, and D. M. Poole. Thanks are also due to Drs. J. V. P. Long, D. A. Smith, and P. J. Turner, and to Cambridge Scientific Instruments Ltd., for permission to reproduce micrographs and for the photograph of the scanning electron microscope.

REFERENCES

1. A. J. DEKKER, 'Solid State Physics', Macmillan, London and New York, 1964.
2. P. R. THORNTON, 'Scanning Electron Microscopy', Chapman and Hall, London, 1968.
3. G. R. BOOKER, in 'Modern Diffraction and Imaging Techniques in Materials Science', S. Amelinckx, R. Gevers, G. Remaut and J. van Landuyt (eds), North-Holland, Amsterdam, 1970, p. 553.
4. G. W. KAMLOTT, *Surf. Sci.*, **25**, 120 (1971).
5. A. V. CREWE, J. WALL and L. M. WELTER, *J. Appl. Phys.*, **39**, 5861 (1968).
6. E. W. WHITE, in 'Scanning Electron Microscopy/1968', IIT Research Institute, Chicago, 1968, p. 89.
7. T. D. McKINLEY, K. F. J. HEINRICH and D. B. WITTRY (eds), 'The Electron Microprobe', John Wiley and Sons, Inc., New York and London, 1966.
8. I. ADLER, 'X-ray Emission Spectrography in Geology', Elsevier, Amsterdam, 1966.
9. J. V. P. LONG, in 'Physical Methods in Determinative Mineralogy', J. Zussman (ed), Academic Press, London and New York, 1967, p. 215.
10. C. W. MEAD, in 'Electron Probe Microanalysis', A. J. Tousimis and L. Marton (eds), 'Advances in Electronics and Electron Physics', Suppl. 6, 1969.
11. W. L. BAUN and E. W. WHITE, *Anal. Chem.*, **41**, 831 (1969).
12. P. DENNY and R. ROY, Proc. XIIth Conf. Clays and Clay Minerals, Atlanta, Ga., 567 (1963).
13. E. W. WHITE, *Amer. Mineral.*, **49**, 196 (1964).
14. G. SPRINGER and J. V. P. LONG, in 'X-ray Optics and X-ray Microanalysis', Academic Press, London and New York, 1963, p. 611.
15. M. P. JONES, J. GAVRILOVIC and C. H. J. BEAVER, *Trans. Inst. Mining and Met.*, **75**, B274 (1966).
16. E. W. WHITE and G. V. GIBBS, *Amer. Mineral*, **54**, 931 (1969).
17. W. REUTER, *Surf. Sci.*, **25**, 80 (1971).
18. P. DUNCUMB, in 'Electron Microscopy and Analysis', Inst. of Physics Conf. Series No. 10, p. 132 (1971).
19. T. R. SWEATMAN and J. V. P. LONG, *J. Petrol.*, **10**, 332 (1969).

20. E. W. WHITE, P. DENNY and S. M. IRVING, Electron Microprobe, Proc. Symp., Washington D.C., 791 (1964).
21. B. JEFFERIES and J. V. P. LONG, in 'Electron Microscopy and Analysis', Inst. of Physics Conf. Series No. 10, p. 150 (1971).
22. G. SPRINGER, in 'X-ray Optics and Microanalysis', Academic Press, London and New York, 1969, p. 424.
23. R. CASTAING, Ph.D. Thesis, Univ. of Paris, Publ. O.N.E.R.A., No. 55, 1951.
24. J .V. P. LONG and S. O. AGRELL, *Miner. Mag.*, **34**, 318 (1965).
25. P. DUNCUMB, in 'X-ray Optics and X-ray Microanalysis', Academic Press, London and New York, 1963, p. 431.
26. D. M. POOLE and J. L. SHAW, in 'X-ray Optics and Microanalysis', Academic Press, London and New York, 1969, p. 319.
27. H. LIEBL, *J. Appl. Phys.*, **38**, 5277 (1967).
28. A. J. SOCHA, *Surf. Sci.*, **25**, 147 (1971).
29. P. A. REDHEAD and R. A. ARMSTRONG, *Nederlands Tijdschrift voor Vacuumtechniek*, **8**, 145 (1970).
30. C. C. CHANG, *Surf. Sci.*, **25**, 53 (1971).
31. J. C. RIVIÈRE, *Phys. Stat. Sol.*, (In press).
32. A. G. JACKSON, M. P. HOOKER, T. W. HAAS, G. J. DOOLEY, and J. T. GRANT, 'Bibliography on Low Energy Electron Diffraction', U.S. Govt. Rept., ARL 69-0003, 1969).
33. B. CARRIERE, J-P. DEVILLE, and S. GOLDSZTAUB, *C.R.Acad. Sci. Paris*, **271B**, 796 (1970).
34. H. POPPA and A. G. ELLIOT, *Surf. Sci.*, **24**, 149 (1971).
35. D. M. HERCULES, *Anal. Chem.*, **42**, 20A (1970).
36. K. SIEGBAHN, *Phil. Trans. Roy. Soc. Lond.*, **A268**, 33 (1970).
37. J. R. LOVERING and J. R. WIDDOWSON, *Lithos.* **1**, 264 (1968).
38. E. W. MÜLLER, *Adv. in Electronics and Electron Physics*, **13**, 83 (1960).
39. K. M. BOWKETT and D. A. SMITH, 'Field-Ion Microscopy', North-Holland, Amsterdam, 1970.
40. H. N. SOUTHWORTH, *Contemp. Phys.*, **11**, 209 (1970).
41. E. W. MÜLLER, J. A. PANITZ, and S. B. McLANE, *Rev. Sci. Instr.*, **39**, 83 (1968).
42. N. C. MACDONALD, in 'Scanning Electron Microscopy 1970', Proc. Fourth Annual SEM Symp. (IIT Res. Inst., Chicago, 1970) p. 89.

CHAPTER 12

Review of Analysis

H. Bennett

British Ceramic Research Association
Queens Road, Penkhull
Stoke-on-Trent ST4 7LQ
England

12.1 INTRODUCTION

This book has been concerned primarily with the use of methods based on the detection and monitoring of various electromagnetic radiations for the elemental and crystallographic analysis of minerals. As an inevitable result, the traditional methods of chemical analysis of such materials have received scant attention, and it is my purpose in this chapter to rectify this omission and to try to bring some perspective into a picture which has been built up rather as pieces in a jig-saw puzzle. In addition, I shall discuss some of the topics which other authors have tended to take for granted, in particular sample preparation and sample decomposition.

In the analysis of minerals we can usually distinguish between two types of activity, to provide the ultimate analysis (to give elemental analysis its rather old-fashioned name) and the structural analysis, to indicate the nature of the separate phases present. A detailed examination of a mineral, a raw material, or a complete ceramic body almost inevitably requires both an evaluation of the proportions of the various elements present and a knowledge of how these elements are combined into chemical compounds, and how they are distributed among the separate phases represented in the sample. Thus, in my own field of ceramics, the technologist certainly needs to know which crystalline phases are present in a body and the nature of the bonding between them.

Most of this chapter will be concerned with elemental analysis, with the emphasis on what, for want of a better term, may be called 'wet analysis'. This does not mean that this form of analysis is the most important, but rather that each of the other techniques has been dealt with in some detail in its own chapter. Also, an attempt will be made to show how information available from the various methods may be correlated to build up a more complete picture of the material under study. Mention will finally be made of some ways in which physicochemical methods are applied in other laboratories.

It is only fair, at this juncture, to put my own position, and thus my prejudices, clearly. In the analysis section of the British Ceramic Research Association, judgements on the applicability of an analytical technique are made on the basis of the needs of a laboratory handling rather more than one thousand samples per annum. Our requirements, in terms of accuracy, are high, in that many analyses are made for referee purposes or for standardisation, while the variety of samples received is wide, potentially covering the whole range of materials which can be called ceramics, plus even a few that cannot. The large through-put of samples does demand use of the most rapid methods available, but our needs, especially from any instrumental technique, must be headed by the proviso that the results obtained shall be no worse than can be achieved by good quality traditional methods, and this is a greater restriction than the instrumental enthusiasts would seem to suggest in the literature.

12.2 SAMPLE PREPARATION

Any analysis starts, logically enough, with the preparation of a small sample of finely divided powder whose composition is identical with the original macro-sample sent to the laboratory. This is not the place to enter into a dissertation on the selection of the bulk sample from its original site, be it quarry, rock-face, stock-pile, production line, etc., and the principles of this procedure have been well outlined elsewhere [1-3] ; reference may be made particularly to the discussion given by Wilson and Wilson [4] on sampling coal. In practice, one of the prime factors which tends to govern the bulk sampling method used is that of cost, and this has tended to make the standard of industrial bulk sampling somewhat less than is desirable, and in many cases it may be positively inadequate. It cannot be too strongly stressed that, no matter how good the subsequent sample preparation and analytical procedures used may be, they will be a complete waste of time and money unless the primary sample is truly representative of the materials which it purports to represent. It follows that economising on the collection of the bulk sample is false economy and that it is worth spending a little more money at this stage, since this may result in a final saving of money by not wasting many times more money in obtaining results which are really quite meaningless, although they will possess the semblance of meaningfulness. The dangers inherent in this situation will be obvious.

But even starting with a well gathered sample of some pounds weight, or perhaps less, supplied to the laboratory, errors can still occur in two distinct ways. In the first place, errors in splitting the sample in order to reduce the final bulk to that of a laboratory sample can result in bias with concentration of one or more of the components either into the laboratory sample or into the discard material, and, in the second place, the process of attrition used in reducing particle sizes will almost certainly create contamination of the sample. It follows that some error is inevitable in virtually all analyses, and the important point lies in recognising this fact and in ensuring that it be kept to acceptable proportions. From the experience of a large number of laboratories, the main danger in all analytical work appears to lie in a failure to recognise the size of the potential errors, a fact which has become all too clear through the results of parallel analyses carried out in different laboratories.

It is convenient to separate the components of this problem into sample splitting and sample grinding for the purpose of the present discussion, but it should be realised that this is not possible in practice. Every individual sample will present different aspects of the problem, so that the sample may be homogeneous and hard, in which case the splitting procedure will present no problems but grinding will be difficult, or it may be heterogeneous and soft, and so requiring care in splitting but easy to grind, or, most complex of all, it may be heterogeneous both in composition and hardness of the separate components, in which case the interactions between the problems of splitting and grinding the sample can be formidable. In the next sections, splitting will be discussed first, followed by grinding.

12.2.1 Sample Splitting

Splitting is normally performed before grinding in order to cut down the amount of material which has to be ground to the final fine size for subsequent analysis. It can best be achieved manually by the process of coning and quartering, a technique which needs considerable skill and care. Basically, the method appears simple, involving thoroughly mixing the sample, forming the mixture into a cone, dividing it vertically into four quarters, removing two opposite quarters for further treatment, and setting the rest aside. The process can be repeated until a sample of the requisite size has been chosen.

Clearly, the method is quite easy to apply when the sample is received as a mixture of small, equi-sized particles since, in this case, even a moderate amount of care will ensure that the half chosen at each stage is truly representative of the composition of the whole. Samples with a wide range of particle sizes present more difficulties, especially if the large, intermediate, and small particles have appreciably different compositions, and it may be necessary to crush the whole sample before splitting to ensure accurate splitting. Major difficulties will arise, however, with a sample in which a coarse sized material is mixed with a fine powder of greatly different chemical composition, without any material of intermediate size, since this will require the greatest of care to ensure that a preponderance of one type of material is not chosen during the quartering procedure. Such a situation may well demand fine grinding of a much greater quantity than is normal, even the whole bulk sample in many cases, before any attempt is made to split the sample.

Rotary and wedge sample splitters, riffles, etc., can also be used with advantage at this stage, but the instruments usually available are so large that they require samples weighing several pounds for effective separation. The Warren Spring laboratory has produced a range of miniature riffles [5] which reduce the size of the sample which can be handled. Alternatively, commercial rotary splitters can be used for larger samples, and smaller versions made specifically to handle the final stages, when the sample size is becoming small. Or, initial use of a rotary splitter may be followed by cone and quarter methods to provide the final sample.

It is important to remember that the errors introduced by poor splitting are statistical in nature and normally involve no addition of new material to the sample, although they may result in removal of a component from the final, analysed sample. For these reasons, splitting errors can be very difficult to identify except by using duplicate samples, which increases the work load unnecessarily. Muller [6] has discussed methods of crushing and grinding minerals in the laboratory.

12.2.2 Sample Grinding

Contamination by grinding is generally a more easily identified error than those introduced by splitting and, although frequently a large error, may actually be less significant than unappreciated sample splitting errors. This does not mean that contamination errors can be ignored but rather that these errors may be more easily spotted and the relevant correction made. Thus, when it is found that a sample of an alumina ceramic containing 99% Al_2O_3 contains 5% SiO_2 after grinding in an agate mortar, it is obvious that the silica has come from the mortar and the sample is worthless.

The degree of contamination introduced during any size reduction step generally increases with the fineness of the subdivision involved, being roughly proportional to the surface area of the final product. For this reason, primary crushing with a steel-jawed jaw-crusher is often permissible, unless the sample involved has a very low iron content. The amount of iron introduced at this stage represents a very small percentage of the total weight of the sample, and it can often, although not always, be reduced by treatment with a magnet. Such treatment is obviously not used with a magnetic sample. A modern jaw-crusher will reduce to a particle size not greater than $\frac{1}{8}$ inch, and the resulting material can often be fed directly, after splitting, to the final grinding stage.

Several materials have been used for the final reduction step, and many of them are now available attached to mechanical grinding units. The choice of material depends on the nature of the material to be ground, and the relevant points to note in the selection of an appropriate material are, firstly, hardness and abrasion resistance and, secondly, brittleness. Cost is also a factor, but again false economies can result from penny-pinching in the choice of grinding material. Comments on some commonly used grinding media are outlined below:

1. Glass	virtually useless, being both soft and brittle.
2. Porcelain	tough but soft; its use is restricted to very soft materials such as clay, and even here hard nodules in the clay can cause contamination.
3. Agate	moderately hard but rather brittle; not advisable for use with hard materials, particularly aluminous compounds, or where the silica content is low and critical.
4. Sintered Alumina	although corundum is theoretically much harder than agate, very few alumina mortars are as resistant to abrasion. Alumina mortars are useful for grinding aluminous materials, but the composition range for which they may be used is rather restricted, in that the inevitable alumina contamination should not be significant.
5. Iron	ordinary iron mortars almost inevitably introduce too much iron contamination when the sample is ground therein, but hardened percussion types can be used over a range of samples, and introduce less iron. The contamination is still relatively large and can be tolerated only on samples where the iron content is quite unimportant.
6. Tungsten Carbide	this is a very popular material for the grinding of rock and ceramic samples generally, and is often hailed as the solution to all grinding problems. Although it is of considerable value, it nevertheless does have its disadvantages. It does add contamination in measurable amounts and, what is more, the contamination is undesirable when chemical methods of analysis are to be

used. The material is usually bonded with cobalt, so that both cobalt and tungsten are introduced into the analysis mixture, and both elements can cause awkward interferences. If XRF methods are to be used the problem may be of lesser significance, but care must be taken to check this out in any single case. The level of contamination introduced into a hard rock or ceramic sample may well be an appreciable fraction of one per cent of the total weight.

7. Boron Carbide — this material is probably the most satisfactory for general use in mortars. It is very hard and tough and, even when grinding very hard alumina ceramics, the maximum contamination amounts to only about ¼% by weight. Moreover, the normal processes of decomposition used in subsequent stages of the analysis usually convert the boron carbide into borate plus carbon dioxide, after which it no longer interferes with the analysis. Boron carbide does suffer from two disadvantages, however, the minor one being that it is brittle and needs to be handled with care, and the more important one being its cost. A small two-inch internal diameter mortar will cost more than £100.

Most samples are amenable to boron carbide treatment but, unless a reasonably large mortar and pestle are available, it will generally be necessary to reduce the initial particle size of the material obtained, for example, from a jaw-crusher, by careful tapping on an iron plate, to give a size which can conveniently be handled in a small mortar. Since such treatment can introduce iron, however, it is often considered necessary to pretreat two samples, one in an iron percussion mortar and the other in either an agate or an alumina mortar, as may be the more appropriate for the given sample. The iron-crushed sample is then used for the overall analysis, and the true iron content of the sample obtained from the agate or alumina-crushed material. Together these data give an accurate value for the ultimate analysis of the sample.

The foregoing sections have drawn attention to a few of the difficulties encountered in the preparation of samples of materials, especially hard materials, for analysis. Fortunately, these difficulties can be minimised, and allowance made for their effects, provided that care is taken at every stage in the procedure, bulk sample collection, splitting, and grinding. Again it may be stressed that if the necessary care, expertise, and thought are not given to preparing the sample for actual analysis so as to obtain a true representation of the original material, no analytical technique however powerful or accurate can give the correct answer.

12.3 SAMPLE DECOMPOSITION TECHNIQUES

One of the first problems faced in the actual analysis of minerals and ceramics is that of decomposing the material. Most substances in these categories require a fusion method for complete decomposition in order to provide the homogeneous solution needed for chemical methods of analysis, and for such instrumental techniques as optical absorption and atomic absorption spectrography. But even nominally non-destructive methods, such as spark source mass spectrography (Chapter 5) and X-ray fluorescence (Chapter 3) normally employ a fusion in a flux such as lithium tetraborate, in order to reduce errors arising from the presence of separate crystalline phases, whenever high accuracy is required.

Many salts and alkalis have been recommended for the decomposition of minerals, in addition to acid attack. Some types of attack are, by their very nature, ideal for certain materials, especially if an efficient selective leaching can be achieved. Thus magnesite, dolomite, and calcium carbonates are most readily attacked by acids, which decompose the carbonaceous compounds and leave only a small residue, usually siliceous material, which can be filtered off, ignited, and brought into solution separately. Similarly, sands, quartzites, and ganisters, which are rich in silica, are very conveniently attacked for many purposes by hydrofluoric acid, and acid fluorides are now finding some use in instrumental techniques when silica is not required since, it must be remembered, such treatment involves removing the silica as the volatile fluoride. Acid fluxes, such as bisulphates, have been used in the past but now find little favour.

Today, in general, the most favoured attacks are achieved with alkali carbonates, hydroxides, and borates. Sodium and potassium hydroxides have been used chiefly when it is desired to take the silica into true solution, the carbonates and borates when a gravimetric silica is to be carried out. The majority of chemical methods, however, use a combination of carbonate and borate as the fluxing agent and various ratios of flux to sample can adequately decompose the wide range of minerals used in the ceramic industries. Normally either lithium or sodium carbonate and tetraborate are used, the relative amounts depending on the material to be decomposed. Thus, some materials, such as aluminous materials including bauxites, kyanites, and andalusites, are best handled as a sintered mass without full fusion, but most substances are best fused. Chrome ores, for example, are attacked by a very fluid melt, high in borate, in which they can be readily swirled and which allow easy oxidation.

The prime requirements for a method of decomposition include the following points:

1. Decomposition of the sample in the molten flux should be rapid and complete. When the whole is molten, a clear melt is useful for indicating complete decomposition.
2. Dissolution of the melt in acid should be fast in order that the transition from an alkaline matrix to an acid solution be rapid. This helps

to keep silica in true solution, and slow dissolution almost inevitably results in flakes of hydrated silica floating in the solution. As before, a clear solution should ideally result after acid attack.

3. The fusion should not cause loss of elements being sought, nor should the elements present in the flux cause interference in any subsequent determinations.

Clearly, these requirements will be met by different flux compositions. Thus, a flux with a high borate content is usually more effective for decomposition than an alkaline flux, high in carbonate, and so item (1) above is often best served with a high borate flux. On the other hand, item (2) demands a flux which produces a considerable amount of effervescence with acid, to help in stirring the melt, and so a high carbonate flux is favoured from this point of view. It follows, therefore, that the blend of carbonate and borate in the flux has to be adjusted to the materials in the sample and to the determinations to be carried out subsequently. Nicholls and Wood have discussed the use of such fluxes in the preparation of samples for spark source mass spectroscopy in Chapter 5, section 5.2.1. More detailed treatments of the use of fluxes in mineral analysis are given in the many standard books on the chemical analysis of these materials [5, 7-9].

12.4 CHEMICAL METHODS OF ANALYSIS

Before attempting to evaluate the various methods of analysis available to the mineral chemist, it is necessary to outline briefly the principles underlying the current techniques of wet chemical analysis. The range of methods now recognised as British Standards, or being prepared for submission as Standard methods, are a far cry from the traditional 'classical method'. It was the practice of determining elements sequentially that made the old method so time-consuming, and the newer methods have concentrated on telescoping and speeding up the processes used so that it is now easy to achieve an accurate analysis of a material, such as a clay, in 1½ days or, by sacrificing a small amount of accuracy, in less than a day. Such an analysis would have needed 4-7 days by the classical method.

In principle, the new methods involve the decomposition of the sample, usually by fusion and acidification, followed by short evaporation. The silica, now in the form of a gel or, in the case of material decomposed by acid attack, as an insoluble residue, is rendered easily filterable by coagulation with polyethylene oxide. A gravimetric silica is determined after ignition, by weighing before and after treatment with hydrofluoric acid.

The residue from this determination is re-fused, dissolved in acid, added to the acidified filtrate from the silica removal step, and the whole diluted to a known volume to provide a working solution for the subsequent elemental determinations.

Alkalis are determined separately, usually by a hydrofluoric acid decomposition followed by a flame photometric finish. Other halogen-containing acids are usually avoided in this connection, because of their interference with the flame photometric determinations. Aluminium salts are added to prevent interference from calcium, and caesium is useful as an ionisation buffer to prevent interference between sodium and potassium. Alkaline earths may be determined either on the main stock solution or on the residue from the alkali determination, depending on the nature of the sample and the anticipated levels of calcium and magnesium. In general, the stock solution would be used for samples which do not yield a complete solution with hydrofluoric acid or where calcium and magnesium contents are high, e.g. carbonate rocks such as magnesite and dolomite.

The main methods chosen for the final determinations of the components in the sample include:

1. Silica. That portion remaining in solution is determined spectrophotometrically with ammonium molydbate either as the yellow compound or, reduced, as the blue compound. The latter has the advantage of not suffering from phosphate interference.
2. Titania. Spectrophotometrically after reaction with hydrogen peroxide.
3. Ferric oxide. Spectrophotometrically with 1,10-phenanthroline after reduction to the ferrous state with hydroxylammonium chloride.
4. Manganic oxide. Spectrophotometrically as permanganate after oxidation with periodate solution.
5. Alumina. Titrimetrically after iron and titanium have been removed by solvent extraction with cupferron/chloroform. The alumina is complexed with an excess of 1,2-diaminocyclohexane-tetra-acetic acid (DCTA) or ethylenediamine-tetra-acetic acid (EDTA), which is then back-titrated with standard zinc solution using dithizone as indicator.
6. Calcium. Titrimetrically with EDTA or ethylene glycol-*bis*-2 aminoethyl-tetra-acetic acid (EGTA), normally after complexing the R_2O_3 oxides present with triethanolamine.
7. Magnesium. Titrimetrically with EDTA or DCTA, again after complexing the R_2O_3 with triethanolamine.
8. Phosphate. Spectrophotometrically as its vanado-molybdate complex.

It will be apparent that gravimetric finishes have almost entirely disappeared from the repertoire of the wet chemical analyst in favour of spectrophotometry and titrimetry, because of the greater speed of these latter techniques. In fact, precipitation as a means of separation is now regarded very much as a last resort, solvent extraction being the method of choice because of the much improved separations and the much greater speed which can be achieved thereby. Two reagents are in common use, cupferron and sodium diethyl dithiocarbamate (DDC), both using chloroform as the organic phase, with the addition of a liquid ion exchange resin for the specific removal of chromate from solution in the analysis of chrome ores and refractories. It may be noted that cupferron will remove iron and titanium from acid solution, and aluminium also from neutral solutions. DDC is used mainly for the removal of manganese and

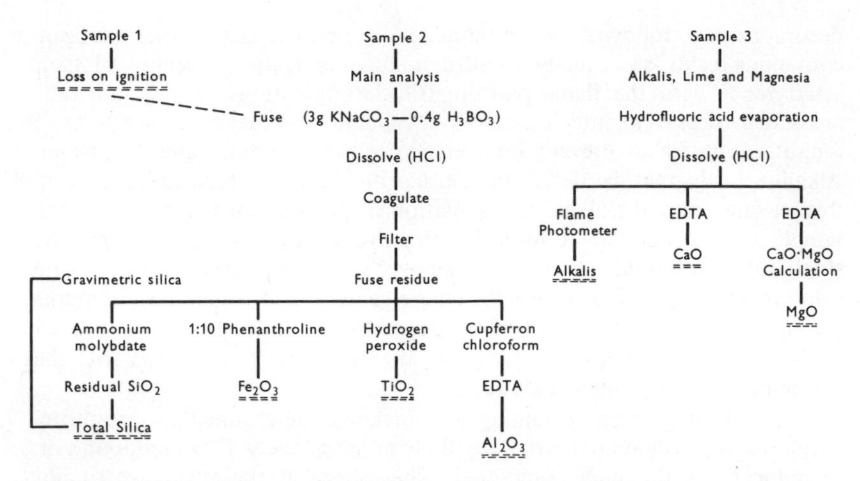

Figure 12.1 Schematic diagram for the method of analysis of high-silica, aluminosilicate and aluminous materials.

occasionally nickel, derived from fusion in a nickel crucible; it also removes iron at the same time.

After these separation steps, and with the appropriate adjustments of pH, the compleximetric finishes for aluminium, calcium, and magnesium are relatively simple to apply, especially with the aid of specific complexing agents such as triethanolamine for the tri-valent cations. Note that EDTA is being replaced by the more specific EGTA and DCTA, the latter being particularly useful for aluminium, which it complexes in the cold thus avoiding interference from chromium.

A typical scheme of analysis useable for materials in the silica/alumina range is shown diagrammatically in figure 12.1.

Note that the loss on ignition is an empirical determination and will include the water content of the material and the amount of carbonate or other volatiles present, as discussed in Chapter 10. Subsequent charts and tables will show how instrumental methods compare with the results obtained from a scheme such as this.

12.4.1 Spectrographic Methods

Optical spectrography is the technique principally used in most laboratories. Spark source mass spectrography is clearly a very powerful technique and capable of giving a very full analysis of the sample being studied, in particular of indicating the presence of unsuspected elements, but its greater cost and the difficulties inherent in its operation, taken in conjunction with the relatively limited range of elements which are normally determined in ceramic or refractory analysis, makes it rather too expensive for routine use, although it is obviously very useful in trouble shooting when all other methods fail. In the majority of cases, arc spectrography will give sufficient information much more cheaply.

In a general laboratory, spectrography tends to find its main use in qualitative or, at best, semi-quantitative analysis. It is an invaluable asset for the preliminary examination of samples of unknown compositions since, in a laboratory liable to need to analyse used refractories, slags and similar materials in which the range of elements present in amounts worth determining is not certain, although the total number of elements involved may be relatively restricted, a spectrographic examination can enable the analyst to plan his scheme of analysis in a logical and efficient manner. Moreover, an arc spectrogram can be prepared quickly and easily and is relatively simple to interpret qualitatively, given a series of standard lines for the elements likely to be present.

If quantitative readings are required, however, the results obtained will depend on the types of samples received by the laboratory. If spectrographic techniques have to be applied to a wide range of samples of markedly different compositions, the accuracy of the results obtained for any given element will inevitably be poor. If, on the other hand, the samples obtained are fairly uniform in composition, and calibration curves can be prepared for this specific use, the results obtained may well be suitable for routine control purposes, particularly if results are required quickly. Fast results can be obtained on a solid sample using some form of pelleting technique, but such results are normally comparable with wet chemical results only for elements present in low concentrations. Control of the purity of sand is a good example of this type of use, when spectrographic analysis is used to monitor the amount of iron and aluminium present. But even here the simple arc spectrograph has lost ground to its automated equivalent, the direct reading spectrograph, since the extra cost of using, developing, and reading spectrographic plates under control conditions can make the employment of the, initially more expensive, direct reader more economical in the long run.

Applications of the Direct Reading Spectrograph

If quantitative determinations are required for concentrations above the minor content level, the necessary accuracies can be obtained only by the use of solution techniques in conjunction with the direct reader. Of course, the need to prepare a solution immediately reduces the gain in time achieved over wet chemical methods, by the use of solid samples, but if the samples can be handled in batches containing comparable elemental compositions, thus avoiding the need to reset the photocell detectors for each sample, considerable savings in time can still be obtained. There is a practical limit to the concentration level which can be handled competitively by the direct reader, however, and about 10% is a probable upper limit for good quality analyses. These limitations define a possible area of activity for the direct reading spectrograph. Its use would appear logically to be confined to the analysis of samples which contain only one major element with the other constituents present at not more than the major impurity level.

Two types of material are obvious candidates for such treatment, high silica materials, including sand, and magnesites. Table 12.1 compares results obtained with a direct reading spectrometer with those obtained by wet chemical analysis. The spectrometer samples were dissolved by first decomposing the sand with hydrofluoric acid, then fusing the residue with lithium carbonate and tetraborate, and finally dissolving in hydrochloric acid. The rotating disc method was used for sample presentation within the instrument.

TABLE 12.1

Analysis of Standard Samples by Direct Reader

Sample	BCS 267	BCS 313*†	BCS 314†	Sand LA	Flint
Al_2O_3					
Chemical	0.85	0.158	0.77	0.10	0.15
D.R. Al 3961	0.88	0.155	0.76	0.08	0.19
CaO					
Chemical	1.75	0.02	1.79	0.13	0.78
D.R. Ca 3968	1.76	0.01	1.86	0.11	0.70
D.R. Ca 3179	1.69	0.01	1.83	0.12	0.71
Fe_2O_3					
Chemical	0.79	0.025	0.48	0.03	0.11
D.R. Fe 2599	0.76	0.023	0.48	0.03	0.11
MgO					
Chemical	0.06	<0.01	0.05	0.01	0.01
D.R. Mg 2802	0.046	0.004	0.048	0.007	0.026
Mn_2O_3					
Chemical	0.17	0.001	<0.01	n.d.	n.d.
D.R. Mn 2576	0.176	<0.002	0.007	0.003	0.003
TiO_2					
Chemical	0.17	0.012	0.19	0.05	0.02
D.R. Ti 3349	0.176	0.011	0.192	0.034	0.008
SiO_2					
Chemical	95.9	99.51	96.32	99.26	97.95
By Difference	96.0	99.6	96.3	99.4	98.24

*2-g sample. † Chemical result of one laboratory only.

The alkali results shown in table 12.1 were determined using an EEL photometer, because the direct reading spectrograph was not capable of determining potassium. Other elements were calibrated for on the direct reader, and determined accurately to ±0.01% of the total sample weight, but these are not reported since no comparative chemical results are available.

For comparison, table 12.2 lists the results obtained from one of the samples (BCS 314) quoted in table 12.1, analysed by wet chemical methods similar to those outlined in figure 12.1, but in the course of an inter-laboratory test involving 9 different analytical laboratories. In comparing the two sets of comparison data, it must be remembered that those quoted in table 12.1 were obtained within the same laboratory, and that errors between laboratories are generally accepted to be about three times greater than those arising within a single laboratory. The results on these high silica materials indicate that the spectrometric values compare very well with those obtained by wet chemical means, although this example probably represents the spectrometer being used under the most favourable conditions, since the impurity levels are very low.

Slightly less favourable conditions apply when the analysis of commercial magnesites, with higher impurity levels, is attempted, as shown in the comparative data listed in table 12.3 for two samples. The chemical values were obtained by the method which is now accepted as British Standard B.S. 1902, Part 2E, 1970 [10] from cooperative work between a group of laboratories, and the direct reader values from one laboratory only by a rotrode technique. In the latter method, the sample is decomposed by fusion in a sodium carbonate/borate mix rather high in acid content, the melt dissolved in sulphuric acid, and the solution sparked with a condensed spark excitation using cobalt as the internal standard.

Despite the fact that the direct reader does not show up in such a favourable light in this latter case, the results are still of adequate accuracy for most industrial purposes. Two large industrial organisations are sufficiently confident of the values so obtained as to utilise the method for their control analyses.

Cost Comparison for the Direct Reading Spectrograph

Experience with the direct reading spectrograph has shown that surprisingly few analyses are needed to justify purchase of the equipment. Where the alternative is full wet chemical analysis by the methods indicated in section 12.4, the break-even point can be as low as 300 analyses per annum. Significant savings can be made with the direct reader when it is possible to analyse samples in batches of sufficient size that a single setting of the calibration will serve for a whole day's work.

The minimum size of batch worth handling is about 5 or 6 samples, and batches of about 12 can be handled in a single day and still provide answers to a high level of accuracy. These figures assume that duplicate solutions are prepared for each sample and 3 or 4 separate sparkings are carried out per solution, with additional sparkings as may be necessary if visual inspection

TABLE 12.2

Co-operative Results on BCS 314 Silica Brick by Standard Chemical Methods

Laboratory	SiO_2	TiO_2	Fe_2O_3	Al_2O_3	CaO	MgO
BCS Figures*	96.2	0.19	0.53	0.77	1.83	0.05
A	95.98	0.22	0.52	0.78	1.80	0.05
	96.06	0.20	0.50	0.78	1.80	0.05
	96.33	0.21	0.53	0.80	1.80	0.05
	95.83	0.19	0.52	0.78	1.83	0.05
	95.89	0.21	0.52	0.79	1.83	0.05
B	96.20	0.21	0.56	0.83	1.94	<0.05
	96.12	0.20	0.52	0.76	1.86	<0.05
C	96.22	0.19	0.51	0.83		
	96.02	0.21	0.52	0.80		
	96.27	0.20	0.53	0.78		
	96.25	0.21	0.54	0.79		
D	95.97	0.21	0.51	0.81	1.90	0.06
	96.09	0.19	0.51	0.81	1.93	0.05
E	96.39	0.18	0.52		2.00	0.07
	96.44	0.20	0.55		1.97	0.05
	96.56					
	96.43					
	96.46					
F	96.44	0.21	0.56	0.85	1.98	0.09
	96.20	0.19	0.60	0.79	1.82	0.06
G	96.34	0.20	0.54	0.88	1.87	<0.01
	96.05	0.18	0.51	0.83	1.80	<0.01
H	95.60					
	95.93					
	96.35					
K	96.24	0.20	0.52	0.80		
	96.29	0.20	0.50	0.81		
	96.15	0.20	0.51	0.78		
	96.16	0.20	0.50	0.78		
	96.27	0.18	0.52	0.77		
	96.19	0.18	0.52	0.79		
	96.26	0.19	0.52	0.78		
Mean	96.21	0.20	0.53	0.80	1.88	0.05
Standard deviation	0.19	0.01	0.02	0.03	0.07	0.02

*British Chemical Standards as supplied by the Bureau of Analysed Samples Ltd.

TABLE 12.3

Comparison of Chemical and Direct Reading Spectrometer Results for Magnesites AN 31 and AN 32

		SiO_2	TiO_2	Fe_2O_3	Al_2O_3	Mn_3O_4	Cr_2O_3	CaO
AN 31								
Chemical	Mean %	2.49	0.04	1.77	0.85	0.10	0.07	2.32
	S.D. %	0.02	0.01	0.03	0.02	0.01	0.01	0.02
D.R.	Mean %	2.56	0.04	1.88	0.86	0.12	0.07	2.33
	S.D. %	0.04	0.004	0.06	0.01	0.005	0.007	0.03
AN 32								
Chemical	Mean %	0.97	0.02	5.44	0.99	0.12	0.75	1.76
	S.D. %	0.02	0.01	0.05	0.04	0.02	0.03	0.04
D.R.	Mean %	0.97	0.03	5.63	0.98	0.14	0.77	1.78
	S.D. %	0.04	0.004	0.06	0.01	0.005	0.007	0.03

reveals a poor sparking. Industrially, much lower standards can be accepted for much routine analysis on control samples where the answers are generally 'known' in advance. Single solutions and two sparkings per solution can often yield acceptable answers, and thus the potential throughput of the instrument is correspondingly enhanced – but only if the laboratory has a need for the extra capacity!

Flame Photometry

This technique enjoyed a brief vogue in the late 1950's. Soon after its general introduction, and particularly after monochromatic-type instruments became widely available, the technique was being used for an extensive range of determinations. Not only were the alkali metals and alkaline earth metals attempted, but also elements such as iron, titanium, manganese, and even aluminium were considered to lie within the range of possible elements determinable by flame photometry. This over-stretching of the technique brought it into some disrepute and resulted, in the event, in a smaller degree of use than the method possibly merited. At the present time, its use is restricted, in most laboratories, to the determination of sodium, potassium, and possibly lithium, with the addition, in a very few laboratories, of calcium and possibly magnesium.

Atomic Absorption Spectrophotometry

This technique is currently in the same danger of over-exploitation as was flame photometry ten or fifteen years ago. Not content with the genuine value of the instrument in the field of trace analysis and for the determination of a range of minor constituents, many of which have given the analyst great difficulties in the past, the enthusiast is trying to convert the instrument into a complete laboratory in itself.

The determination of major constituents in clays, i.e. silica and alumina, by atomic absorption spectrophotometry has received considerable attention from many workers, but accuracies, or even precisions, better than ±1% absolute really have to be worked for. The effort taken to achieve this accuracy, if it were applied to another technique, could well achieve better results in no longer a time.

It is to be hoped that this sort of activity will eventually find its true level, and disappear. Fortunately, the true value of atomic absorption spectrophotometry is sufficiently clear that, unlike flame photometry, it is unlikely to fall from grace but will retain its place in almost any analytical laboratory. Any determination which it does well, and it will determine well about 30 elements in a wide variety of situations, it is likely to do more quickly and/or more simply than other techniques. It may be noted, in passing, that the present emphasis on inorganic pollution may well have arisen only because atomic absorption spectrophotometry has enabled the relevant determinations of very low metal concentrations to be carried out.

12.4.2 X-ray Fluorescence Methods

X-ray fluorescence has hovered around the fringes of accurate mineral analysis for some years, but it is only since about 1970 that fully accurate results have become available on a routine basis. The breakthrough has lain in the use of cast and polished beads to present the sample to the instrument, as detailed in Chapter 3. Before that time, work done with fused beads which were ground and pressed into discs simply failed to match up, in general, to the needs of the analyst in the field of ceramics. Problems associated with sampling accuracy and surface roughness on the disc led to unacceptably high standard deviations on top of insufficient accuracies. Most seriously, XRF methods could not determine the major components, silica and alumina, accurately and such determinations are vital to the ceramics industry, since most aluminosilicates or aluminous materials are sold primarily on the basis of their alumina contents.

The work carried out principally in the steel and glass industries on improving sample presentation methods has resulted in the new and more effective techniques detailed in Chapter 3. Using these methods, wide range calibrations covering almost the whole range of compositions in the silica/alumina refractories may now be prepared. Methods for achieving this include simple calibration against standards, which does impose some limitations on the materials that can be handled; incorporation of lanthanum into the sample disc, to act as a heavy absorber [11] and hence minimise interference from absorption effects by swamping the effect of the other, lighter elements; and a standard multi-dilution technique [12] using the element to be determined as the diluted substance. In this context, one of the advantages of using a fusion method for preparing the sample disc is that it renders possible the use of synthetic standards, and this may well be

TABLE 12.4 (a)

Comparison of Chemical and X-ray Analysis

	SiO_2		Al_2O_3		Fe_2O_3		TiO_2	
	X	C	X	C	X	C	X	C
Bauxite	5.9	5.9	88.8	89.0	1.78	1.74	3.20	3.14
Sillimanite	34.2	34.0	64.0	63.6	0.30	0.30	1.37	1.34
Mullite	25.0	25.0	72.3	71.9	0.57	0.59	0.15	0.15
Firebrick A	50.0	49.9	44.5	44.4	2.60	2.58	1.38	1.35
Firebrick B	57.0	56.7	34.0	33.9	3.30	3.31	1.48	1.48
Ball Clay	52.4	52.6	32.3	31.9	0.98	1.00	1.21	1.19

TABLE 12.4 (b)

Comparison of Chemical and X-ray Analysis

	SiO_2		Al_2O_3		Fe_2O_3		CaO	
	X	C	X	C	X	C	X	C
Kyanite	42.6	42.4	55.0	54.8	0.78	0.75	0.10	0.10
Glaze	49.6	49.3	12.9	12.7	0.35	0.35	10.1	9.9
Stoneware	65.6	65.9	22.3	22.3	22.42	2.40	1.23	1.25
Chinaware	34.7	34.8	14.3	14.4	0.28	0.31	25.7	25.8
Slag	10.8	10.8	2.7	2.9	18.9	19.0*	41.7	41.5
Ore	19.9	20.0	7.3	7.2	35.8	35.5*	3.3	3.3

* Expressed as Fe
X = X-ray analysis
C = Chemical analysis

important since it is not always possible to achieve an adequate spread of standard samples for all the desired elements using natural materials.

Table 12.4 shows sets of results obtained by the simple calibration method, table 12.4a, and by the lanthanum addition method, table 12.4b. Again, in order to make a comparison with the general standard of chemical analysis on this type of material, a set of cooperative figures obtained from 9 laboratories by one of the British Ceramic Research Association's Working Groups are presented in table 12.5.

As before, it should be borne in mind that the results in the two tables are not strictly comparable, since table 12.4 contains data from a single laboratory and table 12.5 from several. Thus, although the X-ray fluorescence results are better than the results obtained by most of the other instrumental methods used and of a standard adequate for the majority of ceramic purposes, they still do not quite reach the standard attainable by

TABLE 12.5

Co-operative Results on BCS 309, Sillimanite by Standard Chemical Methods

Laboratory	SiO_2	TiO_2	Fe_2O_3	Al_2O_3	CaO	MgO
BCS Figures	34.1	1.93	1.53	61.1	0.34	0.17
	33.96	2.08	1.54		0.24	0.10
A	34.04	2.08	1.52		0.22	0.14
	34.14	2.10	1.53		0.21	0.13
B	34.17	2.00	1.56	61.71	0.41	0.15
	34.10	1.85	1.46	61.20	0.37	0.16
	34.05	1.83	1.51	61.20	0.30	0.18
	34.03	1.83	1.49	61.07	0.30	0.16
C	34.07	1.84	1.52	61.14	0.32	0.16
	34.07	1.89	1.48	61.07	0.30	0.16
	34.09	1.87	1.50	61.08	0.32	0.17
D	34.26	1.85	1.47	61.07	0.40	0.14
	34.25	1.85	1.48	61.07	0.40	0.14
E	34.17	1.85	1.50	61.22	0.36	0.16
	34.20	1.87	1.48	61.27	0.33	0.17
F	34.33	1.95	1.60	61.48	0.37	0.38
	34.03	1.90	1.60	61.80	0.27	0.30
G	34.05	1.84	1.50	60.88	0.24	0.13
	34.40	1.84	1.58	61.07	0.24	0.12
	33.91					
H	33.94					
	34.15					
	34.01	1.97	1.55	61.01	0.27	0.20
	33.89	1.98	1.53	60.96	0.28	0.20
	34.01	1.98	1.52	60.92	0.28	0.20
K	34.01	1.98	1.53	60.92	0.29	0.20
	33.95	1.98	1.52	60.98	0.32	0.19
	33.96	1.97	1.53	60.96	0.28	0.18
	34.01	1.98	1.53	60.96	0.30	0.19
Mean	34.08	1.93	1.52	61.14	0.31	0.16
Standard deviation	0.13	0.08	0.04	0.24	0.06	0.03

a good wet chemical method. The standard claimed by workers responsible for the new X-ray fluorescence methods discussed here and in Chapter 3 is ±½% for silica and alumina, a figure which compares with the level of error found in results obtained from different laboratories for the same sample, but greater than the error within a single laboratory.

Nevertheless, despite the slightly lower standard of results obtained using the method, it is clear that X-ray fluorescence analysis is the method of choice today for most general ceramic analyses, and this is probably rapidly becoming equally true in the mineral industry as a whole. The greatly increased speed with which results are available with the technique and its eminent suitability as a routine control instrument even, as shown in Chapter 6, as an on-line or an on-stream instrument operating with a minimum of sample preparation, make it a very attractive proposition to managements with large analytical through-puts, despite the high initial capital cost of the equipment.

12.4.3 Structural Analysis

Thus far, the emphasis in the chapter has been laid on elemental analysis, but much of the content of the book has been concerned with techniques which provide information on the structure and the phase composition of the material under study. Clearly both elemental and structural analyses have their uses, and sometimes it is an understanding of the elemental composition which provides the answer to a problem while on other occasions it is the structure of the material that is the critical factor. Very often, however, both types of information are needed and so, for a complete understanding of the behaviour of a ceramic material such as a refractory brick, it is necessary to have an elemental analysis, a visual appreciation of the texture of the substance, a microscopic examination to reveal the appearance of the various crystalline phases and glassy bonds, X-ray diffraction and differential thermal analytical studies to identify the minerals in the crystalline phases, and an electron probe microanalyser evaluation of the micro-composition and of the bonding within the body.

Thus, all these techniques are complementary and to ask the question 'Which is the most important?' has no real meaning. The important point that should be grasped by all is that much remains to be discovered concerning the reasons why any particular material is useful, or not, and each of the instruments and techniques discussed in the preceding chapters has its own part to play in such studies, together, one may confidently predict, with a few other instruments and techniques that have not yet been invented!

It is not my purpose to go into detail about the use or value of these new instruments, such points having already been adequately covered by people better equipped to deal with these topics, but in one instance I feel capable of contributing to the discussion, and this is in connection with the use of the electron probe micro-analyser. This technique differs from most of the others cited in that it enables elemental analyses to be carried out on volumes so small that they could not possibly be isolated for separate analysis in any other way. In addition, the non-destructive nature of the techniques used makes it possible to gather these data at the same time, and over the same micro-area, as one can obtain structural information about the nature of the crystalline phases present or the presence of grain boundaries, etc. It is, of course, important to remember that almost all other

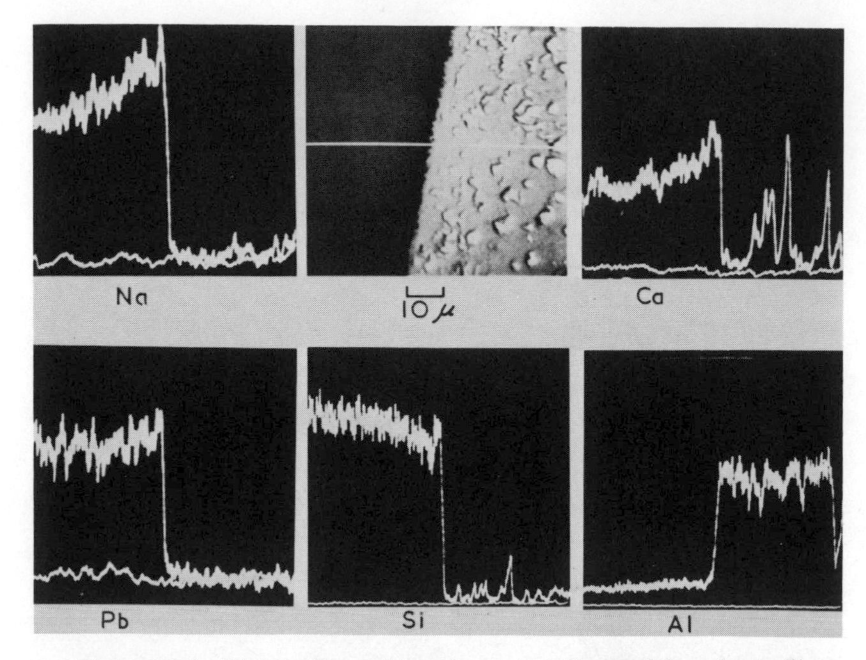

Figure 12.2 Distributions of the elements sodium, calcium, lead, silicon and aluminium across the interface between a lead glaze, on left in centre top, and an impure alumina body, on right, along the line shown.

techniques give average information relating to the whole of the sample without the wealth of detail available from the EPMA. Three examples of the use of EPMA in ceramic research may be cited.

Of these, the first two are rather similar, both being concerned with the interaction between a ceramic material and a glaze which has been applied to it. Figure 12.2 shows the case of an impure alumina body which has been coated with a lead glaze and the glaze subsequently fired on. The top centre photograph shows the electron backscatter image with the alumina body on the right hand side and the glaze on the left, the horizontal white line across the picture marking the line along which the elemental scans were made. The traces clearly show that both the sodium and the calcium contents of the glaze increase as the interface with the alumina substrate is approached, the lead content remains relatively unchanged, but the silicon content decreases. There is virtually no migration of sodium, silicon, lead, or aluminium across the interface, which appears to be very sharp in both the electron micrograph and the elemental traces. The layer of crystals which forms the interface does not appear to be thicker than 1 μm.

The second example, figure 12.3, is taken from a study of the glaze on gehlenite ($2CaO.Al_2O_3.SiO_2$), the glaze again being lead-bearing. Here the body is on the left of the micrograph, with the glaze on the right, and this time the crystalline interface shows up very clearly, being marked by a layer of prismatic and lath-shaped crystals apparently about 10μm long. In fact, two types of crystal have formed, as is clearly shown by the elemental traces, which indicate that the crystals nearer to the gehlenite body are anorthite, with half the lime and twice the silica content of gehlenite, whereas those nearer to the glaze contain a sodium aluminosilicate phase which has a composition identical to that obtained when the same glaze was applied to an anorthite body. Note how, in this case, sodium and lead appear to have migrated from the glaze into the layer of anorthite crystals. Such information assists very materially in elucidating the amount and the type of reaction occurring between the glaze and the substrate body, which can be an important factor in the manufacture of pottery.

The third example, shown in figure 12.4, concerns the texture and structure of an experimentally prepared bone china made by melting together the pre-reacted components. At low magnifications, shown in the top left, it appeared that two immiscible liquids had formed and, at higher magnifications, it became evident that each had separated into two phases

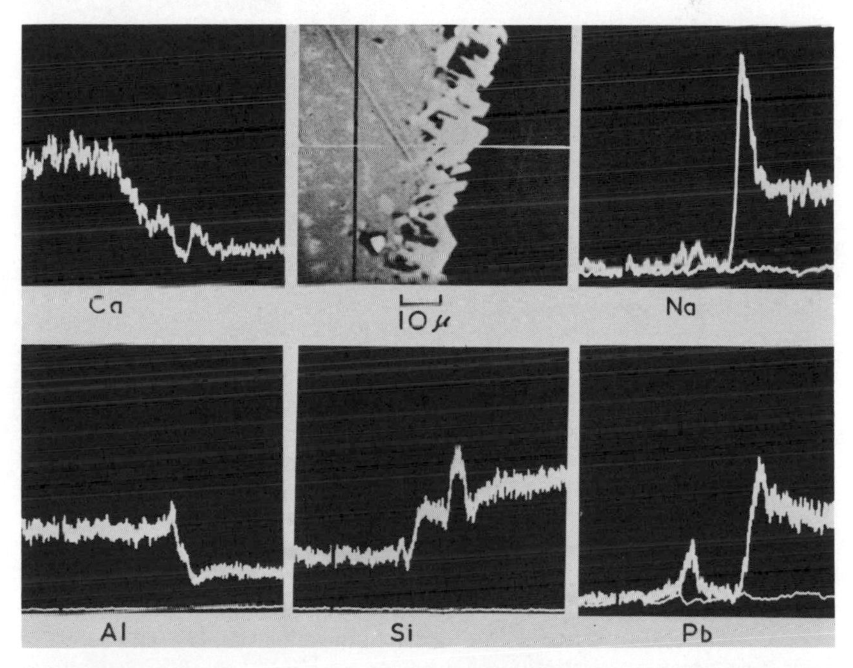

Figure 12.3 Distribution of the elements calcium, sodium, aluminium, silicon and lead across the interface between a gehlenite body, on left in centre top, and a lead glaze, on right, along the line shown. Note the existence of a crystalline layer of reaction products at the line of the interface.

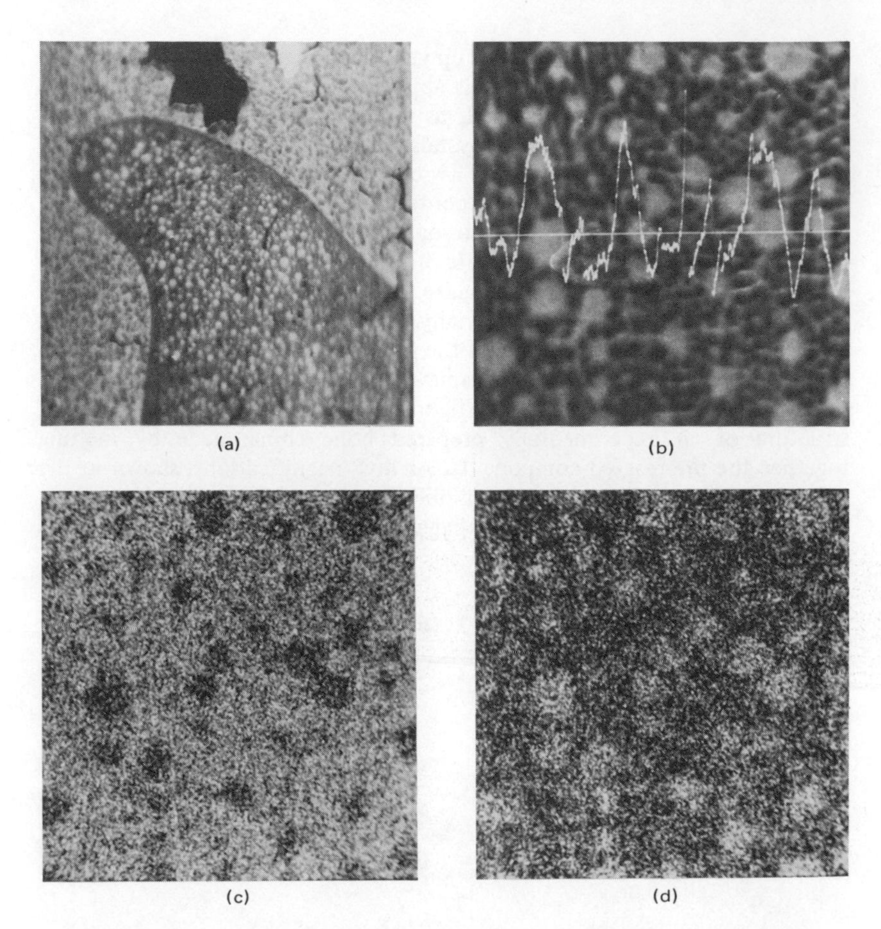

Figure 12.4 Texture of an experimental melted bone china body as revealed in the electron microprobe analyser; (a) low magnification electron image; (b) magnified view of central region of (a) with Al distribution along central line superimposed; (c) and (d) X-ray images for Ca and Si respectively, corresponding to the area shown in (b). Light regions in (c) and (d) indicate high concentrations of the relevant element.

on cooling. The upper right photograph shows part of the darker area at high magnification, with the distribution of aluminium along the central white line superimposed thereupon. The lower photographs show the corresponding X-ray images for calcium and silicon, the light regions indicating a high concentration of the element. The phosphorus image is not shown but it closely parallels the calcium distribution. It is evident that globules of aluminosilicates have separated from the matrix, which is predominantly calcium phosphate, during cooling. Conversely, in the lighter coloured areas, globules of calcium phosphate have separated from an aluminosilicate matrix. Such information is invaluable in the study of the physical and chemical nature of a bone china body.

From these three examples it will be clear that electron probe microanalysis offers an entry into a world of texture and structure which would otherwise be exceedingly difficult to penetrate. It enables the ceramist to study the reactions which take place during sintering, to examine the nature of the bonding in ceramic materials, information which may well prove to be vital in determining the servicability of a body, and to investigate the mechanisms whereby slags may attack refractories. In conjunction with the scanning electron microscope and the methods which White has described (Chapter 11, section 11.3) the field that is opened up is immense and very exciting.

12.5 COST EFFECTIVENESS OF ANALYSIS

It would be very nice, in trying to decide on the purchase of this or that analytical instrument or the adoption of one or another analytical technique, to be able to write down a series of equations relating to the capital costs of the various methods, their running costs, the salary costs involved in the different methods, overheads, depreciation, etc., and arrive at a definitive answer to the question originally posed. Unfortunately, the situation is usually rather too complex to allow of such a simple solution, in fact it is sometimes even difficult to decide what the question asked really ought to be, let alone what the answer should be. The relative economics of the new method and of the method which it replaces, or the importance of the information which the new method can provide, are, of course, prime considerations, but other factors, usually less amenable to quantification, normally enter into an assessment of the true cost of any technique.

In general, an instrumental method will either give information which is not available by any other technique, and we may cite the electron probe microanalysis studies quoted above, or it will give information which is available by an older method, but more rapidly, as in the use of X-ray fluorescence analysis. The reader may have noticed that instruments providing new information generally relate to the structure of the material, whereas those giving information more rapidly relate to the elemental composition.

As far as the first situation is concerned, the primary question that must be asked is whether the new information is really necessary and, if it is, whether the volume of work available is sufficient to warrant the rather large capital outlay normally involved in the purchase of such an instrument, usually of the order of £$N \times 10^4$. Moreover, a sophisticated instrument needs a highly trained operator to obtain the best results from it, and this will result in high running costs in order to obtain the best return from the initial investment. It follows, therefore, that only the largest laboratories will be able to justify the purchase of all the range of instruments discussed in this book, and many laboratories carrying out fairly routine analyses will be able to justify, perhaps, only the purchase of optical microscopic or X-ray diffraction equipment, although they may have occasional need of the information available from the more sophisticated methods. It behoves such a laboratory to investigate the possibility of setting up sharing schemes with other local laboratories to provide a centralised service, which possesses not only the obvious advantages of spreading cost but also has the advantage of making one good operator available to several laboratories instead of to a single laboratory. It is worth remembering that good operators tend to be rather few and far between. Various contract analytical services exist, such as those provided in the United Kingdom by the Harwell Industrial Research Unit, the Fulmer Research Institute, or the British Ceramic Research Association, and most laboratories will find that their local University or Polytechnic is very willing to cooperate and help in solving problems.

When contemplating the rapidity with which results can be made available, it is important to work within the perspective of the local situation. The question which must be asked is whether it matters that moderately, or even highly, accurate results be obtained in 2 hours or 2 minutes instead of in 2 days. Clearly the answer is yes in the case of production control on a continuous process, and so we see the rise in importance of X-ray methods for automatic on-line and on-stream control, discussed in Chapter 6, or the use of neutron activation analysis in determining oxygen in steel, quoted in Chapter 4. In both cases, the high cost of the technique is clearly and amply justified by the resulting increase in productivity. In the laboratory, however, it is often more difficult to decide on the justification for an instrumental method. In particular, it is important to check that the time saved is really saved and does not mean longer tea-breaks but a real increase in through-put. On the other hand, it is necessary to consider the availability of analysts of the right calibre to perform the wet chemical analyses involved, and to consider if it would be better policy to reserve them for the really difficult cases and to use perhaps relatively less skilled operators to carry out the routine work using instrumental techniques. It is also important to remember that, even if the laboratory has a large through-put, the work load may comprise large groups of samples covering a very few types of materials, or it may include a large number of small groups of samples covering a wide range of different materials and problems. Clearly, the former situation lends itself more easily to the application of instrumental methods than does the latter, when the more flexible approach provided by classical methods may be more applicable, and such considerations naturally can materially affect the economics in terms of the actual usage of the equipment involved.

But beyond all these factors there lie some more abstruse considerations which must be taken into account. In particular, it may be pertinent to consider whether you are willing to make the often very basic changes in the character of your laboratory staff that are demanded by the introduction of this or that piece of equipment. Also, it is important to remember that all elemental analyses based on instrumental methods depend, for their ultimate accuracy, on the accuracies of the standard substances used in the calibration procedures, and these depend, in turn, on the accuracy with which the standards have been determined, usually by a wet chemical method. Thus there will always be a need for high calibre analysts to back up the instrumental analyst, and so there will be a need for training grounds for these wet chemical analysts also.

Moreover, the specialist who has been trained to use one technique may be very useful from the point of view of obtaining results with maximum accuracy and despatch, but he may not be the best person to oversee a general analytical laboratory, since he may not have the wide experience of all the available techniques that is so necessary for such a senior position. He may tend to apply his technique to problems where a better solution could be obtained with a different method, or he may try to apply it in situations where it is quite inapplicable, as happened with flame photometry. There are, at present, a large number of senior analysts in industry generally who view the future with some reservations, for this reason. It is not easy to see where the present generation of analysts, who have lived through the analytical revolution of the past quarter century, are going to find replacements with comparable breadths of experience in all methods. Training grounds for general analysts must exist, and it is here that the Research Association and kindred organisations in this country and throughout the world can play a vital role.

Thus, in the British Ceramic Research Association, we tend to be regarded as a training ground for analysts for the ceramics and refractories industry as a whole, and so it is necessary to plan the work of the laboratory to provide a widely based experience for the workers therein, and it is largely for this reason that the laboratory has not embraced X-ray fluorescence methods with the enthusiasm of some other organisations. I strongly believe that there is a case for some laboratories, of the Research Association type, making what may be some sacrifice in immediate efficiency or through-put for the longer term needs of analytical chemistry generally.

12.6 SOME APPLICATIONS AND COMPARISONS

Physicochemical methods are, of course, widely used, and one laboratory which has made a comparative study of some of the techniques currently available is that of General Telephone and Electronics Incorporated, at their Bayside Research Center, Long Island, New York. There they provide an analytical and testing service for all the companies within the group, and handle a wide range of materials from minerals and ceramics to specialised organic compounds. Figure 12.5 shows the range of instruments available in the Center, and provides a useful comparison of the information which each

METHOD	QUALITATIVE ANALYSIS		QUANTITATIVE ANALYSIS		SPECIFI-CITY
	What property is measured?	How good is it for this use?	What property is measured?	How good is it for this use?	To one substance in a mixture
VISIBLE LIGHT MICROSCOPY	Visible appearance, size, shape, color texture	Fair to good	Size and quantities of particles[1]	Excellent	Excellent
SCANNING ELECTRON MICROSCOPY	Visible appearance, size, shape, texture, surface potential, electron-beam-induced current, cathodo-luminescence	Excellent, very high depth of focus	Not applicable	Not applicable	Poor
ELECTRON MICROSCOPY	a. Size, shape, texture b. Electron diffraction angles	Good	a. Size and quantities of particles* b. Amount of electrons diffracted	a. Excellent b. Poor to fair	Excellent
CHEMICAL MICROSCOPY	Specific chemical reactions	Good	Change in physical property, color, etc.	Poor	Fair to good
INFRA-RED ABSORPTION	Wavelength distribution	Good	Absorption at specific wavelength	Good in medium concentration	Fair to good
VISIBLE LIGHT ABSORPTION	Wavelength distribution	Poor	Absorption at specific wavelength	Good in lower, poor in higher ranges	Poor to fair
ULTRAVIOLET ABSORPTION	Wavelength distribution	Poor	Absorption at specific wavelength	Excellent in trace ranges	Poor to fair
FLUORESCENCE SPECTROSCOPY	Wavelength of emitted light	Excellent for line emitters	Intensity of emission at specific wavelength	Excellent	Excellent for line emitters
FLAME PHOTOMETRY	Wavelength of emitted light	Excellent	Intensity of emission at specific wavelength	Fair in low ranges	Good
ATOMIC ABSORPTION SPECTRO-PHOTOMETRY	Not suited for qualitative survey	–	Absorption of specific line emission	Excellent	Excellent
EMISSION SPECTROSCOPY	Wavelength of emitted light	Excellent	Intensity of emission at specific wavelength	Good in trace ranges, fair in higher ranges	Excellent

* Specific particle size measurement also available on Coulter Counter (number frequency of particles). Micromerograph (cumulative weight by diameter) and Isorpta Analyzer (specific surface).

Figure 12.5 Comparison of available methods for determination of chemical composition and phase structure. (Courtesy of Bayside Research Center of General Telephone and Electronics Laboratories Inc.)

TIME OF ANALYSIS (Long: 2 hours) (Short: 1/2 hour)	ACCURACY OF RESULTS	CAPABILITY For multi-component analysis	INSTRUMENT PRICE (High > $20,000) (Low < $10,000)	PRINCIPAL GT&E INSTRUMENTS
Short	As good as sampling accuracy	Fair	Low	Leitz Ortholux Microscope Leitz Panphot Microscope Reichert MEF Microscope B & L Research Metallograph
Short	Not applicable	Not applicable	High	JEOLCO Scanning Electron Microscope
Long	As good as sampling accuracy	Fair	Medium to high	RCA EMU 3F Electron Microscope Philips EM 300 Electron Microscope Philips EM 75 Electron Microscope
Short	Good	Poor	Low	Zeiss Polarizing Microscope
Short	Good	Fair to good	Low to medium	Perkin Elmer 621 IR Spectrophotometer
Short	Good	Fair to good	Low to medium	Perkin Elmer 350 Spectrophotometer Beckman DU Spectrophotometer Cary 14 Spectrophotometer Hitachi 124 Spectrophotometer
Short	Excellent to poor depending on range	Fair to good	Low to medium	Perkin Elmer 350 Spectrophotometer Beckman DU Spectrophotometer Cary 14 Spectrophotometer Hitachi 124 Spectrophotometer
Short to moderate	Good	Good for line emitters	Medium	Norelco X-Ray Generator JACO Spectrometer Beckman DU Spectrophotometer
Short after calibration	Good in proper ranges	Good	Low to medium	Beckman DU Spectrophotometer
Short after calibration	Good to very good	Poor	Low	Instrument Laboratories Model 153 Atomic Absorption Spectrophotometer
Short after calibration	Good in proper ranges	Excellent	High	Jarrel Ash Mark III Ebert Spectrograph Bausch & Lomb Dual Grating Spectrograph Jarrel Ash 1.5-Meter Wide-Angle Wadsworth Spectrograph

METHOD	QUALITATIVE ANALYSIS		QUANTITATIVE ANALYSIS		SPECIFICITY
	What property is measured?	How good is it for this use?	What property is measured?	How good is it for this use?	To one substance in a mixture
X-RAY SPECTROSCOPY	Wavelength of emitted X-radiation	Excellent	Intensity of X-ray emission at specific wavelength	Good in mid-range concentrations	Excellent
ELECTRON PROBE MICRO-ANALYSIS	Characteristic X-radiation, electron back-scatter, secondary electron emission	Excellent	Intensity of X-ray emission at specific wavelength	Very good for micro-analysis of occlusions and micro-segregations	Excellent
X-RAY DIFFRACTION	X-ray diffraction angles	Excellent	Amount of radiation diffracted	Fair	Excellent
MASS SPECTROMETRY	Mass/charge of ionic particles	Excellent	Quantity of particles of specific mass/charge	Excellent (gases, liquids) Good (solids)	Excellent
CHROMATO-GRAPHIC ANALYSIS	Thermal conductivity of gas stream	Good	Conductivity change integrated over time	Very good in bulk, fair to good in trace	Excellent
ACTIVATION ANALYSIS	Energy and decay rate of induced activity	Excellent	Specific energy and amount of induced radio activity	Excellent for trace ranges or for microsamples	Excellent
CLASSICAL† ANALYSIS	–	–	Volume or weight	Excellent	–

† Includes amperometric analysis, gravimetric analysis, combustion analysis, and volumetric analysis.

Figure 12.5 *(continued)*

TIME OF ANALYSIS (Long: 2 hours) (Short: 1/2 hour)	ACCURACY OF RESULTS	CAPABILITY For multi-component analysis	INSTRUMENT PRICE (High > $20,000) (Low < $10,000)	PRINCIPAL GT&E INSTRUMENTS
Short after calibration	Good to very good	Excellent	Medium	Norelco Vacuum Path X-Ray Spectrometer
Moderate to long	Good to very good	Excellent	High	Philips Electronics Electron Probe Microanalyzer
Short	Poor for quantitative use	Fair to good	Medium	Norelco and Siemens X-ray Generators plus many accessories Pailred Automatic Single Crystal Diffractometer
Long (solids) Short (gases and liquids)	Very good (liquids, gases) Fair (solids) in trace range	Unexcelled	Medium (gases, liquids) High (solids)	EAI Quadrupole Mass Analyzer CEC 21-110 and AEI MS 702 Spark Source Solids Mass Spectrographs Hitachi RMU 6E Mass Spectrometer
Short after calibration	Good	Good	Low	Perkin Elmer Model 900 Gas Chromatograph
Short	Excellent in trace ranges	Good to excellent	Medium to high	Technical Measurement Corp. Activatron 211; TII 1017 Multichannel Analyzer with semiconductor detector
Moderate to long	Excellent	–	Low	LECO Combustion Titrimeter; Electro-balances for bulk and microanalysis. Miscellaneous physico-chemical apparatus Perkin Elmer Model 240 Elemental Analyzer

can supply, their speeds and accuracies, their costs, and their general applicabilities. Special attention may be drawn to the 'unexcelled' capability that mass spectrometry has for multicomponent analysis, and the very short analysis times available with activation analysis.

The activation analysis facility is, in fact, an excellent example of automation being used not only to provide answers rapidly but also to make a potentially hazardous procedure very safe indeed. After the sample has been ground, mixed, and made ready for irradiation, a weighed aliquot is placed in a 'rabbit' at the operator's console from where it is passed through a pneumatic tube to the irradiation chamber, irradiated for a pre-determined time at a pre-set flux level, unloaded, passed to the counting chamber, counted for the required elements, and the results displayed all under the control of a mini-computer. As a result, the operator is never in any danger from irradiated material, and the sample is finally discharged automatically. The irradiation unit is an accelerator of the Dynamitron type (Chapter 1, section 1.2.1) and the counting chain uses a Ge(Li) semiconductor detector backed up by a multi-channel analyser, so that a single counting operation suffices to give data for most of the elements of interest.

12.6.1 Laser-Based Instruments

The use of lasers in physicochemical methods of analysis has not been mentioned elsewhere in this book, largely because their use is quite recent and rather limited at present. This does not mean, of course, that they will long continue to be rarities, and it may well be that the next major step in analytical power may come with their wider application. Some uses have been sufficiently developed to warrant mention, moreover, and readers would do well to become aware of the possibilities inherent in the use of laser light both as a source of coherent light and as a high energy source which can be focused very precisely onto the sample under study.

The theory and general use of lasers has been reviewed [13, 14] and suffice it here to remind readers that the light generated from a laser consists of 'pulses' in which all the radiation is in phase, that these are produced as a highly parallel beam with virtually zero dispersion, and that the energy output can be of the order of joules per cm^2 of beam area. The advantages of laser light are thus firstly that a very precise beam of light is available for scanning transparent samples, and secondly that high photon and total energies can be concentrated in a small volume of the sample.

One of the earliest applications, the laser light scattering analyser described by Vand and co-workers [15], uses the very low dispersion of a focused laser beam to investigate defects in crystals. The method depends on the fact that a beam of laser light is very coherent and so it is not scattered by a perfect, defect-free crystal, which is therefore quite transparent and no light can be detected in directions inclined to the beam direction. Defects, however, scatter the laser beam and this scattered light can be detected at angles up to 90° from the beam. In consequence, if the laser beam is directed through the crystal and viewed at right angles, the defects show up clearly against the black background, with a resolution of about 0.1μm. Direct transmission studies can also be made, in which case the defects show up as shadows from the reduction in the energy of the beam due to the scattering process, and the camera or eyepiece can be replaced by a photocell for quantitative work on bulk scattering.

Sherman and Black [16] have independently built a scanned laser infra-red microscope (SLIM) which operates on the same principles as Vand's instrument but uses infra-red radiation at 3.39μm instead of visible light to study the perfection of semiconductors, especially III-V compounds in the current use. A He-Ne laser is focused onto a thin wafer of the sample via two mirrors which vibrate about a vertical and a horizontal axis to scan an area of about $1cm^2$ in a raster pattern. The transmitted radiation is monitored with a lead selenide or indium arsenide detector and the resulting output used to modulate the brightness of a cathode ray tube display scanned synchronously with the beam. Again, defects and homogeneities or the presence of segregated precipitates of infra-red-opaque material can be detected with a resolution approaching 10μm. Other laser emissions, extending from the ultra-violet to the 10.6μm line of the CO_2 laser, could be used for materials other than semiconductors.

A variant of this second technique is afforded by the scanned laser photoluminescence microscope (SLPM) [17] in which the point-to-point luminescence induced in the specimen is scanned instead of the transmitted radiation. The method depends on the energy density in the laser beam being sufficiently high to excite luminescence in situations where a conventional, incoherent light source would be ineffective. The technique has again been principally applied to III-V semiconductors and gives a sensitive measure of the distribution of impurities or of added dopants in the specimen. Narrow band optical filters are used in the detecting system to select specific wavelengths in the luminescent radiation, and hence allow the distribution of specific luminescing sites to be mapped. This technique would appear to be eminently suitable for application to mineral specimens using the correct laser light.

A final application of a laser is in the field of mass spectrometry. Knox [18] has described the use of a beam of high energy radiation from a laser to excite ions from a sample in the ionisation chamber of a mass spectrometer. The advantage that this method has over spark source mass spectrometry is that it can excite molecular groupings from the substance under study, instead of breaking the material down into its constituent atoms, and these groupings can give some indications of the structure of the material from which they have come. Again the application depends on the ability to concentrate energy at a single point and to control the energy input closely. The technique has already shown that the principal ionic species obtained from selenium is the Se_5^+ ion, not the Se^+ ion, showing that the Se_5 group is an important structural unit in this metal. Clearly the spectra obtained from a mineral will be difficult to interpret, but the method would appear to hold out great promise in helping to elucidate some problems of mineral structure.

12.7 CONCLUSION

The preceding chapters of this book have discussed a wide range of methods for the analysis of minerals, from the determination of elemental compositions, through the identification of the chemical compounds present, to detailed examination of the fine structure of the material, and it is a matter for regret that the exigencies of space have prevented me from making a full appraisal of these techniques in this review. For example, the use of the mass spectrometer for the determination of trace elements has received scant mention, not because it is an unimportant technique but because of the utter impossibility of discussing this and the potential importance of trace element analysis to the study of ceramics and of their manufacture within the limitations of this chapter.

As has been mentioned in various places throughout the book, it is always important for the analyst to understand how instrumental methods of analysis interrelate and how a maximum amount of information can be obtained for a minimum cost. Complete characterisation of a sample always depends on obtaining complementary information from several sources and correlating these data to build up the required picture of the substance, but obviously the extent to which this procedure is carried will depend on how well the sample needs to be characterised. The important thing, however, is never to trust to a single method to provide all the data on any aspect of the problem.

Thus, the elemental content of a substance sent for complete analysis and characterisation might first be surveyed using arc spectroscopy or non-dispersive X-ray fluorescence with a Si(Li) detector system, before being subjected to wet chemical, spectroscopic, or X-ray fluorescence analysis, to determine the major elements present, and mass spectrometry or neutron activation analysis for its trace elements. Together, these data will give the complete elemental composition of the sample. Optical microscopic studies, both in transmitted and reflected light, will give some indication of the probable number of phases, crystalline and vitreous, present and may provide at least tentative identifications for some as well as helping with the interpretation of the X-ray powder diffraction patterns obtained from the sample. Armed with the elemental and phase compositions, more detailed studies of how the elements were distributed among the phases could be made with the electron probe microanalyser, or in a scanning electron microscope fitted with an X-ray fluorescence analyser, and the fine structure could be further investigated in thin section using transmission electron microscopy, and selected area electron diffraction patterns could confirm the phase identifications made from the X-ray diffraction patterns or help in their interpretation. Finally, some indication of the atomic groupings present, and further identification of the phases themselves, could be obtained from infra-red spectroscopy and thermogravimetric analysis. Such a study takes time, but it is usually time well spent, especially for the research worker in the field of minerals or materials science.

In this context, it is worth remembering that standards of purity have risen very sharply over the past decade, particularly in the field of semiconductors and solid state electronics. A 'pure' material in this field can be one which has less than 1ppm. of total impurities, and this represents the analytical standard which is required. The standard in the minerals industry has not yet reached this level, although it it tending that way in some specialised ceramics, but the pressure is constantly on to improve detection limits and accuracies, and physicochemical methods can help greatly here since often they may represent the only techniques capable of making the measurements, provided always that they are used sensibly and thoughtfully and not blindly.

ACKNOWLEDGEMENTS

The author thanks his Director of Research, Dr. N. F. Astbury, CBE, for permission to publish this chapter, and is grateful to Mr. A. D. Ambrose of the British Steel Corporation, Tubes Division, for supplying the data used in table 12.4, and to Dr. S. N. Ruddlesden of the British Ceramic Research Association for supplying the photographs reproduced in figures 12.2, 12.3 and 12.4, which are to be published in the 20th Proceedings of the British Ceramic Society.

REFERENCES

1. P. GY, *Revue Ind. miner.*, **38**, 53 (1956).
2. G. M. BROWN, in 'Methods in Geochemistry', A. A. Smales and L. R. Wager (eds), Interscience, New York, 1960, p. 4.
3. D. J. OTTLEY, *Min. and Miner. Eng.*, **2**, 390 (1966).
4. C. L. WILSON and D. W. WILSON, 'Comprehensive Analytical Chemistry', Vol. 1A, Elsevier, London, k960, p. 36.
5. P. G. JEFFERY, 'Chemical Methods of Rock Analysis', Pergamon Press, Oxford, 1970.
6. L. D. MULLER, in 'Physical Methods in Determinative Mineralogy', J. Zussman (ed), Academic Press, London and New York, 1967, p.1.
7. W. F. HILLEBRAND, G. E. F. LUNDELL, H. A. BRIGHT and J.I. HOFFMAN, 'Applied Inorganic Analysis', John Wiley and Sons, Inc., New York and London, 1953.
8. H. BENNETT and R. A. REED, 'Chemical Methods of Silicate Analysis – A Handbook', Academic Press, London and New York, 1971.
9. J. A. MAXWELL, 'Rock and Mineral Analysis', Interscience, New York, 1968.

10. British Standard, B.S. 1902, Part 2E (BSI, 2 Park St., London, 1970).
11. A. D. AMBROSE, Proc. Fourth Ceramic Chemists Conf. (Brit. Ceram. Res. Assoc., Stoke-on-Trent, 1971).
12. D. G. ASHLEY, Proc. Fourth Ceramic Chemists Conf. (Brit. Ceram. Res. Assoc., Stoke-on-Trent, 1971).
13. C. C. EAGLESFIELD, 'Laser Light', Macmillan, London, and St. Martins Press, New York, 1967.
14. D. FISHLOCK (ed), 'A guide to the Laser', Macdonald, London, 1967.
15. V. VAND, E. A. MARGERUM, F. SCHWAB and R. I. HARKER, *U.S. Govt. Res. Devel. Rept.*, **41** (10), 71 (1966).
16. B. SHERMAN and J. F. BLACK, *Appl. Optics,* **9**, 802 (1970).
17. J. F. BLACK, C. J. SUMMERS and B. SHERMAN, *Appl. Phys. Letters,* **19**, 28 (1971).
18. B. E. KNOX, *Adv. Mass Spectrom.,* **4**, 491 (1968).

Subject Index